T0199527

# STATISTICS
# FOR THE
# 21ST CENTURY

# STATISTICS: Textbooks and Monographs

## D. B. Owen, Founding Editor, 1972–1991

*Additional Volumes in Preparation*

# STATISTICS FOR THE 21ST CENTURY

## Methodologies for Applications of the Future

edited by

## C. R. Rao

*Pennsylvania State University*
*University Park, Pennsylvania*

### Gábor J. Székely

*Bowling Green State University*
*Bowling Green, Ohio*

*and*

*Alfréd Rényi Institute*
*of Mathematics*
*Hungarian Academy of Sciences*
*Budapest, Hungary*

CRC Press
Taylor & Francis Group
Boca Raton London New York

CRC Press is an imprint of the
Taylor & Francis Group, an **informa** business

First published in 2000 by Marcel Dekker

Published in 2020 by CRC Press
Taylor & Francis Group
6000 Broken Sound Parkway NW, Suite 300
Boca Raton, FL 3487-2742

First issued in paperback 2020

© 2000 by Taylor & Francis Group, LLC
CRC Press is an imprint of Taylor & Francis Group, an Informa business

No claim to original U.S. Government works

ISBN-13: 978-0-367-57903-6 (pbk)
ISBN-13: 978-0-8247-9029-5 (hbk)

This book contains information obtained from authentic and highly regarded sources. Reasonable efforts have been made to publish reliable data and information, but the author and publisher cannot assume responsibility for the validity of all materials or the consequences of their use. The authors and publishers have attempted to trace the copyright holders of all material reproduced in this publication and apologize to copyright holders if permission to publish in this form has not been obtained. If any copyright material has not been acknowledged please write and let us know so we may rectify in any future reprint.

Except as permitted under U.S. Copyright Law, no part of this book may be reprinted, reproduced, transmitted, or utilized in any form by any electronic, mechanical, or other means, now known or hereafter invented, including photocopying, microfilming, and recording, or in any information storage or retrieval system, without written permission from the publishers.

For permission to photocopy or use material electronically from this work, please access www.copyright.com (http://www.copyright.com/) or contact the Copyright Clearance Center, Inc. (CCC), 222 Rosewood Drive, Danvers, MA 01923, 978-750-8400. CCC is a not-for-profit organization that provides licenses and registration for a variety of users. For organizations that have been granted a photocopy license by the CCC, a separate system of payment has been arranged.

**Trademark Notice:** Product or corporate names may be trademarks or registered trademarks, and are used only for identification and explanation without intent to infringe.

**Visit the Taylor & Francis Web site at**
**http://www.taylorandfrancis.com**

**and the CRC Press Web site at**
**http://www.crcpress.com**

Library of Congress Cataloging-in-Publication Data

Statistics for the 21st century : methodologies for applications of the future / edited by C. R. Rao., Gabor J. Székely.
   p. cm. — (Statistics, textbooks, and monographs ; v. 161)
   ISBN 0-8247-9029-4
   1. Mathematical statistics. I. Title: Statistics for the twenty-first century.
II. Rao, C. Radhakrishna (Calyampudi Radhakrishna) III. Székely, Gabor J. IV. Lukacs Symposium (8th : 1998 : Bowling Green University)
V. Series.

QA276.16 .S844 2000
519.5—dc21                              99-057753

# Preface

*Statistics for the 21st Century* is a collection of articles, most of which were presented at the Eighth Lukacs Symposium, held at Bowling Green State University, and some invited from selected authors. Statistics has a long antiquity but a short history. It has come into academia as a separate discipline only in the second quarter of the 20th century. During this period, some major contributions have been made to the theory and methodology of statistics motivated by practical problems in biological and social sciences and industry. The main architect was R. A. Fisher. We have reproduced in this volume two previously published articles on his contributions, one by B. Efron and another by one of us (C. R. Rao). No doubt much of the core theory in statistics was motivated by Fisher's fundamental contributions to statistical inference. There are, however, controversies over some of Fisher's works, not all of which have been resolved, and which will be topics for discussion for a long time to come. We have decided to put on record Fisher's contributions in the present volume with the hope that future generations of statisticians will continue the debate on Fisherian concepts and methods.

We have requested the authors to describe the current state of statistical methodology in different areas of application and to mention what, in their opinion, should be the future lines of research in statistics. We want to thank all the authors for their excellent surveys of the existing statistical methods and for the insights they provided for future deliberations.

There is some concern among statisticians that not much attention is being given to interdisciplinary research, which nurtured the growth of statistics in the first half of the 20th century, and that advances in computer science, artificial intelligence and machine learning are not fully exploited in the analysis of data for extracting information. New concepts are emerging, such as data mining and machine learning, advocated by researchers in applied areas, which seem to brush aside the core methodology of statistics as model-based and inadequate for the analysis of the massive data sets we are acquiring today. In the words attributed to J. Tukey, the bulk of current statistical research appears to be "finding exact solutions to wrong problems instead of approximate solutions to right problems." We hope that, as we enter the 21st century, we will critically evaluate what has been achieved so far and what needs to be done in the future in terms of statistical education and research to make statistics a viable subject addressed to solving critical societal problems.

There is no index in this book. Our volume is more descriptive than systematic and we are certain that an index would be inappropriate and misleading, and no reader would benefit from it. We wish to close by acknowledging the outstanding work of Kata Vantso, Gabor Sipotzy, and Csaba Borotzky as technical editors.

C. R. Rao
Gábor J. Székely

# Contents

# Contributors

**Alan Agresti**   Department of Statistics, University of Florida, Gainesville, Florida

**G. Jogesh Babu**   Department of Statistics, Pennsylvania State University, University Park, Pennsylvania

**Anil K. Bera**   Department of Economics, University of Illinois, Champaign, Illinois

**Carles M. Cuadras**   Department of Statistics, University of Barcelona, Barcelona, Spain

**Bradley Efron**   Department of Statistics, Stanford University, Stanford, California

**John D. Emerson**   Department of Mathematics and Computer Science, Middlebury College, Middlebury, Vermont

**V. T. Farewell**   Department of Statistical Science, University College, London, England

**Andreas Fieger**   Statistical Institute, Ludwig-Maximilians Universität München, Munich, Germany

**Josep Fortiana**   Department of Statistics, University of Barcelona, Barcelona, Spain

**D. A. S. Fraser**   Department of Statistics, University of Toronto, Toronto, Ontario, Canada

**Janos Galambos**   Department of Mathematics, Temple University, Philadelphia, Pennsylvania

**Leo A. Goodman**   Department of Statistics, University of California, Berkeley, California

**Christian Heumann**   Statistical Institute, Ludwig-Maximilians Universität München, Munich, Germany

**Satish Iyengar**   Department of Statistics, University of Pittsburgh, Pittsburgh, Pennsylvania

**G. D. Johnson**   Center for Statistical Ecology and Environmental Statistics, Department of Statistics, Pennsylvania State University, University Park, Pennsylvania

**Lucien Le Cam**   Department of Statistics, University of California, Berkeley, California

**G. S. Maddala**   Department of Economics, Ohio State University, Columbus, Ohio

**V. Mandrekar**   Department of Statistics and Probability, Michigan State University, East Lansing, Michigan

**Frederick Mosteller**   Department of Statistics, Harvard University, Cambridge, Massachusetts

**W. L. Myers**   School of Forest Resources and Environmental Resources Research Institute, Pennsylvania State University, University Park, Pennsylvania

**P. K. Pathak**   Department of Statistics and Probability, Michigan State University, East Lansing, Michigan

**G. P. Patil**   Center for Statistical Ecology and Environmental Statistics, Department of Statistics, Pennsylvania State University, University Park, Pennsylvania

**Shyamal D. Peddada**   Division of Statistics, University of Virginia, Charlottesville, Virginia

**C. R. Rao**   Department of Statistics, Pennsylvania State University, University Park, Pennsylvania

**N. Reid**   Department of Statistics, University of Toronto, Toronto, Ontario, Canada

**Alan D. Rogol**   Department of Pediatrics and Pharmacology, University of Virginia Health Sciences Center, Charlottesville, Virginia

**D. A. Sprott**   Department of Statistics and Actuarial Science, University of Waterloo, Waterloo, Ontario, Canada, and Centro de Investigación en Matemáticas, Guanajuato, México

**J. N. Srivastava**   Department of Statistics, Colorado State University, Fort Collins, Colorado

**Gábor J. Székely**   Department of Mathematics and Statistics, Bowling Green State University, Bowling Green, Ohio, and Alfréd Rényi Institute of Mathematics, Hungarian Academy of Sciences, Budapest, Hungary

**C. Taillie**   Center for Statistical Ecology and Environmental Statistics, Department of Statistics, Pennsylvania State University, University Park, Pennsylvania

**Helge Toutenburg**   Statistical Institute, Ludwig-Maximilians Universität München, Munich, Germany

**Grace L. Yang**   Department of Mathematics, University of Maryland, College Park, Maryland

**Cleo Youtz**   Department of Statistics, Harvard University, Cambridge, Massachusetts

**Juan Zhang**   Roth Associates, Rockville, Maryland

# Challenges for Categorical Data Analysis in the Twenty-first Century

Alan Agresti[1]

Department of Statistics
University of Florida
Gainesville, Florida

**Abstract.** As we approach the millenium, the state-of-the-art in categorical data analysis, as in all branches of statistics, is vastly different than at the start of this century. This article focuses on some of the primary developments of the past ten to twenty years and discusses areas likely to see significant progress early in the next century. Topics on which I focus are methods for ordered categorical responses, repeated measurement and clustered data, and small samples and sparse data. As the literature evolves, the variety of options for data analysis continues to increase dramatically. As a consequence, perhaps the greatest challenge of the next century is maintaining adequate communication among statisticians and between statisticians and other scientists.

## 1. Introduction

The beginning of this century was an auspicious one for statistical methodology pertaining to categorical data. In 1900 Karl Pearson proposed chi-squared tests of fit for multinomial distributions and for the hypothesis of independence in two-way contingency tables, and G. Udny Yule proposed the odds ratio and some related measures of association. After this, the development of categorical data methodology proceeded at a rather slow

---

[1] This work was partially supported by a NIH grant.

pace in the first two-thirds of this century, although notable contributions were made by such luminaries as R. A. Fisher, William Cochran, Maurice Bartlett, and Jerzy Neyman. The mid-to-late sixties were a turbulent period politically in world events but also generated an explosion of new ideas and fundamental changes in the cultural fabric of Western society. Whether such stimuli extended to the statistical arena is debateable, but this certainly marked the beginning of a period of tremendous progress in methodology for categorical data.

There is not space here to review all the major developments since then, but a few highlights are as follows: Leo Goodman was an ever-present force throughout the sixties and, indeed, the rest of the century; see, for instance, Goodman (1978) for a compilation of some of his early publications on loglinear, logit, and latent class models, his work on the latter showing one of the first uses of the EM algorithm. Influential work also occurred at Harvard by William Cochran, Frederick Mosteller, and their students, and at the University of North Carolina by Gary Koch and his students and colleagues. For instance, Grizzle, Starmer and Koch (1969) provided a regression approach to analyzing contingency tables using weighted least squares. Ultimately perhaps most influential was the Nelder and Wedderburn (1972) introduction of the generalized linear model (GLM), which unified maximum likelihood (ML) fitting of logit models for binary data and loglinear models for count data with regression models for continuous responses. As these new methods were blossoming, Rao (1973) presented an elegant theoretical framework for asymptotic theory for parametric models for contingency tables. As the century nears its end, we are witnessing a continuing explosion in the development of methods for categorical data. My article focuses on some of the primary developments of recent years and on possibly fruitful areas for progress in the twenty-first century.

Predicting the future is always dangerous. However, I am sure most statisticians would agree that, regardless of the statistical area, much of future research will focus on computationally-intensive methods. This will happen in an increasing variety of venues, but some of the obvious ones are: (1) Generalized linear mixed models, in which it is currently a challenge to estimate well regression parameters and variance components, (2) Bayesian alternatives to frequentist methods, (3) "Data mining" methods for huge data sets with possibly large numbers of variables, and (4) Complex sampling schemes and data structures, including the handling of missing data. In the twentieth century statistical science has emphasized estimation and significance testing, but in the future these (especially testing) may be less important than description, model selection, and prediction.

I will discuss future challenges for statistics with reference to methods for three types of categorical data that have achieved much progress in the past 10-20 years — small-samples and sparse data, ordered categorical data, and repeated measurement and other forms of clustered data. I review some of the main ideas and suggest areas for future work. With the development of advanced methods, the challenges also increase of maintaining adequate communication among statisticians and between statisticians and other scientists. This may be the greatest challenge of all for the twenty-first century.

## 2. Small-sample 'exact' inferential methods

In this era of ever-increasing computer power, the tendency seems to be to collect increasingly large data sets. Nonetheless, in some applications small sample sizes are the norm. Examples include biomedical studies with rare conditions or with constraints on sample size due to ethical considerations. In addition, as data sets become larger there is the concomitant tendency to measure more variables, with the consequence of sparse data sets and questionnable relevance of large-sample theorems having regularity conditions requiring the parameter space to be fixed as the sample size increases. 'Exact' inferential methods provide an alternative to large-sample methods for contingency tables with small samples or sparse data.

The development of computational algorithms for exact inferential analyses has been a major advance of the past decade in contingency table analysis. With exact methods, one can guarantee that the size of a test is no greater than some prespecified level and that the coverage probability for a confidence interval is at least the nominal level. The release of *StatXact* (Cytel Software 1995) now makes a wide variety of exact methods easy to use. At the same time, the 'exactness' refers only to inference being based on probability distributions that do not depend on unknown parameters, and there is no unique way to do this.

The exact approach most fully developed for contingency tables is *exact conditional inference*, which eliminates nuisance parameters by conditioning on their sufficient statistics. These methods have the advantage of versatility, applying to exponential family linear models that use the canonical link function, such as loglinear models for Poisson responses and logit models for binomial responses. The best known conditional method is Fisher's exact test of independence in $2 \times 2$ tables, in which unknown row and/or column probabilities are eliminated by conditioning on row and column marginal counts. Traditionally regarded as a small-sample method, exact conditional inference is increasingly applied to sparse data problems with large tables. In recent years, its applications include tests of independence in $r \times c$ contingency tables that are sensitive to trends for ordered categories, tests of conditional independence in stratified tables, tests comparing several odds ratios, tests for parameters in logit models, and goodness-of-fit tests for models for square contingency tables. Although the literature focuses mainly on testing, confidence intervals for odds ratios are possible through inversion of tests for non-null parameter values. Survey articles about the exact approach include Agresti (1992), Mehta (1994), and Mehta and Patel (1995).

For contingency table problems, in eliminating nuisance parameters this approach conditions on certain margins that are typically not fixed by the sampling scheme. For instance, in testing independence in a two-way table, conditioning on all row and column marginal totals yields a hypergeometric distribution for the cell counts. When the sampling scheme is multinomial or independent binomial rather than hypergeometric, however, the restriction of the sample space to samples having exactly the same response margins as the observed one may seem artificial. By contrast, an exact *unconditional* approach conditions only on margins of the contingency table naturally fixed by the sampling design. This unconditional approach eliminates nuisance parameters using a "worst-case" scenario; the P-value is a tail probability maximized over all possible values for the nuisance parameters. For instance, consider testing equality of binomial parameters $p_1$ and $p_2$ for two independent binomial samples in a $2 \times 2$ table, using a test statistic $T$ with observed value $t$. One finds $P_p(T \geq t)$ for a given value $p$ of the unknown common parameter value under the null, and then defines the P-value to be $\sup_p[P_p(T \geq t)]$.

The unconditional approach also has critics, and some advocates of the conditional approach have often been quite vociferous in opposing it (e.g., Yates 1984). Some statisticians question whether response distributions for parameter values that are highly different than those suggested by the observed data should contribute to computations of P-values. Berger and Boos (1994) proposed an adaptation that addresses this weakness by calculating the supremum over a confidence region for the nuisance parameter and adjusting the P-value accordingly. In the two binomial case, letting $C_\gamma$ denote a $100(1 - \gamma)\%$ confidence interval for $p$, their P-value is $\sup_{p \in C_\gamma}[P_p(T \geq t)] + \gamma$, where $\gamma$ is very small, such as .001. Most of the literature on the unconditional approach refers to $2 \times 2$ tables, but a recent technical report by Freidlin and Gastwirth proposes unconditional versions of the Mantel-Haenszel test for several $2 \times 2$ tables.

For any given way of approaching exact inference, as in large-sample methods there are variations in ways to conduct it. For instance, one can define P-values in different ways, one can use different test statistics (e.g., likelihood-ratio, Wald, score), and one can construct confidence intervals by inverting two separate one-tailed tests or a single two-tailed test.

Although the development of exact methodology has seen considerable progress, certain analyses are still computationally infeasible and likely to be so for some time. For instance, even for the conceptually simple task of testing independence in a two-way table, the number of tables in the conditional reference set grows exponentially in the number of rows and columns and the sample size, and it is easy to display tables that cannot be handled with existing software. Another recent advance is an ever-increasing variety of methods for accurate approximation of exact methods. These include simple Monte Carlo (e.g., Agresti et al. 1979), Monte Carlo with importance sampling (e.g., Mehta et al. 1988, Booth and Butler 1998), Markov Chain Monte Carlo (MCMC, Forster et al. 1996), the iterated bootstrap (Presnell 1996), and saddlepoint approximations (Strawderman and Wells 1998). The Forster et al. paper applies the MCMC approach to a wide variety of logit and loglinear models, and other work by these authors addresses models for square tables that are awkward to handle with alternative methods.

The literature on approximating exact methods applies mainly to the conditional approach. The unconditional approach is much more computationally intensive, and the development of approximate exact inference of this form is an interesting challenge for future work. Most of the literature deals with testing, but in practice interval estimation provides more informative results. Exact algorithms are usually much slower for interval estimation than for testing a null parameter value, so this is a computational challenge. Also, the existing literature pays little attention to power studies.

As discussed in Section 6, discreteness can cause the performance of exact methods to be conservative. In extreme cases, a relevant conditional distribution can be degenerate. Consider, for instance, logistic regression. For subject $i$, the response $y_i = 0$ or 1, and denote the explanatory variables by $\mathbf{x}_i = (x_{i0}, x_{i1}, \cdots, x_{ik})$, with $x_{i0} = 1$. Then, for the model

$$\log[P(Y_i = 1)/P(Y_i = 0)] = \sum_j \beta_j x_{ij}$$

the sufficient statistic for $\beta_j$ is $T_j = \sum_i y_i x_{ij}$. Exact inference for $\beta_j$ uses the distribution of $T_j$, conditional on $\{T_i, i \neq j\}$. For instance, for $H_0 : \beta_j = 0$ and $H_a : \beta_j > 0$, the P-value is

$$P(T_j \geq t_j | \{T_i = t_i, i \neq j\}).$$

If any predictor is continuous, this probability may equal 1, since the observed data configuration $\{y_i\}$ may be the only one that has the fixed values of all the sufficient statistics for the nuisance parameters.

Another future challenge is to develop an *approximate* exact inference, perhaps based on approximate conditioning, in-which the extreme discreteness does not occur. Pierce and Peters (1998) provide a promising start in this direction. Basically, they suggest applying the best of the higher-order asymptotic methods (such as saddlepoint approximations) as for continuous data, incorporating the approximate conditioning by making no continuity correction. Also, adjustments of exact methods that use the mid-P value (*half* the probability of the observed test statistic plus the probability of more extreme results) are less susceptible to discreteness problems. This seems to be a reasonable compromise between the conservativeness of exact methods and the uncertain adequacy of a large-sample method, but it is not sufficiently helpful when the conditional distribution is degenerate. For cases in which exact mid P-value calculation is infeasible, Monte Carlo or saddlepoint approximations to the mid P-value are likely to perform very well.

The exact conditional approach does not apply to GLMs with non-canonical link, because of the lack of reduction in sufficient statistics. Examples include probit models for binary data and cumulative logit models for ordinal responses. It also does not apply for GLMs with random effects. Open problems include exact and approximately exact methods for handling models for which the conditional approach does not apply. Random effects approaches may become a more common way of eliminating some nuisance parameters, such as when the number of parameters grows with the sample size. Methods based on permutation arguments are also increasingly feasible.

## 3. Modeling ordinal responses

In applications with a categorical response, logit models have occupied a central place for some time, both for binary and for multiple-category responses. Perhaps surprisingly, as of twenty years ago little attention had been paid to modeling ordinal responses. Although articles by S. Haberman and by G. Simon in 1974 showed how logit and loglinear models could incorporate the ordering, the models did not receive much attention until being further developed in landmark papers by Goodman (1979) and McCullagh (1980).

Currently, the most popular logit model for ordinal responses applies the logits to cumulative probabilities. For a *c*-category response variable $Y$ and a set of predictors $\mathbf{x}$ with corresponding effect parameters $\beta$, this *cumulative logit* model has form

$$\text{logit}[P(Y \leq j)] = \alpha_j - \beta\mathbf{x}, \ j = 1, \ldots, c - 1.$$

The model applies simultaneously to all $c - 1$ cumulative probabilities, and it assumes an identical effect of the predictors for each cumulative probability. It and related models with link functions such as the probit have several appealing properties, such as (1) one can motivate the model with a regression model for an underlying continuous response, (2) if the model holds for a particular set of response categories, it holds with the same effects when the response scale is collapsed in any way, and (3) it is unnecessary to assign scores to the response categories.

Another popular model uses *adjacent-category* logits, $\log[P(Y = j)/P(Y = j+1)]$, $j = 1, \ldots, c-1$. In practice, cumulative logit and adjacent-category logit models usually provide similar substantive results. They tend to fit well in similar situations, such as when the response distribution at different predictor values has a shift in location but not in dispersion. Although McCullagh (1980) proposed a general model for cumulative probabilities that has dispersion as well as location effects, it does not seem to have been used much. A helpful future contribution would be to extend this general model to other ordinal logits and to show practitioners how to use and interpret such a model.

The ordinal logit models are like ordinary regression models in distinguishing between response and explanatory variables. *Association models*, on the other hand, describe association between categorical variables and treat those variables symmetrically (Goodman 1979). The most popular association model assumes that expected frequencies $\{\mu_{ij}\}$ in a two-way contingency table satisfy

$$\log \mu_{ij} = \lambda + \lambda_i^X + \lambda_j^Y + \beta u_i v_j.$$

The case $\beta = 0$ corresponds to statistical independence. The *linear-by-linear association* model fixes monotone row scores $\{u_i\}$ and column scores $\{v_j\}$, the *row effects* model fixes $\{v_j\}$ but treats $\{u_i\}$ as parameters, and the *column effects* model fixes $\{u_i\}$ but treats $\{v_j\}$ as parameters. These are all loglinear models, unlike the *row and column effects* (RC) model that treats both sets of scores as parameters, and for which ML estimation is more difficult. The model with equally-spaced scores for $Y$ relates to logit models for adjacent-category logits, and the models generalize to include covariates (Becker and Clogg 1989). A related literature on correspondence analysis models and equivalent canonical correlation models (Goodman 1986) uses an association term of form $\beta u_i v_j$ to model the difference between the expected count and its independence value rather than the difference between the log expected count and its independence value. In association models, unlike cumulative logit models, the quantitative treatment of ordinal variables requires assigning scores.

Despite recent advances in ordinal logit and loglinear modeling, in practice the most common approach to añalyzè ordinal response variables is to assign scores to categories and use regression or ANOVA methods with ordinary least squares. This modeling approach has the advantage of simplicity, particularly when it is sufficient to summarize effects in terms of location. Its limitations are due not only to a nonnormal response but also to a tendency for the variance to depend on the mean; for instance, less variability tends to occur when the mean is near the high end or low end of the ordinal scale. Alternatively, one can fit such regression models assuming a multinomial rather than normal distribution for the response (Grizzle et al. 1969), which addresses the lack of constant variance. This model is a reasonable compromise for statisticians to use when dealing with clients who do not have the mathematical background to understand a logit or loglinear model but who can understand a trend in means. With it, one sacrifices the ability to make predictions about proportions falling in particular response categories, but often this is of secondary importance.

For further details about models described in this section, see Agresti (1996). A future challenge is the development of simpler ways of describing effects in loglinear and logit models for ordinal data. Parameters in the models relate naturally to odds ratios, but multiplicative effects on an odds are difficult for many users to understand. In particular, even though standardized measures of association such as R-squared have well-known limitations, they are helpful for letting users assess magnitudes of effects, and more attention could be paid to developing such measures for these models. Another area that could use further study is the consequence of using various models when others are more appropriate. For instance, if a cumulative logit model truly holds, in what cases does the use of ordinary regression suffer in terms of efficiency of parameter estimation and validity of interpretation, and how does this depend on the number of response categories? There is also scope for further work with multinomial modeling, for nominal, ordinal and mixed responses. Areas for potential work include random effects in discrete choice modeling, and modeling with multiple random effects such as in hierarchical models. Generally, there is a need for the comparison of various ways of handling correlation and missing data for repeated measurement of multinomial responses. Finally, software exists for fitting many ordinal models (such as PROC LOGISTIC in SAS), but there are still some surprising gaps (e.g., no procedure in SPSS) that may have limited their application.

## 4. Models for repeated categorical measurement

The most active area of new research in categorical data analysis in the past decade has been the modeling of clustered data, such as occur in longitudinal studies and other forms of repeated measurement. A variety of ways now exist of modeling the correlation structure among responses in the same cluster.

Two major types of models differ in terms of whether they focus on *subject-specific* or *population-averaged* effects. The former refer to conditional distributions at the subject level, whereas the latter refer to marginal distributions that result from averaging over subjects in the population. For instance, consider the simplest case in which this occurs, matched-pairs data with a binary response. Denote the two responses for subject or cluster $s$ by $(Y_{s1}, Y_{s2})$. The subject-specific model

$$\text{logit}[P(Y_{s1} = 1)] = \alpha_s, \quad \text{logit}[P(Y_{s2} = 1)] = \alpha_s + \beta$$

assumes a common effect $\beta$ for each subject. With it, the odds that $Y_{s2} = 1$ are $\exp(\beta)$ times the odds that $Y_{s1} = 1$. By contrast, the marginal model

$$\text{logit}[P(Y_1 = 1)] = \alpha, \quad \text{logit}[P(Y_2 = 1)] = \alpha + \beta$$

refers to the marginal proportions of responses in the two categories. According to it, the population odds that $Y_2 = 1$ are $\exp(\beta)$ times the population odds that $Y_1 = 1$. Because the subject-specific probability is a nonlinear function of $\beta$, the effect sizes differ in the two models.

Most subject-specific models represent subject (or cluster) effects by a random effects term and then assume that repeated responses given that effect are independent. For binary responses, the basic model has a logit link with a linear predictor that contains a random effect having a normal distribution. For instance, the model for the response $Y_{st}$ corresponding to the $t$th observation for subject $s$ might have form

$$\text{logit}P(Y_{st} = 1) = \alpha_s + \beta\mathbf{x}_t$$

where $\mathbf{x}_t$ is a vector of predictors for that response and $\{\alpha_s\}$ are *iid* from a normal distribution with unknown parameters. Random effects models can handle a variety of departures from standard models, including subject heterogeneity, unobserved covariates, and other forms of overdispersion. Subject-specific models themselves imply certain patterns of effects at the population-averaged level, but the implied model may have a complex form. For instance, a logit subject-specific model does not imply a marginal model of logit form. Approximate relationships exist between population-averaged and subject-specific parameters, with population-averaged effects being attenuated, more so when stronger within-subject correlation occurs (e.g., Neuhaus et al. 1991).

Regardless of the choice of model, ML estimation is difficult, because of the lack of a natural multivariate family of distributions for handling correlations among categorical responses. For mixed models, integrating out the random effect to obtain the likelihood function is complex, requiring an approximation such as numerical integration. Not surprisingly, MCMC has been applied here. One such approach is essentially Bayesian with improper priors, but this has the danger of potentially improper posteriors (Natarajan and McCulloch 1995). A promising alternative method is a Monte Carlo EM algorithm that uses random sampling to construct Monte Carlo approximations at the E step (Booth and Hobert 1998). This approach has the advantage that the Monte Carlo error can be accurately assessed at each iteration, and one can accurately reproduce the ML estimates with sufficiently many iterations.

Similarly, ML is difficult for marginal models. The model applies to the marginal distributions of the multivariate response rather than the joint distribution to which components of a multinomial likelihood refer. A weighted least squares approach is simple but has the severe limitation of categorical predictors with nonsparse data (Koch et al. 1977). Recently, generalized estimating equations (GEE) have provided a popular way to estimate parameters in marginal models. With this approach, it is unnecessary to specify fully the joint distribution. One specifies models (such as a logit model if the response is binary) only for marginal distributions and uses a working guess for the correlation structure (Liang and Zeger, 1986). With marginal models, since the association structure is not the primary focus, it is regarded as a nuisance. Estimates of model parameters are consistent even if the correlation structure is misspecified. The GEE methodology, originally specified for univariate marginal distributions such as the binomial and Poisson, extends to multinomial categorical responses. For ordinal responses, for instance, see Lipsitz et al. (1994). The GEE approach is appealing for marginal modeling because of its relative simplicity, but it has limitations. Since the model does not specify the joint distribution, it lacks a likelihood function and hence inference relies on Wald methods. It is unclear whether a particular choice of mean and variance function and correlation structure would be implied by any meaningful joint distribution for multinomial probabilities.

In principle, one can also specify and fit marginal models using maximum likelihood with a likelihood specified for a joint distribution. Multivariate logistic models have been defined that have a one-to-one correspondence between joint cell probabilities and parameters of marginal models as well as higher-order parameters of the joint distribution (e.g., Fitzmaurice and Laird 1993, McCullagh and Glonek 1995). This correspondence is awkward to specify for more than a few dimensions. Alternatively, one can treat a marginal model as a set of constraint equations and use methods of maximizing likelihoods subject to constraints (Lang and Agresti 1994). In this approach, it is possible also to model simultaneously the joint distribution. For instance, for repeated ordinal responses one might use a cumulative logit model for the marginal distributions and a model assuming a simple common odds ratio structure for the pairwise conditional associations. This approach also becomes computationally more difficult as the size of the joint distribution and the number of covariates increase. Generally, ML is difficult for marginal models except sometimes when the complete data form a contingency table. An important problem for the future is to find ways of extending ML to handle large multivariate problems.

The modeling of clustered correlated data is likely to be an active area of research in coming years. The GEE approach may lose some popularity as ML becomes computationally more feasible for marginal models and as modeling with random effects becomes more fully developed. The class of generalized linear mixed models is certain to see substantial work and further generalization in the near future. For instance, many applications have several random effects terms, and it is sensible to allow them to be correlated. To illustrate, one might observe a vector of responses, some continuous and some binary or counts, repeatedly in a longitudinal study. One might assume that, conditionally on a vector of random effects, this is a vector of normal, binomial, and Poisson counts, and that that vector of random effects has a multivariate normal distribution. Another extension is to allow generalized *additive* mixed models. For multinomial responses, the random effects literature is also incomplete, so far considering primarily cumulative logit and probit models. Time series models for categorical responses have also received relatively little attention. ML computations for all such models with correlated responses are complex, and model diagnostics and ways of handling missing data are of vital importance and need much more development.

## 5. Other areas

Many areas not addressed here are likely to see considerable future work. For instance, sample size and power considerations are difficult for categorical data analyses. The number of parameters can be quite large, and even with focusing on a single parameter in a model the nonconstant variance implies that one must guess marginal response proportions as well as the size of the effect. The multiplicity of parameters is also a complication for Bayesian modeling. Model specification can involve numerous components, especially for hierarchical models that also assume distributions for prior parameters. Model selection is also a sticky issue, and model averaging is likely to become more common. In population size estimation with capture-recapture modeling, for instance, it is not necessarily wise to choose the most parsimonious model that fits well, as it may lead to severe underprediction. Finally, new methods are unlikely to find much application unless accompanied by user-friendly software. Eventually one would hope for a "super-GLM" program that can handle a variety of strategies for discrete and continuous data, including ML fitting of marginal models, GEE methods, and mixed models for various random effects distributions. Even with binary data, such goals currently require a variety of software, and much more basic software needs exist that are not nearly as ambitious.

## 6. The challenge of fragmentation

As we approach the millennium, the state-of-the-art in categorical data analysis, as in all branches of statistics, is vastly different than at the start of this century. The variety of options for analyzing data continues to increase dramatically, and both statisticians and statistically-literate scientists have a serious challenge in choosing among them. Although some will disagree, to me there does not seem to be a single theoretical framework with the potential to unify us in providing a clear way to approach statistical analysis.

With the continuing explosion in research literature, the danger exists of increasing fragmentation among statisticians in ways of approach modeling and data analysis. For instance, even for a certain choice of response variable, probability distribution, and link function in a GLM, modeling approaches may vary according to the (1) handling of dependence due to factors such as clustering, (2) choice of a semiparametric or fully parametric approach, (3) choice of method of estimation (Bayes, frequentist), confidence interval and test type (likelihood-ratio, score, Wald; approximate or exact), (4) method of accounting for missing data, and (5) way of describing the effects in the model. Indeed, confusion about the "best" method can occur even for the seemingly simplest of data structures.

For example, how should one construct a confidence interval for a proportion or for a simple measure of association based on proportions, such as the odds ratio? 'Exact' methods exist, but their performance can be unsatisfactory for small samples because of discreteness, and they rarely can achieve a desired size. For interval estimation, for instance, the actual coverage probability can be much larger than the nominal confidence level and is unknown (Neyman 1935). The implication is conservativeness: As the relevant distribution becomes more highly discrete, exact tests lose power and exact confidence intervals tend to be overly wide. As a consequence, it is not obvious when one should use such a method instead of a large-sample approximate method.

To illustrate, I first consider interval estimation for a binomial parameter, $p$. The most commonly cited exact method, called the Clopper-Pearson interval, is based on inverting two single-tailed binomial tests. For nominal 95% confidence based on binomial outcome $x$ with index $n$ and sample proportion $\hat{p} = x/n$, the endpoints are the values of $p$ that satisfy

$$\sum_{k=x}^{n} \binom{n}{k} p^k (1-p)^{n-k} = .025 \text{ and } \sum_{k=0}^{x} \binom{n}{k} p^k (1-p)^{n-k} = .025.$$

An approximate interval can be based on large-sample normality of $\hat{p}$. This interval has endpoints that are the values of $p$ that satisfy $1.96 = (\hat{p} - p)/\sqrt{p(1-p)/n}$. The test inverted for this interval is the score test, and the interval itself follows directly from the general method for confidence interval construction given in Rao's (1947) classic paper. Because of the conservativeness of the exact interval, for most purposes the approximate method is actually better, even for small $n$. To illustrate, I compare them by reporting, as a function of $n$, the mean of the actual coverage probabilities for all $p$ in the [0, 1] parameter space. The mean for the nominal 95% Clopper-Pearson interval is .990 when $n = 5$, .980 when $n = 15$, and .973 when $n = 30$. By contrast, the means for the score interval are .955, .953, .952. Figure 1 plots the mean coverage probabilities for the Clopper-Pearson exact interval (E), the interval based on the score test (S), and the interval based on inverting the large-sample Wald test (W) with estimated standard error (i.e., solving $1.96 = (\hat{p} - p)/\sqrt{\hat{p}(1-\hat{p})/n}$. For further details, see Agresti and Coull (1998). Better methods exist of forming exact intervals for $p$, but the conservativeness issue persists.

Similarly, Cornfield's exact confidence interval for the odds ratio in a $2 \times 2$ table, based on inverting two separate exact tests for non-null values of the odds ratio, can also be very conservative. I illustrate for the odds ratio based on two independent binomial samples. For cell counts $\{a, b, c, d\}$, a large-sample 95% confidence interval based on the delta method is $\exp[\log(ad/bc) \pm 1.96(a^{-1} + b^{-1} + c^{-1} + d^{-1})^{1/2}]$. I randomly sampled 10,000 pairs of binomial parameters $(p_1, p_2)$ from the uniform distribution over the unit square and evaluated coverage probabilities of nominal 95% intervals, for binomial samples of size 10 each for each combination of parameters. The mean coverage probability was .977 for the large-sample method and .986 for the exact; the minimum coverage probability was .941 for the large-sample method and .970 for the exact. For each pair of parameters, I computed expected lengths conditional on the event that all four cell counts are positive, so that both intervals have finite length. Because of tail behavior, differences in coverage probability can translate to large differences in expected interval length. With each of the 10,000 probability pairs oriented so that $p_1 \geq p_2$, the median of the expected lengths was 59 for the large-sample method and 228 for the exact.

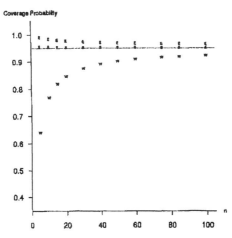

Figure 1. Mean coverage probability as a function of sample size for the nominal 95% exact (E), score (S), and wald (W) intervals, when $p$ has a uniform $(0,1)$ distribution

Situations exist, of course, in which it is imperative to guarantee a lower bound on the actual coverage probability and/or in which large-sample methods are highly unreliable, such as several degree-of-freedom chi-squared tests of fit and one-sided inference with highly unbalanced data sets. In addition, the conservativeness problem diminishes as the sample size and table size increase. Thus, exact methods have an important niche, and it is not the purpose of these examples to criticize them. The complications due to discreteness point out, though, that the choice of a statistical method may not be obvious even for experienced statisticians with simple problems.

The diversity of ways of handling even simple data sets such as 2×2 tables makes the goal of having statisticians agree about the analysis of more complex problems seem quite challenging. In addition, the problems of fragmentation alluded to above cause other obvious challenges, such as that of keeping up-to-date with new ideas across the field. At the same time, I do not want to sound too alarmist. Although more and more ways exist of analyzing data, we must keep in mind that in most cases different methods provide the same substantive conclusions. Distinctions such as

(Bayesian / frequentist) and (exact conditional / exact unconditional) that may seem fundamentally important to us do not seem all that striking to an outsider.

## 7. The challenge of communication

For any statistical subject, the continuing dramatic improvements in computing power have fueled the development of ever more complex methodology. A major challenge for research statisticians is to explain this methodology to those who are not specialists in that area and who could benefit from the methods. Statisticians should not underestimate this challenge. For instance, models for ordinal responses still seem to have little use in many disciplines, such as the social sciences, in which ordinal responses are common. Whereas it may not be crucial for scientists to understand the technical details of how to obtain parameter estimates and their standard errors, it is important for them to know what the estimate means. The task for the statistician in explaining model interpretation and differences among potential models, especially when the model is nonlinear in the mean response or has any random effect structure, is not simple. Even basic distinctions such as (probability / odds), (population-averaged / subject-specific), and (maximum likelihood / quasi likelihood) are confusing to many.

This challenge is part of a much broader one of improved communication of all types between statisticians and students and other faculty in academia and co-workers in industry and government. In terms of the impact of our field, this is a bigger challenge than any mentioned above in this article about the development of new methodology. In recent years, for instance, statistics departments at several major universities have faced threat of closure. In the U. S. statistics departments uniformly seem to be having difficulty attracting quality American students to graduate programs. We as statisticians may be much to blame for the poor regard we often have from administrators, academicians in other departments, and students.

As statistical methodology achieves a reasonable level of maturity, it seems to me that our profession should shift some emphasis from statistical research to collaborative research and the innovative application of statistical methods to problems in science and public affairs. More joint appointments in academia, and different expectations for tenure and promotion, would be helpful in this regard. We should also pay more attention to providing service through quality teaching of introductory statistics courses, through short courses, expository seminars, and course modules on recent advances, and through writing survey papers for journals in other disciplines. This is the work that will likely have the greatest impact on how our profession is viewed by others in the twenty-first century. Finally, statistics departments should expend more energy in expanding and improving the undergraduate curriculum for statistics majors and in attracting capable undergraduate majors. This will help to give the field more visibility and to channel better students to graduate programs.

It is perhaps all too easy to be a curmudgeon in making remarks about the future of statistical science. However, if you have any doubt about how well we have succeeded so far, simply think of the many times you have been introduced to others as a statistician. Probably the most common response you have had, as I have, is something to the effect of "I had to take a statistics course once. It was (choose one) horrible / traumatic / incomprehensible / irrelevant." There has been real progress in recent years in the development of textbooks that emphasize ideas and applications in presenting statistical science, but much can be improved. The statistics service course should not be watered-down mathematical statistics; even the undergraduate math stat course that enrolls engineers and business students as well as our own majors needs stronger emphasis on statistical modeling and using the likelihood function and less emphasis on optimality theorems that have limited practical relevance. The magnitude of this challenge is reinforced by the fact that four times as many college students take introductory statistics from mathematics departments as from statistics departments (see *AmStat News*, April 1998, p. 28), and countless more students take it from departments of psychology, sociology, business, education, . . . .

Of course, as our field becomes more fragmented, communication among statisticians is also more of a challenge. I believe that many statisticians could substantially improve their performance in this area, especially in verbal communication. In seminar presentations, for instance, many speakers seem to misjudge seriously how much an audience can absorb from one talk. How many listeners really care that much about the technical details rushed through as time is expiring? What are we expected to learn from figures in which axes and plots are unlabeled or poorly labeled, from tables that contain hundreds of numbers, from P-values, powers, and mean squared errors in the tables reported to four or more decimal places?

Recently I reread several papers given at a conference in 1967 on the future of statistics (Watts 1967). It is sobering to see how the overwhelming majority of concerns voiced then are still relevant today. In particular, in the preface Donald Watts noted

> Although the theme of the conference was The Future of
> Statistics, only one speaker addressed himself directly to that

topic. Significantly the other speakers and panelists ... all independently chose to discuss one important implicit feature which
profoundly affects statistics. That feature is *communication*. It
is vital to the future of statistics that we statisticians be able to
communicate among ourselves, but *it is equally vital that we be
able to communicate with others.*

## References

[1] Agresti, A., (1992), A survey of exact inference for contingency tables
(with discussion). *Statistical Science*, **7**, 131–177.

[2] Agresti, A. (1996) *An Introduction to Categorical Data Analysis*, New
York, Wiley.

[3] Agresti, A., and Coull, B. C. (1998) Approximate is better than 'exact' for interval estimation of binomial parameters, *The American
Statistician*, to appear (May).

[4] Agresti, A., Wackerly, D., and Boyett, J. (1979), Exact conditional
tests for cross classifications: Approximation of attained significance
levels. *Psychometrika*, **44**, 75–83.

[5] Becker, M. P., and Clogg, C. C. (1989), Analysis of sets of two-way
contingency tables using association models. *Journal of the American
Statistical Association*, **84**, 142-151.

[6] Berger, R. L., and Boos, D. D. (1994), *P* values maximized over a
confidence set for the nuisance parameter. *Journal of the American
Statistical Association*, **89**, 1012–1016.

[7] Booth, J., and Butler, R. (1998) Monte Carlo approximation of exact
conditional tests for log-linear models, to appear in *Biometrika*.

[8] Booth, J., and Hobert, J. (1998) Maximizing generalized linear mixed
model likelihoods with an automated Monte Carlo EM algorithm, to
appear in *Journal of the Royal Statistical Society*, B.

[9] Clopper, C. J., and Pearson, E. S. (1934), The use of confidence or
fiducial limits illustrated in the case of the binomial. *Biometrika*, **26**,
404–413.

[10] Cytel Software, (1995) *StatXact 3 for Windows*, Cambridge, MA.

[11] Fitzmaurice, G. M. and Laird, N. M. (1993), A likelihood-based
method for analysing longitudinal binary responses. *Biometrika*, **80**,
141–151.

[12] Forster, J. J., McDonald, J. W., and Smith, P. W. F. (1996), Monte
Carlo exact conditional tests for log-linear and logistic models. *Journal of the Royal Statistical Society, Series B*, **58**, 445–453.

[13] Goodman, L. A. (1978) *Analyzing Qualitative/Categorical Data*,
Cambridge, MA, Abt books

[14] Goodman, L. A. (1979), Simple models for the analysis of association in cross-classifications having ordered categories. *Journal of the American Statistical Association*, **74**, 537–552.

[15] Goodman, L. A. (1986), Some useful extensions of the usual correspondence analysis approach and the usual log-linear models approach in the analysis of contingency tables. *International Statistical Review*, **54**, 243–309.

[16] Grizzle, J. E., Starmer, C. F., and Koch, G. G. (1969), Analysis of categorical data by linear models. *Biometrics*, **25**, 489–504.

[17] Koch, G. G., Landis, J. R., Freeman, J. L., Freeman, D. H., and Lehnen, R. G. (1977), A general methodology for the analysis of experiments with repeated measurement of categorical data. *Biometrics*, **33**, 133–158.

[18] Lang, J. and Agresti, A. (1994), Simultaneously modeling joint and marginal distributions of multivariate categorical responses. *Journal of the American Statistical Association*, **89**, 625–632.

[19] Liang, K. Y., and Zeger, S. L. (1986), Longitudinal data analysis using generalized linear models. *Biometrika*, **73**, 13–22.

[20] Lipsitz, S. R., Kim, K., and Zhao, L. (1994), Analysis of repeated categorical data using generalized estimating equations. *Statistics in Medicine*, **13**, 1149–1163.

[21] McCullagh, P. (1980), Regression models for ordinal data (with discussion). *Journal of the Royal Statistical Society, B*, **42**, 109–142.

[22] McCullagh, P., and Glonek, G. F. V. (1995), Multivariate logistic models. *Journal of the Royal Statistical Society, B*, **57**, 533–546.

[23] Mehta, C. R. (1994), The exact analysis of contingency tables in medical research. *Statistical Methods in Medical Research*, **3**, 135–156.

[24] Mehta, C. R., and Patel, N. R. (1995), Exact logistic regression: Theory and examples. *Statistics in Medicine*, **14**, 2143–2160.

[25] Mehta, C. R., Patel, N. R., and Senchaudhuri, P. (1988), Importance sampling for estimating exact probabilities in permutational inference. *Journal of the American Statistical Association*, **83**, 999–1005.

[26] Neuhaus, J. M., Kalbfleisch, J. D., and Hauck, W. W. (1991), A comparison of cluster-specific and population-averaged approaches for analyzing correlated binary data. *International Statistical Review*, **59**, 25–35.

[27] Natarajan, R., and McCulloch, C. E. (1995), A note on the existence of the posterior distribution for a class of mixed models for binomial responses. *Biometrika*, **82**, 639–643.

[28] Neyman, J. (1935), On the problem of confidence limits. *Annals of*

*Mathematical Statistics*, **6**, 111–116.

[29] Pierce, D. A., and Peters, D. (1998) Improving on exact tests by approximate conditioning, unpublished manuscript.

[30] Presnell, B. (1996), Bootstrap unconditional P-values for the sign test with ties and the 2 × 2 matched-pairs trial. *Journal of Nonparametric Statistics*, **7**, 47–55.

[31] Rao, C. R. (1947), Large sample tests of statistical hypotheses concerning several parameters with applications to problems of estimation. *Proc. Cambridge Phil. Soc.*, **44**, 50–57.

[32] Rao, C. R. (1973) *Linear Statistical Inference and its Applications*, 2nd ed. New York, Wiley.

[33] Strawderman, and Wells, M. (1998) Approximately exact inference for the common odds ratio in several 2×2 tables, *Journal of the American Statistical Association*, **93**

[34] Yates, F. (1984), Tests of significance for 2×2 contingency tables. *Journal of the Royal Statistical Society, Series A*, **147**, 426–463.

# Consistency and Accuracy of the Sequential Bootstrap

G. Jogesh Babu[1] and C. R. Rao[2]

Department of Statistics
Pennsylvania State University
University Park, Pennsylvania

P. K. Pathak

Department of Statistics and Probability
Michigan State University
East Lansing, Michigan

**Abstract.** The object of this paper is to present a brief account of the sequential bootstrap from a survey sampling point of view. This sequential resampling scheme entails resampling from the observed sample sequentially (with replacement) until a preassigned number of distinct original observations appear. This approach stems from the observation made by Efron in 1983 that the usual bootstrap samples are supported on approximately $.632n$ of the original data points. We outline a number of approaches that can be employed to study the theoretical as well as the empirical properties of the sequential bootstrap. Our investigation shows that there is a great potential for sequential bootstrap in applications often encountered in practice.

[1] Supported by NSA grant MDA 904-97-1-0023 and NSA grant DMS-9626189.

[2] Supported by the Army Research Office under Grant DAAH04-96-1-0082.

## 1. Introduction

In this study we have examined the resampling procedure for bootstrap from a survey sampling point of view. Given an observed sample, resampling for the bootstrap involves $n$ repeated trials of simple random sampling with replacement (SRSWR). It is well known that SRSWR does not yield samples that are equally informative ([10]) as it results in different number of distinct observations in different bootstrap samples. Our investigation shows that this randomness in the information content of the bootstrap samples is unnecessary. Stemming from the observation made by Efron ([6]) that the usual bootstrap samples are supported on approximately $k = (1-e^{-1})n \sim .632n$ of the original data points, we have proposed a sequential resampling procedure which keeps the information content of each sample at a constant level without affecting its correctness properties.

Let $S = (x_1, \ldots, x_n)$ denote a random sample from a distribution $F$, and let $\theta(F)$ be a given parameter of interest. As an illustrative example we take $\theta(F)$ to be the population mean:

$$(1.1) \qquad \mu(F) = \int x \, dF.$$

Let $F_n$ denote the empirical distribution function based on $S$ so that $F_n(x) = n^{-1} \Sigma \delta_{x_i}(x)$ in which $\delta_{x_i}$ denotes the delta function at the $i$th sample observation $x_i$, $1 \leq i \leq n$. We also assume for simplicity in the exposition that the population variance $\sigma^2 = \int (x - \mu)^2 dF = 1$. Consider now the plug in estimator $\mu_n := \mu(F_n)$ of $\mu(F)$ so that

$$(1.2) \qquad \mu_n = \frac{1}{n} \sum_{i=1}^{n} x_i$$

with the corresponding pivot:

$$\pi_n = \sqrt{n}(\mu_n - \mu) = \sqrt{n}(\bar{x} - \mu)$$

in which $\bar{x}$ is the sample mean and $\mu$ is the population mean.

The central limit theorem entails that the sampling distribution of $\pi_n$ can be approximated by the standard normal distribution. On the other hand, the bootstrap furnishes an alternative approach based on resampling to estimating the sampling distribution of $\pi_n$ from $S$ more precisely. Generally speaking the central limit approximation is accurate to $o(1)$, while resampling approximation is accurate to order $o(\frac{1}{\sqrt{n}})$. For example, let $G_n$ denote the distribution of $\pi_n$. Then the central limit theorem yields that

$$\|G_n - \Phi\|_\infty : = \sup_x |G_n(x) - \Phi(x)|$$

$$(1.3) \qquad\qquad = o(1)$$

where $\Phi(x) = \frac{1}{\sqrt{2\pi}} \int_{-\infty}^{x} e^{-t^2/2} dt$, while the bootstrap approximation captures the skewness of the distribution $G_n$ in the following sense:

$$(1.4) \qquad \|G_n - H_n\|_\alpha = o(1/\sqrt{n})$$

in which $H_n$ represents a one-term Edgeworth expansion for $G_n$.

The usual bootstrap procedure to approximating the distribution $G_n$ is based on resampling. Given the observed sample $S$, one selects a simple random sample with replacement (SRSWR) of size $n$ from the observed sample $S$. Let $S_n^* = (x_1^*, \ldots, x_n^*)$ be an SRSWR sample drawn from $S$ in this manner. Let $F_n^*$ be the empirical distribution based on $S_n^*$. Let

$$(1.5) \qquad \begin{aligned} \pi_n^* &= \sqrt{n}(\mu(F_n^*) - \mu(F_n))/\sigma(F_n) \\ &= \sqrt{n}(\mu_n^* - \mu_n)/s_n, \quad \text{say}, \end{aligned}$$

denote the pivot based on $S_n^*$, where $s_n^2 = \frac{1}{n}\sum_1^n (x_i - \bar{x})^2$. Then for large $n$, the conditional distribution of $\pi_n^*$ given $S$ is close to that of $\pi_n$. In practice, this conditional distribution of $\pi_n^*$ is approximated by the observed frequency distribution (ensemble) of $\pi_n^*$ obtained by repeated resamplings from $S$ by SRSWR of size $n$ a large number of times of order $O(1000)$. This observed frequency distribution is referred to as the bootstrap distribution of the original pivot $\pi_n$.

We can now examine what may be viewed as a certain drawback of this method of resampling. It is well known that owing to the with replacement nature of SRSWR, not all of the observations in an observed bootstrap sample $S_n^*$ will be based on distinct observations from $S$. In fact, the information content of $S_n^*$, the set of distinct observations from $S$, is a random variable. Let $\nu_n$ denote the number of distinct observations in $S_n^*$. Then

$$(1.6) \qquad \begin{aligned} E(\nu_n) &= n[1 - (1 - \frac{1}{n})^n] \\ &= n(1 - e^{-1}) + O(1) \end{aligned}$$

$$(1.7) \qquad \simeq n(.632)$$

$$(1.8) \qquad V(\nu_n) = ne^{-1}(1 - e^{-1}) + O(1)$$

$$(1.9) \qquad \simeq n(.32)n$$

so that

$$(1.10) \qquad SD(\nu_n) \simeq (.48)\sqrt{n}.$$

Thus on the average, the usual bootstrap utilizes approximately 63% of the information from the original sample, the rest of the 37% of the data in it represents repetitious information. The $2\sigma$-limits for $\nu_n$ are approximately $(.63)n \pm 2(.48)\sqrt{n} \simeq (.63)n \pm \sqrt{n}$. For example, if $n = 100$, $\nu_n$ ranges from a lower limit of 53 to an upper limit of 73 in approximately 95% of the bootstrap samples. This sort of randomness in the information content of bootstrap samples can be eliminated by adopting a sequential approach of the kind described in the following sections.

## 2. A Sequential Resampling Approach

To select a bootstrap sample, draw sample units (observations) from $S$ sequentially by SRSWR until there are $(m + 1) \approx n(1 - e^{-1}) + 1$ distinct sample units (original observations from $S$) in the observed bootstrap sample; discard the last observation. The recorded sequential bootstrap sample has the form:

$$(2.1) \qquad S_N^* = (x_1^*, x_2^*, \ldots, x_n^*)$$

in which the number of distinct observations is precisely $[n(1 - e^{-1})]$. The size $N$ of the bootstrap sample is now a random variable with $E(N) \approx n$. The pivot based on $S_N^*$ is

$$(2.2) \qquad \begin{aligned} \pi_N^* &= \sqrt{N}(\mu(F_N^*) - \mu(F_n))/\sigma(F_n) \\ &= \sqrt{N}(\mu_N^* - \mu_n)/s_n, \quad \text{say} \end{aligned}$$

in which $F_N^*$ denotes the empirical distribution based on the sequential bootstrap sample $S_N^*$.

The simplest approach to studying the sequential bootstrap is to compare the pivot $\pi_N^*$ (based on the sequential approach) with the pivot $\pi_n^*$ (based on the bootstrap of constant size $n$). It is easily seen that the random sample size $N$ admits the following representation in terms of independent (geometric) random variables:

$$(2.3) \qquad N = I_1 + I_2 + \ldots + I_m$$

in which $m = [n(1 - e^{-1})]$; $I_1 = 1$, and for each $k$, $2 \leq k \leq m$,

$$(2.4) \qquad P(I_k = j) = \left(1 - \frac{k-1}{n}\right)\left(\frac{k-1}{n}\right)^{j-1}$$

for $j = 1, 2, \ldots$. Therefore

$$E(N) = n[\frac{1}{n} + \frac{1}{(n-1)} + \ldots + \frac{1}{(n-m+1)}]$$

(2.5)
$$= n + O(1).$$

Similarly

(2.6)
$$V(N) = \sum_{k=1}^{m} \frac{n(k-1)}{(n-k+1)^2} = n(e-2) + O(1).$$

Thus

(2.7)
$$\frac{E(N-n)^2}{n^2} = \frac{(e-2)}{n} + O\left(\frac{1}{n^2}\right)$$

so that $(N/n)$ tends to 1 in probability. Further an analysis similar to that of [9] shows that Hajek's disparity between $\pi_n^*$ and $\pi_N^*$ satisfies:

$$\frac{E(\pi_N^* - \pi_n^*)^2}{\text{Var}(\pi_n)} \leq k\sqrt{\frac{\text{Var}(N)}{n^2}}$$

(2.8)
$$= O\left(\frac{1}{\sqrt{n}}\right).$$

This inequality implies that $\pi_N^*$ and $\pi_n^*$ are asymptotically equivalent, thus providing a simple justification that the sequential bootstrap provides its asymptotic correctness (consistency).

A second approach for consistency of the sequential bootstrap can be based on the so-called Mallows' metric (see, for example, [5]). Suppose that we define a sequence of distribution functions $\{G_n : n \geq 1\}$ to converge in $M$-sense to a distribution function $G$ iff (a) the sequence $\{G_n : n \geq 1\}$ converges weakly to $G$, and (b) $\int x^2 \, dG_n$ converges to $\int x^2 \, dG$. Then it is easily seen that this convergence is induced by the following metric ([8]):

(2.9)
$$d^2(F, G) = \inf_{x \sim F, y \sim G} E(x - y)^2$$

where the infimum is with respect to all random variables $x$ and $y$ such that $x$ has the distribution $F$ and $y$ has the distribution $G$.

Under the metric (2.9), it is easily shown that

(2.10)
$$d(\pi_n^*, \pi_n) \leq d(F_n, F)$$

so that

$$d(\pi_n^*, \Phi) \le d(\pi_n^*, \pi_n) + d(\pi_n, \Phi)$$
(2.11)
$$\le d(F_n, F) + d(\pi_n, \Phi).$$

The first term on the right converges to zero by the law of large numbers, while the second term does so by the central limit theorem. This establishes the consistency of the usual constant size bootstrap. The added complication in the sequential case is that the sample size $N$ is now a random variable and for consistency to go through we need to show that $d(\pi_N^*, \Phi)$ can be made arbitrarily small. Based on techniques similar to those of [9] and [10], it can be shown that

(2.12)
$$d(\pi_N^*, \pi_n^*) = O(n^{-1/4}).$$

Consistency of the pivot $\pi_N^*$ now follows as a consequence of (2.11), (2.12) and the triangle inequality for metric spaces.

A limitation of the preceding two approaches is that they apply only to linear statistics and cannot be easily extended to more general statistics such as the sample quantiles. A third and more general approach is to treat pivots $\pi_n^*$ and $\pi_N^*$ as random signed measures and study their convergence in the functional sense. In this functional setting, we have established the following key results ([11]).

**Theorem 2.1** *Let $F$ be a distribution function in $R'$. Let $F_n^*$ and $F_N^*$ denote empirical distribution functions based respectively on the usual and the sequential bootstrap samples. Then*

(2.13)
$$\|F_N^* - F_n^*\|_\infty = O_p(n^{-\frac{3}{4}})$$

(2.14)
$$\|\sqrt{N}(F_N^* - F_n^*) - \sqrt{n}(F_n^* - F_n)\|_\infty = O_p(n^{-\frac{1}{4}})$$

Moreover if $F$ is uniform on $[0.1]$, the sequence of stochastic processes

$$\{\sqrt{N}(F_N^*(t) - F_n(t)) : 0 \le t \le 1\}$$

converge weakly to the standard Brownian bridge $B(t) := w(t) - tw(1)$, where $w(t)$ is the standard Wiener process. More generally, if $F$ is any continuous distribution function then this limit is $(BoF)(t) = B(F(t)), -\infty < t < \infty$. Results such as these imply consistency of the sequential bootstrap for statistics of the form $\theta_n = T(F_n)$ in which $T$ is a compactly differentiable functional and includes functionals such as the quantiles.

The three different approaches of this section show that the distance between the sequential and the usual bootstrap is at most of the order of $O(n^{-\frac{1}{4}})$. Although this entails the consistency of the sequential bootstrap, it does not guarantee its second order correctness. To do so, one needs to capture the skewness of the pivot $\pi_N^*$. There are two approaches for accomplishing it.

## 3. Second Order Correctness of the Sequential Bootstrap

The proof of the second order correctness of the sequential bootstrap requires Edgeworth type expansions for dependent random variables. A simple rigorous justification of such expansion is unavailable in the literature. Along the lines of [7], we first outline an approach based on the computation of cumulants. This approach assumes that a formal Edgeworth expansion is valid for pivot under sequential bootstrap.

Let $N_i$ denote the number of times the $i$th observation $x_i$ from the original sample appears in the sequential bootstrap sample, $1 \leq i \leq n$. Then

$$(3.1) \qquad N = N_1 + N_2 + \ldots + N_n$$

in which $N_1, N_2, \ldots$ are exchangeable random variables.

The second order correctness of the sequential bootstrap for statistics such as the sample sum is closely related to the behavior of the moments of the random variables $N_i$'s, $1 \leq i \leq n$. Among other things the asymptotic distribution of each $N_i$ should be Poisson with mean 1. In fact it can be shown that

$$(3.2) \quad E(N_1 - 1)^{k_1} \ldots (N_l - 1)^{k_l} = \prod_{i=1}^{l} (e^\Delta - 1 - \Delta)(x - 1)^{k_i} + O\left(\frac{1}{n}\right)$$

where $\Delta$ is the difference operator with unit increment.

It follows from (3.2) that to the order of $O(1/n)$, the $N_i$'s are asymptotically independent. This implies that the Hall-Mammen ([7]) type conditions for the second order correctness of the sequential bootstrap hold. This approach is based on the tacit assumption that formal Edgeworth type expansions go through in the sequential bootstrap.

A second approach without such assumption entails the following modification of the sequential bootstrap. It is based on the Poisson distribution.

## Poisson Resampling Scheme:

The original sample $(x_1, \ldots, x_n)$ is assumed to be from $\mathbb{R}^k$ for greater flexibility. Let $\alpha_1, \ldots, \alpha_n$ denote $n$ independent observations from $P(1)$, the Poisson distribution with unit mean. If there are exactly $m = [n(1 - e^{-1})]$ non-zero values among $\alpha_1, \ldots, \alpha_n$, take

$$(3.3) \qquad S_N^* = \{(x_1, \alpha_1), \ldots, (x_n, \alpha_n)\},$$

otherwise reject the $\alpha$'s and repeat the procedure. This is the conceptual definition. The sample size $N$ of the Poisson resampling scheme admits the representation:

$$(3.4) \qquad N = \alpha_1 + \alpha_2 + \ldots + \alpha_n$$

in which $\alpha_1, \ldots, \alpha_n$ are iid Poisson random variable with mean $\lambda = 1$ and with the added restriction that exactly $m$ of the $\alpha$'s are non-zero, i.e. $I_{\{\alpha_1 > 0\}} + \ldots + I_{\{\alpha_n > 0\}} = m$. Further it can be shown that the moment generating function $M_N(t)$ of $N$ is:

$$(3.5) \qquad M_N(t) = \left[ \frac{\left(e^{(e^t - 1)} - e^{-1}\right)}{(1 - e^{-1})} \right]^m$$

so that the distribution of $N$ can be viewed as that of $m$ iid random variables with a common moment generating function:

$$(3.6) \qquad m(t) = \frac{\left(e^{(e^t - 1)} - e^{-1}\right)}{(1 - e^{-1})}.$$

It is clear that $m(t)$ is the moment generating function of the Poisson distribution with mean $\lambda = 1$ and censored at $x = 0$. This representation of $M_N(t)$ provides a practical way of implementing this scheme. To implement Poisson resampling scheme, first assign at random $(n - m)$ $\alpha$'s to zero and then to the remaining $m$ $\alpha$'s assign values independently chosen from the Poisson distribution with mean $\lambda = 1$ and censored at $x = 0$.

This modification of the sequential bootstrap enables us to develop a rigorous proof of the second order correctness in the sequential case. The techniques we use are similar to those of [1], [3] and [4].

Let $\{Y_j : j \geq 1\}$ be a sequence of iid Poisson random variables with mean $\lambda = 1$. Let

$$(3.7) \qquad V_n^2 = \frac{1}{n} \sum_{j=1}^{n} (x_j - \bar{x})(x_j - \bar{x})'$$

$$(3.8) \qquad c_j = V_n^{-1}(x_j - \bar{x})$$

$$(3.9) \qquad P_n(x) = \frac{1}{6n} \sum_{j=1}^{n} ((c_j x)^3 - 3(c_j^2)(c_j x))$$

$$(3.10) \qquad N = \sum_{j=1}^{n} y_j$$

$$(3.11) \qquad T_n = \sum_{j=1}^{n} I_{\{y_j > 0\}}$$

$$(3.12) \qquad \bar{Y} = \frac{N}{n}$$

$$(3.13) \qquad U_n = \frac{1}{\sqrt{n}} \sum_{j=1}^{n} y_j V_n^{-1}(x_j - \bar{x})$$

$$(3.14) \qquad \tilde{\Psi}_n(x) = (1 + \frac{1}{\sqrt{n}} P_n(x)) \frac{e^{-\|x\|^2/2}}{(\sqrt{2\pi})^k}.$$

For a real-valued function $h$, let

$$(3.15) \qquad M_h = \sup |h(x)|(1 + \|x\|)^{-3}$$

and for any $\delta > 0$, let

$$(3.16) \qquad \omega(h, \delta; x) = \sup_{\|x - z\| < \delta} |h(x) - h(z)|$$

$$(3.17) \qquad \omega(h, \delta) = \int \omega(h, \delta; x) \phi(x) \, dx$$

where $\phi$ denotes the standard normal density.

In terms of the foregoing terminology, the following results furnish a rigorous justification for the second order correctness of the sequential bootstrap. These results are consequences of a technical result on conditional Edgeworth expansions for weighted means of multivariate random vectors presented in [2].

**Theorem 3.1** *Let $x_1, x_2, \ldots$ be i.i.d. random vectors with mean $\mu$ and covariance matrix $\Sigma$. Let $H$ be a 3-times continuously differentiable function in a neighborhood of $\mu$. Suppose that $x_1$ has a strongly non-lattice* distribution and $E(\|x_1\|^3) < \infty$. Let $l(y)$ denote the vector of first order partial derivatives at $y$, and $l(\mu) \neq 0$. If $m - n(1 - e^{-1})$ is bounded, then for almost all sample sequences $x_1, x_2, \ldots$, we have

$$\sup_z \sqrt{n} \left| P\left( \frac{\sqrt{N}(H(\frac{1}{N}\sum_{i=1}^n x_i y_i) - H(\bar{x}))}{\sqrt{l'(\bar{x})V_n^2 l(\bar{x})}} \leq z \middle| T_n = m; x_1, \ldots, x_n \right) \right.$$
$$\left. - P\left( \sqrt{n}\left( H(\bar{x}) - H(\mu) \right) \leq z\sqrt{l'(\mu)\Sigma l(\mu)} \right) \right| \to 0,$$

as $n \to \infty$.

The next result is more suitable for applications to studentized statistics.

**Theorem 3.2** Let $\{x_n\}$ be as in Theorem 3.1. Suppose the function $H$ is three times continuously differentiable in a neighborhood of the origin and $H(0) = 0$. If $m - n(1 - e^{-1})$ is bounded, then for almost all sample sequences $x_1, x_2, \ldots$, we have

$$\sup_z \sqrt{n} \left| P\left( \frac{\sqrt{N}H(\frac{1}{N}\sum_{i=1}^n (x_i - \bar{x})y_i)}{\sqrt{l'(0)V_n^2 l(0)}} \leq z \middle| T_n = m; x_1, \ldots, x_n \right) \right.$$
$$\left. - P\left( \sqrt{n}H(\bar{x} - \mu) \leq z\sqrt{l'(0)\Sigma l(0)} \right) \right| \to 0,$$

as $n \to \infty$.

It is easily seen that the second order correctness of the sequential bootstrap pivot such as

$$\pi_N^* = \sqrt{N}(\sum_{i=1}^n (x_j - \bar{x})y_j)/s_n$$

given that $T_n := \left( \sum_{j=1}^n I_{\{y_j > 0\}} \right) = m$ follows from Theorem 3.2. The one-term correction captures the skewness of the underlying distribution. For further details the reader is referred to [2].

### References

[1] Babu, G.J. and Bai, Z.D. (1996), Mixtures of global and local Edgeworth expansions and their applications. *J. Multivariate Analysis*, **59**, 282–307.

[2] Babu, G.J., Pathak, P.K. and Rao, C.R. (1998) *Second order corrections of the sequential bootstrap*, submitted for publication.

[3] Babu, G.J. and Singh, K. (1989), On Edgeworth expansions in the mixture cases. *Annals of Statistics*, **17**, 443–447.

[4] Bai, Z.D. and Rao, C.R. (1992), A note on Edgeworth expansion for ratio of sample means. *Sankhya*, **54 A**, 309–322.

[5] Bickel, P.J. and Freedman, D.A. (1981), Some asymptotic theory for the bootstrap. *Annals of Statistics*, **9**, 1196–1217.

[6] Efron, B. (1983), Estimating the error rate of a prediction rule: Improvement on cross-validation. *J. Amer. Statist. Assoc.*, **78**, 316–331.

[7] Hall, P. and Mammen, E. (1994), On general resampling algorithms and their performance in distribution estimation. *Annals of Statistics*, **24**, 2011–2030.

[8] Mallows, C.L. (1972), A note on asymptotic joint normality. *Annals of Statistics*, **43**, 508–515.

[9] Mitra, S.K. and Pathak, P.K. (1984), The nature of simple random sampling. *Annals of Statistics*, **12**, 1536–1542.

[10] Pathak, P.K. (1964), Sufficiency in sampling theory. *Annals of Mathematics Statistics*, **35**, 795–808.

[11] Rao, C.R., Pathak, P.K., Koltchinskii, V.I. (1997), Bootstrap by sequential resampling. *Journal of Statistical Planning and Inference*, **64**, 257–281.

# Hypothesis Testing in the 20th Century with a Special Reference to Testing with Misspecified Models

Anil K. Bera

Department of Economics,
University of Illinois,
Champaign, Illinois

**Abstract.** This century and the history of modern statistics began with Karl Pearson's, [181], (1900) goodness–of–fit test, one of the most important breakthroughs in science. The basic motivation behind this test was to see whether an assumed probability model adequately described the data at hand. Then, over the first half of this century we saw the developments of some general principles of testing, such as Jerzy Neyman and Egon Pearson's, [159], (1928) likelihood ratio test, Neyman's, [155], (1937) smooth test, Abraham Wald's test in 1943 and C.R. Rao's score test in 1948. All these tests were developed under the assumption that the underlying model is correctly specified. Trygve Haavelmo, [99], (1944) termed this underlying model as the priori admissible hypothesis. Although Ronald Fisher, [80], (1922) identified the "Problem of Specification" as one of the most fundamental problems in statistics much earlier, Haavelmo was probably the first to draw the attention to the consequences of misspecification of the priori admissible hypothesis on the standard hypothesis testing procedures. We will call this the type-III error. In this paper, we will deal with a number of ways that an assumed probability model can be misspecified, and discuss how some of the standard tests could be modified to make them valid under various misspecification.

## 1. Introduction

Science in each century comes with its pioneers and their innovations. The twentieth century is no exception. In broader science it started with Max Planck's theory of quantum physics and Sigmund Freud's publication of *The Interpretation of Dreams*. Statistics, though it has been used for centuries, is a relatively new science. Statistics in this century started with the Pearson, [181], (1900)[1] goodness–of–fit test, which is regarded as one of the 20 most important scientific breakthroughs of this century along with advances and discoveries like the theory of relativity, the IQ test, hybrid corn, antibiotics, television, the transistor and the computer [see Hacking, [101], (1984)]. Although there is always a danger in personalizing history, we can take Pearson, [181], (1900) as the starting point of modern statistics [see Mahalanobis, [148], (1933)]. For any-one encountering the current statistical scene for the first time, it is almost impossible to guess how the statistical world looked 100 years ago. Only a handful of academicians were actively engaged in statistical research at the turn of the century. During this century, however, statistics has developed from its infancy into a mature science. Nowadays, statistics is being applied quite successfully in almost all disciplines. Psychometrics, biometrics, econometrics and sociometrics are all now regarded as established fields. The main objectives of statistical techniques can be broadly divided into two interrelated areas, namely, estimation and testing. As the title suggests, this paper is concerned only with testing. Over the first half of this century, we saw the development of some general principles of testing, such as Neyman and Pearson's, [159], (1928) likelihood ratio (LR) test, Neyman's, [155], (1937) smooth test, Wald's, [217], (1943) test and Rao's, [186], (1948) score (RS) or the Lagrange multiplier (LM) test. In the next section we will review these tests and their influence in econometric model specification tests. Compared with statistics, econometrics is a relatively new discipline. Work in econometrics began in the 1920s mainly due to the initiatives of Ragnar Frisch and Jan Tinbergen. The new discipline got off to a slow start and was not welcomed by many eminent economists, including John Maynard Keynes. Keynes' very skeptical view was expressed in his review of Tinbergen's book, *Statistical Testing of Business–Cycle Theories* [see Hendry, [109], (1980)]. Substantial progress in econometrics was made in the late 1940s and the early 1950s following the works of Chernoff, Haavelmo, Koopmans, Rubin and Simon from the then Cowles Commission for Research in Economics. They for-

---

[1] Here, I follow the popular, though technically inaccurate custom of including the year 1900 in the twentieth century.

mulated the simultaneous equation systems and discussed their identification and estimation problems. This led to another ambitious project, the development of large scale econometric models under the guidance of Lawrence Klein. During the 1950s and 1960s, numerous econometric models were formulated, estimated and used for policy purposes. Their success was phenomenal. However, the main problem in these model-building exercises was that issues of specification and testing were systematically ignored, while attention focussed on the estimation. This was quickly recognized by econometricians, and during the late 1970s and 1980s substantial progress was made in developing a number of model specification tests.

All the test principles discussed in Section 2 were developed under the assumption that the underlying probability model is correct. In section 3, we outline some of the possible ways a probability model can be misspecified, and then discuss how the some of the misspecified test can be modified to make them valid under misspecification. We close the paper in Section 4 with some concluding remarks. At the outset, we would like to mention three features of this paper. First, we try to explore how the developments in *econometrics* are intertwined with those in *statistics*. Second, and closely related, many of the recent advances in testing in econometrics may be traced back to much earlier developments. Third, along with pure academic discussions, we also try to provide glimpses of the personalities involved, and to put some faces behind in our day–to–day statistical and econometric tools. We hope that these features will provide a sense of historical development, both academic and personal, that has happened in this field area over the last 100 years.

## 2. A Brief History of Testing

Although the systematic use of hypothesis testing began after the publication of Pearson, [181], (1900), the history of statistical testing is, indeed, very long. In the introduction to their paper, Neyman and Pearson, [160], (1933), pages 289-290 wrote:

"The problem of testing statistical hypothesis is an old one. Its origin is usually connected with the name of *Thomas Bayes*, who gave the well known theorem on the probabilities a *posteriori* of the possible 'causes' of a given event."

Although Bayes, [20], (1763) did not not explicitly mention testing, the paper posed the fundamental problem of what we can say about a "parameter" after observing the "data"; more specifically (in modern notation), what is $Pr[a < \theta < b | Y = y]$ given $Y \sim \text{Binomial}(n, \theta)$. According to

Lehmann, [138], (1986), page 126 hypothesis testing developed gradually over time and he cited the writings of Gavarret, [85], (1840), Lexis, [141], (1875), [142], (1877) and Edgeworth, [73], (1885) in the nineteenth century.

## 2.1 Pearson's (1900) Goodness–of–Fit Test and R.A. Fisher

To describe the Pearson, [181], (1900) $\chi^2$ test let us consider the multinomial distribution with $p$ classes and let the probability of an observation belonging to the $j$-th class be $\theta_j (\geq 0)$, $j = 1, 2, \ldots, p$, so that $\sum_{j=1}^{p} \theta_j = 1$. We are interested in testing $H_0$: $\theta_j = \theta_{j0}$, $j = 1, 2, \ldots, p$, where $\theta_{j0}$'s are known constants. Let $n_j$ denote the observed frequency of the $j$-th class, with $\sum_{j=1}^{p} n_j = n$. Pearson, [181], (1900) suggested the goodness-of-fit test statistic

$$(2.1) \qquad P = \sum_{j=1}^{p} \frac{(n_j - n\theta_{j0})^2}{n\theta_{j0}} = \sum \frac{(O - E)^2}{E},$$

where O and E denote observed and expected frequencies.

Pearson, [181], (1900) demonstrated that under $H_0$, $P$ is asymptotically distributed as $\chi^2_{p-1}$. The basic motivation behind this test was to see whether an assumed probability model adequately described the data at hand, and it can be used after fitting any probability model not just within the multinomial model described above. To outline Pearson's approach, let us define the difference between the observed and theoretical (expected) frequencies as errors:

$$(2.2) \qquad e_j = n_j - n\theta_j, \qquad j = 1, 2, \ldots, p.$$

By definition $\sum_{j=1}^{p} e_j = 0$; therefore, effectively we have $(p-1)$ errors. By denoting $e = (e_1, e_2, \ldots, e_{p-1})'$, it can be easily shown that

$$E(e) = 0$$

and

$$(2.3) \qquad V(e) = n[\operatorname{diag}(\theta_1, \theta_2, \ldots, \theta_{p-1}) - \theta\theta'],$$

where $\theta = (\theta_1, \theta_2, \ldots, \theta_{p-1})'$. For large $n$, $e$ converges in distribution to normal and, therefore,

$$(2.4) \qquad e'V(e)^{-1}e d \to \chi^2_{p-1}.$$

where $d \to$ denotes convergence in distribution. Let us define $e_{j0} = n_j - n\theta_{j0}$, $j = 1, 2, \ldots, p-1$ and $e_0 = (e_{10}, e_{20}, \ldots, e_{(p-1)0})'$. Then

(2.5) $$P = e_0' V_0^{-1} e_0,$$

where $V_0$ is $V(e)$ in (2.3) but evaluated at $\theta_j = \theta_{j0}$. Since both $\theta_j$ and $\theta_{j0}$ are simply constants, under $H_0$,

(2.6) $$e_0' V_0^{-1} e_0 d \to \chi_{p-1}^2.$$

Here, we are testing $H_0 : \theta_j = \theta_{j0}$, $j = 1, 2, \dots, p$, which is a simple hypothesis. If the null hypothesis is composite with some nuisance parameters $\eta$, then Fisher, [80], (1922) demonstrated that the degrees of freedom should be reduced by the number of (nuisance) parameters. Karl Pearson, however, insisted that estimating parameters should not make any difference in large samples. Fisher's result can be stated, in a general form, as: under $H_0$, Pearson's $P(\eta)$ can be decomposed asymptotically as

(2.7) $$P(\eta) = P(\hat{\eta}) + (\hat{\eta} - \eta)' V(\hat{\eta})^{-1} (\hat{\eta} - \eta),$$

where $\hat{\eta}$ is an efficient estimator of $\eta$ [see Bera and Bilias, [22], (1999)]. The two components in the right hand side of (2.7) can also be shown to be independent. Now under certain regularity conditions

(2.8) $$(\hat{\eta} - \eta)' V(\hat{\eta})^{-1} (\hat{\eta} - \eta) d \to \chi_q^2,$$

where $q$ is the dimension of $\eta$. Given $P(\eta)d \to \chi_{p-1}^2$, it follows that $P(\hat{\eta})d \to \chi_{p-q-1}^2$.

Fisher, [81], (1922) and Yule, [229], (1922) that supported Fisher's results were published in the same issue of the *Journal of Royal Statistical Society*. These papers angered Karl Pearson, and his reply [Pearson, [183], (1922)] appeared shortly after in *Biometrika*. Although Fisher wanted to respond, The Royal Statistical Society declined to publish anything more on the debate, and this ultimately led to Fisher's resignation from the society. For an account of this Pearson–Fisher debate see Box, [42], (1978) pages 84-88, Agresti, [1], (1996) pages 259-260 and Hald, [102], (1998), Section 27.4 and for general treatment on Pearson's goodness-of-fit test see Lancaster, [134], (1969), Rayner and Best, [190], (1989), Barnard, [16], (1992) and Greenwood and Nikulin, [96], (1996).

In spite of the shortcomings of Pearson's work, the importance of the $\chi^2$ test and its profound impact on the development of statistics is beyond question. In fact, Fisher himself recognized its value, along with the Karl Pearson's work on the system of frequency curves and method of moment estimation. In Fisher, [80], (1922), page 314, he stated: "We may instance the development by *Pearson* of a very extensive system of skew curves, the elaboration of a method of calculating their parameters, and the preparation of the necessary tables, a body of work which has enormously extended the power of modern statistical practice, and which has been, by pertinacity and inspiration alike, practically the work of a single man. Nor is the introduction of the Pearsonian system of frequency curves the only contribution which their author has made to solution of problems of specification: of even greater importance is the introduction of an objective criterion of goodness-of-fit."

After Karl Pearson, Ronald Fisher took the lead in developing modern statistical methodologies and the underlying theories. It can be said that Fisher took up where Pearson left off [see Neyman, [157], (1967)]. When he was a third year undergraduate at Gonville and Caius college, Cambridge, he suggested the maximum likelihood estimation procedure as an alternative to Pearson's, [180], (1894) method of moment estimation [Fisher, [79], (1912)]. In 1922 he published a monumental paper, Fisher, [80], (1922) which introduced such basic concepts as consistency, efficiency, sufficiency, and even the term "parameter" with its present meaning [see Stephen Stigler's comment on Savage, [201], (1976)]. The intervening period 1913-1920, however, was years in the "wilderness" for Fisher [Box, [42], (1978), Chapter 2]. After being rejected for the military due to poor eyesight, he did some odd jobs and even planned to try farming. During this period he also came in contact with William Gosset (Student), and this was the beginning of a lifelong mutual friendship between them. Near the end of this period was also the start of a lifelong enmity between Karl Pearson and Fisher. In 1919 when Fisher was offered a temporary job at the Rothamsted Experimental Station, and by 1922 he had attained the position of the Chief Statistician. We provide this background because of the importance of Fisher, [80], (1922) [see also Fisher, [82], (1925)]. Hald, [102], (1998), page 713 succinctly summarized the paper by saying: "For the first time in the history of statistics a framework for a frequency-based general theory of parametric statistical inference was clearly formulated."

In this paper ([80], page 313) Fisher divided the statistical problems into three types:

(1) Problems of Specification. These arise in the choice of the mathematical form of the population.

(2) Problems of Estimation. These involve the choice of methods of calculating from a sample statistical derivatives, or as we shall call them statistics, which are designed to estimate the values of the parameters of the hypothetical population.

(3) Problems of Distributions. These include discussions of the distribution of statistics derived from samples, or in general any functions of quantities whose distribution is known.

Fisher, however, put little emphasis on the problem of *specification* as he immediately stated (see [80], page 314) after his above formulation:

"As regards problems of specification, these are entirely a matter for the practical statistician, for these cases where the qualitative nature of the hypothetical population is known do not involve any problems of this type. In other cases we may know by experience what forms are likely to be suitable, and the adequacy of our choice may be tested a *posteriori*."

Although he mentioned *testing*, Fisher occupied himself mostly with the problems of *estimation* and *distribution*. One of the fundamental contributions of Fisher, [80], (1922), [82], (1925) was providing the foundation of maximum likelihood procedure and studying its efficiency relative to the method of moments.

Here, we should mention that Fisher was not the first to formulate the general statistical problem into the above three broad categories. Emile Borel mentioned these in his book, *Eléments de la Théorie des Probabilités*, (1909) [see Neyman, [157], (1967), page 1457]. Even before that, Pearson, [182], (1902) had given the same categories in somewhat different language [see Hald, [102], (1998), pages 710-711].

## 2.2 Neyman and Pearson's (1928, 1933) LR and MP Tests

The idea for the Neyman and Pearson, [159], (1928) likelihood ratio (LR) test is hidden in Fisher's treatment of likelihood function. To discuss the LR and other tests we introduce some notation. Suppose there are $n$ independent observations, $y_1, y_2, \ldots, y_n$ with identical density function $f(y; \theta)$, where $\theta$ is a $p \times 1$ parameter vector with $\theta \in \Theta \subset \Re^p$. It is assumed that $f(y; \theta)$ satisfies the regularity conditions stated in Rao, [187], (1973), page 364 and Serfling, [207], (1980), page 144. The log-likelihood function, the score function, and the information matrix are then defined, respectively, as

$$(2.9) \qquad l(\theta) = \sum_{i=1}^{n} \ln f(y_i; \theta)$$

$$(2.10) \qquad s(\theta) = \frac{\partial l(\theta)}{\partial \theta}$$

$$(2.11) \qquad \mathcal{I}(\theta) = -E \left[ \frac{\partial^2 l(\theta)}{\partial \theta \partial \theta'} \right].$$

Suppose the hypothesis to be tested is $H_0$: $h(\theta) = c$, where $h(\theta)$ is an $r \times 1$ vector function of $\theta$ with $r \leq p$ and $c$ is a known constant vector. It is assumed that $H(\theta) = \partial h(\theta)/\partial\theta$ has full column rank, i.e., rank$[H(\theta)] = r$. We denote the maximum likelihood estimator (MLE) of $\theta$ by $\hat{\theta}$, and by $\tilde{\theta}$, the restricted MLE of $\theta$, i.e., $\tilde{\theta}$ is obtained by maximizing the log-likelihood function $l(\theta)$ subject to the restriction $h(\theta) = c$.

When the null hypothesis is true, we would expect $\hat{\theta}$ and $\tilde{\theta}$ to be close, and likewise $l(\theta)$ and $l(\tilde{\theta})$. Based on this intuitive idea and using Fisher's likelihood principle, Neyman and Pearson, [159], (1928) formulated their LR test as

$$(2.12) \qquad\qquad LR = 2[l(\hat{\theta}) - l(\tilde{\theta})].$$

Wilks, [225], (1938) derived the asymptotic distribution of the LR statistic. This statistic is applicable in most of the testing problems, and has been found to be very useful. However, there seems to be a lack of a general principle of constructing optimal tests as Egon Pearson, [176], (1966) later noted: "We were also careful to state that while this likelihood ratio procedure led to 'good' tests, we had no grounds for claiming them as 'best'. Indeed, we pointed out that acceptance of this or any other process of reasoning must be a matter of individual preference, depending ultimately on an appeal 'to the way we think'."

Neyman and Pearson, [160], (1933) laid the foundation of the theory of hypothesis testing. For the first time the concept of "risk" was introduced in the context of testing. Student urged them to consider some alternative hypothesis that would "explain the occurance of the sample with a more reasonable probability." And this led Neyman and Pearson to define the concepts of "power function" and uniformly most powerful (UMP) test. The problem then was to find a test that would maximize the power subject to the condition that the rejection probability under the tested (null) hypothesis has some preassigned value, say $\alpha$. A general solution was obtained when both the null and alternative hypotheses were simple, and this solution is the Neyman-Pearson (N-P) fundamental Lemma [for more historical account see, Pearson, [176], (1966), Reid, [192], (1982), pages 78-104 and Lehmann, [140], (1992)]. Simply stated the Lemma is as follows:

Suppose we want to test $H_0 : \theta = \theta_0$ against $H_1 : \theta = \theta_1$ based on the sample $y_1, y_2, \ldots, y_n$. Let $L(\theta) = \Pi_{i=1}^{n} f(y_i, \theta)$ denote the likelihood function. Then the critical region defined by

$$(2.13) \qquad\qquad \omega(y) = \{y | L(\theta_1) > k_\alpha L(\theta_0)\},$$

where $k_\alpha$ is such that $Pr[\omega(y)|H_0] = \alpha$, provides most powerful (MP) test for $H_0$ against $H_1$. Here by "y" we mean the values of the sample $(y_1, y_2, \ldots, y_n)$.

The importance of N-P lemma in econometrics and statistics can never be over-emphasized. This lemma laid the foundation of the *theory* of hypothesis testing. It also provided the logical basis for the LR test. If an MP test maximizes power uniformly in $\theta \in \Theta_1 \subset \Theta$, the test is called UMP test.

The *first* formal specification test in econometrics, the Durbin-Watson (DW) test for autocorrelation has its foundation in the UMP test principle via a theorem of Anderson, [6], (1948) [see Durbin and Watson, [72], (1950)]. Unfortunately, an UMP test exists rarely, and, therefore, it is necessary to restrict optimal tests to a suitable subclass by requiring the tests to satisfy other criteria, such as, local optimality, unbiasedness and invariance. For the N-P lemma, there is only one side condition, i.e., the size($\alpha$) of the test. Once the test is restricted, there will be more than one side condition, and one needs to use the generalized N-P lemma given in Neyman and Pearson, [161], (1936).

When an UMP test does not exist, there is not a single region which is best for all alternatives. We, therefore, try to find a critical region that is good for alternatives close to the null hypothesis, called local alternatives, hoping that the critical region will also be good for alternatives away from the null. For the critical region $\omega(y)$ let us define the power function as

$$(2.14) \qquad \gamma(\theta) = Pr[\omega(y)|\theta \neq \theta_0] = \int_{\omega(y)} L(\theta)dy.$$

Assuming that $\gamma(\theta)$ admits Taylor series expansion, we have

$$(2.15) \qquad \gamma(\theta) = \gamma(\theta_0) + (\theta - \theta_0)\gamma^{'}(\theta_0) + \frac{(\theta - \theta_0)^2}{2}\gamma^{''}(\theta^*),$$

where $\theta^*$ is a value in between $\theta$ and $\theta_0$. If we consider local alternatives of the form $\theta = \theta_0 + \delta/\sqrt{n}$, $0 < \delta < \infty$, the third term will be of order $O(n^{-1})$. To obtain highest power, we need to maximize,

$$(2.16) \qquad \gamma^{'}(\theta_0) = \frac{\partial}{\partial \theta}\gamma(\theta)\Big|_{\theta=\theta_0} = \int_{\omega(y)} \frac{\partial}{\partial \theta}L(\theta_0)dy,$$

for $\theta > \theta_0$, assuming the regularity conditions that allow differentiation under the sign of integration. Using the generalized N-P lemma, it is easy to see that the locally most powerful (LMP) test for $H_0 : \theta = \theta_0$ vs $H_1 : \theta > \theta_0$ should be based on the critical region defined by

$$(2.17) \qquad \frac{\partial}{\partial \theta}\ln L(\theta_0) = \frac{\partial}{\partial \theta}l(\theta_0) = s(\theta_0) > k,$$

where $k$ is a constant such that the size of the test is $\alpha$. This is the basis of the Rao, [186], (1948) score test [see Rao and Potti, [188], (1946)]. As we will have occasion to discuss later, Rao's Score (RS) test has been very useful to econometrics for developing various model diagnostic procedures.

The LMP test in (2.17) can also be obtained directly from the N-P lemma. We can write the critical region $\omega(y)$ in (2.13) as

$$(2.18) \qquad \frac{\ln L(\theta_1) - \ln L(\theta_0)}{\theta_1 - \theta_0} > k_1,$$

where $k_1$ is a constant such that the size of the test is equal to $\alpha$. Now as $\theta_1 \to \theta_0$, it is clear the critical region defined by (2.18) converges to that of (2.17) [see Kendall and Stuart, [122], (1973), page 181 and Gourieroux and Monfort, [98], (1995), page 32.

It is clear from the expansion (2.15) that LMP test does not exist for the two sided alternatives $\theta > \theta_0$ and $\theta < \theta_0$. To find an optimal test for such alternatives, let us now add unbiasedness to our requirements. A test $\omega(y)$ of size $\alpha$ is unbiased for $H_0 : \theta \in \omega_0$ against $H_1 : \theta \in \omega_1$, if $E[\omega(y)|H_0] \leq \alpha$ and $E[\omega(y)|H_1] \geq \alpha$. Let us again consider the expansion of $\gamma(\theta)$ around $\theta_0$

$$(2.19) \qquad \gamma(\theta) = \gamma(\theta_0) + (\theta - \theta_0)\gamma'(\theta_0) + \frac{(\theta - \theta_0)^2}{2}\gamma''(\theta_0) + \circ(n^{-1}).$$

Unbiasedness requires that $\gamma'(\theta_0) = 0$, since "power" should be minimum at $\theta_0$. To maximize the local power, we need to maximize $\gamma''(\theta_0)$ for both $\theta > \theta_0$ and $\theta < \theta_0$, and this will lead to the locally most powerful unbiased (LMPU) test. Neyman and Pearson, [161], (1936), page 9 call the corresponding critical region type-A region. Using generalized N-P lemma, it can be shown that the type-A critical region (or the LMPU test) is given by

$$(2.20) \qquad \frac{\partial^2}{\partial \theta^2}l(\theta_0) + \left[\frac{\partial}{\partial \theta}l(\theta_0)\right]^2 > k_1 \frac{\partial}{\partial \theta}l(\theta_0) + k_2,$$

where $k_1$ and $k_2$ are such that the test is unbiased and has size $\alpha$. In contrast to the LMP test in (2.17), we see that the LMPU test requires both the first and second derivatives of the log-likelihood function. However, for a certain class of distributions, the second derivatives can be expressed in terms of the first derivatives (score function), so that the LMPU test will essentially be based on score.

For many testing problems, the score function evaluated at the null hypothesis is identically zero. To illustrate this, consider the following simple example from Cox and Hinkley, [56], (1974), page 117. Let

$$(2.21) \qquad y_i \sim IIDN(0, (1 + \theta^2 z_i)), \qquad\qquad i = 1, 2, \ldots, n$$

where $z_i$'s are given positive constants. We are interested in testing $H_0$ : $\theta = 0$, that is, $y_i$ has constant variance. The log-likelihood function is given by:

$$(2.22) \qquad l(\theta) = const - \frac{1}{2} \sum_{i=1}^{n} \ln(1 + \theta^2 z_i) - \frac{1}{2} \sum_{i=1}^{n} y_i^2 / (1 + \theta^2 z_i),$$

and the score with respect to $\theta$ is

$$(2.23) \qquad \frac{\partial}{\partial \theta} l(\theta) = -\theta \sum_{i=1}^{n} \left[ \frac{z_i}{(1 + \theta^2 z_i)} - \frac{z_i y_i^2}{(1 + \theta^2 z_i)^2} \right].$$

It is clear that $\frac{\partial}{\partial \theta} l(\theta) = 0$ under $H_0 : \theta = 0$.

In econometrics such situations occur quite often [for more on this see Bera, Ra and Sarkar, [37], (1998)]. Kiefer, [124], (1982), Waldman, [218], (1982), Schmidt and Lin, [202], (1984) and Rockinger, [194], (1994) provided some examples in which the score is identically zero. Lee and Chesher, [137], (1986) have offered a comprehensive treatment of this problem, and one of the examples they considered is the stochastic production frontier model of Aigner, Lovell and Schmidt, [2], (1977), Cox and Hinkley, [56], (1974) proposed an easy solution to this kind of problem by reparametarization. Let us again consider their example given in (2.21) and assume $\theta^2 = \gamma$. Then, we have

$$(2.24) \qquad y_i \sim IIDN(0, 1 + \gamma z_i), : : \quad i = 1, 2, \ldots, n.$$

The corresponding log-likelihood and score functions are given by, respectively,

$$(2.25) \qquad l(\gamma) = const - \frac{1}{2} \sum_{i=1}^{n} \ln(1 + \gamma z_i) - \frac{1}{2} \sum_{i=1}^{n} \frac{y_i^2}{(1 + \gamma z_i)}$$

and

$$(2.26) \qquad \frac{dl(\gamma)}{d\gamma} = -\frac{1}{2} \sum_{i=1}^{n} \left[ \frac{z_i}{1 + \gamma z_i} - \frac{z_i y_i^2}{(1 + \gamma z_i)^2} \right].$$

Under the null hypothesis $H_0 : \gamma = 0$, the score function is reduced to

(2.27)
$$\left. \frac{dl(\gamma)}{d\gamma} \right|_{\gamma=0} = \frac{1}{2} \sum_{i=1}^{n} z_i(y_i^2 - 1)$$

A test can be based on the above quantity. In fact, it is easy to show that [see Godfrey, [92], (1988, page 92)]

(2.28)
$$\frac{\sum_{i=1}^{n} z_i(y_i^2 - 1)}{\sqrt{2 \sum_{i=1}^{n} z_i^2}} d \to N(0, 1)$$

Using (2.23) and (2.27), we see that

(2.29)
$$\left. \frac{\partial^2}{\partial\theta^2} l(\theta) \right|_{\theta=0} = 2 \left. \frac{\partial l(\gamma)}{\partial \gamma} \right|_{\gamma=0}.$$

Therefore, the test in (2.28) is essentially based on the *second order* derivative of the original log-likelihood function (2.22), This suggests a general approach that if the first-order derivatives (score) are zero we can use the second-order derivatives of the log-likelihood function to construct the test. Lee and Chesher, [137], (1986) provided further justification for this approach. In the LMP test, we check whether the score function is close to zero [see (2.17)], examining whether the log-likelihood function reaches "maximum" at the restricted MLEs. At the maximum the second derivative should be negative definite. Therefore, a reasonable procedure would be to reject $H_0$ if the second derivative is positive definite or has a very high value. If we go back to (2.20), we see that it provides an automatic procedure to test $H_0$ when the score vanishes. The type-A critical region reduces to

(2.30)
$$\frac{\partial^2}{\partial\theta^2} l(\theta_0) > k_2.$$

This is exactly the test advocated in Cox and Hinkley, [56], (1974) and Lee and Chesher, [137], (1986), The moral of the story is that, when the score is zero, we cannot have an LMP test, but we can base an LMPU test on the second derivative of the log-likelihood function using Neyman and Pearson, [161], (1936) approach.

## 2.3 Neyman's (1937) Smooth Test

Neyman suggested this test to rectify some of the drawbacks of the Pearson, [181], (1900) goodness-of-fit test discussed in Section 2.1. He noted that in Pearson $\chi^2$ test it was not clear how the class intervals should be determined and that the corresponding distributions under the alternative hypothesis were not "smooth." By smooth alternatives Neyman meant those densities that have few intersections with the null density function and that are close to null. Suppose we want to test the null hypothesis ($H_0$) that $f(y, \theta)$ is the true density function for the random variable $Y$, i.e, $H_0 : Y \sim f(y, \theta)$. For simplicity we will also denote this as $f(Y|H_0)$. The specification of $f(y, \theta)$ will be *different* depending on the problem at hand. Let us denote the alternative hypothesis as $H_1 : Y \sim g(y)$. Using a probability integral transformation, Neyman, [155], (1937) first transformed any hypothesis testing problem of this type to testing only *one kind of hypothesis*. Let us define a new random variable $Z$ by

$$(2.31) \qquad z = \int_{-\infty}^{y} f(Y|H_0)dY.$$

The density function of $Z$ is given by

$$
\begin{aligned}
h(z) &= g(y)\frac{dy}{dz} \\
&= \frac{g(y)}{f(y|H_0)}.
\end{aligned}
$$

(2.32)

If $H_0$ is true, $f(y|H_0) = g(y)$, i.e.,

$$(2.33) \qquad h(z) = 1, : : \qquad 0 < z < 1.$$

If we denote by $H_0^*$ as the hypothesis that Z follows uniform distribution in the interval (0,1), then it is clear that $H_0$ and $H_0^*$ are equivalent. Therefore, instead of testing $H_0$ whether the probability density of $Y$ is $f(y; \theta)$, we may just test whether $Z$ has uniform distribution.

Neyman was not the first to use the idea of probability integral transformation to reformulate the original hypothesis testing problem $H_0$. As discussed by Egon Pearson, [175], (1938), Fisher, [83], (1932) and Karl Pearson, [184], (1933) developed the same idea. They did not, however, construct any formal test statistic. As for the specific alternative to hypothesis $H_0^*$, Neyman suggested to consider the following smooth alternative $H_1^*$. He specified the density of $Z$ as

$$(2.34) \qquad h(z) = C(\delta) \exp\left[\sum_{j=1}^{r} \delta_j \pi_j(z)\right],$$

where $C(\delta)$ is the constant of integration that depends on the $\delta_j$ values, and $\pi_j(z)$ are orthogonal polynomials satisfying

$$\int_0^1 \pi_j(z)\pi_k(z)dy = 1 \text{ for } j = k$$

(2.35)
$$= 0 \text{ for } j \neq k.$$

Using generalized N-P lemma, Neyman derived a locally most powerful symmetric unbiased test for the hypothesis $H_0^* : \delta_1 = \delta_2 = \ldots = \delta_r = 0$. The test is symmetric in the sense that the asymptotic power of the test depends only on the "distance" $\sum_{j=1}^r \delta_j^2$ between $H_0^*$ and $H_1^*$. The test statistic is

(2.36)
$$\psi_r^2 = \sum_{j=1}^r \frac{1}{n}\left[\sum_{i=1}^n \pi_j(z_i)\right]^2,$$

which, under $H_0^*$ or $(H_0)$, asymptotically follows a central $\chi_r^2$, and under $H_1^*$ follows a non-central $\chi_r^2$ with non-centrality parameter $\sum_{j=1}^r \delta_j^2$.

Neyman's approach requires the computation of the probability integral transformation variable $Z$. It is, however, easy to recast the testing problem in terms of the original variable $Y$. Writing the relation (2.31) as $z = F(y)$ and defining $q_j(y) = \pi_j(F(y))$ we can obtain the orthogonal polynomials with respect to $f(y, \theta)$ from (2.35) as

$$\int \pi_j(F(y))\pi_k(F(y))dz = \int q_j(y)q_k(y)f(y; \theta)dy$$

$$= 1 \text{ for } j = k$$

(2.37)
$$= 0 \text{ for } j \neq k.$$

Then, from (2.34) the density function under the alternative hypothesis in terms of $Y$ is given by

$$h(y) = h(F(y))\frac{dz}{dy}$$

(2.38)
$$= \left(C(\delta, \theta) \quad \exp\left[\sum_{j=1}^r \delta_j q_j(y)\right]\right)f(y; \theta)$$

Under this formulation the test statistic $\psi_r^2$ reduces to

(2.39)
$$\psi_r^2 = \sum_{j=1}^r \frac{1}{n}\left[\sum_{i=1}^n q_j(y_i)\right]^2$$

which has the same asymptotic distribution as before.

Unlike Neyman's earlier work with Egon Pearson, the smooth test did not have much impact for some time. There is an amusing anecdote even before the paper was published as described by Reid, [192], (1982), page 149. In 1937, W.E. Deming was preparing publication of Neyman's lectures by the United States Department of Agriculture. On one occasion Neyman misspelled *smooth* when referring to this 1937 paper. "I don't understand the reference to 'Smouth' ", Deming wrote Neyman, "Is that name of a statistician?". When in 1970 George Roussas wrote, "During the years 1934–38 Neyman made four fundamental contributions to the science of statistics" nominating Neyman for an honorary degree [Reid, [192], (1982) page 157], he did not include the smooth test. Barton, [17], (1953), [18], (1955), [19], (1956) wrote a series of papers on smooth test, applied the results to discrete data and generalized the test to the composite null hypothesis situation [see also Hamdan, [105], (1962), [106], (1964)]. Neymans's original formulation of the test assumed a fully specified null hypothesis. A major impetus for the smooth test came from the works of Kopecky and Pierce, [130], (1979) and Thomas and Pierce, [215], (1979). While Neyman's statistic for testing a simple null hypothesis has limiting $\chi^2$ distribution, Barton's, [19], (1956) test involves a mixture $\chi^2$ distribution. For the composite hypothesis, Thomas and Pierce, [215], (1979) developed a natural modification of Neyman's test that has limiting $\chi^2$ distribution. Koziol, [131], (1979), [133], (1987) also suggested similar tests. Rayner and Best, [190], (1989) give an excellent review of smooth tests and related procedures. They also elaborated on many interesting, little known results. For example, Pearson $\chi^2$ goodness-of-fit test discussed in Section 2.1 is a categorized form of the Neyman's smooth test. To see this result, let us write the alternative class probabilities as

$$(2.40) \qquad q_k = C(\delta) \, \exp\left[\sum_{j=1}^{r} \delta_j h_{ik}\right] \theta_{k0},$$

for $k = 1, 2, \ldots, p$ [see the description above equation (2.1)]. In (2.40), $h_{jk}$ are values taken by a random variable $H_j$ with $Pr(H_j = h_{jk}) = \theta_{k0}$, $k = 1, 2, \ldots, p$; $j = 1, 2, \ldots, r$. These $h_{jk}$ are also orthonormal with respect to the multinomial probabilities $\theta_{k0}$. Rayner and Best, [190], (1989), pages 57–60 showed that the smooth test for testing $H_0 : \delta_1 = \delta_2 = \ldots = \delta_r = 0$ is same as the Pearson statistic $P$ in (2.1) with $r = p - 1$.

Smooth-type tests can be viewed as a compromise between the omnibus test procedures such as Pearson $\chi^2$ with generally low power in all directions and specific tests with power directed only towards certain alternatives. Neyman's original test has been extended in various directions and the resulting procedures have been very useful [see, for instance, Rayner and Best, [190], (1989), [191], (1990), LaRiccia, [135], (1991), Ingott and Ledwina, [116], (1996) and Fan, [76], (1996) ]. There are, however, two major issues that linger in connected with Neyman's original suggestion. First, is the choice of the function, for example, $q_j(y)$ in (2.38) for testing $f(y, \theta)$ as the null hypothesis. The orthonormal functions suggested by Rayner and Best, [189], (1986) and Koziol, [132], (1986), [133], (1987) seem to be a good choice. Thomas and Pierce, [215], (1979) used powers of the distribution function of $Y$ and found them theoretically attractive. The second problem is deciding on suitable values of $r$. Neyman, [155], (1937), page 134 himself felt that, "there will be no need to go beyond the fourth order test," and suggested to take $r = 3$ or 4. He added, however, that this was "only an opinion and not any mathematical result." Thomas and Pierce, [215], (1979), page 442 found that in their experience $r = 2$ may be a better choice. Ledwina, [136], (1994) and Kallenburg and Ledwina, [120], (1995), [121], (1997) suggested a data-driven version smooth test in which they selected the value of $r$ using Schwarz's, [203], (1978) information criterion.

We have seen no formal application of Neyman's smooth test in econometrics. However, the probability integral transformation has been used in analysis of duration models and Bayesian econometrics. One of the earliest interactions between statisticians and econometricians in connection with the smooth test dates from the 1942–43 MIT Statistics Seminar list; Lawrence Klein gave a seminar on "Neyman's Smooth Test of Goodness–of–Fit" [see *Statistical Science*, November 1991 issue]. Klein joined the MIT graduate program in September 1942 after studying with Neyman's group in statistics at Berkeley, and he wanted to draw the attention to this paper since it has not been published in an easily accessible journal. Unfortunately, Klein's effort did not bring Neyman's test into econometrics. However, some of the test in econometrics can be easily be given a smooth-test interpretation. The test for normality derived in Jarque and Bera, [118], (1980) and Bera and Jarque, [25], (1981) can be viewed as smooth tests [see Rayner and Best, [190], (1989), page 90]. Tests suggested in Smith, [210], (1989) and Cameron and Trivedi, [48], (1990) are also in the spirit of Neyman, [155], (1937), In a recent paper, Diebold, Gunther and Tay, [69], (1998) suggested the use of probability integral transformed series for density forecast evaluation. They traced the origin of the procedure to Rosenblatt, [196], (1952) but as we mentioned earlier this is related to the ideas of Fisher, [83], (1932), Pearson, [184], (1933) and, more formally, of Neyman, [155], (1937). It seems that the potential for application of smooth test in econometrics is very high and is almost an unexplored area. It could be used particularly to develop more effective econometric model specification tests.

## 2.4 Wald (1943), Rao's (1948) Score and Neyman's (1959) $C(\alpha)$ Tests

Let us go back to the framework of Section 2.2 and consider testing the general hypothesis $H_0 : h(\theta) = c$ where $\theta$ is a $p \times 1$ parameter vector, $h(\theta)$ is an $r \times 1$ vector function of $\theta$ and $c$ is a known vector. As mentioned earlier, if $H_0$ is true, the "distance" between the restricted and unrestricted MLEs, $\bar{\theta}$ and $\hat{\theta}$, should be small. For the Wald $(W)$ test the distance is measured through $h(\theta)$. Since $h(\bar{\theta}) = c$ by construction, the test is based on $h(\hat{\theta})$ and its algebraic form is

$$(2.41) \qquad W = (h(\hat{\theta}) - c)' [H(\hat{\theta})' \mathcal{I}(\bar{\theta})^{-1} H(\hat{\theta})]^{-1} (h(\hat{\theta}) - c),$$

where $H(\theta) = \frac{\partial}{\partial \theta} h(\theta)$ and $\mathcal{I}(\theta)$ is the information matrix. We might recall that for the LR test the distance is measured through the log-likelihood function $l(\theta)$. Under $H_0$, asymptotically $W$ is distributed as $\chi_r^2$.

Now consider a simple case, say $p \neq 1$ and test $H_0 : \theta = \theta_0$ against $H_1 : \theta > \theta_0$. As stated in (2.17), a locally most power test for $H_0$ is based on the score function and is given by

$$(2.42) \qquad\qquad s(\theta_0) > k,$$

where $k$ is so determined that the size of the test is equal to a preassigned value $\alpha$. This is the basis of the Rao's, [186], (1948) score (RS) test.

Generalization of this to the $p \geq 2$ case is not trivial. There will be scores for each individual parameter, and the problem is how to combine them in an "optimal" way. Rao proceeded as follows. Let $H_0 : \theta = \theta_0$, where now $\theta = (\theta_1, \theta_2, \ldots, \theta_p)'$ and $\theta_0 = (\theta_{10}, \theta_{20}, \ldots, \theta_{p0})'$, and the (local) alternative hypothesis be as $H_1 : \theta = \theta_\delta$, where $\theta_\delta = (\theta_{10} + \delta_1, \theta_{20} + \delta_2, \ldots, \theta_{p0} + \delta_p)'$. Proportionate change in the log-likelihood function for moving from $\theta_0$ to $\theta_\delta$ is given by

$$(2.43) \qquad\qquad \sum_{j=1}^{p} \delta_j \frac{\partial l(\theta_0)}{\partial \theta_j} = \delta' s(\theta_0),$$

where $\delta = (\delta_1, \delta_2, \ldots, \delta_p)'$. If the $\delta$'s were known a test could be based on

$$(2.44) \qquad\qquad \frac{[\delta' s(\theta_0)]^2}{\delta' \mathcal{I}(\theta_0) \delta},$$

which under $H_0$ will be asymptotically distributed as $\chi_1^2$. Then Rao posed
the question: what linear function would yield the *maximum discrimination*? Therefore, his quest for optimality was in the form of maximizing
something close to the non-centrality parameter of a $\chi^2$ test. It is easy to
see that (2.44) is maximized at $\delta = \mathcal{I}(\theta_0)^{-1} s(\theta_0)$. This leads to the test
statistic

$$(2.45) \qquad \sup_{\delta} \frac{[\delta' s(\theta_0)]^2}{\delta' \mathcal{I}(\theta_0)\delta} = s(\theta_0)' \mathcal{I}(\theta_0)^{-1} s(\theta_0)$$

When the null hypothesis is composite, as stated earlier, namely, $H_0$:
$h(\theta) = c$ with $r \leq p$ restrictions, Rao suggested to use

$$(2.46) \qquad RS = s(\bar{\theta})' \mathcal{I}(\bar{\theta})^{-1} s(\bar{\theta})$$

where $\bar{\theta}$ is the restricted MLE of $\theta$.

To find the asymptotic distribution of RS, Rao, [186], (1948) showed
(for $p = 2$) that $s(\theta)' \mathcal{I}(\theta)^{-1} s(\theta)$ can be decomposed as

$$(2.47) \qquad s(\theta)' \mathcal{I}(\theta)^{-1} s(\theta) a = s(\bar{\theta}) \mathcal{I}(\bar{\theta})^{-1} s(\bar{\theta}) + (\bar{\theta} - \theta)' \mathcal{I}(\theta)(\bar{\theta} - \theta)$$

where $a =$ denotes asymptotic equivalence meaning that equality holds after neglecting $o_p(1)$ terms [for a proof for general $p$, see Bera and Bilias, [21],
(1999)]. This result is similar to Fisher's decomposition of Pearson's statistics as given in (2.7), Since scores are asymptotically normally distributed
with mean zero and variance $\mathcal{I}(\theta)$, the left hand side of (2.47) is asymptotically distributed as $\chi_p^2$. It can be shown that $\bar{\theta}$ satisfying $r$ restrictions
is asymptotically normally distributed with mean $\theta$ and variance whose
$g$-inverse is $\mathcal{I}(\theta)$. Therefore, $(\bar{\theta} - \theta)' \mathcal{I}(\theta)(\bar{\theta} - \theta)d \to \chi_{p-r}^2$, and it can be
shown that the two terms in the right hand side of (2.47) are asymptotically
independent. Therefore, under $H_0$, $RSd \to \chi_r^2$.

In terms of our "distance" interpretation, the intuitive idea behind
the score test is that, if $H_0$ is true, $s(\bar{\theta})$ is expected to be close to zero by
virtue of the fact that $s(\hat{\theta}) = 0$. From this we can see the duality between
the Wald and score tests. At the unrestricted MLE $\hat{\theta}$, $s(\hat{\theta}) = 0$, and the
Wald test checks whether $h(\hat{\theta})$ is away from $c$. On the other hand, at the
restricted MLE $\bar{\theta}$, $h(\bar{\theta}) = c$ is satisfied and the score tests verifies whether
$s(\bar{\theta})$ is close to zero.

Econometricians use the score form of the test; however, they call it
the Lagrange multiplier (LM) test. The terminology LM came from the
three articles Aitchison and Silvey, [3], (1958), [4], (1960) and Silvey, [208]

(1959), where an LM interpretation of (2.46) was given. Note that $\bar{\theta}$ can be obtained from the solution to the equation

$$(2.48) \qquad s(\bar{\theta}) - H(\bar{\theta})\bar{\lambda} = 0,$$

along with $h(\bar{\theta}) = c$, where $\lambda$'s are the Lagrange multipliers. Therefore, we have $s(\bar{\theta}) = H(\bar{\theta})\bar{\lambda}$. Given that $H(\bar{\theta})$ has full rank, $s(\bar{\theta}) = 0$ is equivalent to $\bar{\lambda} = 0$, i.e., the Lagrange multipliers vanish. These multipliers can be interpreted as the implicit cost (shadow prices) of imposing the restrictions. It can be shown that

$$(2.49) \qquad \bar{\lambda} = \frac{\partial l(\bar{\theta})}{\partial c},$$

that is, the multipliers give the rate of change of the maximum attainable value with respect to the change in the constraints. If $H_0 : h(\theta) = c$ is true and $l(\bar{\theta})$ gives the optimal value, $\bar{\lambda}$ should be close to zero. Given this "economic" interpretation in terms of multipliers, it is not surprising that the econometricians prefer the term LM rather than RS. In terms of Lagrange multipliers, (2.46) can be expressed as

$$(2.50) \qquad RS = LM = \bar{\lambda}' H(\bar{\theta})' \mathcal{I}(\bar{\theta})^{-1} H(\bar{\theta})\bar{\lambda}.$$

It is clear from (2.46) and (2.50) that the RS form of the test is much easier to compute, and this explains its popularity among econometricians. However, they have given it a name that is closer to economists' way of thinking. In an interview Rao was asked about his favorite publication among his many books and papers [see DeGroot, [67], (1987)]. Part of Rao's reply was: "In 1947 ... I introduced two general asymptotic test criteria called score tests for simple and composite hypotheses as alternative to Wald's tests. I find that my score test for composite hypotheses has become entrenched in the econometrics literature under a fancier name, the Lagrange Multiplier test."

In econometrics, Byron, [46], (1968) was probably the first to apply the RS test. He used Silvey's, [208], (1959) LM version along with the LR statistic for testing homogeneity and symmetry restrictions in demand equations. It took another decade for econometricians to realize the potential of the RS test. In this regard, the work of Breusch and Pagan, [44], (1980) has been the most influential. It collected relevant research reported in the statistics literature, presented the RS test in a general framework in the context of econometric model evaluation, and discussed many applications. Since the late seventies, econometricians have applied the score principle to a variety of econometric testing problems and studied the properties of the resulting tests. Now the $RS$ tests are the most common items in the econometricians' kit of testing tools. Famous tests such as Moran's, [152], (1950) I statistic for spatial dependence, Cox's, [52], (1961), [53],(1962) non-nested test statistics and White's, [223], (1982) information matrix test can also be viewed as score tests. Bera and Ullah, [33], (1991) and Anselin and Bera, [13], (1998) discussed many other applications of the score test to econometric models and demonstrated that many of the old and new econometric tests could be given a score test interpretation [see also Amemiya, [5], (1985), McAleer, [149], (1987), Godfrey, [92], (1988), Davidson and MacKinnon, [64], (1993) and a number of papers in Maddala et al., [147], (1995) and Bera and Mukherjee, [29], (1999)]. The following examples illustrate the versatility and usefulness of the RS test.

Let us consider again the Pearson goodness-of-fit test but now with multinomial probability framework. The likelihood function is given by

(2.51)
$$L(\theta) = n! \prod_{j=1}^{p} \frac{\theta_j^{n_j}}{n_j!},$$

and we test $H_0 : \theta_j = \theta_{j0}$, $j = 1, 2, \ldots, p$, where $\theta_{j0}$ are known constants. The scores and information matrix are, respectively,

(2.52)
$$\frac{s(\theta)}{((p-1) \times 1)} = \begin{bmatrix} \dfrac{n_1}{\theta_1} - \dfrac{n_p}{\theta_p} \\ \vdots \\ \dfrac{n_{p-1}}{\theta_{p-1}} - \dfrac{n_p}{\theta_p} \end{bmatrix}$$

and

$$\frac{\mathcal{I}(\theta)}{[(p-1) \times (p-1)]} = n \left[ \text{diag}\left( \frac{1}{\theta_1}, \ldots, \frac{1}{\theta_{p-1}} \right) + \frac{1}{\theta_p} 11' \right],$$

where $1 = (1, 1, \ldots, 1)'$ is a $(p-1) \times 1$ vector of ones. Since $\theta_j$ and $\theta_{j0}$ should sum to 1, effectively, we have only $(p-1)$ parameters, and $(p-1)$ restrictions, namely, $H_0$: $\theta_j = \theta_{j0}$, $j = 1, 2, \ldots, p-1$. That is why $s(\theta)$ has dimension $(p-1) \times 1$, and $\mathcal{I}(\theta)$ is a $(p-1) \times (p-1)$ matrix. Using the above expressions, it is easy to see that the score statistic

(2.53)
$$s(\theta_0)' \mathcal{I}(\theta_0)^{-1} s(\theta_0) = \sum_{j=1}^{p} \frac{(n_j - n\theta_{j0})^2}{n\theta_{j0}},$$

where $\theta_0 = (\theta_{10}, \ldots, \theta_{p0})'$ [see Rao, [187], (1973), page 442 and Cox and Hinkley, [56], (1974), page 316]. Therefore, the RS statistic is the same as Pearson's $P$ given in (2.1),

For this particular problem, the other two test statistics LR and $W$ are given by

$$(2.54) \qquad LR = 2 \sum_{i=1}^{p} n_j \ln \left( \frac{n_j}{n\theta_{j0}} \right) = \sum O \ln \left( \frac{O}{E} \right)$$

and

$$(2.55) \qquad W = \sum_{j=1}^{p} \frac{(n_j - n\theta_{j0})^2}{n_j} = \sum \frac{(O - E)^2}{O}.$$

It is quite a coincidence that Pearson, [181], (1900) suggested a score test mostly based on intuitive grounds almost 50 years before Rao, [186], (1948), To see another coincidence, consider the Neyman's smooth statistic $\psi_r^2$. It can be shown that $\psi_r^2$ in (2.39) is a score statistic for testing $H_0 : \delta_1 = \delta_2 = \ldots = \delta_r = 0$ within the class of alternative densities given by (2.38), [see Rayner and Best, [190], (1989), Sections 4.2 and 6.1]. In a way, Neyman's smooth test is the first formally derived RS test. Given the connection between the smooth and score tests, it is not surprising that Pearson's test is nothing but a categorized version of the smooth test, as noted earlier. We can also relate some of the seemingly unrelated tests principles suggested after 1948 to Rao's score principle.

Roy, [197], (1953) proposed a heuristic method of constructing a test of any hypothesis $H_0$ that can expressed as an intersection of a class of hypotheses (finite or infinite), Suppose $H_0 = \bigcap_{j \in J} H_{0j}$, where $J$ is an index set. Roy's union-intersection (UI) method gives the rejection region for $H_0$ as the union of rejection regions for all $H_{0j}, j \in J$. Let us consider testing $H_{0\delta} : \delta' \theta_0 = \delta' \theta_0$ against $H_{1\delta} : \delta' \theta \neq \delta' \theta_0, \delta \in \Re^p$. Let $H_0 = \bigcap_{\delta \in \Re^p} H_{0\delta}$ and $H_1 = \bigcap_{\delta \in \Re^p} H_{1\delta}$. If $T_\delta$ is the LR statistic for testing $H_{0\delta}$ against $H_{1\delta}$, then

$$(2.56) \qquad T = \sup_{\delta \in \Re^p} T_\delta$$

is the LR statistic for testing $H_0$ against $H_1$ [see Hochberg and Tamhane, [111], (1987), pages 28-35]. This is the same principle used by Rao, [186], (1948) to convert a multivariate problem into series of univariate problems [see equation (2.45)]. Note that under $H_0$ the $\delta$ values are not identified. In many testing problems some nuisance parameters take the role of $\delta$. For example, while testing for no structural change, the particular change point ($\delta$) is not identified under the null hypothesis. Davies, [65], (1977) was the first to consider such a problem in a general context and proposed a test procedure when the test statistics have the normal distribution at any fixed value of the unidentified parameter. Davies, [66], (1987) extended his results to the case where the test statistic follows $\chi^2$ distribution and suggested a quick method to approximate the upper bound of the significance level. This approach has found many applications in econometrics, for example, see, Watson and Engle, [220], (1985), Bera and Higgins, [24], (1992), Andrews and Ploberger, [9], (1994), Teräsvirta, [214], (1994), Bera and Ra, [31], (1995), [32], (1997), Hansen, [107], (1996), González–Rivera, [95], (1998) and Lin and Teräsvirta, [143], (1999), Roy and Mitra, [198], (1956) derived Pearson's $\chi^2$ statistic from Roys's, [197], (1953) UI principle. They followed an approach similar to that of Rao, [186], (1948): first they obtained an MP test against a simple alternative using the N-P lemma and then maximized the test statistic to derive a test against the composite alternative. This is related to our earlier result that the Pearson $\chi^2$ is a special case of the RS test.

Apart from its computational simplicity and local asymptotic efficiency, the RS test has advantages from some other theoretical and practical points of view. For example, consider the case of testing, when under the null hypothesis the parameter values lie on the boundary of the parameter space. In this context, the standard theory associated with tests based on MLEs will not be valid. The large sample properties of MLEs and the associated tests in boundary situations have been examined by Chernoff, [50], (1954), Moran, [153], (1971), Chant, [49], (1974) and Self and Liang, [204], (1987). One important general result of their investigation is that the $W$ and LR tests in the boundary situation will not follow their usual asymptotic $\chi^2$ distribution, while the asymptotic properties of the RS test are not altered. As a result it has been argued that the RS test is particularly suitable for testing hypotheses at the boundary of the parameter space. It may be appropriate to note that most of testing exercises on the boundary lead to one-sided testing problems. There has been an increased awareness of the potential one-sided nature of many other testing problems in econometrics [see, for example, Farebrother, [77], (1986), Hillier, [110], (1986), King and Smith, [125], (1986), Kodde and Palm, [127], (1986), Rogers, [195], (1986) and Wolak, [226], (1987)]. Since economic theory often provides information on signs of various parameters that are used to improve power of relevant tests, the resulting testing exercises become one–sided. In this connection, one might mention the work of King and Wu, [126], (1997), who tested a simple null hypothesis $H_0 : \theta = 0$ against a multi-parameter one-sided alternative given by $H_1 : \theta_1 \geq 0, \theta_2 \geq 0, \ldots, \theta_p \geq 0$ with at least one inequality being a strict inequality, where $\theta_i$ is the $i^{th}$–component of the $p \times 1$ parameter vector $\theta$. They constructed a locally most mean powerful (LMMP) test of the preceding testing exercise. It may be noted that Sengupta and Vermeire, [206], (1986) introduced the class of a locally most mean powerful unbiased (LMMPU) test for testing $H_0 : \theta = 0$ against $H_1 : \theta \neq 0$. These tests are so constructed that they maximize the mean curvature of the power hyper-surface in the neighborhood of $\theta = 0$. Comparisons between the LMMPU tests and other tests like the RS test have been made in Mukherjee and Sengupta, [154], (1993).

Sometimes the $W$ test can have some computational advantage. For models in which the null hypothesis imposes nonlinear restrictions on $\theta$, the unrestricted MLE $\hat{\theta}$ is easier to calculate. Examples of such cases can be found in the dynamic specification test of Sargan, [200], (1980), the rational expectation hypothesis tests of Wallis, [219], (1980) and Hoffman and Schmidt, [112], (1981) and the test of market efficiency under rational expectation as described in Baillie et al., [15], (1983). Although the $W$ test may be easier, particularly in this kind of situation, it runs into a serious problem. Gregory and Veal, [97], (1985) and Vaeth, [216], (1985) pointed out that numerical value of the $W$ statistic is not invariant to the algebraically equivalent forms of the null hypothesis. However, the RS test is invariant to different equivalent forms of nonlinear restrictions. The main problem with the $W$ test is that it uses a wrong "metric" and is not invariant to changes in coordinates. For more on this topic see Breusch and Schmidt, [45], (1988), Phillips and Park, [185], (1988), Ferrari and Cribari-Neto, [78], (1993) and Critchley et al., [58], (1996).

To construct the LR, $W$ and RS statistics, it is necessary to calculate the restricted and/or unrestricted MLEs. In some cases, however, MLEs may be computationally difficult to obtain while consistent, though inefficient, estimates are easily available. When $\sqrt{n}$- consistent estimators are used there is an attractive way to construct a *score type* test developed by Neyman, [156], (1959). In the literature this is known as the $C(\alpha)$ test. Regarding this unusual name, Neyman, [158], (1979) wrote:

"With reference to many questions I heard, I must explain the origin of the symbol $C(\alpha)$. Here $\alpha$ refers to the possibility of prescribing an arbitrarily chosen level of significance $\alpha$. The letter $C$ refers to Harald Cramér, whose work I greatly admire. Particularly I value greatly Cramér's book, the *Mathematical Methods of Statistics*, which first published in 1946, had a very strong and beneficial influence on our discipline; it made mathematical statistics considerably more mathematical than it was before. In 1959 a jubilee volume was published in honor of Cramér. The basic theory of $C(\alpha)$ tests is published in that volume. Originally, I thought of calling these tests the Cramér tests. This idea was abandoned because the use of Cramér's name could have been interpreted as loading Cramér with the responsibility for all the deficiencies of these particular tests ...... My intension was, and is, merely to honor Cramér."

Neyman, however, was disappointed that the paper did not attract much attention as he confessed to Reid, [192], (1982), page 252.

"Presumably the reason was that I made a mistake in publishing it in the Festschrift .... My motivation was that I have a great respect for Cramér and also quite warm affection [but], as is well known, not many people read the Festschrifts!"

Although by 1959, Neyman was sixty-five years old, his $C(\alpha)$ paper is in the same class as the series of papers with Egon Pearson and the smooth test paper. Neyman was fond of referring the $C(\alpha)$ test as his "last performance" [Reid, [192], (1982), pages 286-287]. The fundamental contribution of this paper is the *derivation* of an asymptotically optimal test in the presence of nuisance parameter.

To follow Neyman, [156], (1959), let us partition $\theta$ as $\theta = [\theta_1', \theta_2]$ where $\theta_2$ is a scalar and test $H_0 : \theta_2 = \theta_{20}$. Therefore, $\theta_1$ is the nuisance parameter with dimension $(p-1) \times 1$. Neyman restricted the search for an optimal statistic within the class of functions that satisfy the regularity conditions given in Cramér, [57], (1946), page 500. A function $g(.; \theta)$ belonging to this class is called Cramér function. Suppose under $H_0 : \theta_2 = \theta_{20}$, $E[g(y; \theta_1, \theta_{20})] = g_1(\theta_1, \theta_{20})$ and $V[g(y; \theta_1, \theta_{20})] = \sigma^2(\theta_1, \theta_{20})$. Then by Lindberg–Lévy central limit theorem

$$(2.57) \qquad Z_n(\theta_1, \theta_{20}) = \frac{1}{\sqrt{n}} \sum_{i=1}^{n} \left[ \frac{g(y_i; \theta_1, \theta_{20}) - g_1(\theta_1, \theta_{20})}{\sigma(\theta_1, \theta_{20})} \right]$$
$$d \to N(0, 1),$$

If $\theta_1$ were known $Z_n(\theta_1, \theta_{20})$ could be used for testing $H_0$. To avoid confusion let us denote $\sqrt{n}$-consistent estimator of $\theta$ under $H_0$ as $\theta^+ = (\theta_1^{+'}, \theta_{20})'$. To find the optimal test Neyman proceeded in two steps. In the *first* he asked the question how we select $Z_n(\theta_1, \theta_{20})$ [or $g(y, \theta)$] so that

$$(2.58) \qquad Z_n(\theta_1, \theta_{20}) - Z_n(\theta_1^+, \theta_{20}) = o_p(1),$$

Neyman, [156], (1959), Theorem 1 proved that the necessary and sufficient condition for (2.58) to hold is that

$$(2.59) \qquad Cov[g(y; \theta_1, \theta_{20}), s_{1j}(y; \theta_1, \theta_{20})] = 0,$$

where $s_{1j} = \frac{\partial l(\theta)}{\partial \theta_{1j}}$, i.e., the score for the $j^{th}$ component of $\theta_1$, $j = 1, 2, \ldots$, $p - 1$. In other words, the function $g(y; \theta)$ should be orthogonal to $s_1 =$

$\frac{\partial l(\theta)}{\partial \theta_1}$. Let us assume $g(y; \theta_1, \theta_{20})$ is already a normed Cramér function. Starting from a normed Cramér function let us construct

$$(2.60) \qquad \bar{g}(y; \theta_1, \theta_{20}) = g(y; \theta_1, \theta_{20}) - \sum_{j=1}^{p-1} b_j s_{1j}(\theta_1, \theta_{20}),$$

where $b_j$ are the regression coefficients of regressing $g(y; \theta_1, \theta_{20})$ on $s_{11}, s_{12}, \ldots, s_{1p-1}$. Denote by $\sigma^2 (\theta_1, \theta_{20})$ the minimum variance of $\bar{g}(y; \theta_1, \theta_{20})$, and define

$$(2.61) \qquad g^*(y; \theta_1, \theta_{20}) = \frac{\bar{g}(y; \theta_1, \theta_{20})}{\sigma(\theta_1, \theta_{20})}.$$

Note that $g^*(y; \theta_1, \theta_{20})$ is also a normed Cramér function, and the covariance between $g^*(y; \theta_1, \theta_{20})$ and $s_{1j}(\theta_1, \theta_{20})$ is also zero, $j = 1, 2, \ldots, p - 1$. Therefore, a class of the $C(\alpha)$ test can be based on $Z_n(\theta_1^+, \theta_{20}) = \frac{1}{\sqrt{n}} \sum_{i=1}^{n} g^*(y_i; \theta_1^+, \theta_{20})$. Neyman's *second* step was to find the starting function $g(y; \theta)$. Theorem 2 of Neyman, [156], (1959) states that under the sequence of local alternatives $H_{1n} : \theta_2 = \theta_{20} + \frac{\delta}{\sqrt{n}}, 0 < \delta < \infty, Z_n(\theta_1^+, \theta_{20})$ is distributed as asymptotically normal with mean $\delta \rho \sigma_2$ and variance unity where

$$(2.62) \qquad \rho = \mathrm{Corr} \left[ \bar{g}(y; \theta_1, \theta_{20}), \frac{\partial l(\theta)}{\partial \theta_2} \right],$$

$$\text{and } \sigma_2 = V \left[ \frac{\partial l(\theta)}{\partial \theta_2} \right].$$

The asymptotic power of the test will be purely guided by $\rho$, and to maximize the power we should select the function $g(y; \theta)$ such that $\rho = 1$, i.e., the optimal choice should be $g(y; \theta) = \frac{\partial l(\theta)}{\partial \theta_2} = s_2(\theta_1, \theta_{20})$ say, score for the testing parameter $\theta_2$. Therefore, from (2.60), an asymptotically and locally optimal test should be based on the part of the score for the parameter tested that is orthogonal to the score for the nuisance parameter, namely,

$$(2.63) \qquad s_2(\theta_1^+, \theta_{20}) - \sum_{j=1}^{p-1} b_j s_{1j}(\theta_1^+, \theta_{20}),$$

It can be shown that if we use test function $g(y; \theta)$ other than $s_2(\theta)$, the asymptotic efficiency of the resulting test would be $\rho^2$ [see Neyman and

Scott, [162], (1967), page 478. Therefore, any test function with $|\rho| > 0$ will have some power. The optimal test, however, will result only when the function $g(y, \theta)$ coincide with $s_2(\theta)$. In (2.63) $b_j, j = 1, 2, \ldots, p-1$ are now regression coefficient of regressing $s_2(\theta_1^+, \theta_{20})$ on $s_{11}, s_{12}, \ldots, s_{1p-1}$, and we can express (2.63) as

$$(2.64) \qquad s_2(\theta^+) - \mathcal{I}_{21}(\theta^+)\mathcal{I}_{11}^{-1}(\theta^+)s_1(\theta^+) = s_2^*, \text{say},$$

where $\mathcal{I}_{ij}$ are the appropriate blocks of the information matrix $\mathcal{I}(\theta)$ corresponding to $\theta_1$ and $\theta_2$. $s_2^*(\theta)$ is called the *effective* score for $\theta_2$ and its variance, $\mathcal{I}_{22}^*(\theta) = \mathcal{I}_{22} - \mathcal{I}_{21}\mathcal{I}_{11}^{-1}(\theta)\mathcal{I}_{12}(\theta)$ is termed effective information [see, for instance, Hall and Mathiason, [104], (1990)]. Note that, since $s_2^*(\theta)$ is the residual score obtained from running a regression of $s_2(\theta)$ on $s_1(\theta)$, it will be orthogonal to the score for $\theta_1$. The operational form of Neyman's $C(\alpha)$ test is

$$(2.65) \qquad C(\alpha) = s_2^*(\theta^+)'\mathcal{I}_{22}^*(\theta^+)^{-1}s_2^*(\theta^*)$$

Bera and Bilias, [21], (1999) derived this test using the Rao, [186], (1948) framework. If we replace the $\sqrt{n}$-consistent estimator $\theta^+$ by the restricted MLE, then $s_2(\theta^+)^*$ and $\mathcal{I}_{22}^*(\theta^+)$ reduces to $s_2(\tilde{\theta})$ and $\mathcal{I}_{22}(\tilde{\theta})$, respectively and the $C(\alpha)$ test becomes the standard RS test.

Neyman's $C(\alpha)$ provides an attractive way to take account of the nuisance parameter $\theta_1$. The test can be viewed as adaptive in the sense of Stein, [211], (1956) because it is independent of the value of the nuisance parameter. Apart from requiring only $\sqrt{n}$-consistent estimator, the $C(\alpha)$ test has many advantages. The test can be computed by artificial regression [see Davidson and MacKinnon, [63], (1991) and Bera and Bilias, [21], (1999)]. The non-invariance of the $W$ test for nonlinear restrictions when it has some computational advantages was mentioned earlier. Dagenais and Dufour, [59], (1991) and Dufour and Dagenais, [70], (1992) have studied the invariance properties of the $C(\alpha)$ test showing that the $C(\alpha)$ statistic is invariant to the equivalent reformulation of the null hypothesis and giving conditions under which it is invariant to the re-parameterization of the model space and to the transformation of the model variables.

After Neyman, [156], (1959) no new fundamental test principle has been suggested. While hypothesis testing is an active area of research, what we have seen more refinements of the old tests and studies of their asymptotic and finite sample properties. Econometricians have developed a number of test procedures, such as those in Durbin, [71], (1970), Hausman, [108], (1978), Newey, [163], (1985), Tauchen, [213], (1985) and White,

[223], (1982), but these tests could be put in the framework of the basic test principles that we just covered.        –    ·

## 3. Misspecified Models

As we mentioned in Section 2.1 Karl Pearson, [182], (1902) and Ronald Fisher, [81], (1922) drew attention to the problem of *specification* in statistical modeling. A formal mention of *misspecification* occurred even a little earlier. Pearson, [180], (1894) proposed the method of moment estimation technique to fit a density to the crab measurements data set brought to him by W.F.R. Weldon. Pearson's, [179], (1893) own class of frequency distribution was not broad enough to fit the data, and he dissected the asymmetrical "double humped" density into a mixture of *two* normal densities. Francis Galton who handled the paper for the Royal Society, had his doubts about Pearson's methodology. In a letter dated November 25, 1893, Galton pointed out the possibility of "misspecification" in Pearson's model [see, Stigler, [212], (1986), pages 332-333]: "My misgivings, rightly or wrongly based, about the practical application of your methods are (2.1) if there be really 3 or more components and if the given curve be dissected into only 2, then neither of the 2 calculated components can be right......
it seems to me that observed curves of frequency are never so exact in contour as to lend themselves to exact & minute treatment." This is possibly one of the earliest statement about model misspecification in statistics. Subsequently, many others expressed the problem in a variety of ways.

### 3.1 Pearson, Student and Fisher Debate on Effects of Non-normality

The issue of misspecification and its consequences was clearly exposed in a somewhat acrimonious debate among Egon Pearson, Student and Fisher. One of the landmarks in the history of statistics is Fisher's book *Statistical Methods for Research Workers*. The book, however, has not received a single good review [see, Box, [42], (1978), pages 130-131]. Egon Pearson, [170], (1926) reviewed the first edition in *Science Progress*, and the second edition in *Nature* in 1929. Pearson's, [173], (1929) review prompted a chain of twenty-eight letters. The contentious issue was the assumption of normality on which many of Fisher's methods were based. Pearson, [173], (1929) wrote: "A large number of the tests developed are based upon the assumption that the population sampled is of 'normal' form. That this is the case may be gathered from a very careful reading of the text, but the point is not sufficiently emphasized. It does not appear reasonable to lay stress on the 'exactness' of tests, when no means whatever are given of appreciating how rapidly they become in-exact as the population sampled diverges from normality. That the tests, for example, connected with the analysis of variance are far more dependent on normality than those involving 'Student's' $z$ (or $t$) distribution is almost certain, but no clear indication of the need for caution in their application is given to the worker. It would seem wiser in the long run, even in a text-book, to admit the incompleteness of theory in this direction, rather than risk giving the reader the impression that the solution of all his problems has been achieved."

As Egon Pearson, [177], (1990), pages 95–101, page 108 later recalled, Student was sympathetic with Pearson's view that more attention should be given to the "normality" assumption, and Student was also anxious to prevent controversy among Fisher and the two Pearsons, all of whom were his good friends, though in different ways. In a letter to *Nature* (July 20, 1929) Student "gently" prodded Fisher to address Egon Pearson's concern: "The *question of the applicability* of normal theory to non–normal material is, however, of considerable importance and merits attention both from the mathematicians and from those of us whose province it is to apply the results of his labour to practical work .... We should all of us, however, be grateful to Dr. Fisher if he would show us elsewhere on theoretical grounds what *sort* of modification of his tables we require to make when the samples with which we are working are drawn from populations which are neither symmetrical nor mesokurtic."

In his reply to *Nature*, August 17, 1929, Fisher, "quite unusually," disagreed with Student and ruled out the possibility of modifying his tables to adapt for non-normality, saying:

"The theoretical reasons may be made most clear by ignoring the limits of practical possibility, and supposing that an army of computers had extended the existing tables some two hundred fold, with the view of providing tests of significance for all distributions conforming to the Pearsonian system of frequency curves. The system of tests of significance so produced would then be exposed to criticism from three different angles."

The three difficulties Fisher cited are the equal possibility of distributions outside the Pearsonian system, problems that would arise from the sampling errors when estimating parameters, and that for Pearsonian curves quite other statistics would be required, not just the mere revision of the familiar statistics appropriate to normal distribution. Here, Fisher points out the real difficulties in carrying out a systematic attack on the robustness problem. Having been trained in Karl Pearson's tradition, Egon Pearson had long recognized the existence of real non–normal distributions and since 1926 he had been quietly working on this problem [see, Pearson, [177], (1990), page 100]. Apparently, Pearson had the last word on the ongoing debate in the pages of *Nature* (October 19, 1929):

"A-normality may arise in an infinite number of ways; it is only humanly possible to explore a few of these, but each fresh piece of information makes us more certain of the strength or weakness of our tools. By representing populations by some variable system of mathematical curves, it is possible to examine certain typical forms of deviation from the normal. But it certainly is not claimed that such an exploration would be exhaustive."

If we read carefully, it is quite apparent that Egon Pearson and Fisher were not really far apart in their views. Both agreed that how difficult it is define "non-normality" and even more difficult to investigate the effects of non-normality in a systematic way. These difficulties are compounded even further if we want to define "misspecification" within a broad framework. Leo Tolstoy opened his classic novel *Anna Karenina* with. "All happy families resemble one another, but each unhappy family is unhappy in its own way." Possibly, we could paraphrase this to characterize misspecified model by saying, "All well-specified models resemble one another, but each misspecified model is misspecified in its own way." Therefore, it is quite natural that for a long time there was no *general* treatment of misspecification either in the statistics or econometrics literature.

Egon Pearson [171], (1928), [172], (1929), [174], (1931) established that Fisher's $z$ test for comparing two variances and the related tests were much more sensitive to non-normality than others. A more exhaustive exploration on the robustness of "normal theory" was possible only after the introduction of the electronic computer, and the results were reported in Pearson and Please, [178], (1975) [see also Pearson, [177], (1990), pages 100–101]. Geary, [89], (1946), [90], (1947) and Gayen, [86], (1949), [87], (1950), [88], (1950) in a series of papers investigated theoretically the effects of non-normality and supported Egon Pearson's earlier numerical results. Box, [40], (1953) took this analysis further and showed that the sensitivity of the test on variances is even greater when the number of variances to be compared exceeds two. Over the years there have been many studies assessing the consequences of hypothesis testing when the normality assumption is not valid. Hotelling, [113], (1947), [114], (1961) demonstrated marked effects of non-normality on the sample standard deviations, correlation coefficients, and the $t$-test. Box and Watson, [41], (1962) studied the effects of non-normality on testing for regression coefficients. Even the $\chi^2$ test about the variance of the linear model is not asymptotically valid in the presence of non-normality [see, for example, Arnold, [14], (1980)]. Also, if the disturbance term in a linear regression model can characterized by a stable distribution with the index parameter $\alpha < 2$, then usual hypothesis testing procedures will not be valid [see Blattberg and Sargent, [39], (1971)].

## 3.2 Neyman's (1937) and Haavelmo's (1944) Approaches to Misspecification

Along with the suggestion of the smooth test, Neyman, [155], (1937) carried out a thorough study of power when the alternative hypothesis is misspecified. As given in (2.36), the statistic $\psi_r^2$ tests $H_0 : \delta_1 = \delta_2 = \ldots = \delta_r = 0$ against the alternative $h(z)$ stated in (2.34), Neyman called $\psi_r^2$ the $r^{th}$-order smooth test. For his misspecification analysis he considered powers of the test for three cases: (1) a higher-order test is used when the true distribution is lower-order, (2) a lower-order test than the required one is used and (3) when the true distribution does not belong to the smooth family. For Case-1, Neyman established that there will be loss of power and stated ([155], page 193) that, if we test $H_0$ by means of $r^{th}$-order smooth test calculating $\psi_r^2$ for $r > 1$, then the falsehood of the hypothesis tested will be less frequently detected than if we applied the test of a lower-order, say, $\psi_{r-1}^2$. For Case-2, he proved that the lower-order test will be biased; however, this bias will disappear as the number of observations increases since the test components are orthogonal to each other. The main result is that for large $n$, smooth test or any order $r$ is indifferent to the presence of $\delta_{r+1}, \delta_{r+2}, \ldots$ etc. ([155], page 198). For the last case, by expanding the true probability density Neyman put the situation as a special form of Case-2, and commented that a smooth test of the fourth or, at most, fifth order will have a fair chance of detecting the falsehood of the hypothesis tested. Rayner and Best, [190], (1989), pages 52–53 provide a discussion on the effective order and power comparison for the smooth test. They suggest that a sequence of $\psi_r^2$ tests could be applied to a data set to identify the effective order of the alternative generating the data.

Haavelmo, [99], (1944), in this pioneering work, put Neyman's analysis in a general framework. He was probably the first to do a systematic analysis of the problems that arise when the alternative hypothesis used to construct a test deviates from the data generating process (DGP). Haavelmo was clearly influenced by Neyman's approach to statistics as he stated in his Nobel Lecture, December 7, 1989 [see Havelmo, [100], (1997)]:

"For my own part I was lucky enough to be able to visit the United States in 1939 on a scholarship .... I then had the privilege of studying with the world famous statistician Jerzy Neyman in California for a couple of months. At that time, young and naive, I thought I knew something about econometrics. I exposed some of my thinking on the subject to Professor Neyman. Instead of entering into a discussion with me, he gave me two or three exercises for me to work out. He said he would talk to me when I had done these exercises. When I met him for that second talk, I had lost most of my illusions regarding the understanding of how to do econometrics. But Professor Neyman also gave me hopes that there might be other more fruitful ways to approach the problem of econometric methods than those which had so far caused difficulties and disappointments."

While discussing the problem of testing economic relations, Haavelmo, [99], (1944), pages 65-66 stated: "Whatever be the principles by which we choose a "best" critical region of size $\alpha$, the essential thing is that a test is always developed with respect to a *given fixed* set of possible *alternatives* $\Omega^0$. If, on the basis of some general principles, a "best" test, or region, $W_0'$ say, is developed for testing a given hypothesis $P \in \omega^0$ with respect to a set $\Omega^0$, of a priori admissible hypotheses, and if we shift the attention to *another a priori admissible set*, $\Omega'$, also containing $\omega^0$, the same general principle will, usually, lead to *another* "best" critical region, say $W_0''$. In other words, if a test is developed on the basis of a given set of a priori admissible hypotheses, $\Omega^0$, the test is, in general, valid only for this set, $\Omega^0$."

In testing any economic relations, specification of the priori admissible hypotheses, $\Omega^0$, is of fundamental importance. According to Haavelmo, a test is not robust if there are some alternatives in $\Omega'$ for which the test has poor power, where $\Omega'$ may be obtained by extending (or changing) $\Omega^0$ to include new (or different) alternatives.

Very often it is difficult to interpret the results of a test applied to a misspecified model. For example, while testing the significance of some of the regression coefficients in the linear regression models, the results are not easily interpretable when a nonlinear model is appropriate one [see White, [222], (1980), Bera and Byron, [23], (1983) and Byron and Bera, [47], (1983)]. This is due to the fact that under the linear regression model the "allowable" alternatives include only the system of regression equations of the same (linear) form, but with regression coefficients that are different from zero [see Haavelmo, [99], (1944), page 66].

Typically, the alternative hypothesis may be misspecified in three different ways. The first is what we shall call "complete misspecification". In this case, the set of assumed alternatives, $\Omega^0$, and the DGP, $\Omega'$ say, are mutually exclusive, i.e., $(\Omega^0 - \omega^0) \bigcap (\Omega' - \omega^0) = \emptyset$. In the second case the alternative is under-specified in that it is a subset of a more general model representing the DGP, i.e., $\Omega^0 \subset \Omega'$. This leads to the problem of "under-testing" which one has to guard against when performing "one–directional" tests [or "fewer-directional" tests than actually required]. The last is "over-testing," which results from over-specification, that is, when $\Omega^0 \supset \Omega'$. This is more likely when "multi-directional" joint tests are applied based on an overparamerterized alternative model. [For a detailed discussion of the concepts of under-testing and over-testing, see Bera and

Jarque, [26], (1982)]. In both under-testing and over-testing some loss of power is to be expected. Kendall and Stuart, [122], (1973), page 462 call the loss of sensitivity due to over-testing "dilution" of the test.

In the statistics and econometrics literature, most emphasis has been put on the minimization of type–I and type–II error probabilities. There are, however, only a few papers that seriously consider the consequences and remedies of misspecifying the priori admissible hypothesis–which we may call the type–III error. Bera and Jarque, [26], (1982) reported some Monte Carlo results on the estimated power of the some of the well-known one–directional and multi-directional specification tests under different kinds of misspecification (see also the references cited there for other related research), On the basis of their Monte Carlo results, Bera and Jarque, [26] (1982), page 71 concluded that under-testing resulted in considerable loss of power while the effect of over-testing was less severe. Godfrey, [92], (1988), page 79, Pagan and Wickens, [169], (1989), page 993, and Pagan, [168], (1990) highlighted the importance of this issue and Wooldridge, [227], (1990) developed some robust, regression-based specification tests. Kopecky and Pierce, [130], (1979), Davidson and MacKinnon, [60], (1985), [62], (1987) and Saikkonen, [199], (1989) provided some analytical treatment of the problem. In the context of a duration model, Jaggia and Trivedi, [117], (1994) investigated the performance of various tests under misspecification.

To demonstrate the non-robustness of a significance test let us utilize an example from Haavelmo, [99], (1944), also investigated in Bera and Yoon, [34], (1991):

$$(3.1) \qquad y_t = b + kt + \varepsilon_t \qquad\qquad (t = 1, 2, \ldots, n),$$

$$E(y_t) = b + kt,$$

$$E(\varepsilon_t) = 0, E(e_t^2) = \sigma^2,$$

$$f(y_t) = \frac{1}{\sqrt{2\pi}\sigma} e^{-\frac{1}{2\sigma^2}(y_t - b - kt)^2},$$

where $\sigma^2$ is assumed to be known. We use 't' instead of 'i' since this is a time-series model. Let $H_0 : k = 0$ be the hypothesis to be tested. The following joint probability specifies $\Omega^0$, the set of admissible hypotheses:

$$(3.2) \qquad f(y_1, y_2, \ldots, y_n) = \frac{1}{(\sqrt{2\pi}\sigma)^n} e^{-\frac{1}{2\sigma^2}\sum_t (y_t - b - kt)^2},$$

with $-\infty < k < \infty$ and $-\infty < b < \infty$, and under $H_0$, this reduces to $\omega^0$, namely

$$(3.3) \qquad f(y_1, y_2, \ldots, y_n) = \frac{1}{(\sqrt{2\pi}\sigma)^n} e^{-\frac{1}{2\sigma^2}\sum_t (y_t - b)^2}.$$

Using standard notation, the test will be based on

$$(3.4) \qquad \hat{k} = \frac{\sum (t - \bar{t})(y_t - \bar{y})}{\sum (t - \bar{t})^2},$$

which has the following sampling distribution

$$(3.5) \qquad \hat{k} \sim N(k, \frac{\sigma^2}{\sum (t - \bar{t})^2}),$$

The critical region for this test is

$$(3.6) \qquad \left| \frac{\hat{k}}{\sqrt{\sigma^2 / \sum (t - \bar{t})^2}} \right| > 1.96,$$

at 5% level of significance. The power function, $\gamma(k)$, can then be written as

$$(3.7) \qquad \gamma(k) = 1 - \left[ \Phi(1.96 - \frac{k}{s}) - \Phi(-1.96 - \frac{k}{s}) \right],$$

where $\Phi(.)$ is the distribution function of the standard normal distribution and $s = \sqrt{\sigma^2 / \sum (t - \bar{t})^2}$.

Haavelmo perturbed the priori admissible hypotheses $\Omega^0$ by making $\varepsilon_t$ dependent, more specifically,

$$(3.8) \qquad \varepsilon_t = \frac{1}{\sqrt{2}}(\nu_t + \nu_{t-1}),$$

where $\nu_t \sim IIDN(0, \sigma^2)$, and studied the change in the power function $\gamma(k)$. However, under this setup the correlation coefficient between $\varepsilon_t$ and $\varepsilon_{t-1}$ is fixed at $1/2$. More generally let us assume

$$(3.9) \qquad \varepsilon_t = \rho \varepsilon_{t-1} + \nu_t,$$

with $|\rho|$ and $\nu_t \sim IIDN(0, \sigma^2(1 - \rho^2))$. Therefore, $\varepsilon_t$ will have the same variance $\sigma^2$ as in (3.1), The sampling distribution of $\hat{k}$ with $\varepsilon_t$'s following the $AR(1)$ of (3.9) is given by

$$(3.10) \qquad \hat{k} \sim N(k, s^2 C)$$

where

$$C = 1 + \frac{1}{\sum (t - \bar{t})^2} \left[ 2 \sum_{t < t'} (t - \bar{t})(t' - \bar{t}) \rho^{(t' - t)} \right].$$

At 5% significance level, the new power power function, $\gamma(k|\rho)$ say, of the test based on the critical region (3.6) may now be expressed as

$$(3.11) \qquad \gamma(k|\rho) = 1 - \left[ \Phi\left(1.96\frac{s}{s'} - \frac{k}{s'}\right) - \Phi\left(-1.96\frac{s}{s'} - \frac{k}{s'}\right)\right],$$

where $s' = s\sqrt{C}$. Obviously, the two power functions defined by (3.7) and (3.11) coincide when $\rho = 0$. Using the same numerical value for $n$ and $\sigma$ as in Haavelmo's example, we obtain plots of $\gamma(k|\rho)$ for different values of $\rho$. These plots are presented in Figure 1. From the figure it is easy to observe what happens to size and power of the above test based on the set of priori admissible hypotheses $\Omega^0$ specified by (3.2) when, in fact, $\varepsilon_t$'s are serially correlated. For instance when $\rho = 0.8$ the true type I error probability could be as high as 0.48, and convergence of power to 1 for distant alternatives is quite slow.

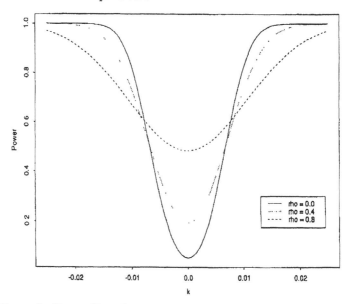

Figure 1. Power Functions

This example is very simple and more than half a century old. In spite of that, it is at the heart of the central problem in model specification tests. It clearly demonstrates that we should study the properties of our commonly used one-directional specification tests for certain alternatives not contained in the priori admissible hypotheses $\Omega^0$, since it is quite possible that some outside scheme is a true one having serious consequences for our inference [Haavelmo, [99], (1944), page 181].

As we discussed earlier a model can be misspecified in a variety of ways. Godfrey and Orme, [93], (1999) classified different kinds of misspecification in three broad groups: (i) parametric, (ii) distributional and (iii) higher-order moments. However, we will combine the last two groups into one and simply call it distributional misspecification. Our two types of misspecification, parametric and distributional, are not really mutually exclusive. Consider, for instance Haavelmo's example. It can be viewed either as a case of parametric misspecification since the standard test for $H_0 : k = 0$ puts $\rho = 0$, or as a distributional misspecification by ignoring the dependence structure in the distribution of $\varepsilon_t$. The two broad categories of are only for convenience.

### 3.3 Testing with Locally Misspecified Alternatives

In the previous section, we discussed behavior of a test statistic, when the model is misspecified, using an example. We now set up a general theoretical framework and study analytically the distribution of a one–directional test under local parametric misspecification. Using that distribution, we suggest a robust specification test under misspecified alternatives. This discussion is based on Bera and Yoon, [34], (1991), [35], (1993), Although we concentrate on the RS test, our analysis could be extended to the LR and $W$ tests.

Consider a general statistical model represented by the log-likelihood $l(\gamma, \psi, \phi)$ where $\gamma$, $\psi$, and $\phi$ are parameter vectors with dimensions $(m \times 1)$,$(r \times 1)$ and $(q \times 1)$, respectively. That is our $\theta = (\gamma', \psi', \phi')'$ and $p = m + r + q$. We follow Saikkonen's, [199], (1989) notation whenever possible. Suppose that one's primary interest is in model diagnostics or in specification search such that $l_0(\gamma)$ is the null model with possible alternatives, $l_1(\gamma, \psi)$, $l_2(\gamma, \phi)$, and $l(\gamma, \psi, \phi)$. Let us assume, as in Saikkonen (1989), that the following relations are true: $l_0(\gamma) = l_1(\gamma, \psi_*) = l_2(\gamma, \phi_*)$; $l_1(\gamma, \psi) = l(\gamma, \psi, \phi_*)$; and $l_2(\gamma, \phi) = l(\gamma, \psi_*, \phi)$, where $\psi_*$ and $\phi_*$ are known parameter values.

We will focus on the one–directional test for $H_0 : \psi = \psi_*$ in the alternative model $l_1(\gamma, \psi)$ ignoring the nuisance parameter $\phi$. Typically $\psi_* = 0$, representing zero restriction, and MLE of $\gamma$ under $H_0$, $\bar{\gamma}$ say, is readily available. In this situation the RS test is the preferred approach, and it is, locally optimal if the alternative correctly represents the DGP. Let $RS_\psi$ be the RS test statistic for $H_0$. Since $l_1(\gamma, \psi) = l(\gamma, \psi, \phi_*)$, we can express the score vector and the information matrix needed for $RS_\psi$ conveniently using $\theta$ and $l(\theta)$. Imposing the standard regularity conditions on $l(\theta)$, let $s_\psi(\theta) = \frac{\partial l(\theta)}{\partial \psi}$ and $\mathcal{I}(\theta) = -E\left(\frac{1}{n}\frac{\partial^2 l(\theta)}{\partial \theta \partial \theta'}\right)$. Not e that now our $\mathcal{I}(\theta)$ is $1/n$ of our earlier definition in (2.11); this will make our expressions for non-centrality parameters easier. $RS_\psi$ for testing $H_0$ based on $l_1(\gamma, \psi)$ can be written as

$$(3.12) \qquad\qquad RS_\psi = \frac{1}{n} s_\psi(\bar\theta)' \mathcal{I}_{\psi \cdot \gamma}^{-1}(\bar\theta) s_\psi(\bar\theta),$$

where $\bar\theta = (\bar\gamma', \psi_*', \phi_*')'$, $\mathcal{I}_{\psi \cdot \gamma}(\theta) = \mathcal{I}_\psi(\theta) - \mathcal{I}_{\psi\gamma}(\theta) \mathcal{I}_\gamma^{-1}(\theta) \mathcal{I}_{\gamma\psi}(\theta)$, $\mathcal{I}_\psi(\theta) \equiv \mathcal{I}_{\psi\psi}(\theta) = -E\left(\frac{1}{n} \frac{\partial^2 l(\theta)}{\partial\psi\partial\psi'}\right)$, and $\mathcal{I}_{\psi\gamma}(\theta) = -E\left(\frac{1}{n} \frac{\partial^2 l(\theta)}{\partial\psi\partial\gamma'}\right)$, etc. Given correct specification, $RS_\psi$ has well-known asymptotic distribution under the null and a sequence of local alternatives. This may be summarized as as follows:

CASE 1. *Correct Specification*
    Consider testing $H_0 : \psi = \psi_*$ in $l_1(\gamma, \psi)$, where $l_1(\gamma, \psi)$ represents the true model. Under $H_0$,

$$(3.13) \qquad\qquad RS_\psi d \to \chi_r^2(0)$$

Under the local alternative, $H_1 : \psi = \psi_* + \xi/\sqrt{n}$,

$$(3.14) \qquad\qquad RS_\psi d \to \chi_r^2(\lambda_1),$$

where $\lambda_1 = \lambda_1(\xi) = \xi' \mathcal{I}_{\psi \cdot \gamma} \xi$ and $\xi \neq 0$.
    Here $\chi_r^2(\lambda_1)$ stands for the non-central chi–square distribution with $r$ degrees of freedom and non–centrality parameter $\lambda_1$. Note also that the argument of $\mathcal{I}_{\psi \cdot \gamma}$ is suppressed such that $\mathcal{I} \equiv \mathcal{I}(\theta_*)$ where $\theta_* = (\gamma_0', \psi_*', \phi_*')'$ with $\gamma_0$ denoting the true value of $\gamma$.
    Let us now consider the case of misspecification. Suppose the true log-likelihood function is $l_2(\gamma, \psi)$, so that the alternative $l_1(\gamma, \psi)$ becomes misspecified completely. Using the sequence of *local* DGP $\phi = \phi_* + \delta/\sqrt{n}$ ($\delta \neq 0$), Davidson and MacKinnon, [62], (1987) and Saikkonen, [199], (1989) obtained the asymptotic distribution of $RS_\psi$ under $l_2(\gamma, \phi)$. The result may be stated as:

CASE 2. *Complete Misspecification*
    Consider testing $H_0 : \psi = \psi_*$ in $l_1(\gamma, \psi)$ where $l_2(\gamma, \phi)$ represents the true model. Under $l_2(\gamma, \phi)$ with $\phi = \phi_* + \delta/\sqrt{n}$,

$$(3.15) \qquad\qquad RS_\psi d \to \chi_r^2(\lambda_2),$$

where $\lambda_2 \equiv \lambda_2(\delta) = \delta' \mathcal{I}_{\phi\psi \cdot \gamma} \mathcal{I}_{\psi \cdot \gamma}^{-1} \mathcal{I}_{\psi\phi \cdot \gamma} \delta$ and $\mathcal{I}_{\psi\phi \cdot \gamma} = \mathcal{I}_{\psi\phi} - \mathcal{I}_{\psi\gamma} \mathcal{I}_\gamma^{-1} \mathcal{I}_{\gamma\phi} = \mathcal{I}_{\phi\psi \cdot \gamma}'$.
    Using this asymptotic distribution of the misspecified test $RS_\psi$, Davidson and MacKinnon, [62], (1987) and Saikkonen, [199], (1989) investigated

the power properties of $RS_\psi$ in the direction of the $\phi$ parameter. In particular, Saikkonen, [199], (1989) explicitly computed the asymptotic relative efficiency of $RS_\psi$ with respect to the optimal test (based on the true model), $RS_\phi$ say, for the general case, $r \neq q$. It should be noted that the asymptotic distribution of $RS_\psi$ was obtained under $l_2(\gamma, \phi)$, which is, in fact, $l(\gamma, \psi_*, \phi)$ by construction. Thus, one can interpret $l_2(\gamma, \phi)$ as the log-likelihood of the 'null' model 'contaminated' locally by the nuisance parameter $\phi = \phi_* + \delta/\sqrt{n}$. In other words, the situation can be treated as hypothesis testing in the presence of the nuisance parameter. An immediate effect of this parameter is that, even asymptotically, the size of the test as it is apparent from the non-centrality parameter $\lambda_2$, is not correct, unless $\delta(\neq 0)$ belongs to the null space of $\mathcal{I}_{\psi\phi\cdot\gamma}$ or $\mathcal{I}_{\psi\phi\cdot\gamma}$ itself is zero.

Using the result (3.15), Bera and Yoon, [35], (1993) modified $RS_\psi$ so that the resulting test is robust to the presence of $\phi$. The modified statistic is given by

$$RS_\psi^* = \frac{1}{n} \left[ s_\psi(\tilde\theta) - \mathcal{I}_{\psi\phi\cdot\gamma}(\tilde\theta)\mathcal{I}_{\psi\cdot\gamma}^{-1}(\tilde\theta)s_\phi(\tilde\theta) \right]'$$
$$\left[ \mathcal{I}_{\psi\cdot\gamma}(\tilde\theta) - \mathcal{I}_{\psi\phi\cdot\gamma}(\tilde\theta)\mathcal{I}_{\phi\cdot\gamma}^{-1}(\tilde\theta)\mathcal{I}_{\phi\psi\cdot\gamma}(\tilde\theta) \right]^{-1}$$
(3.16)
$$\left[ s_\psi(\tilde\theta)\mathcal{I}_{\psi\phi\cdot\gamma}(\tilde\theta)\mathcal{I}_{\psi\cdot\gamma}^{-1}(\tilde\theta)s_\phi(\tilde\theta) \right].$$

Under the regularity conditions, and when $H_0 : \psi = \psi_*$ is true

(3.17) $$RS_\psi^* d \to \chi_r^2(0),$$

This new test essentially adjusts the mean and the variance of the standard $RS_\psi$. It should be noted that $RS_\psi^*$ has the same asymptotic null distribution as the $RS_\psi$ based on the correct specification, thereby producing as asymptotically correct size test under the locally misspecified alternative $l_2(\gamma, \phi)$. Three points are worth noting. First, $RS_\psi^*$ requires estimation only under the joint null, that is $\psi = \psi_*$ and $\phi = \phi_*$. Given the full specification of the model $l(\gamma, \psi, \phi)$, it is, of course, possible to derive a RS test for $\psi = 0$ in the presence of $\phi$. However, that requires MLE of $\phi$ which could be difficult to obtain in some cases. Second, when $\mathcal{I}_{\psi\phi\cdot\gamma} = 0$, $RS_\psi^* = RS_\psi$. This is a very simple condition to check in practice. And if this condition is true, $RS_\psi$ is an asymptotically valid test in the *local* presence of $\phi$. Finally, Bera and Yoon, [35], (1993) showed that for local misspecification $RS_\psi^*$ is asymptotically equivalent to Neyman's $C(\alpha)$ test discussed in Section 2.4, and therefore, shares its optimality properties.

To illustrate the usefulness of the adjusted score statistic $RS_\psi^*$, let us consider the tests developed in Anselin et al., [12], (1996) for the mixed regressive–spatial autoregressive model with a spatial autoregressive disturbance:

$$y = \phi W y + X\gamma + u,$$
(3.18)
$$u = \psi W u + \varepsilon,$$
$$\varepsilon \sim N(0, \sigma^2 I)$$

In this model, y is an $(n \times 1)$ vector of observations on a dependent variable recorded at each of n locations, $X$ is an $(n \times m)$ matrix of exogenous variables, and $\gamma$ is a $(m \times 1)$ vector of parameters, $\phi$ and $\psi$ are scalar spatial parameters and W is a observable spatial weights matrix with positive elements, associated with the spatially lagged dependent variable and the spatial autoregressive disturbance. This spatial weight matrix represents "degree of potential interaction" between neighboring locations and are scaled so that the sum of the row elements in each matrix is equal to one.

We are interested in the testing $H_0 : \psi = 0$, in the local presence of the spatial log dependence parameter $\phi$. Using the results in Anselin, [10], (1988), Chapter 6, Anselin et al., [12], (1996) derived the $RS_\psi^*$ statistic as

(3.19)      $$RS_\psi^* = \frac{\left[ \tilde{u} W \tilde{u}/\tilde{\sigma}^2 - T(n\mathcal{I}_{\phi\cdot\gamma})^{-1}\tilde{u}'Wy/\tilde{\sigma}^2 \right]^2}{T[1 - T(n\mathcal{I}_{\phi\cdot\gamma})]^{-1}},$$

where $\tilde{u} = y - X\tilde{\gamma}$ are the ordinary least square residuals, $\tilde{\sigma}^2 = \tilde{u}'\tilde{u}/n$,

$$\mathcal{I}_{\phi\cdot\gamma} = \frac{1}{n\tilde{\sigma}^2}[(W X\tilde{\gamma})'M(W X\tilde{\gamma}) + T\tilde{\sigma}^2],$$

and $T = tr[(W' + W)W]$. And the conventional test $RS_\psi$ is given by

(3.20)      $$RS_\psi = \frac{[\tilde{u}'W\tilde{u}/\tilde{\sigma}^2]^2}{T}.$$

A comparison of (3.19) with (3.20) clearly reveals that $RS_\psi^*$ modifies the standard $RS_\psi$ by correcting the asymptotic mean and variance of the score. Similarly we can find $RS_\phi$ and $RS_\phi^*$.

Anselin and Florax, [11], (1995) and Anselin et al., [12], (1996) provided simulation results on the finite sample performance of the adjusted and unadjusted score and the related tests. The adjusted tests $RS_\psi^*$ and $RS_\phi^*$ performed remarkably well. They had reasonable empirical sizes, remaining within the confidence interval in all cases. In terms of power they performed exactly the way they were supposed to. For instance, when the data were generated under $\phi > 0$, $\psi = 0$, although $RS_\phi$ had most power, the powers of $RS_\phi^*$ were very close to that of $RS_\phi$. That is, the price paid for adjustments that were not needed turned out to be small. The real superiority $RS_\phi^*$ was revealed when $\psi > 0$ and $\phi = 0$. It yielded low rejection frequencies even for $\psi = 0.9$ whereas $RS_\phi$ rejected the true null $H_0 : \phi = 0$ too frequently. When $\phi > 0$, the power function of $RS_\phi^*$ was seen to be almost unaffected by the values of $\psi$, even for those far away from zero (global misspecification), Similarly, $RS_\psi^*$ also performed well [see also Anselin and Bera, [13], (1998)]. For other applications and discussions on the adjusted score test see Bera et al., [37], (1998) and Godfrey and Orme, [93], (1999). The benefit of considering several departures simultaneously in model testing has been demonstrated by McGuirk et al., [150], (1993) and Godfrey and Veal, [94], (1998).

### 3.4 Testing with Distributional Misspecification

In this section we discuss hypothesis testing when one does not have confidence in the maintained *probability model* itself. In particular, we investigate how the asymptotic distribution of the classical tests, namely LR, $W$ and RS tests, are affected by distributional misspecification, and outline procedures to adjust the tests whenever possible so that they remain valid.

Let the true data generating process (DGP) be described by the unknown density $g(y)$ and $f(y; \theta)$ be our assumed distribution. We assume that $g(y)$ and $f(y, \theta)$ satisfy some appropriate regularity conditions, such as those given by White, [223], (1982) and Kent, [123], (1982). We define the Fraser, [84], (1965) information of $f(y, \theta)$ under $g(y)$ as

$$(3.21) \qquad I_F(\theta) = E_g[\ln f(y, \theta)] = \int \ln f(y, \theta) g(y) dy,$$

where $E_g[.]$ denotes expectation under $g(y)$. Let $\theta_g$ be the value of $\theta$ that maximizes $I_F(\theta)$, i.e.,

$$(3.22) \qquad I_F(\theta_g) = \sup_{\theta \in \Theta} I_F(\theta)$$

Under our current set up, we interpret $\hat{\theta}$ as the quasi-maximum likelihood estimator (QMLE) that maximizes the assumed likelihood function $l(\theta)$. Note that $\hat{\theta}$ is a natural estimator of $\theta_g$, just as $l(\theta)/n$ is a natural "estimator" of $I_F(\theta)$. White, [223], (1982) showed that under regularity conditions $\hat{\theta}$ always exists and it converges almost surely to $\theta_g$ [see also

Huber, [115], (1967)]. Note that White, [223], (1982) defined $\theta_g$ that minimizes the Kullback–Leibler information criterion

(3.23) $\qquad I_{KL}(\theta) = E_g \left[ \ln \frac{g(y)}{f(y;\theta)} \right] = \int \ln \left[ \frac{g(y)}{f(y;\theta)} \right] g(y) dy.$

However, maximizing $I_F(\theta)$ or minimizing $I_{KL}(\theta)$ gives the same solution. If the model $f(y;\theta)$ is correctly specified, i.e., $g(y) = f(y;\theta_0)$ for some $\theta_0 \in \Theta$ and for all $y$, then $\theta_g = \theta_0$, so that $\hat{\theta}$ is consistent for the true parameter.

Let us define a function $b(\theta)$ as

(3.24) $\qquad b(\theta) = E_g \left[ \frac{\partial ln f(y;\theta)}{\partial \theta} \right] = \int \frac{\partial ln f(y;\theta)}{\partial \theta} g(y) dy.$

Given our definition of $\theta_g$ in (3.22), it is easy to see that $b(\theta_g) = 0$. And this is the counterpart of the standard result that $E_f [\frac{\partial ln f(y;\theta)}{\partial \theta}] = 0$ where $E_f[.]$ denotes expectation under $f(y;\theta)$. Another standard result is the information matrix equality, namely,

(3.25) $\qquad E_f \left[ \frac{\partial ln f(y;\theta)}{\partial \theta} \cdot \frac{\partial ln f(y;\theta)}{\partial \theta'} \right] = E_f \left[ -\frac{\partial^2 ln f(y;\theta)}{\partial \theta \partial \theta'} \right].$

The counterpart of this equality does not hold under distributional misspecification and when the expectation is taken with respect to $g(y)$. Let us define

$$J(\theta_g) = E_g \left[ \frac{\partial ln f(y;\theta)}{\partial \theta} \cdot \frac{\partial ln f(y;\theta)}{\partial \theta'} \right]$$

(3.26) $\qquad K(\theta_g) = E_g \left[ -\frac{\partial^2 ln f(y;\theta)}{\partial \theta \partial \theta'} \right].$

$J(\theta_g)$ and $K(\theta_g)$ are, in general, different as illustrated by the following example in White, [223], (1982). Suppose we take $f(y;\theta) \equiv N(\mu, \sigma^2)$, and let the DGP satisfies $E_g(y) = \mu$, $E_g(y - \mu)^2 = \sigma^2$, $E_g(y - mu)^3 = \mu_3$ and $E_g(y - \mu)^4 = \mu_4$. Then, it is easy to show that

(3.27) $\qquad J(\theta_g) = \begin{bmatrix} \frac{1}{\sigma^2} & \frac{\mu_3}{2\sigma^6} \\ \frac{\mu_3}{2\sigma^6} & \frac{\mu_4}{4\sigma^8} - \frac{1}{4\sigma^4} \end{bmatrix}$

and

(3.28) $\qquad K(\theta_g) = \begin{bmatrix} \frac{1}{\sigma^2} & 0 \\ 0 & \frac{1}{2\sigma^2} \end{bmatrix}.$

Hence, $J(\theta_g) = K(\theta_g)$ if and only if $\mu_3 = 0$ and $\mu_4 = 3\sigma^4$. Clearly, for this example, normality of the DGP is a sufficient, but not necessary, condition for $J(\theta_g)$ and $K(\theta_g)$ to be equal.

Under distributional misspecification, our earlier tests, LR, $W$ and RS will no longer be asymptotically distributed as $\chi_r^2$ as in Section 2. The main reason is the divergence between the matrices $J$ and $K$ under misspecification ( we drop the arguments of $K$ and $J$ whenever convenient ). And this also changes the asymptotic distribution of the QMLE $\hat{\theta}$, as follows:

$$(3.29) \qquad \sqrt{n}(\hat{\theta} - \theta_g)d \to N[0, K^{-1}JK^{-1}].$$

When there is no misspecification, $K = J$, and then asymptotic variance is
$K(\theta_g)^{-1}J(\theta_g)K(\theta_g)^{-1} = n\mathcal{I}(\theta_g)^{-1}$, in terms of our information matrix defined in (2.11). Therefore, we can view $J(\theta_g)K(\theta_g)^{-1}$ as the correction factor to the variance formula. Although the result (3.29) is derived in Huber, [115], (1976) and White, [223], (1982) under a general framework, this was also mentioned in Koopmans et al., [129], (1950), pages 148–150 in the context of simultaneous equation system.

The "sandwich" variance formula $K^{-1}JK^{-1}$ has been found to be useful in many contexts. For example, consider the linear regression model with heteroskedastic errors:

$$(3.30) \qquad y_t = x_t'\beta + \varepsilon_t,$$

where $\varepsilon_t$'s are independent with $E(\varepsilon_t) = 0$ and $V(\varepsilon_t) = \sigma_t^2$, $t = 1, 2, \ldots, n$. Let us write the general variance covariance matrix of $\varepsilon = (\varepsilon_1, \varepsilon_2, \ldots, \varepsilon_n)'$ as $V(\varepsilon) = \Sigma$. For the heteroskedastic case $\Sigma = diag(\sigma_1^2, \sigma_2^2, \ldots, \sigma_n^2)$. It is easy to show that

$$(3.31) \qquad \plim_{n \to \infty} \left[ \bar{\sigma}^2 \left( \left( \frac{X'X}{n} \right) \right)^{-1} \right] \neq nV(\bar{\beta}),$$

i.e., the standard variance formula for $V(\bar{\beta})$ does not provide a consistent estimator. Here $\bar{\beta} = (X'X)^{-1}X'y$ and $\bar{\sigma}^2 = \bar{\varepsilon}'\bar{\varepsilon}/n$ with $\bar{\varepsilon} = (\bar{\varepsilon}_1, \bar{\varepsilon}_2, \ldots, \bar{\varepsilon}_n) = y - X\bar{\beta}$ as the OLS residual vector. However, The sandwich-type formula suggested by Eicker, [74], (1963), [75], (1967) and White, [221], (1980) gives a *consistent* estimator for correct variance of $\bar{\beta}$, namely,

$$(3.32) \qquad \plim_{n \to \infty} \left( \frac{X'X}{n} \right)^{-1} \frac{X'\bar{\Sigma}X}{n} \left( \frac{X'X}{n} \right)^{-1} = \lim_{n \to \infty} nV(\bar{\beta}),$$

where $\bar{\Sigma} = \mathrm{diag}(\bar{\varepsilon}_1^2, \bar{\varepsilon}_2^2, \ldots, \bar{\varepsilon}_n^2) = \bar{\Sigma}_0$ (say), This variance formula can be used to draw asymptotically valid inferences in the face of an unknown form of heteroskedasticity. By exploiting this remarkably simple technique, Davidson and MacKinnon, [61], (1985) suggested heteroskedasticity–robust tests for many situations. Wooldridge, [227], (1990), [228], (1991) further generalized this technique and proposed convenient regression-based specification tests [see also MacKinnon, [146], (1992)].

We can even see the same approach when $\varepsilon_t$'s are also dependent. The most popular method is to use the Newey and West, [164], (1987) form of positive, semi–definite heteroskedasticity and autocorrelation consistent covariance matrix. For this case we need to replace $\bar{\Sigma}$ by $\bar{\Sigma}_0 + \bar{\Sigma}_1$, so that

$$X'\bar{\Sigma}X = X'\bar{\Sigma}_0X + X'\bar{\Sigma}_1X$$

$$(3.33) \qquad = \sum_{t=1}^{n} \bar{\varepsilon}_t^2 x_t x_t' + \sum_{j=1}^{L} \sum_{t=j+1}^{n} \bar{\varepsilon}_t \bar{\varepsilon}_{t-j}(x_t x_{t-j}' + x_{t-j} x_t')$$

The second term in (3.33) takes care of the serial dependence and the summation is truncated at $L$ since for stationary $\varepsilon_t$, since correlations at higher-order lags will be very small. To ensure positive semi-definiteness of $X\bar{\Sigma}X$, Newey and West, [164], (1987) suggested weighting the second term by $\omega_j = 1 - [j/(L+1)]$, $j = 1, 2, \ldots, L$ so that for their estimator

$$(3.34) \qquad X\bar{\Sigma}X = \sum_{t=1}^{n} \bar{\varepsilon}_t^2 x_t x_t' + \sum_{j=1}^{L} \sum_{t=j+1}^{n} \omega_j \bar{\varepsilon}_t \bar{\varepsilon}_{t-j}(x_t x_{t-j}' + x_{t-j} x_t')$$

Choices of the truncation length $L$ and the weights $\omega_j$ are the major problems, and these issues have received wide attention in the econometrics literature. For example, see Andrews, [7], (1991), Andrews and Monahan, [8], (1992), Newey and McFadden, [166], (1994), Newey and West, [165], (1994), Den Haan and Levin, [68], (1997) and Robinson and Velasco, [193], (1997).

Going back to the Haavelmo's example in Section 3.2, we see that the standard test in (3.6) fails to take account of the dependence in $\varepsilon_t$. When asked about his contribution to linear models, Geoff Watson mentioned the Durbin and Watson, [72], (1950) test but pointed out this inherent problem [see Beran and Fisher, [38], (1998), page 91]:

"What do I do if I have a regression and find the errors don't survive the Durbin-Watson test? What do I actually do? There is no robust method. You'd like to use a procedure that would be robust against errors no matter what the covariance matrix is. Most robustness talk is really about outliers, long-tail robustness. Dependence robustness is largely un-touched."

One way to carry out a serial correlation-robust test for $H_0 : k = 0$ would be to use the Newey and West, [164], (1987) formula for the variance of $\hat{k}$ in (3.4), We carried out a simple numerical exercise to illustrate the usefulness of using robust variance formula. Data were generated using (3.9) for $\rho = 0.4$ and $n = 100$. Power functions for different cases are drawn in Figure 2. The dotted line is the same as in Figure 1, which depicts the power of the standard tests. As we see without adjusting the variance, the test has very high type-I error probability, and hence, its higher power really does not indicate that superiority of the test. The solid line is for the known $\rho$ case. This has the ideal size–power combination. In practice, however, let alone $\rho$, the true autoregressive structure such as (3.9) will hardly be known. The modified test using Newey and West, [164], (1987) formula with $L = 6$, seems to perform quite well as we can see from the dashed-lined power curve. Though it is not as good as the known $\rho$ case, we should note that the modified test does not use any knowledge of the dependence structure of $\varepsilon_t$.

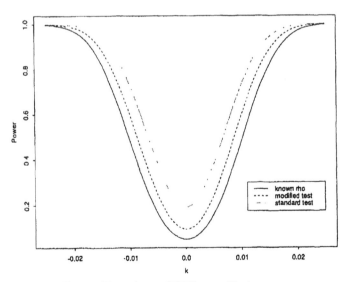

Figure 2. Power Functions of Different Tests

To see how the distribution of the standards tests will change and how we need to modify them in face of distributional misspecification, let us now consider the Wald test for testing $H_0 : h(\theta_g) = c$. From the Taylor series expansion

$$(3.35) \qquad \sqrt{n}h(\hat{\theta})a = \sqrt{n}h(\hat{\theta}_g) + \sqrt{n}H(\theta_g)(\hat{\theta} - \theta_g)$$

we have, under $H_0$,

$$(3.36) \qquad \sqrt{n}(h(\hat{\theta}) - c)d \to N[0, H'BH],$$

where $B = K^{-1}JK^{-1}$. Therefore, our earlier form of the Wald statistic (2.41) will not be valid. In light of (3.36), it should be modified as

$$(3.37) \qquad W^* = n[h(\hat{\theta}) - c]'[H(\hat{\theta})'B(\hat{\theta})H(\hat{\theta})]^{-1}[h(\hat{\theta}) - c],$$

and it will then have the standard asymptotic $\chi_r^2$ distribution [for details see Kent, [123], (1982) and White, [223], (1982), and for an easy exposition see Pace and Salvan, [167], (1997)]. In a similar fashion a robust form of the RS statistic (2.46) can be obtained as

$$(3.38) \quad RS^* = \frac{1}{n}s(\tilde{\theta})'K(\tilde{\theta})^{-1}H(\tilde{\theta})[H(\tilde{\theta})'B(\tilde{\theta})H(\tilde{\theta})]^{-1}H(\tilde{\theta})'K(\tilde{\theta})^{-1}s(\tilde{\theta})$$

Some of the adjusted score tests in the econometrics literature can be put in the framework of $RS^*$. For example, let us consider the test for heteroskedasticity under the framework of (3.30), where we now explicitly specify $V(\varepsilon_t) = \sigma_t^2 = \sigma^2 + \delta'z_t$, where $\delta$ is a $r \times 1$ vector and $z_t$'s are fixed. Assuming normality of $\varepsilon_t$, the RS test for the homoskedasticity hypothesis $H_0 : \delta = 0$ is given by [see Godfrey, [91], (1978) and Breusch and Pagan, [43], (1979)]

$$(3.39) \qquad\qquad RS = \frac{v'Z(Z'Z)^{-1}Z'v}{2\bar{\sigma}^4},$$

where $v_t = \bar{\varepsilon}_t^2 - \bar{\sigma}^2$, $v = (v_1, v_2, \dots, v_n)'$ and $Z = (z_1, z_2, \dots, z_n)'$. The factor "$2\bar{\sigma}^4$" is the consequence of the normality assumption, and, therefore, the test in (3.39) will not be valid even asymptotically if $\varepsilon_t$'s are not distributed as normal. The robust form of the RS from (3.38) can be derived as

$$(3.40) \qquad\qquad RS^* = \frac{v'Z(Z'Z)^{-1}Z'v}{(v'v/n)}.$$

This is the same modification suggested by Koenker, [128], (1981). Note that the modification amounts to replacing "$2\bar{\sigma}^4$" by a robust estimate of $V(\varepsilon_t^2)$, namely by $\frac{1}{n}\sum_{t=1}^{n}(\bar{\varepsilon}_t^2 - \bar{\sigma}^2)^2$. For other recent applications of this approach see Lucas, [144], (1998) and Bera and Premaratne, [30], (1999).

While we obtain these modified statistics $W^*$ and $RS^*$ by adjusting the variances of $h(\hat{\theta})$ and $s(\bar{\theta})$, respectively, a similar adjustment for the LR statistic (2.12) is not possible. Kent, [123], (1982) showed that under distributional misspecification the LR statistic is asymptotically distributed as a weighted sum of $r$ independent $\chi_1^2$ variables where the weights are the eigenvalues of the correction factor $JK^{-1}$. Therefore, there is no obvious "variance" adjustment for the LR test. Without misspecification $JK^{-1}$ reduces to an identity matrix and the eigenvalues are all equal to unity resulting in the asymptotic distribution of LR as standard $\chi_r^2$. White, [223], (1982) quite ingeniously used this factor $JK^{-1}$ to suggest a test for correct specification. If $f(y;\theta)$ is correctly specified, then we should expect $J(\hat{\theta})K(\hat{\theta})^{-1}$ to be close to an identity matrix or $J(\hat{\theta}) - K(\hat{\theta})$ to be close to a null matrix. White's, [223], (1982) information matrix (IM) test is based on comparing the matrices $J(\hat{\theta})$ and $K(\hat{\theta})$ element by element.

Consider a vector $d(y, \theta)$ defined by

$$(3.41) \qquad d(y, \theta) = vech\left[\frac{\partial^2 ln f(y;\theta)}{\partial\theta\partial\theta'} + \frac{\partial ln f(y;\theta)}{\partial\theta}\cdot\frac{\partial ln f(y;\theta)}{\partial\theta'}\right],$$

where "vech" denotes an operator that stacks distinct elements of a symmetric matrix. Thus, $d(y, \theta)$ is a vector with $p(p+1)/2$ elements, since $\theta$ is a $p \times 1$ vector. The actual IM test utilizes a sample average of $d(y,\theta)$, namely

$$(3.42) \qquad \bar{d}(\hat{\theta}) = \frac{1}{n}\sum_{i=1}^{n} d(y_i; \hat{\theta})$$

If the probability model is correctly specified, $\bar{d}(\hat{\theta})$ should take small values. Chesher, [51], (1984) provided an interesting interpretation of the IM test. He considered the problem of testing for random parameters and assumed that $\theta$ has mean, $\bar{\theta}$, and variance $\Gamma$, a $p \times p$ matrix. The null hypothesis that $\theta$ is non-stochastic is equivalent to $H_0 : \Gamma = 0$. For $\Gamma$ close to a null matrix, the marginal density of $y$ can be approximated by

$$(3.43) \qquad f^*(y; \bar{\theta}, \Gamma) = f(y; \bar{\theta})[1 + tr\{(F_2 + F_1 F_1')\Gamma],$$

where $F_1 = \frac{\partial ln f}{\partial\theta}$ and $F_2 = \frac{\partial^2 ln f}{\partial\theta\partial\theta'}$. It is clear that under $H_0 : \Gamma = 0$, $f^*(y, \theta, 0) = f(y; \theta)$. Chesher, [51], (1984) showed that the score vector for testing $H_0$ is precisely equal to $n\bar{d}(\hat{\theta})$. Thus the IM test has a score test interpretation [see also Cox, [54], (1983)]. In fact, as Bera and Bilias, [21], (1999) discussed, "all" moment-type-tests are score tests under a suitably defined density function. Application of the IM test principle to standard regression and other econometric models have resulted in some simple and interesting specification tests, for example, see Smith, [209], (1985), Hall, [103], (1987), Bera and Lee, [27], (1993), White, [224], (1994), Bera and Zuo, [36], (1996) and Bera and Mallick, [28], (1999).

## 4. Epilogue

The paper has reviewed what has happened over this century in testing statistical hypotheses with some emphasis on testing with misspecified models and viewed the progress with an historical focus. It is by no means exhaustive. We have not discussed many tests nor considered some of the specification concerns raised, for example, in Cox, [55], (1990), Lehmann, [139], (1990), Maasoumi, [145], (1990) and Monfort, [151], (1996), we have said nothing about finite sample issues and pretesting problems. Each of these issues would require more than a full-length paper.

We began the paper with Pearson, [181], (1900), which in a way foretold the advances to come in this century. Will there be another paper of similar influence to inaugurate the new century or millennium? Possibly, not. Given the vastness of the field we have now lost those "sharp moments of birth." The basic motivation behind Pearson's test was to see whether an assumed model fits the data adequately. In the light of recent advancements in nonparametrics techniques and developments of robust test procedures such as those discussed here, we probably no longer require explicit specification of the underlying probability model to carry out statistical inference. Quite strangely, because of these developments in adaptive procedures, possibly we will now have less use for the Pearson's goodness-of-fit test. In a way that measures the advancements in statistics and testing over this century, and signals a bright future for the next.

## Acknowledgements

I would to extend my sincere thanks to Gábor Székely for his endless patience; without his urging this paper would have never been written. I am also thankful to C.R. Rao for his encouragement. Gamini Premaratne did a superb job as a research assistant and helped me tremendously to complete this paper. Yulia Kotlyarova drew the two figures. Janet Fitch helped me to improve the exposition. I am most grateful to them. I however, retain the responsibility for any remaining errors. Financial supports from the Research Board and the Office of Research, College of the Commerce and

Business Administration of the University of Illinois at Champaign-Urbana are gratefully acknowledged.

### References

[1] Agresti, A. (1996), *An Introduction to Categorical Data Analysis*, Wiley, New York.

[2] Aigner, D. J., Lovell C. A. K. and Schmidt P., (1977), Formulation and estimation of stochastic frontier production function model. *J. Econometrics*, **6**, 21–37.

[3] Aitchison, J. and Silvey, S.D. (1958), Maximum-likelihood estimation of parameters subject to restraints. *Ann. Math. Statist.*, **29**, 813–828.

[4] Aitchison, J. and Silvey, S.D. (1960), Maximum-likelihood estimation procedures and associated tests of significance. *J. Roy. Statist. Soc.*, **B22**, 154–171.

[5] Amemiya, T. (1985), *Advanced Econometrics*, Harvard University press, Cambridge.

[6] Anderson, T.W. (1948), On the theory of testing serial correlation. *Skand. Aktuarietidskr.*, **31**, 88–116.

[7] Andrews, D.W.K. (1991), Heteroskedasticity and autocorrelation consistent covariance matrix estimation. *Econometrica*, **59**, 817–858.

[8] Andrews, D.W.K. and Monahan, J.C. (1992), An improved heteroskedasticity and autocorrelation consistent covariance matrix estimator. *Econometrica*, **60**, 953–966.

[9] Andrews, D.W.K. and Ploberger, W. (1994), Optimal tests when a nuisance parameter is present only under the alternative. *Econometrica*, **62**, 1383-1414.

[10] Anselin, L. (1988), *Spatial Econometrics: Methods and Models*, Kluwer, Dordrecht.

[11] Anselin, L. and Florax, R. (1995), Small sample properties of tests for spatial dependence in regression models: Some further results. In: L. Anselin and R. Florax Eds., *New Directions in Spatial Econometrics*, Springer Verlag, Berlin, 21-74.

[12] Anselin, L. Bera, A.K., Florax, R. and Yoon, M.J. (1996), Simple diagnostic tests for spatial dependence. *Regional Science and Urban Economics*, **26**, 77–104.

[13] Anselin, L. and Bera, A.K. (1998), Spatial dependence in linear regression models with an introduction to spatial econometrics. In: A. Ullah and D.E.A. Giles, Eds., *Handbook of Applied Economic Statistics*, Marcel Dekker, New York, 237–289.

[14] Arnold, S.F. (1980), Asymptotic validity of $F$ tests for the ordinary linear model and the multiple correlation model. *J. Amer. Statist. Assoc.*, **75**, 890–894.

[15] Baillie, R.T., Lippens, R.E. and McMahon, P.C. (1983), Testing rational expectation and efficiency in the foreign exchange market. *Econometrica*, **51**, 553–563.

[16] Barnard, G.A. (1992), Introduction to Pearson (1990) "On the criterion that a given system of deviation from the probable in the case of a correlated system of variables is such that it can be reasonably supposed to have arisen from random sampling." In: S. Kotz and N.L. Johnson, Eds., *Breakthroughs in Statistics*, Volume II. Springer-Verlag, New York, 1–11.

[17] Barton, D.E. (1953), On Neyman's smooth test of goodness of fit and its power with respect to a particular system of alternatives. *Skand. Aktuarietidskr*, **36**, 24–63.

[18] Barton, D.E. (1955), A form of Neyman's $\chi^2$ test of goodness of fit applicable to grouped and discrete data. *Skand. Aktuarietidskr*, **38**, 1–16.

[19] Barton, D.E. (1956), Neyman's $\chi^2$ test of goodness of fit when the null hypothesis is composite. *Skand. Aktuarietidskr*, **39**, 216–245.

[20] Bayes, Rev. T. (1763), An essay toward solving a problem in the doctrine of chances. *Phil. Trans. Roy. Soc.*, **53**, 370–418.

[21] Bera, A.K. and Billias, Y. (1999), Rao's score, Neyman's $C(\alpha)$ and Silvey's LM test: An essay on historical developments and some new results. *J. Statist. Plan. and Inf.*, forthcoming.

[22] Bera, A.K. and Bilias, Y. (1999), A brief history of estimation. Mimeo.

[23] Bera, A.K. and Byron, R. (1983), A note on the effects of linear approximation on hypothesis testing. *Economics Letters*, **12**, 251–254.

[24] Bera, A.K. and Higgins, M. (1992), A test for conditional heteroskedasticity in time series models. *J. Time Series Analysis*, **13**, 501–519.

[25] Bera, A.K. and Jarque, C.M. (1981), An efficient large-sample test for normality of observations and regression residuals. Working Paper in Economics and Econometrics, Number 40, The Australian National University, Canberra.

[26] Bera, A.K. and Jarque, C.M. (1982), Model specification tests: A simultaneous approach. *J. Econometrics*, **20**, 59–82.

[27] Bera, A.K., and Lee, S. (1993), Information matrix test, parameter heterogeneity and ARCH: A synthesis. *Review of Economic Studies*, **60**, 229–240.

[28] Bera, A.K. and Mallick, N.C. (1999), Information matrix tests for the composed error frontier model. Office of Research working Paper Number 99-0102, University of Illinois.

[29] Bera, A.K. and Mukherjee, R. (1999), 50 years of Rao's score test. A special issue of the. *J. Stat. Plan. Inf.*, forthcoming.

[30] Bera, A.K. and Premaratne, G. (1999), Adjusting the tests for skewness and kurtosis for distributional misspecification. Mimeo.

[31] Bera, A.K. and Ra, S-S. (1995), A test for the presence of conditional heteroskedsticity within ARCH–M framework. *Econometric Reviews*, **14**, 473–485.

[32] Bera, A.K. and Ra, S-S. (1997), Testing for regression coefficient stability. *J. Quant. Econ.*, **13**, 17–35.

[33] Bera, A.K. and Ullah, A. (1991), Rao's score test in econometrics. *J. Quantit. Econom.*, **7**, 189–220.

[34] Bera, A.K. and Yoon, M.J. (1991), Specification testing with misspecified alternatives. Bureau of Economic and Business Research Faculty Working Paper 91-0123, University of Illinois.

[35] Bera, A.K. and Yoon, M.J. (1993), Specification testing with locally misspecified alternatives. *Econometric Theory*, **9**, 649–658.

[36] Bera, A.K., and Zuo, X-L. (1996), Specification test for linear regression model with ARCH process. *J. Stat. Plan. Inf.*, **50**, 283–308.

[37] Bera, A.K., Ra, S-S. and Sarkar, N. (1998) Hypothesis testing for some non-regular cases in econometrics. In: S. Chakravarty, D. Coondoo and R. Mukherjee, Eds. *Econometrics: Theory and Practice*, Allied Publishers, New Delhi, 319-351.

[38] Beran, R.J. and Fisher, N.I. (1998), A conversation with Geoff Watson. *Stat. Science*, **13**, 75–93.

[39] Blattberg, R. and Sargent, T. (1971), Regression with non–Gaussian stable disturbance: Some sampling results. *Econometrica*, **39**, 501–510.

[40] Box, G.E.P. (1953), Non–normality and tests on variances. *Biometrika*, **40**, 318–335.

[41] Box, G.E.P. and Watson, G.S. (1962), Robustness to non–normality of regression tests. *Biometrika*, **49**, 93–106.

[42] Box, J.F. (1978), *R.A. Fisher, The Life of a Scientist*. Wiley, New York.

[43] Breusch, T.S. and Pagan, A.R. (1979), A simple test for heteroscedasticity and random coefficient variation. *Econometrica*, **47**, 1287–1294.

[44] Breusch, T.S. and Pagan, A.R. (1980), The Lagrange multiplier test and its applications to model specification in econometrics. *Rev. Econ. Stud.*, **47**, 239–253.

[45] Breusch, T.S. and Schmidt, P. (1988), Alternative forms of the Wald test: How long a piece of string?. *Comm. Stat. Theor. Method.*, **17**, 2789–2795.

[46] Byron, R.P. (1968), Methods for estimating demand equations using prior information: A series of experiments with Australian data. *Austr. Econ. Papers*, **7**, 227–248.

[47] Byron, R. and Bera, A.K. (1983), Least squares approximation to unknown regression functions: A comment. *Int. Econ. Rev.*, **24**, 255–260.

[48] Cameron, A.C. and Trivedi, P.K. (1990), Conditional moment tests and orthogonal polynomials, Working Paper in Economics, Number 90-051, Indiana university.

[49] Chant, D. (1974), On asymptotic tests of composite hypotheses in nonstandard conditions. *Biometrika*, **61**, 291–298.

[50] Chernoff, H. (1954), On the distribution of likelihood ratio. *Ann. Math. Statist.*, **25**, 573–578.

[51] Chesher, A.D. (1984), Testing for neglected heterogeneity. *Econometrica*, **52**, 865–872.

[52] Cox, D.R. (1961), Tests of separate families of hypotheses. In: *Proceedings of the Fourth Berkeley Symposium on Mathematical Statistics and Probability*, Vol.11, University of California Press, Berkeley, California.

[53] Cox, D.R. (1962), Further results on tests of separate families of hypotheses. *J. Roy. Statist. Soc.*, **B24**, 406–424.

[54] Cox, D.R. (1983), Some remarks on over-dispersion. *Biometrika*, **70**, 269–274.

[55] Cox, D.R. (1990), Role of models in statistical analysis. *Statistical Science*, **5**, 169–174.

[56] Cox, D.R. and Hinkley D.V., (1974), *Theoretical Statistics*, Chapman and Hall, London.

[57] Cramér, H. (1946), *Mathematical Methods of Statistics*. Princeton University Press, Princeton, New Jersey.

[58] Critchley, F., Marriott, P. and Salmon, M. (1996), On the differential geometry of the Wald test with nonlinear restrictions. *Econometrica*, **64**, 1213–1222.

[59] Dagenais, M.G. and Dufour, J-M. (1991), Invariance, nonlinear models, and asymptotic tests. *Econometrica*, **59**, 1601–1615.

[60] Davidson, R. and MacKinnon, J.G. (1985), The interpretation of test statistics. *Canadian Journal of Economics*, **18**, 38–57.

[61] Davidson, R. and MacKinnon, J.G. (1985), Heteroskedasticity-robust tests in regression directions. *Annales de l' INSÈÈ,*, **59/60**, 183–218.

[62] Davidson, R. and MacKinnon, J.G. (1987), Implicit alternatives and the local power of test statistics. *Econometrica*, **55**, 1305–1329.

[63] Davidson, R. and MacKinnon, J.G. (1991), Artificial regression and $C(\alpha)$ tests. *Economics Letters*, **35**, 149–153.

[64] Davidson, R. and MacKinnon, J.G. (1993), *Estimation and Inference in Econometrics*. Oxford University Press, Oxford.

[65] Davies, R.D. (1977), Hypothesis testing when a nuisance parameter is present only under the alternative. *Biometrika*, **64**, 247–254.

[66] Davies, R.D. (1987), Hypothesis testing when a nuisance parameter is present only under the alternative. *Biometrika*, **74**, 33–43.

[67] DeGroot, M.H. (1987), A conversation with C.R. Rao. *Statist. Sc.*, **2**, 53–67.

[68] Den Haan, W.J. and Levin, A.T. (1997), A practitioner's guide to robust covariance estimation. In: G.S. Maddala and C.R. Rao, Eds., *Handbook of Statistics, Volume 15*, Elsevier Science, Amsterdam, 299–342.

[69] Diebold, F.X. and Gunther, T.A. and Tay, A.S. (1998), Evaluating density forecasts with application to financial risk management. *Int. Eco. Rev.*, **39**, 863–905.

[70] Dufour, J-M. and Dagenais, M.G. (1992), Nonlinear models, rescaling and test invariance. *J. Stat. Plann. Inf.*, **32**, 111–135.

[71] Durbin, J. (1970), Testing for serial correlation in least squares regression when some of the regressors are lagged dependent variables. *Econometrica*, **38**, 410–421.

[72] Durbin, J. and Watson, G.S. (1950), Testing for serial correlation in least squares regression I. *Biometrika*, **37**, 409–428.

[73] Edgeworth, F.Y. (1885), *Methods of Statistics*, Jubilee Volume of the Statistical Society, London.

[74] Eicker, F. (1963), Asymptotic normality and consistency of the least squares estimators for families of linear regression. *Ann. Math. Statist.*, **34**, 447–456.

[75] Eicker, F. (1967), Limit theorems for regression with unequal and dependent errors. In: *Proceedings of the Fifth Berkeley Symposium in Mathematical Statistics and Probability*, University of California Press, Berkeley, 59–82.

[76] Fan, J. (1986), Test of significance based on wavelet thresholding and Neyman's truncation. *J. Amer. Statist. Ass.*, **91**, 674–688.

[77] Farebrother, R.W. (1986), Testing linear inequality constraints in the standard model. *Communications in Statistics, Series A*, **15**, 7–31.

[78] Ferrari, S.L.P. and Cribari-Neto, F. (1993), On the corrections to the Wald test of non-linear restrictions. *Econom. Lett.*, **42**, 321–326.

[79] Fisher, R.A. (1912), On an absolute criterion for fitting frequency curves. *Messenger of Mathematics*, **41**, 155–160.

[80] Fisher, R.A. (1922), On the mathematical foundations of theoretical statistics. *Phil. Trans. Roy. Soc.*, **A222**, 309–368.

[81] Fisher, R.A. (1922), On the interpretation of $\chi^2$ from contingency tables, and the calculation of $P$. *J. Roy. Stat. Soc.*, **85**, 87–94.

[82] Fisher, R.A. (1925), Theory of statistical estimation. *Proc. Camb. Phil. Soc.*, **22**, 700–725.

[83] Fisher, R.A. (1932), *Statistical Methods for Research Workers*, Oliver and Boyd, Edinburgh.

[84] Fraser, D.A.S. (1965), On information in statistics. *Ann. Math. Statist.*, **36**, 890–896.

[85] Gavarret, J. (1840), *Principles Généraux de Statistique Médicale*, Paris.

[86] Gayen, A.K. (1949), The distribution of 'Student's' $t$ in random samples of any size drawn from non–normal universes. *Biometrika*, **36**, 353–369.

[87] Gayen, A.K. (1950), The distribution of the variance ratio in random samples of any size drawn from non–normal universes. *Biometrika*, **37**, 236–255.

[88] Gayen, A.K. (1950), Significance of the difference between the means of two non-normal samples. *Biometrika*, **37**, 399–408.

[89] Geary, R.C. (1946), The distribution of student's ratio for non-normal samples. *J. Roy. Statist. Soc. Suppl.*, **3**, 178–184.

[90] Geary, R.C. (1947), Testing for normality. *Biometrika*, **21**, 259–286.

[91] Godfrey, L.G. (1978), Testing for multiplicative heteroscedasticity. *J. Econometrics*, **8**, 227–236.

[92] Godfrey, L.G. (1988), *Misspecification Tests in Econometrics, The Lagrange Multiplier Principle and Other Approaches* . Cambridge University Press, Cambridge.

[93] Godfrey, L.G. and Orme, C.D. (1999), On improving the robustness and reliability of Rao's score test. *J. Stat. Plann. Inf.*, forthcoming.

[94] Godfrey, L.G. and Veal, M.R. (1998), Robust test statistics for regression models. Mimeo.

[95] González–Rivera, G. (1998), Smooth transition GARCH models. *Studies in Nonlinear Dynamics and Econometrics*, forthcoming.

[96] Greenwood, P.E. and Nikulin, M.S. (1996), *A Guide to Chi-Squared Testing*, John Wiley & Sons, New York.

[97] Gregory, A.W. and Veal, M.R. (1985), Formulating Wald tests of nonlinear restrictions. *Econometrica*, **53**, 1465–1468.

[98] Gourieroux, C. and Monfort, A. (1995), *Statistics and Econometric Models 2*, Cambridge University Press, Cambridge.

[99] Haavelmo, T. (1944), The probability approach in econometrics. *Supplements to Econometrica*, **12**, .

[100] Haavelmo, T. (1997), Econometrics and the welfare state: Nobel lecture, December 1989. *Ame. Econ. Rev.*, **87**, 13–15.

[101] Hacking, I. (1984), Trial by number. *Science 84*, **5**, 67–70.

[102] Hald, A. (1998), *A History of Mathematical Statistics from 1750 to 1930*. Wiley, New York.

[103] Hall, A. (1987), The information matrix test for the linear model. *Rev. Econ. Stud.*, **54**, 257-263.

[104] Hall, W.J. and Mathiason, D. (1990), On large-sample estimation and testing in parametric models. *Internat. Statist. Rev.*, **58**, 77–97.

[105] Hamdan, M.A. (1962), The power of certain smooth tests of goodness of fit. *Aust. J. Statist.*, **4**, 25–40.

[106] Hamdan, M.A. (1964), A smooth test of goodness of fit based on the Walsh functions. *Aust. J. Statist.*, **6**, 130–136.

[107] Hansen, B. (1996), Hypothesis testing when a nuisance parameter is identified only under the alternative hypothesis. *Econometrica*, **64**, 413-430.

[108] Hausman, J.J. (1978), Specification tests in econometrics. *Econometrica*, **46**, 1215–1272.

[109] Hendry, D.E. (1980), Econometrics–alchemy or science?. *Economica*, **47**, 387–406.

[110] Hillier, G.H. (1986), Joint test for zero restrictions on non-negative regression coefficients. *Biometrika*, **73**, 657–383.

[111] Hochberg, Y. and Tamhane, A.C. (1987), *Multiple Comparison Procedures*, John Wiley & Sons, New York.

[112] Hoffman, D.L. and Schmidt, P. (1981), Testing the restriction implied by the rational expectation hypothesis. *J. Econometrics*, **15**, 265–287.

[113] Hotelling, H. (1947), Effects of non-normality at high significance levels (abstract). *Ann. Math. Statist*, **18**, 608–609.

[114] Hotelling, H. (1961), The behavior of some standard statistical tests under nonstandard conditions. *Proceedings of the 4th Berkeley Symposium on Mathematical Statistics and Probability*, Volume 1, University of California Press, Berkeley, 319–359.

[115] Huber, P.J. (1967), The behavior of maximum likelihood estimates under nonstandard conditions. In:¯*Proceedings of the Fifth Berkeley Symposium in Mathematical Statistics and Probability*, University of California Press, Berkeley.

[116] Inglot, T. and Ledwina, T. (1996), Asymptotic optimality of data-driven Neyman's test for uniformity. *Ann. Stat.*, **24**, 1982–2019.

[117] Jaggia, S. and Trivedi, P.K. (1990), Joint and separate score tests of state dependence and unobserved heterogeneity. *J. Econometrics*, **60**, 272–291.

[118] Jarque, C.M. and Bera, A.K. (1980), Efficient tests for normality, homosedasticity and serial independence of regression residuals. *Economics Letters*, **6**, 255–259.

[119] Jarque, C.M. and Bera, A.K. (1987), Test for normality of observations and regression residuals. *Internat. Statist. Rev.*, **55**, 163–172.

[120] Kallenberg, W.C.M. and Ledwina, T. (1995), On the data-driven Neyman's test. *Prob. Math. Stat.*, **15**, 409–426.

[121] Kallenberg, W.C.M. and Ledwina, T. (1997), Data-driven smooth tests when the hypothesis is composite. *J. Amer. Statist. Ass.*, **92**, 1094–1104.

[122] Kendall, M.G. and Stuart, A. (1973), *The Advanced Theory of Statistics 2*, Griffin, London.

[123] Kent, J.T. (1982), Robust properties of likelihood ratio tests. *Biometrika*, **69**, 19–27.

[124] Kiefer, N.M. (1982), A remark on the parameterization of a model for heterogeneity. Working Paper, No. 278, Department of Economics, Cornell University.

[125] King, M.L. and Smith, M.D. (1986), Joint one–sided tests of linear regression coefficients. *J. Econometrics*, **32**, 367–383.

[126] King, M.L. and Wu, P.X. (1997), Locally optimal one–sided tests for multiparameter hypotheses. *Econometric Reviews*, **16**, 131–156.

[127] Kodde, D.A. and Palm, F.C. (1986), Wald criteria for jointly testing equality and inequality restrictions. *Econometrica*, **54**, 1243–1248.

[128] Koenker, R. (1981), A note on Studentizing a test for heteroscedasticity. *J. Econometrics*, **17**, 107–112.

[129] Koopmans, T.C., Rubin, H. and Leipnik, R.B. (1950), Measuring the equation systems of dynamic economics. In: T.C. Koopmans, Ed., *Statistical Inference in Dynamic Economic Models*, Cowles Commision for Research in Economics, Monograph No. 10, John Wiley, New York, 53–244.

[130] Kopecky, K.J. and Pierce, D.A. (1979), Efficiency of smooth goodness of fit tests. *J. Amer. Statist. Ass.*, **74**, 393–397.

[131] Koziol, J.A. (1979), A smooth test for bivariate independence. *Sankhya*, Series B, **41**, 260–269.

[132] Koziol, J.A. (1986), Assessing multivariate normality a compendium. *Commun. Statist. -Theor. Meth.*, **15**, 2763–2783.

[133] Koziol, J.A. (1987), An alternative formulation of Neyman's smooth goodness of fit tests under composite alternatives. *Metrika*, **34**, 17–24.

[134] Lancaster, H.O. (1969), *The Chi-Squared Distribution*. Wiley, New York.

[135] LaRiccia, V.N. (1991), Smooth goodness of fit tests: A quantile function approach. *J. Amer. Statist. Ass.*, **86**, 427–431.

[136] Ledwina, T. (1994), Data-driven version of Neyman's smooth test of fit. *J. Amer. Stat. Ass.*, **89**, 1000–1005.

[137] Lee, L. F. and Chesher, A. (1986), Specification testing when score test statistics are individually zero. *J. Econometrics*, **31**, 121–149.

[138] Lehmann, E.L. (1986), *Testing Statistical Hypotheses*. John Wiley & Sons, New York.

[139] Lehmann, E.L. (1990), Model specification: The views of Fisher and Neyman, and later developments. *Stat. Sci.*, **5**, 160–168.

[140] Lehmann, E.L. (1992), Introduction to Neyman and Pearson(1933) "On the problem of the most efficient tests of Statistical hypothesis," In : S. Kotz and N.L. Johnson, Eds., *Breakthroughs in Statistics*, Volume I, Springer-Verlag, New York, 67–72.

[141] Lexis, W. (1875), *Einleitung in die Theorie der Bevölkerungsstatistik.* Strassburg.

[142] Lexis, W. (1877), *Zur Theorie der Massenerscheinungen in der Menschlichen Gesellschaft.* Freiburg.

[143] Lin, C.F.J. and Teräsvirta, T. (1999), Testing parameter constancy in linear models against stochastic stationary parameters. *J. Econometrics*, **90**, 193–213.

[144] Lucas, A. (1998), Inference on cointegrating ranks using LR and LM tests based on pseudo-likelihoods. *Econometric Reviews*, **17**, 185–214.

[145] Maasoumi, E. (1990), How to live with misspecification if you must. *J. Econometrics*, **44**, 67–86.

[146] MacKinnon, J.G. (1992), Model specification tests and artificial regressions. *J. Economics Literature*, **30**, 102–146.

[147] Maddala, G.S., P.C.B. Phillips and T.N. Srinivasan, Eds., (1995), *Advances in Econometrics and Quantitative Economics*, Blackwell, Oxford.

[148] Mahalanobis, P.C. (1933), Editorial. *Sankhya*, **1**, 1.–4.

[149] McAleer, M. (1987), Specification tests for separate models: A survey. In: M.L. King and D.E.A. Giles, Eds., *Specification Analysis in the Linear Model*, Routledge and Kegan Paul, London, 146–196.

[150] McGuirk, A.M., Driscoll, P. and Alwang, J. (1993), Misspecification testing: A comprehensive approach. *Amer. J. Agr. Econ.*, **75**, 1044–1055.

[151] Monfort, A. (1996), A reappraisal of misspecified econometric models. *Econometric Theory*, **12**, 597–619.

[152] Moran, P.A.P. (1950), A test for the serial independence of residuals. *Biometrika*, **37**, 178–181.

[153] Moran, P.A.P. (1971), Maximum likelihood estimation in nonstandard conditions. *Proc. Camb. Phil. Soc.*, **70**, 441–445.

[154] Mukherjee, R. and Sengupta, A. (1993), Comparison between the locally most mean power unbiased and Rao's tests in the multiparameter case. *J. Multivariate Analysis*, **45**, 9–24.

[155] Neyman, J. (1937), "Smooth test" for goodness of fit. *Skand. Akturarietidskr*, **20**, 150–199.

[156] Neyman, J. (1959), Optimal asymptotic test of composite statistical hypothesis. In: U. Grenander, Ed., *Probability and Statistics, the Harald Cramér Volume*, Almqvist and Wiksell, Uppsala, 213–234.

[157] Neyman, J. (1967), R.A. Fisher (1890–1962): An appreciation. *Science*, **156**, 456–1460.

[158] Neyman, J. (1979), $C(\alpha)$ tests and their use. *Sankhya*, **A41**, 1–21.

[159] Neyman, J. and Pearson, E.S. (1928), On the use and interpretation of certain test criteria for purpose of statistical inference. *Biometrika*, **20**, 175–240.

[160] Neyman, J. and Pearson, E.S. (1933), On the problem of the most efficient tests of statistical hypothesis. *Phil. Trans. Roy. Soc. Ser. A.*, **231**, 289–337.

[161] Neyman, J. and Pearson, E.S. (1936), Contribution to the theory of testing statistical hypothesis I: Unbiased critical regions of type $A$ and type $A_1$. *Statistical Research Memoirs*, **1**, 1–37.

[162] Neyman, J and Scott, E. (1967), On the use of $C(\alpha)$ optimal tests of composite hypothesis. *Bulletin of the International Statistical Institute*, **41**, 477–497.

[163] Newey, W. (1985), Maximum likelihood specification testing and conditional moment tests. *Econometrica*, **53**, 1047–1070.

[164] Newey, L.K. and West, K.D. (1987), A simple, positive semi-definite, heteroskedasticity and autocorrelation consistent covariance matrix. *Econometrica*, **55**, 703–708.

[165] Newey, L.K. and West, K.D. (1994), Automatic lag selection in co-variance matrix estimation. *Rev. Econom. Stud.*, **61**, 631–653.

[166] Newey, L.K. and McFadden, D.L. (1994), Large sample estimation and hypothesis testing. In: R.F. Engle and D.L. McFadden, Eds., *Handbook of Econometrics, Volume IV*, Elsevier Science, Amsterdam, 2112–2245.

[167] Pace, L. and Salvan, A. (1997), *Principles of Statistical Inference: From a Neo-Fisherian Perspective*. World Scientific, New Jersey.

[168] Pagan, A.R. (1990), Evaluating models: A review of L.G. Godfrey, misspecification tests in econometrics. *Econometric Theory*, **6**, 273–281.

[169] Pagan, A.R. and Wickens, M.R. (1989), A survey of some recent econometric methods. *Economic Journal*, **99**, 962–1025.

[170] Pearson, E. (1926), Review of Statistical Methods for Research Workers (R.A. Fisher), First Edition. *Sci. Prog.*, **20**, 733–734.

[171] Pearson, E. (1928), The distribution of frequency constants in small samples from symmetric populations (Preliminary Notice). *Biometrika*, **20**, 356–360.

[172] Pearson, E. (1929), The distribution of frequency constants in small samples from non–normal symmetrical and skew populations. *Biometrika*, **21**, 259–286.

[173] Pearson, E. (1929), Review of Statistical Methods for Research Workers (R.A. Fisher), Second Edition. *Nature*, **123**, 866.

[174] Pearson, E. (1931), The analysis of variance in cases of non-normal variation. *Biometrika*, **23**, 114–133.

[175] Pearson, E. (1938), The probability integral transformation for testing goodness of fit and combining independent tests of significance. *Biometrika*, **30**, 134–148.

[176] Pearson, E. (1966), The Neyman-Pearson story: 1926-34, Historical sidelights on an episode in Anglo-Polish collaboration. In: F.N. David, Ed., *Research Papers in Statistics, Festschrift for J. Neyman*, Wiley, New York, 1–23.

[177] Pearson, E. (1990), *'Student': A Statistical Biography of William Sealy Gosset*. Edited by R.L. Plackett and G.A. Barnard, Clarendon Press, Oxford.

[178] Pearson, E. and Please, N.W. (1975), Relation between the shape of population distribution and the robustness of four simple test statistics. *Biometrika*, **62**, 223–241.

[179] Pearson, K. (1893), Asymmetrical frequency curves. *Nature*, **48**, 615–616.

[180] Pearson, K. (1894), Contribution to the mathematical theory of evolution. *Phil. Trans. Roy. Soc. London, Ser. A*, **185**, 71–110.

[181] Pearson, K. (1900), On the criterion that a given system of deviations from the probable in the case of a correlated system of variables is such that it can be reasonably supposed to have arisen from random sampling. *Phil. Mag. Ser. 5*, **50**, 157–175.

[182] Pearson, K. (1902), On the systematic fitting of curves to observations and measurements. *Biometrika*, **1**, 265–363.

[183] Pearson, K. (1922), On the $\chi^2$ test of goodness of fit. *Biometrika*, **14**, 186–191.

[184] Pearson, K. (1933), On a method of determining whether a sample of size n supposed to have been drawn from a parent population having a known probability integral has probably been drawn at random. *Biometrika*, **25**, 379–410.

[185] Phillips, P.C.B. and J.Y.Park (1988), On the formulation of Wald tests of nonlinear restrictions. *Econometrica*, **56**, 1065–1083.

[186] Rao, C.R. (1948), Large sample tests of statistical hypotheses concerning several parameters with applications to problems of estimation. *Proc. Camb. Phil. Soc.*, **44**, 50–57.

[187] Rao, C.R. (1973), *Linear Statistical Inference and its Applications*. John Wiley and Sons, New York.

[188] Rao, C.R. and Poti, S.J. (1946), On locally most powerful tests when alternative are one sided. *Sankhya*, **7**, 439.

[189] Rayner, J.C.W. and Best D.J. (1986), Neyman-type smooth tests for location-scale families. *Biometrika*, **73**, 437–446.

[190] Rayner, J.C.W. and Best D.J. (1989), *Smooth Tests of Goodness of Fit*, Oxford University Press, New York.

[191] Rayner, J.C.W. and Best D.J. (1990), Smooth tests of goodness of fit: An overview. *Int. Stat. Rev.*, **58**, −17.

[192] Reid, C. (1982), *Neyman − From Life*. Springer-Verlag, New York.

[193] Robinson, P.M. and Velasco, C. (1997), Autocorrelation-robust inference. In: G.S. Maddala and C.R. Rao, Eds,*Handbook of Statistics, Volume 15*, Elsevier Science, Amsterdam, 267–298.

[194] Rockinger, M. (1994), Switching regressions of unexpected macro-economic events explaining the French stock index. Discussion Paper, HEC–School of Management, Paris.

[195] Rogers, A.J. (1986), Modified Lagrange multiplier tests for problems with one–sided alternatives. *J. Econometrics*, **31**, 341–361.

[196] Rosenblatt, M. (1952), Remarks on a multivariate transformation. *Ann. Math. Stat.*, **23**, 470–472.

[197] Roy, S. N. (1953), On a Heuristic method of test construction and its use in multivariate analysis. *Ann. Math. Statist.*, **24**, 220–238.

[198] Roy, S.N. and Mitra, S.K. (1956), An introduction to some non-parametric generalization of analysis of variance and multivariate analysis. *Biometrika*, **43**, 361–376.

[199] Saikkonen, P. (1989), Asymptotic relative efficiency of the classical test statistics under misspecification. *J. Econometrics*, **42**, 351–369.

[200] Sargan, J.D. (1980), Some tests for dynamic specification for a single equation. *Econometrica*, **48**, 879–897.

[201] Savage, L.J. (1976), On rereading R.A. Fisher. *Ann. Stat.*, **4**, 441–500.

[202] Schmidt, P. and Lin T. F., (1984), Simple tests of alternative specifications in stochastic frontier model. *J. Econometrics*, **24**, 349–361.

[203] Schwarz, G. (1978), Estimating the dimension of a model. *Ann. Statist.*, **6**, 461–464.

[204] Self, S.G. and Liang, K.Y. (1987), Asymptotic properties of maximum likelihood estimators and likelihood ratio tests under nonstandard conditions. *J. Amer. Statist. Assoc.*, **82**, 605–610.

[205] Sengupta, A.A. (1991), A review of optimality of multivariate tests. *Statistics and Probability Letters*, **12**, 527–535.

[206] Sengupta, A. and Vermeire, L. (1986), Locally optimal tests for multiparameter hypotheses. *J. Amer. Statist. Assoc.*, **81**, 819–825.

[207] Serfling, R.J. (1980), *Approximation Theorems of Mathematical Statistics*, John Wiley and Sons, New York.

[208] Silvey, S.D. (1959), The Lagrange multiplier test. *Ann. Math. Statist.*, **30**, 389–407.

[209] Smith, R.J. (1985), Some tests for misspecification in bivariate limited dependent variable models. *Annales de l' INSÉÉ*, **59/60**, 97–122.

[210] Smith, R.J. (1989), On the use of distributional mis-specification checks in limited dependent variable models. *Economic Journal*, **99**, 178–192.

[211] Stein, C. (1956), Efficient nonparametric testing and estimation. In: J. Neyman Ed., *Proceedings of 3rd Berkeley Symposium on Mathematical Statistics and Probability 1*, Berkeley, University of California Press, 187–195.

[212] Stigler, S.M. (1986), *The History of Statistics, The Measurement of Uncertainty Before 1900*. Harvard University Press, Cambridge.

[213] Tauchen, G. (1985), Diagnostic testing and evaluation of maximum likelihood models. *J. Econometrics*, **30**, 415–443.

[214] Teraäsvirta, T. (1994), Specification, estimation, and evaluation of smooth transition autoregressive models. *J. Amer. Statist. Assoc.*, **89**, 208–218.

[215] Thomas, D.R. and Pierce, D.A. (1979), Neyman's smooth goodness of fit test when the hypothesis is composite. *J. Amer. Statist. Ass.*, **74**, 441–445.

[216] Vaeth, M. (1985), On the use of Wald's test in exponential families. *Int. Statist. Rev.*, **53**, 199–214.

[217] Wald, A. (1943), Tests of statistical hypothesis concerning several parameters when the number of observation is large. *Trans. Amer. Math. Soc.*, **54**, 426–482.

[218] Waldman, D.M. (1982), A stationary point for the stochastic frontier likelihood. *J. Econometrics*, **18**, 275–279.

[219] Wallis, K.F. (1980), Econometric implications of the rational expectation hypothesis. *Econometrica*, **48**, 49–73.

[220] Watson, M.W. and Engle, R.F. (1985), Testing for regression coefficient stability with a stationary AR(1) alternative. *Rev. Econ. and Stat.*, **67**, 341–346.

[221] White, H. (1980), A heteroskedasticity-consistent covariance matrix estimator and a direct test for heteroskedasticity. *Econometrica*, **48**, 817–838.

[222] White, H. (1980), Using least squares to approximate unknown regression functions. *International Economic Review*, **21**, 149–170.

[223] White, H. (1982), Maximum likelihood estimation of misspecified models. *Econometrica*, **50**, 1–25.

[224] White, H. (1994), *Estimation, Inference and Specification Analysis*, Cambridge University Press.

[225] Wilks, S.S. (1938), The large-sample distribution of the likelihood ratio for testing composite hypotheses. *Ann. Math. Statist.*, **9**, 60–62.

[226] Wolak, E.A. (1987), An exact test for multiple inequality constraints in the linear regression model. *J. Amer. Statist. Ass.*, **82**, 782–793.

[227] Wooldridge, J.M (1990), A unified approach to robust, regression-based specification tests. *Econometric Theory*, **6**, 17–43.

[228] Wooldridge, J.M. (1991), On the application of robust, regression based diagnostics to models of conditional means and conditional variances. *J. Econometrics*, **47**, 5–46.

[229] Yule, G.U. (1922), On the application of the $\chi^2$ method to association and contingency tables, with experimental illustration. *J. Roy. Stat. Soc.*, **85**, 95–104.

# The Importance of Geometry in Multivariate Analysis and Some Applications

Carles M. Cuadras[1] and Josep Fortiana
Department of Statistics
University of Barcelona
Barcelona, Spain

**Abstract.** Geometrical concepts, including distance functions between observations, geometric variabilities and proximity functions, are used to develop some new aspects of multivariate analysis. These include the influence of principal components in comparing populations, the detection of atypical observations in discrimination with mixed variables, and the construction of orthogonal expansions for a continuous random variable. Some illustrations are given using two well-known data sets.

## 1. Introduction

Multivariate Analysis is mainly based on results proceeding from three mathematical areas: matrix calculus, distribution theory and metric geometry. This last subject is fundamental in methods such as multidimensional scaling and correspondence analysis, where the notion of distance function plays a basic role.

Following the utility of previous results by J.C. Gower, J.C. Lingoes, C.R. Rao and others, Cuadras (1989) proposed using distances in other areas of statistics, and continued this approach in several papers. A list of applications is given below. See Cuadras, Fortiana and Oliva (1995) for a general perspective.

---

[1] Work supported in part by CGYCIT PB-96-1004-C02-01 and 1997SGR-00183

93

*Regression and prediction*: Cuadras and Arenas (1990), Cuadras, Arenas and Fortiana (1996).

*Discriminant analysis*: Cuadras (1992), Cuadras, Fortiana and Oliva (1997).

*Constructing probability densities*: Cuadras (1992), Cuadras, Atkinson and Fortiana (1997), Cuadras and Fortiana (1997).

*Goodness-of-fit*: Cuadras and Fortiana (1994), Fortiana and Grane (1997).

*Orthogonal expansions*: Cuadras and Fortiana (1995, 1996).

*Multidimensional scaling*: Cuadras (1997), Cuadras and Fortiana (1997), Cuadras, Fortiana and Greenacre (1998).

There are some key geometrical concepts related to a distance, common for these topics, which are described in Section 2. The aim of this paper is to propose new applications of these concepts. Section 3 interprets an inequality concerning Mahalanobis distances, and shows that the means of several populations can follow a geometrical direction other than that of the observable variables. Section 4 presents an approach for detecting outlying observations in discrimination with mixed data. Section 5 gives a general procedure for obtaining orthogonal expansions for a continuous random variable from a geometrical point of view.

## 2. Main geometrical concepts

Let $\mathbf{X}$ be a random vector with a probability density $f(\mathbf{x})$, with respect to a suitable measure $\lambda$, and support $S$. Let $\delta(\mathbf{x}_1\mathbf{x}_2\mathbf{x})$ be a distance function between observations of $\mathbf{X}$. There are some useful concepts related to $\delta$.

The *geometric variability* of $\mathbf{X}$ with respect to $\delta$ is defined by

$$(1) \qquad V_\delta(\mathbf{X}) = \frac{1}{2}\int_{S\times S} \delta^2(\mathbf{x}_1,\mathbf{x}_2)f(\mathbf{x}_1)f(\mathbf{x}_2)\lambda(d\mathbf{x}_1)\lambda(d\mathbf{x}_2).$$

The *proximity function* of an observation $\mathbf{x}_0$ of $\mathbf{X}$ to the population $\Pi$ represented by $\mathbf{X}$ is defined by

$$(2) \qquad D^2(\mathbf{x}_0, \Pi) = \int_S \delta^2(\mathbf{x}_0,\mathbf{x})f(\mathbf{x})\lambda(d\mathbf{x}) - V_\delta(\mathbf{X}).$$

Related to distance $\delta$ is the *symmetric function* $G$ defined by

$$(3) \qquad G(\mathbf{x}_1, \mathbf{x}_2) = -\frac{1}{2}\left[\delta^2(\mathbf{x}_1,\mathbf{x}_2) - D^2(\mathbf{x}_1,\Pi) - D^2(\mathbf{x}_2,\Pi)\right],$$

which satisfies

(4) $\qquad \delta^2(\mathbf{x}_1, \mathbf{x}_2) = G(\mathbf{x}_1, \mathbf{x}_1) + G(\mathbf{x}_2, \mathbf{x}_2) - 2G(\mathbf{x}_1, \mathbf{x}_2).$

Suppose that we have two populations $\Pi_1, \Pi_2$, represented by two random vectors $\mathbf{X}, \mathbf{Y}$ with the same support $S$. The *distance* between $\Pi_1, \Pi_2$ is defined by

(5) $\quad \Delta^2(\Pi_1, \Pi_2) = \displaystyle\int_{S \times S} \delta^2(\mathbf{x}, \mathbf{y}) f(\mathbf{x}) g(\mathbf{y}) \lambda(d\mathbf{x}) \lambda(d\mathbf{y}) - V_\delta(\mathbf{X}) - V_\delta(\mathbf{Y}),$

where $g(\mathbf{y})$ is the density of $\mathbf{Y}$. See Rao (1982).

These functions have interesting properties, especially if the metric space $(S, \delta)$ can be represented in a Euclidean (or Hilbert) space $L$, i.e., there exists $\phi : S \to L$ such that

$$\delta^2(\mathbf{x}_1, \mathbf{x}_2) = \|\phi(\mathbf{x}_1) \text{-} \phi(\mathbf{x}_2)\|^2,$$

where $\|\cdot\|$ is the standard norm in $L$.

- If $\delta$ is the ordinary Euclidean distance and $\Sigma = \mathrm{cov}(\mathbf{X})$, then

$$V_\delta(\mathbf{X}) = \mathrm{tr}(\Sigma).$$

- If $E(\|\phi(\mathbf{X})\|^2)$ and $\|E(\phi(\mathbf{X}))\|$ exists then

$$V_\delta(\mathbf{X}) = E(\|\phi(\mathbf{X})\|^2) \text{-} \|E(\phi(\mathbf{X}))\|^2.$$

- The proximity function is the squared distance from $\phi(\mathbf{x}_0)$ to the $\delta$-mean $E(\phi(\mathbf{X}))$

(6) $\qquad D^2(\mathbf{x}_0, \Pi) = \|\phi(\mathbf{x}_0) \text{-} E(\phi(\mathbf{X}))\|^2.$

- Affine transformations of the squared distance $\delta^2 \to a\delta^2 + b$ give $V_\delta(\mathbf{X}) \to aV_\delta(\mathbf{X}) + b/2$ and $D^2(\mathbf{x}, \Pi) \to aD^2(\mathbf{x}, \Pi) + b/2.$
- By suitable choices of $a$ and $b$, we may transform $\delta$ and generate the probability density

$$f_\delta(\mathbf{x}) = \exp(-D^2(\mathbf{x}, \Pi)).$$

- A measure of the divergence between $f_\delta$ generated by $\delta$ and the true density $f$ is

$$I(f \parallel f_\delta) = V_\delta(\mathbf{X}) - H(f) \geq 0,$$

where $I(f \parallel f_\delta)$ is the Kullback-Leibler divergence and $H(f)$ is the Shannon entropy. A choice of a distance $\delta$ may be good if the geometric variability is close to $H(f)$.

- The distance between the two populations can be expressed as

(7) $$\Delta^2(\Pi_1, \Pi_2) = \|E(\phi(\mathbf{X})) - E(\phi(\mathbf{Y}))\|^2.$$

- The problem of allocating one observation $\mathbf{x}$ to $\Pi_1$ or $\Pi_2$ can be solved by using the rule

(8)     allocate $\mathbf{x}$ to $\Pi_i$ if $D^2(\mathbf{x}, \Pi_i) = \min\{D^2(\mathbf{x}, \Pi_1), D_2(\mathbf{x}, \Pi_2)\}$.

That is, from (6), $\mathbf{x}$ is allocated to the nearest population.

See Cuadras, Atkinson and Fortiana (1997) and Cuadras, Fortiana and Oliva (1997) for proofs, details and applications. It is worth noting that, if the chosen distance can be estimated from a sample, then it is easy to find estimators for the geometric variability and the proximity function, without knowing the probability density.

## 3. When is Mahalanobis greater than Pearson?

### 3.1 Two populations

Let $Y$ be a response variable and consider the multiple regression of $Y$ on $\mathbf{X} = (X_1, \ldots, X_p)$. If $R$ is the multiple correlation coefficient, one expects the inequality

(9) $$R^2 \leq r_1^2 + \cdots + r_p^2,$$

where $r_i$, $i = 1, \cdots, p$, are the simple correlations between $Y$ and $\mathbf{X}$. This should be so as the right-hand part of (9) expresses a lack of redundancy between the explanatory variables. However, the opposite inequality

(10) $$R^2 > r_1^2 + \cdots + r_p^2$$

can arise with real data. This inequality was studied and discussed by several authors (Bertrand, Cuadras, Hamilton, Holder, Freund, Routledge).

Cuadras (1993) showed that (10) holds when $Y$ is mainly correlated with the principal components (PCs) of $\mathbf{X}$ obtained from the correlation matrix $\mathbf{R}$, with the smallest variances. This quirk in regression can occur and these PCs can be important in predicting $Y$. Jolliffe (1982) gives some real data examples. See also Cuadras (1995) and Rao (1994).

The above inequality admits a distance version. Suppose that $\Pi_i = (\mu_i, \Sigma), i = 1, 2$, are two populations and consider the inequality

$$(11) \qquad (\mu_1 - \mu_2)'\Sigma^{-1}(\mu_1 - \mu_2) > (\mu_1 - \mu_2)'\Sigma_0^{-1}(\mu_1 - \mu_2),$$

where $\Sigma_0 = \text{diag}\Sigma$. That is, Mahalanobis distance between $\Pi_1$ and $\Pi_2$ is greater than K. Pearson distance. Redundancy between variables, contained in $\Sigma$ but absent in $\Sigma_0$, leads us to expect the opposite inequality.

The above inequality can also be justified using principal components. Let $\mathbf{u}_i, \lambda_i$ be an eigenvector, eigenvalue of $\Sigma$. Then

$$(12) \qquad \Sigma u_i = \lambda_i \mathbf{u}_i \quad \text{and} \quad \Sigma^{-1} = \mathbf{u}_i/\lambda_i.$$

The corresponding PC is $Y_i = \mathbf{u}_i'\mathbf{X}$. We can suppose that the ordered eigenvalues $\lambda_1 \geq \cdots \geq \lambda_p$ satisfy

$$\lambda_1 > \sigma_{ii} > \lambda_p > 0, \quad i = 1, \ldots, p.$$

Then for any vector $\gamma$

$$(13) \qquad \frac{1}{\lambda_1}\gamma'\gamma < \gamma'\Sigma_0^{-1}\gamma \quad \text{and} \quad \frac{1}{\lambda_p}\gamma'\gamma > \gamma'\Sigma_0^{-1}\gamma.$$

Now suppose that $\mathbf{u}_i$ follows the same direction as the line joining $\mu_1, \mu_2 \in R^p$,

$$\mathbf{u}_i = \mu_1 - \mu_2.$$

Then from (12)

$$(14) \qquad (\mu_1 - \mu_2)'\Sigma^{-1}(\mu_1 - \mu_2) = (\mu_1 - \mu_2)'(\mu_1 - \mu_2)/\lambda_i,$$

and from (13)

$$(\mu_1 - \mu_2)'\Sigma^{-1}(\mu_1 - \mu_2) < (\mu_1 - \mu_2)'\Sigma_0^{-1}(\mu_1 - \mu_2) \text{if} \quad \lambda_i = \lambda_1,$$
$$(\mu_1 - \mu_2)'\Sigma^{-1}(\mu_1 - \mu_2) > (\mu_1 - \mu_2)'\Sigma_0^{-1}(\mu_1 - \mu_2) \text{if} \quad \lambda_i = \lambda_p.$$

Thus, the inequality (11) holds if $\mu_1, \mu_2$ follows the direction of the last PC. In general, we can find this inequality when $\overline{\mu_1\mu_2}$ follows the direction of a PC with small variance. This is a multivariate version of Simpson's paradox.

## 3.2 More than two populations

When we have $g$ populations $(\mu_i, \Sigma), i = 1, \ldots, g$, a generalization of (11) is

$$V_M = \frac{1}{2g^2} \sum_{i,j=1}^{g} \mu'_{ij} \Sigma^{-1} \mu_{ij} > \overline{V_K} = \frac{1}{2g^2} \sum_{i,j=1}^{g} \mu'_{ij} \Sigma_0^{-1} \mu_{ij},$$

where $\mu_{ij} = \mu_i - \mu_j$ and $V_M, V_K$ are the geometric variabilities considering the $g$ populations. This inequality indicates that the vector means tend to follow the direction of the PCs with smallest variance obtained from $\Sigma$.

**Example 1.** Consider the Iris setosa, I.versicolor and I. virginica data used by R.A. Fisher to illustrate discriminant analysis. The variables are: $X_1, X_2$=sepal length, sepal width,   $X_3, X_4$=petal length, petal width. For the sepal variables we find

$$V_M = 2.16 > V_K = 1.14,$$

and for the petal variables:

$$V_M = 9.89 < V_K = 14.55 .$$

As it is illustrated in Figures 1 and 2, the bivariate data $X_1, X_2$ follow a direction orthogonal to the means, while for $X_3, X_4$, data and means follow the same direction.

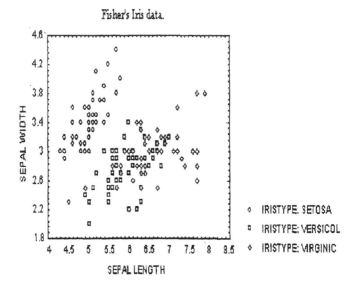

Figure 1. The main direction of the data in each population is orthogonal to the direction of the means ($V_M > V_K$).

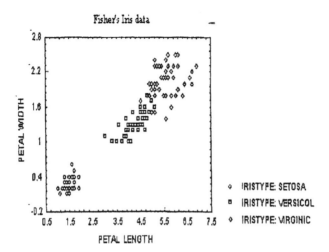

Figure 2. The first principal component and the means follow the same direction $(V_M < V_K)$.

### 3.3 A generalization

Let us suppose that **X** has density $f(x; \theta)$, where $\theta \in \Theta$ is a q-dimensional vector. If $\mathbf{x}_1, \ldots, \mathbf{x}_n$ are iid as **X**, the *efficient scores* are the column vectors

$$\mathbf{Z}_i(\theta) = \frac{\partial}{\partial \theta} \log f(\mathbf{x}_i; \theta).$$

Let

$$\mathbf{V}_\theta = \frac{1}{\sqrt{n}} \sum_{i=1}^{n} \mathbf{Z}_i(\theta).$$

Then $E(\mathbf{V}_\theta) = 0$ and $\mathbf{V}_{\widehat{\theta}} = 0$, where $\widehat{\theta}$ is the ML estimator of $\theta$. The asymptotic distribution of $\mathbf{V}_\theta$ is $N(\mathbf{0}, \mathbf{F}_\theta)$, where $\mathbf{F}_\theta = E(\mathbf{Z}_i(\theta)\mathbf{Z}_i'(\theta))$ is the information matrix and, asymptotically,

$$\mathbf{V}_\theta' \mathbf{F}_\theta^{-1} \mathbf{V}_\theta \sim \chi_q^2.$$

Consider the composite null hypothesis

$$H_0 : \theta \in \Theta_o \subset \Theta.$$

Let $\theta^*$ the restricted ML estimator. Under $H_0$ the proximity of $\theta^*$ to $\widehat{\theta}$ implies the proximity of $\mathbf{V}_{\theta^*}$ to $\mathbf{V}_{\widehat{\theta}} = 0$. The Mahalanobis distance between $\mathbf{V}_{\theta^*}$ and $\mathbf{0}$ is

$$S_c = \mathbf{V}_{\theta^*}' \mathbf{F}_{\theta^*}^{-1} \mathbf{V}_{\theta^*} .$$

This is Rao's efficient score test. Its asymptotic distribution is $\chi^2_{q-s}$, where $s = \dim(\Theta_o)$. It can be proved that

$$-2 \log \Lambda a = S_c,$$

where $\Lambda = L(\theta^*)/L(\widehat{\theta})$ is the likelihood ratio. See Rao (1973).

Suppose that $\mathbf{X}=(X_1, \ldots, X_p)$, the composite hypothesis has a univariate version (e.g., linear hypothesis in MANOVA), $\Lambda$ is the likelihood ratio for $\mathbf{X}$ and $\Lambda_i, i = 1, \ldots, p$ are the likelihood ratios for $X_i, i = 1, \ldots, p$, respectively. Then the above distance interpretation provides a generalization of (11):

$$-2 \log \Lambda > -2 \log \Lambda_1 - \cdots - 2 \log \Lambda_p.$$

This inequality was considered by Cuadras (1996).

## 4. Typicality in Discrimination using distances

Suppose that $\mathbf{X}$ is a $p$-dimensional random vector, $\Pi_i$, $i = 1, 2$ , are two known populations and $\Pi_3$ is a third population, possibly unknown. Typicality can be described as a test to determine whether an observation $\mathbf{x}$ of $\mathbf{X}$ is typical of a mixture of $\Pi_1$ and $\Pi_2$ or belongs to $\Pi_3$. When $\Pi_i \sim N_p(\mu_i, \Sigma)$, $i = 1, 2, 3$, the appropriate hypotheses to be tested are

$$H_0 : \mathbf{x} \in N_p(\alpha\mu_1 + (1 - \alpha)\mu_2, \Sigma), 0 \le \alpha \le 1,$$
$$H_1 : \mathbf{x} \in N_p(\mu_3, \Sigma).$$

For testing the hypotheses $H_0'(\alpha = 1)$, $H_0''(\alpha = 0)$, $H_0(0 \le \alpha \le 1)$, Rao (1973) proposes the statistics

(15)
$$
\begin{aligned}
U_1(\mathbf{x}) &= ((\mathbf{x}\text{-}\mu_1)'\Sigma^{-1}(\mu_2 - \mu_1))^2/\Delta^2(\Pi_1, \Pi_2) \sim \chi_1^2 \\
U_2(\mathbf{x}) &= ((\mathbf{x}\text{-}\mu_2)'\Sigma^{-1}(\mu_2 - \mu_1))^2/\Delta^2(\Pi_1, \Pi_2) \sim \chi_1^2 \\
W(\mathbf{x}) &= (\mathbf{x}\text{-}\mu_1)'\Sigma^{-1}(\mathbf{x} - \mu_1) - U_1(\mathbf{x}) \qquad \sim \chi_{p-1}^2.
\end{aligned}
$$

where

(16)
$$\Delta^2(\Pi_1, \Pi_2) = (\mu_2\text{-}\mu_1)'\Sigma^{-1}(\mu_2 - \mu_1)$$

is the Mahalanobis distance. These tests work with $p$-variate normal data when $\Sigma$ is known or estimated from large samples. If $W(\mathbf{x})$ is significant, then $\mathbf{x}$ is atypical and could not belong to the predefined populations, otherwise $U_1(\mathbf{x})$ and $U_2(\mathbf{x})$ can be used to decide whether $\mathbf{x}$ comes from $\Pi_1$ or $\Pi_2$. See also Bar-Hen and Daudin (1997). Let us present a typicality test for general data using distances and proximity functions.

The squared distance between two populations $\Pi_1, \Pi_2$, obtained from distances between observations, is $\Delta^2(\Pi_1, \Pi_2)$ defined in (5). This distance reduces to (16) when $\delta$ is the Mahalanobis distance, while the proximity functions $D_i^2(\mathbf{x}) = D^2(\mathbf{x}, \Pi_i)$ are

$$D_i^2(\mathbf{x}) = (\mathbf{x}\text{-}\mu_i)'\Sigma^{-1}(\mathbf{x} - \mu_i), \quad i = 1, 2.$$

Suppose now that we have multivariate mixed data and a suitable distance $\delta$ between observations giving general proximity functions $D_1^2(\mathbf{x})$, $D_2^2(\mathbf{x})$. The allocation using proximity functions of one observation $\mathbf{x}$ to $\Pi_1$ or $\Pi_2$, is solved by means of rule (8). This distance-based rule reduces to the linear discriminant rule when $\delta$ is the Mahalanobis distance.

We next follow a similar geometric approach in typicality. The distance-based versions for $P_1(\mathbf{x}) = (\mu_2\text{-}\mu_1)' \Sigma^{-1}(\mathbf{x} - \mu_1)$ and $P_2(\mathbf{x}) = (\mu_2\text{-}\mu_1)'\Sigma^{-1}(\mathbf{x}\text{-}\mu_2)$, see (15), when $\delta$ is a general distance, are

$$P_1(\mathbf{x}) = [\Delta^2(\Pi_1, \Pi_2) + D_1^2(\mathbf{x}) - D_2^2(\mathbf{x})]/2,$$
$$P_2(\mathbf{x}) = [\Delta^2(\Pi_1, \Pi_2) + D_2^2(\mathbf{x}) - D_1^2(\mathbf{x})]/2.$$

The test whether $\mathbf{x}$ comes from the convex linear combination $\alpha\Pi_1 + (1-\alpha)\Pi_2$, i.e., from a population with $\delta$-mean $\alpha E(\phi(\mathbf{X}))+(1-\alpha)E(\phi(\mathbf{Y}))$, $0 \leq \alpha \leq 1$, and the other two related tests, may be performed by using the statistics

$$H_0'(\alpha = 1) : U_1(\mathbf{x}) = (P_1(\mathbf{x}))^2/\Delta^2(\Pi_1, \Pi_2),$$
$$H_0''(\alpha = 0) : U_2(\mathbf{x}) = (P_2(\mathbf{x}))^2/\Delta^2(\Pi_1, \Pi_2),$$
$$H_0(0 \leq \alpha \leq 1) : W(\mathbf{x}) = D_1^2(\mathbf{x}) - U_1(\mathbf{x}) = D_2^2(\mathbf{x}) - U_2(\mathbf{x}).$$

$W(\mathbf{x})$ significant means that $\mathbf{x}$ comes from a different population $\Pi_3$, that is, $\mathbf{x}$ is one outlier for the populations $\Pi_1, \Pi_2$.

These statistics have a geometrical interpretation. Consider the triangle $\bar{\mathbf{x}}, \mathbf{a}, \mathbf{b}$, where $\bar{\mathbf{x}} = \phi(\mathbf{x})$, $\mathbf{a}=E(\phi(\mathbf{X}))$, $\mathbf{b}=E(\phi(\mathbf{Y}))$. Then $U_1(\mathbf{x})$, $U_2(\mathbf{x})$ are the (squared) projections of $\bar{\mathbf{x}}$ on the sides $\{\bar{\mathbf{x}},\mathbf{a}\}$ and $\{\bar{\mathbf{x}},\mathbf{b}\}$ while $W(\mathbf{x})$ is the height or orthogonal distance from $\bar{\mathbf{x}}$ to $\{\mathbf{a},\mathbf{b}\}$.

The sampling distributions of $U_1, U_2$ and $W$ can be difficult to find for mixed data and may be obtained by resampling methods. Nevertheless, we follow a procedure proposed by Cuadras and Fortiana (1994), consisting in correlating the sample with the underlying distribution and its orthogonal expansion. This is explained and illustrated in the next sections.

## 5. Orthogonal expansions

### 5.1 General expansions from distances

Orthogonal expansions appear in many fields of statistics: analysis of variance, factorial analysis, time series, goodness of fit, etc. In fact, any random variable can be expanded as the sum of a countable sequence of uncorrelated random variables. One general expansion can be obtained using distances.

Let $X$ be a continuous random variable with support $I = [a, b]$ and probability density $f(x)$ with respect to Lebesgue measure. Let $\delta(x, x')$ be a distance function between the observations $x, x'$ of $X$ and consider the eigendecomposition of $G(x, x')$, see (3),

$$(17) \qquad f(x)^{1/2} G(x, x') f(x')^{1/2} = \sum_{n \geq 1} \lambda_n u_n(x) u_n(x'),$$

where $\{u_n, n \geq 1\}$ is a countable complete orthonormal set. That is

$$G(x, x') = \sum_{n \geq 1} c_n(x) c_n(x'),$$

where

$$(18) \qquad c_n(x) = \sqrt{\lambda_n} f(x)^{-1/2} u_n(x).$$

Then, if $X'$ is iid as $X$, we obtain the following expansions

$$G(X, X') = \sum_{n \geq 1} X_n X_n',$$

$$G(X, \alpha) = \sum_{n \geq 1} c_n(\alpha) X_n, \quad \alpha \in I,$$

$$G(X, X) = \sum_{n \geq 1} X_n^2,$$

where $\{X_n = c_n(X), n \geq 1\}$ is a sequence of centered and uncorrelated random variables. That is, this countable set of variables satisfy:

- $E(X_n) = 0$.
- $\text{Var}(X_n) = \lambda_n$.
- $\text{cov}(X_m, X_n) = 0$, if $m \neq n$.
- The sequence of variances $\{\lambda_n, n \geq 1\}$ give a decomposition of the geometric variability

$$V_\delta(X) = \sum_{n \geq 1} \lambda_n.$$

The proof is as follows.

$E(X_n) = 0$ is a consequence of $E_X G(X, x) = 0, x \in I$.

The covariances are

$$\text{cov}(X_m, X_n) = \int_I c_m(x) c_n(x) f(x) dx$$

$$= (\lambda_n \lambda_m)^{1/2} \int_I u_m(x) u_n(x) dx,$$

$$= \begin{cases} 0, & m \neq n, \\ \lambda_n, & m = n. \end{cases}$$

The geometric variability can be expressed as

$$V_\delta(X) = \frac{1}{2} \left( EG(X, X) + EG(X', X') - 2EG(X, X') \right)$$

$$= EG(X, X)$$

$$= \sum_{n \geq 1} E(X_n^2).$$

Finally, it can be proved that $\{X_n, n \geq 1\}$ can be interpreted as principal components.

## 5.2 A particular expansion

A useful expansion can be obtained by using the distance

$$\delta(x, x') = \sqrt{|x - x'|}.$$

From $|x - x'| = x + x' - 2 \min\{x, x'\}$, we can work with the symmetric function $K(x, x') = \min\{x, x'\}$ instead of $G(x, x')$. Note that $K(x, x')$ also satisfies (4) and the relation between $K$ and $G$ is

$$G(x, x') = K(x, x') - E_X K(X, x') - E_{X'} K(x, X') + E_{XX'} K(X, X'),$$

$$= K(x, x') - \varphi(x) - \varphi(x') + \mu,$$

where $X, X'$ are iid. We have $\varphi(a) = a, \varphi(x) = G(x, a) - \mu$ and $\mu = G(a, a) - a$, assuming $a < -\infty$. Hence

$$\min\{x, x'\} = a + G(x, x') - G(x, a) - G(a, x') + G(a, a)$$

$$= a + \sum_{n \geq 1} (c_n(x) - c_n(a))(c_n(x') - c_n(a)).$$

Writing $h_n(X) = c_n(X) - c_n(a) = c_n(X) - \mu_n$, where $\mu_n = E(h_n(X))$, we obtain the expansion

$$\min\{X, X'\} = a + \sum_{n \geq 1} h_n(X) h_n(X'),$$

and consequently

(19)
$$X = a + \sum_{n \geq 1} h_n(b) h_n(X),$$

(20)
$$X = a + \sum_{n \geq 1} h_n^2(X),$$

where $\{h_n(X), n \geq 1\}$ is a countable set of uncorrelated random variables.

In general, including the case $a = \infty$, expanding $\min\{x, x'\} - \min\{y, y'\}$ and setting $y = x_0, y' = b$, we find

$$\min\{X, X'\} = x_0 + \sum_{n \geq 1} (h_n(X) h_n(X') - h_n(x_0) h_n(b)).$$

**Example 2.** If $U$ follows the uniform $(0, 1)$ distribution the sequence of uncorrelated variables is

$$U_n = \frac{\sqrt{2}}{n\pi}(1 - \cos n\pi U), \quad n \geq 1.$$

Expansion (19) is

$$U = \frac{4}{\pi^2} \sum_{n \geq 1} (1 - \cos(2n - 1)\pi U)/(2n - 1)^2.$$

Expansion (20) can be written as

$$U = \sum_{n \geq 1} \frac{U_n^2}{n^2 \pi^2},$$

which is formally analogous to the expansion of the limit distribution of the Cramér-von Mises statistic used in goodness-of-fit. See Cuadras and Fortiana (1993, 1994) and Fortiana and Cuadras (1997).

If $X$ follows the exponential distribution with mean 1 and $J_0, J_1$ are the Bessel function of the first kind, then the orthogonal sequence is

$$X_n = 2(J_0(\xi_n \exp(-X/2)) - J_0(\xi_n))/\xi_n J_0(\xi_n),$$

where $\xi_n$ is the n-th positive root of $J_1$. Expansion (19) is then

$$(21) \quad X = \sum_{n \geq 1} 4(1 - J_0(\xi_n))(J_0(\xi_n \exp(-X/2) - J_0(\xi_n))/\xi_n^2 J_0(\xi_n)^2.$$

See Cuadras and Lahlou (1998) for other expansions.

## 6. One statistical illustration

To illustrate the distance-based discriminant rule (8), Cuadras, Fortiana and Oliva (1997) used a cancer data set, taken from Krzanowski (1980). and consisting of 11 mixed measurements (7 continuous, 2 binary and 2 categorical). This rule uses Gower (1971) similarity coefficient for mixed variables and is applied to allocate individuals in two groups of tumours: benign ($N_1 = 78$) and malignant ($N_2 = 59$).

To ascertain whether or not $\mathbf{x}$ is atypical, we consider the statistic introduced in Section 4

$$W(\mathbf{x}) = D_1^2(\mathbf{x}) - \frac{1}{4}(\Delta^2 + D_1^2(\mathbf{x}) - D_2^2(\mathbf{x}))^2/\Delta^2$$

$$= D_2^2(\mathbf{x}) - \frac{1}{4}(\Delta^2 + D_2^2(\mathbf{x}) - D_1^2(\mathbf{x}))^2/\Delta^2,$$

where $\Delta^2 = \Delta^2(\Pi_1, \Pi_2)$. $W(\mathbf{x})$ significant may be interpreted as $\mathbf{x}$ comes from another different population (e.g., the individual may not have this tumour).

The distribution of $W(\mathbf{x})$ for this cancer data seems to be exponential and taking the sample of $N = N_1 + N_2 = 137$ individuals, we compute the maximal Hoeffding correlation $r_N^+$ between the sample and the distribution. This correlation is the maximum correlation that one can reach considering all the bivariate distributions with two univariate marginals: the theoretical distribution (exponential) and the empirical distribution. We obtain $r_N^+ = 0.9285$. Similar maximal correlations can be found by relating the sample with the components of the orthogonal expansion (21).

Individual 11 of the malignant group gives one extreme $W$ value and may be atypical, so it is removed. The computations for the $N - 1 = 136$ remaining individuals give:

| Correlation | Theoretical | Sample |
|---|---|---|
| $\rho^+$ | 1 | 0.9732 |
| $\rho_1$ | −0.9488 | −0.9038 |
| $\rho_2$ | 0.1895 | 0.1188 |
| $\rho_3$ | −0.1934 | −0.1915 |
| $\rho_4$ | 0.0806 | 0.0909 |

The fit to the exponential distribution is quite good. Assuming this distribution, individual 11 is clearly atypical. See Cuadras and Fortiana (1998).

References

[1] Bar-Hen, A. and Daudin, J.-J. (1997), A test of a special case of typicality in linear discriminant analysis. *Biometrics*, **53**, 39–48.

[2] Cuadras, C.M. (1989) Distance analysis in discrimination and classification using both continuous and categorical variables. In: Y. Dodge (Ed.), *Statistical Data Analysis and Inference*, 459–473. Elsevier Science Publishers B.V. (North–Holland), Amsterdam.

[3] Cuadras, C.M. and Arenas, C. (1990), A distance based regression model for prediction with mixed data. *Communications in Statistics A. Theory and Methods*, **19**, 2261–2279.

[4] Cuadras, C.M. (1992a), Probability distributions with given multivariate marginals and given dependence structure. *J. of Multivariate Analysis*, **42**, 51–66.

[5] Cuadras, C.M. (1992b), Some examples of distance based discrimination. *Biometrical Letters*, **29**, 3–20.

[6] Cuadras, C.M. (1993), Interpreting an inequality in multiple regression. *The American Statistician*, **47**, 256–258.

[7] Cuadras, C.M. (1995) Increasing the correlations with the response variable may not increase the coefficient of determination: a PCA interpretation. In: E. Tiit, T. Kollo and H. Niemi, (Eds.), *New Trends in Probability and Statistics. Vol 3. Multivariate Statistics and Matrices in Statistics*, 75–83, VSP/TEV, The Netherlands.

[8] Cuadras, C.M. (1996), Discussion of "The relation between theory and applications in statistics" by D.R. Cox.. *TEST*, 4, 253–261.

[9] Cuadras, C.M. (1998a) Multidimensional Dependencies in Ordination and Classification. In: K. Fernandez-Aguirre and A. Morineau, (Eds.), *Analyses Multidimensionnelles des Donnees*, 15–25, CISIA, Saint Mande (France).

[10] Cuadras, C.M. (1998b), Comments on Some Cautionary Notes on the Use of Principal Components Regression. *The American Statistician*, **52**, 371.

[11] Cuadras, C.M., Arenas, C. and Fortiana, J. (1996), Some computational aspects of a distance–based model for prediction. *Communications in Statistics. Simulation and Computation*, **25(3)**, 593–609.

[12] Cuadras, C.M., Atkinson, R.A. and Fortiana, J. (1997), Probability densities from distances and discriminant analysis. *Statistics and Probability Letters*, **33**, 405–411.

[13] Cuadras, C.M. and Fortiana, J. (1993) Continuous metric scaling and prediction. In: C.M. Cuadras and C.R. Rao (Eds.), *Multivariate*

*Analysis, Future Directions 2*, 47–66. Elsevier Science Publishers B.V. (North–Holland), Amsterdam.

[14] Cuadras, C.M. and Fortiana, J. (1994) Ascertaining the underlying distribution of a data set. In: R. Gutierrez and M.J. Valderrama (Eds.), *Selected Topics on Stochastic Modelling*, 223–230. World-Scientific, Singapore.

[15] Cuadras, C.M. and Fortiana, J. (1995), A continuous metric scaling solution for a random variable. *J. of Multivariate Analysis*, **52**, 1–14.

[16] Cuadras, C.M. and Fortiana, J. (1996) Weighted continuous metric scaling. In: Gupta, A.K. and V.L. Girko (Eds.), *Multidimensional Statistical Analysis and Theory of Random Matrices*, 27–40. VSP, Zeist, The Netherlands.

[17] Cuadras, C.M., Fortiana, J. and Oliva, F. (1996) Representation of statistical structures, classification and prediction using multidimensional scaling. In: W. Gaul and D. Pfeifer, (Eds.), *From Data to Knowledge*, 20–31. Springer-Verlag, Berlin.

[18] Cuadras, C.M., Fortiana, J. and Oliva, F. (1997), The proximity of an individual to a population with applications in discriminant analysis. *Journal of Classification*, **14**, 117–136.

[19] Cuadras, C.M. and Fortiana, J. (1997) Continuous scaling on a bivariate copula. In: Viktor Benes and Josef Stepan, (Eds), *Distributions with given marginals and moment problems*, 137–142. Kluwer Academic Pub., Dordrecht.

[20] Cuadras, C.M. and Fortiana, J. (1997) Visualizing categorical data with related metric scaling. In: J. Blasius and M. Greenacre, (Eds.) *Visualization of Categorical Data*, 365–376, Academic Press, N. York.

[21] Cuadras, C.M. and Fortiana, J. (1998) Typicality in discriminant analysis with mixed variables. *Data Science, Classification and Related Methods, VI Conference IFCS-98*, 82–85, Rome.

[22] Cuadras, C.M., Fortiana, J. and Greenacre, M.J. (1999) Continuous extensions of matrix formulations in Correspondence Analysis, with applications to the FGM family of distributions. In: D.S.G. Pollock, R.D.H. Heijmans and A. Satorra, (Eds.), *Innovations in Multivariate Statistical Analysis*, Kluwer Academic Pub., Dordrecht, in press.

[23] Cuadras, C.M. and Lahlou, Y. (1998) An orthogonal expansion for the logistic distribution. *Mathematics Preprint Series, No. 260*, University of Barcelona.

[24] Fortiana, J. and Cuadras, C.M. (1997), A family of matrices, the discretized Brownian Bridge and distance-based regression. *Linear Algebra and its Applications*, **264**, 173–188.

[25] Fortiana, J. and Grane, A. (1998) Goodness-of-fit tests based on the maximum correlation and its orthogonal expansion. *Working paper*.

[26] Gower, J. C. (1971), A general coefficient of similarity and some of its properties. *Biometrics*, **27**, 857–874.

[27] Jolliffe, I.T. (1982), A note on the use of principal components in regression. *Applied Statistics*, **31**, 300–303.

[28] Krzanowski, W.J. (1980), Mixtures of continuous and categorical variables in discriminant analysis. *Biometrics*, **36**, 493–499.

[29] Rao, C.R. (1973) *Linear Statistical Inference and Its Applications*, New York, Wiley.

[30] Rao, C.R. (1982), Diversity and dissimilarity coefficients: a unified approach. *Theoretical Population Biology*, **21**, 24–43.

[31] Rao, C.R. (1994), Statistics: an essential technology in environmental research and management. *Envir. and Ecol. Statist.*, **1**, 7–20.

# R.A. Fisher in the 21st Century*

Bradley Efron
Department of Statistics
Stanford University
Stanford, California

**Abstract.** Fisher is the single most important figure in 20th Century statistics. This talk examines his influence on modern statistical thinking, trying to predict how Fisherian we can expect the 21st Century to be. Fisher's philosophy is characterized as a series of shrewd compromises between the Bayesian and Frequentist viewpoints, augmented by some unique characteristics that are particularly useful in applied problems. Several current research topics are examined with an eye toward Fisherian influence, or the lack of it, and what this portends for future statistical developments. Based on the 1996 Fisher lecture, the article closely follows the text of that talk.

## 1. Introduction

Even scientists need their heroes, and R.A. Fisher was certainly the hero of 20th Century statistics. His ideas dominated and transformed our field to an extent a Caesar or an Alexander might have envied. Most of this happened in the second quarter of the century, but by the time of my own education Fisher had been reduced to a somewhat minor figure in American academic statistics, with the influence of Neyman and Wald rising to their high water mark.

---

* Invited Paper Presented at the 1996 R. A. Fisher Lecture. This paper was previously published in: *Statistical Science* **13**(2), May 1998; pp. 95–114. Reprinted with the permission of the Institute of Mathematical Statistics.

There has been a late 20th Century resurgence of interest in Fisherian statistics, in England where his influence never much waned, but also in America and the rest of the statistical world. Much of this revival has gone unnoticed because it is hidden behind the dazzle of modern computational methods. One of my main goals here will be to clarify Fisher's influence on modern statistics. Both the strengths and limitations of Fisherian thinking will be described, mainly by example, finally leading up to some speculations on Fisher's role in the statistical world of the 21st Century.

What follows is basically the text of the Fisher lecture presented to the August 1966 Joint Statistical meetings in Chicago. The talk format has certain advantages over a standard journal article. First and foremost, it is meant to be absorbed quickly, in an hour, forcing the presentation to concentrate on main points rather than technical details. Spoken language tends to be livelier than the gray prose of a journal paper. A talk encourages bolder distinctions and personal opinions, which are dangerously vulnerable in a written article but appropriate I believe for speculations about the future. In other words, this will be a broad-brush painting, long on color but short on detail.

These advantages may be viewed in a less favorable light by the careful reader. Fisher's mathematical arguments are beautiful in their power and economy, and most of that is missing here. The broad brush-strokes sometimes conceal important areas of controversy. Most of the argumentation is by example rather than theory, with examples from my own work playing an exaggerated role. References are minimal, and not indicated in the usual author-year format but rather collected in annotated form at the end of the text. Most seriously, the one-hour limit required a somewhat arbitrary selection of topics, and in doing so I concentrated on those parts of Fisher's work that have been most important to me, omitting whole areas of Fisherian influence such as randomization and experimental design. The result is more a personal essay than a systematic survey.

This is a talk (as I will now refer to it) on Fisher's influence, not mainly on Fisher himself or even his intellectual history. A much more thorough study of the work itself appears in L.J. Savage's famous talk and essay, "On Rereading R.A. Fisher," the 1971 Fisher lecture, a brilliant account of Fisher's statistical ideas as sympathetically viewed by a leading Bayesian (Savage, 1976). Thanks to John Pratt's editorial efforts, Savage's talk appeared, posthumously, in the 1976 *Annals of Statistics*. In the article's discussion, Oscar Kempthorne called it the best statistics talk he had ever heard, and Churchill Eisenhart said the same. Another fine reference is Yates and Mather's introduction to the 1971 five-volume set of Fisher's collected works. The definitive Fisher reference in Joan Fisher Box's 1978 biography, 'The Life of a Scientist.'

It is a good rule never to meet your heroes. I inadvertently followed this rule when Fisher spoke at the Stanford Medical School in 1961, without notice to the Statistics Department. The strength of Fisher's powerful personality is missing from this talk, but not I hope the strength of his ideas. Heroic is a good word for Fisher's attempts to change statistical thinking, attempts that had a profound influence on this century's development of statistics into a major force on the scientific landscape. "What about the next century?" is the implicit question asked in the title, but I won't try to address that question until later.

## 2. The Statistical Century

Despite its title, the greater portion of the talk concerns the past and the present. I am going to begin by looking back on statistics in the 20th Century, which has been a time of great advancement for our profession. During the 20th Century statistical thinking and methodology have become the scientific framework for literally dozens of fields including education, agriculture, economics, biology, and medicine, and with increasing influence recently on the hard sciences such as astronomy, geology, and physics.

In other words, we have grown from a small obscure field into a big obscure field. Most people and even most scientists still don't know much about statistics except that there is something good about the number .05 and perhaps something bad about the bell curve. But I believe that this will change in the 21st Century and that statistical methods will be widely recognized as a central element of scientific thinking.

The 20th Century began on an auspicious statistical note with the appearance of Karl Pearson's famous $\chi^2$ paper in the spring of 1900. The groundwork for statistics' growth was laid by a pre-World War II collection of intellectual giants: Neyman, the Pearsons, Student, Kolmogorov, Hotelling, and Wald, with Neyman's work being especially influential. But from our viewpoint at the century's end, or at least from my viewpoint, the dominant figure has been R.A. Fisher. Fisher's influence is especially pervasive in statistical applications, but it also runs through the pages of our theoretical journals. With the end of the century in view this seemed like a good occasion for taking stock of the vitality of Fisher's legacy and its potential for future development.

A more accurate but less provocative title for this talk would have been 'Fisher's Influence on Modern Statistics'. What I will mostly do is examine some topics of current interest and assess how much Fisher's ideas have or have not influenced them. The central part of the talk concerns six research areas of current interest that I think will be important during the next couple of decades. This will also give me a chance to say something about the kinds of applied problems we might be dealing with soon, and whether or not Fisherian statistics is going to be of much help with them.

First though I want to give a brief review of Fisher's ideas and the ideas he was reacting to. One difficulty in assessing the importance of Fisherian statistics is that it's hard to say just what it is. Fisher had an amazing number of important ideas and some of them, like randomization inference and conditionality, are contradictory. It's a little as if in economics Marx, Adam Smith, and Keynes turned out to be the same person. So I am just going to outline some of the main Fisherian themes, with no attempt at completeness or philosophical reconciliation. This and the rest of the talk will be very short on references and details, especially technical details, which I will try to avoid entirely.

In 1910, two years before the 20 year old Fisher published his first paper, an inventory of the statistics world's great ideas would have included the following impressive list: Bayes theorem, least squares, the normal distribution and the central limit theorem, binomial and Poisson methods for count data, Galton's correlation and regression, multivariate distributions, Pearson's $\chi^2$, and Student's $t$. What was missing was a core for these ideas. The list existed as an ingenious collection of ad hoc devices. The situation for statistics was similar to the one now faced by computer science. In Joan Fisher Box's words, "The whole field was like an unexplored archaeological site, its structure hardly perceptible above the accretions of rubble, its treasures scattered throughout the literature."

There were two obvious candidates to provide a statistical core: 'objective' Bayesian statistics in the Laplace tradition of using uniform priors for unknown parameters, and a rough frequentism exemplified by Pearson's $\chi^2$ test. In fact, Pearson was working on a core program of his own through his system of Pearson distributions and the method of moments.

By 1925, Fisher had provided a central core for statistics — one that was quite different and more compelling than either the Laplacian or Pearsonian schemes. The great 1925 paper already contains most of the main elements of Fisherian estimation theory: consistency, sufficiency, likelihood, Fisher information, efficiency, and the asymptotic optimality of the maximum likelihood estimator. Partly missing is ancillarity, which is mentioned but not fully developed until the 1934 paper.

The 1925 paper even contains a fascinating and still controversial section on what Rao has called the second order efficiency of the maximum likelihood estimate (MLE). Fisher, never really satisfied with asymptotic results, says that in small samples the MLE loses less information than competing asymptotically efficient estimators, and implies that this helps solve the problem of small-sample inference. (At which point Savage wonders why one should care about the amount of information in a point estimator.)

Fisher's great accomplishment was to provide an optimality standard for statistical estimation — a yardstick of the best it's possible to do in any given estimation problem. Moreover, he provided a practical method, maximum likelihood, that quite reliably produces estimators coming close to the ideal optimum even in small samples.

Optimality results are a mark of scientific maturity. I mark 1925 as the year statistical theory came of age, the year statistics went from an ad hoc collection of ingenious techniques to a coherent discipline. Statistics was lucky to get a Fisher at the beginning of the 20th Century. We badly need another one to begin the 21st, as will be discussed near the end of the talk.

## 3. The Logic of Statistical Inference

Fisher believed that there must exist a logic of inductive inference that would yield a correct answer to any statistical problem, in the same way that ordinary logic solves deductive problems. By using such an inductive logic the statistician would be freed from the a priori assumptions of the Bayesian school.

Fisher's main tactic was to logically reduce a given inference problem, sometimes a very complicated one, to a simple form where everyone should agree that the answer is obvious. His favorite target for the 'obvious' was the situation where we observe a single normally distributed quantity $x$ with unknown expectation $\theta$,

$$(1) \qquad \sim N(\theta, \sigma^2),$$

the variance $\sigma^2$ being known. Everyone agrees, says Fisher, that in this case, the best estimate is $\hat{\theta} = x$ and the correct 90% confidence interval for $\theta$ (to use terminology Fisher hated), is

$$(2) \qquad \hat{\theta} \pm 1.645\,\sigma.$$

Fisher's inductive logic might be called a theory of types, in which problems are reduced to a small catalogue of obvious situations. This had been tried before in statistics, the Pearson system being a good example, but never so forcefully nor successfully. Fisher was astoundingly resourceful at reducing problems to simple forms like (1). Some of the devices he invented for this purpose were sufficiency, ancillarity and conditionality, transformations, pivotal methods, geometric arguments, randomization inference, and

asymptotic maximum likelihood theory. Only one major reduction principle has been added to this list since Fisher's time, invariance, and that one
is not in universal favor these days.

Fisher always preferred exact small-sample results but the asymptotic
optimality of the MLE has been by far the most influential, or at least the
most popular, of his reduction principles. The 1925 paper shows that in
large samples the MLE $\hat{\theta}$ of an unknown parameter $\theta$ approaches the ideal
form (1),

$$\hat{\theta} \to N(\theta, \sigma^2),$$

with the variance $\sigma^2$ determined by the Fisher information and the sample size. Moreover, no other 'reasonable' estimator of $\theta$ has a smaller
asymptotic variance. In other words, the maximum likelihood method automatically produces estimator that can reasonably be termed 'optimal,'
without ever invoking Bayes theorem.

Fisher's great accomplishment triggered a burst of interest in optimality results. The most spectacular product of this burst was the Neyman-
Pearson lemma for optimal hypothesis testing, followed soon by Neyman's
theory of confidence intervals. The Neyman-Pearson lemma did for hypothesis testing what Fisher's MLE theory did for estimation, by pointing
the way toward optimality.

Philosophically, the Neyman-Pearson lemma fits in well with Fisher's
program: using mathematical logic it reduces a complicated problem to
an obvious solution without invoking Bayesian priors. Moreover, it is a
tremendously useful idea in applications, so that Neyman's ideas on hypotheses testing and confidence intervals now play a major role in day-to-
day applied statistics.

However, the success of the Neyman-Pearson lemma triggered new developments, leading to a more extreme form of statistical optimality that
Fisher deeply distrusted. Even though Fisher's personal motives are sus-
pect here, his philosophical qualms were far from groundless. Neyman's
ideas, as later developed by Wald into decision theory, brought a qualitatively different spirit into statistics.

Fisher's maximum likelihood theory was launched in reaction to the
rather shallow Laplacian Bayesianism of the previous century. Fisher's
work demonstrated a more stringent approach to statistical inference. The
Neyman-Wald decision theoretic school carried this spirit of astringency
much further. A strict mathematical statement of the problem at hand,
often phrased quite narrowly, followed by an optimal solution became the
ideal. The practical result was a more sophisticated form of frequentist
inference having enormous mathematical appeal.

Fisher, caught I think by surprise by this flanking attack from his right, complained that the Neyman-Wald decision theorists could be *accurate* without being *correct*. A favorite example of his concerned a Cauchy distribution with unknown center

$$(3) \qquad f_\theta(x) = \frac{1}{\pi\left[1 + (x - \theta)^2\right]}.$$

Given a random sample $\mathbf{x} = (x_1, x_2, \ldots, x_n)$ from (3), decision theorists might try to provide the shortest interval of the form $\hat{\theta} \pm c$ that covers the true $\theta$ with probability .90. Fisher's objection, spelled out in his 1934 paper on ancillarity, was that $c$ should be different for different samples $\mathbf{x}$ depending upon the correct amount of information in $\mathbf{x}$.

The decision theory movement eventually spawned its own counter-reformation. The neo-Bayesians, led by Savage and DeFinetti, produced a more logical and persuasive Bayesianism, emphasizing subjective probabilities and personal decision making. In its most extreme form the Savage-DeFinetti theory directly denies Fisher's claim of an impersonal logic of statistical inference. There has also been a post-war revival of interest in objectivist Bayesian theory, Laplacian in intent but based on Jeffreys' more sophisticated methods for choosing objective priors, which I shall talk more about later on.

Very briefly then, this is the way we arrived at the end of the 20th Century with three competing philosophies of statistical inference: Bayesian, Neyman-Wald frequentist, and Fisherian. In many ways the Bayesian and frequentist philosophies stand at opposite poles from each other, with Fisher's ideas being somewhat of a compromise. I want to talk about that compromise next because it has a lot to do with the popularity of Fisher's methods.

## 4. Three Competing Philosophies

The chart in Figure 1 shows four major areas of disagreement between the Bayesians and the frequentists. These are not just philosophical disagreements. I chose the four categories because they lead to different behavior at the data-analytic level. For each category I have given a rough indication of Fisher's preferred position.

### 4.1 Individual decision-making versus scientific inference

Bayes theory, and in particular Savage-DeFinetti Bayesianism (the kind I'm focusing on here, though later I'll also talk about the Jeffreys' brand of objective Bayesianism), emphasizes the individual decision-maker, and it has been most successful in fields like business where individual decisions are paramount. Frequentists aim for universal acceptance of their inferences. Fisher felt that the proper realm of statistics was scientific inference, where it is necessary to persuade all or at least most of the world of science that you have reached the correct conclusion. Here Fisher is far over to the frequentist side of the chart (which is philosophically accurate but anachronistic, since Fisher's position predates both the Savage-Definetti and Neyman-Wald Schools).

## 4.2 Coherence versus optimality

Bayesian theory emphasizes the coherence of its judgments, in various technical ways but also in the wider sense of enforcing consistency relationships between different aspects of a decision- making situation. Optimality in the frequentist sense is frequently incoherent. For example, the Uniform Minimum Variance Unbiased (UMVU) estimate of $\exp\{\theta\}$ does not have to equal $\exp\{$the UMVU of $\{\theta\}$, and more seriously there is no simple calculus relating the two different estimates. Fisher wanted to have things both ways, coherent and optimal, and in fact maximum likelihood estimation does satisfy

$$\exp\{\hat{\theta}\} = \widehat{\exp\{\theta\}}.$$

The tension between coherence and optimality is like the correctness/ accuracy disagreement concerning the Cauchy example (3), where Fisher argued strongly for correctness. The emphasis on correctness, and a belief in the existence of a logic of statistical inference, moves Fisherian philosophy toward the Bayesian side of Figure 1. Fisherian practice is a less clear story. Different parts of the Fisherian program don't cohere with each other and in practice Fisher seemed quite willing to sacrifice logical consistency for a neat solution to a particular problem, for example switching back and forth between frequentist and non-frequentist justifications of the Fisher information. This kind of case-to-case expediency, which is a common attribute of modern data analysis, has a frequentist flavor. I have located the Fisherian stars for this category a little closer to the Bayesian side of Figure 1, but spreading over a wide range.

## 4.3 Synthesis versus analysis

Bayesian decision-making emphasizes the collection of information across all possible sources, and the synthesis of that information into the final inference. Frequentists tend to break problems into separate small pieces that can be analyzed separately (and optimally). Fisher emphasized the use of all available information as a hallmark of correct inference, and in this way he is more in sympathy with the Bayesian position.

In this case Fisher tended toward the Bayesian position both in theory and in methodology: maximum likelihood estimation and its attendant theory of approximate confidence intervals based on Fisher information are superbly suited to the combination of information from different sources. [On the other hand, we have this quote from Yates and Mather: "In his own work Fisher was at his best when confronted with small self-contained sets of data ... He was never much interested in the assembly and analysis of large amounts of data from varied sources bearing on a given issue." They blame this for his stubbornness or the smoking-cancer controversy. Here as elsewhere we will have to view Fisher as a lapsed Fisherian.]

## 4.4 Optimism versus pessimism

This last category is more psychological than philosophical, but it is psychology rooted in the basic nature of the two competing philosophies. Bayesians tend to be more aggressive and risk-taking in their data analyses. There couldn't be a more pessimistic and defensive theory than minimax, to choose an extreme example of frequentist philosophy. It says that if anything can go wrong it will. Of course a minimax person might characterize the Bayesian position as 'If anything can go right it will'.

Fisher took a middle ground here. He scorns the finer mathematical concerns of the decision theorists ('Not only does it take a cannon to shoot a sparrow, but it misses the sparrow!'), but he fears averaging over the states of nature in a Bayesian way. One of the really appealing features of Fisher's work is its spirit of reasonable compromise, cautious but not overly concerned with pathological situations. This has always struck me as the right attitude toward most real-life problems, and it's certainly a large part of Fisher's dominance in statistical applications.

Looking at Figure 1, I think it is a mistake trying too hard to make a coherent philosophy out of Fisher's theories. From our current point of view they are easier to understand as a collection of extremely shrewd compromises between Bayesian and frequentist ideas. Fisher usually wrote as if he had a complete logic of statistical inference in hand, but that didn't stop him from changing his system when he thought up another landmark idea.

DeFinetti, as quoted by Cifarelli and Regazzini, puts it this way: "Fisher's rich and manifold personality shows a few contradictions. His common sense in applications on one hand and his lofty conception of scientific research on the other lead him to disdain the narrowness of a genuinely objectivist formulation, which he regarded as a *wooden attitude*. He professes his adherence to the objectivist point of view by rejecting the errors of the Bayes-Laplace formulation. What is not so good here is his mathematics, which he handles with mastery in individual problems but rather cavalierly in conceptual matters, thus exposing himself to clear and sometimes heavy criticism. From our point of view it appears probable that many of Fisher's observations and ideas are valid provided we go back to the intuitions from which they spring and free then from the arguments by which he thought to justify them."

| BAYES | FISHER | FREQUENTIST |
|---|---|---|
| 1. Individual (personal decisions) | *** | Universal (world of science) |
| 2. Coherent (correct) | ************* | Optimal (accurate) |
| 3. Synthetic (combination) | **** | Analytic (separation) |
| 3. Optimistic (aggressive) | ***** | Pessimistic (defensive) |

Figure 1. Four major areas of disagreement between Bayesian and frequentist methods. For each one I have inserted a row of stars to indicate, very roughly, the preferred location of Fisherian inference.

Figure 1 describes Fisherian statistics as a compromise between the Bayesian and frequentist schools, but in one crucial way it is not a compromise: in its ease of use. Fisher's philosophy was always expressed in very practical terms. He seemed to think naturally in terms of computational algorithms, as with maximum likelihood estimation, analysis of variance, and permutation tests. If anything is going to replace Fisher in the 21st Century it will have to be a methodology that is equally easy to apply in day to day practice.

## 5. Fisher's Influence on Current Research

There are three parts to this talk: past, present, and future. The past part, which you have just seen, didn't do justice to Fisher's ideas, but the subject here is more one of influence than ideas, admitting of course that the influence is founded on the ideas' strengths. So now I am going to discuss Fisher's influence on current research once again.

What follows are several (actually six) examples of current research topics that have attracted a lot of attention recently. No claim of completeness is being made here. The main point I'm trying to make with these examples is that Fisher's ideas are still exerting a powerful influence

on developments in statistical theory, and that this is an important indication of their future relevance. The examples will gradually get more
speculative and futuristic, and will include some areas of development *not*
satisfactorily handled by Fisher — holes in the Fisherian fabric — where we
might expect future work to be more frequentist or Bayesian in motivation.

The examples will also allow me to talk about the new breed of applied problems statisticians are starting to see, the bigger, messier, more
complicated data sets that we will have to deal with in the coming decades.
Fisherian methods were fashioned to deal with the problems of the 1920's
and 1930's. It is not a certainty that they will be equally applicable to the
problems of the 21st Century — a question I hope to shed at least a little
light upon.

### 5.1 Fisher information and the bootstrap

This first example is intended to show how Fisher's ideas can pop
up in current work, but be difficult to recognize because of computational
advances. First, here is a very brief review of Fisher information. Suppose
we observe a random sample $x_1, x_2, \ldots, x_n$ from a density function $f_\theta(x)$
depending on a single unknown parameter $\theta$,

$$f_\theta(x) \rightarrow x_1, x_2, \ldots, x_n .$$

The Fisher information in any one $x$ is the expected value of minus th e
second derivative of the log density,

$$i_\theta = \mathbf{E}_\theta \left\{ -\frac{\partial^2}{\partial \theta^2} \log f_\theta(x) \right\} ,$$

and the total Fisher information in the whole sample is $n i_\theta$.

Fisher showed that the asymptotic standard error of the MLE is in-
versely proportional to the square root of the total information,

$$(4) \qquad\qquad se_\theta(\hat{\theta}) \doteq \frac{1}{\sqrt{n i_\theta}} ,$$

and that no other consistent and sufficiently regular estimation of $\theta$ —
essentially no other asymptotically unbiased estimator — can do better.

A tremendous amount of philosophical interpretation has been attached to $i_\theta$, concerning the meaning of statistical information, but in
practice Fisher's formula (4) is most often used simply as a handy estimate
of the standard error of the MLE. Of course, (4) by itself cannot be used
directly because $i_\theta$ involves the unknown parameter $\theta$. Fisher's tactic,
which seems obvious but in fact is quite central to Fisherian methodology,

is to *plug in* the MLE $\hat{\theta}$ for $\theta$ in (4), giving a usable estimate of standard error,

$$(5) \qquad\qquad \hat{se} = \frac{1}{\sqrt{ni_\theta}} \, .$$

Here is an example of formula (5) in action. Figure 2 shows the results of a small study designed to test the efficacy of an experimental anti-viral drug. A total of $n = 20$ AIDS patients had their CD4 counts measured before and after taking the drug, yielding data

$$x_i = (\text{before}_i, \text{ after}_i), \qquad \text{for } i = 1, 2, \ldots, 20 \, .$$

The Pearson sample correlation coefficient was $\hat{\theta} = 0.723$. How accurate is this estimate?

If we assume a bivariate normal model for the data,

$$(6) \qquad\qquad N_2(\mu, \Sigma) \to x_1, x_2, x_3 \ldots, x_{20} \, ,$$

the notation indicating a random sample of 20 pairs from a bivariate normal distribution with expectation vector $\mu$ and covariance matrix $\Sigma$, then $\hat{\theta}$ is the MLE for the true correlation coefficient $\theta$. The Fisher information for estimating $\theta$ turns out to be $i_\theta = 1/(1 - \theta^2)^2$ (after taking proper account of the 'nuisance parameters' in (6) — one of those technical points I am avoiding in this talk) so (6) gives estimated standard error

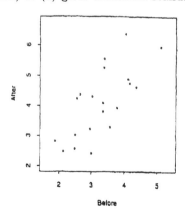

Figure 2. The CD4 data; 20 AIDS patients had their CD4 counts measured before and after taking an experimental drug; correlation coefficient $\hat{\theta} = 0.723$.

$$\widehat{se} = \frac{\left(1 - \hat{\theta}^2\right)}{\sqrt{20}} = 0.107\,.$$

Here is a bootstrap estimate of standard error for the same problem, also assuming that the bivariate normal model is correct. In this context the bootstrap samples are generated from model (6), but with estimates $\hat{\mu}$ and $\hat{\Sigma}$ substituted for the unknown parameters $\mu$ and $\Sigma$ :

$$N\left(\hat{\mu}, \hat{\Sigma}\right) \rightarrow x_1^*, x_2^*, x_3^* \ldots, x_{20}^* \rightarrow \hat{\theta}^*\,,$$

where $\hat{\theta}^*$ is the sample correlation coefficient for the bootstrap data set $x_1^*, x_2^*, x_3^* \ldots, x_{20}^*$.

This whole process was independently repeated 2000 times, giving 2000 bootstrap correlation coefficients $\hat{\theta}^*$. Figure 3 shows their histogram.

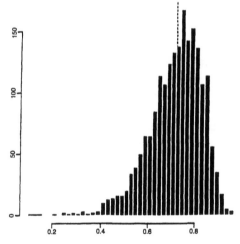

Figure 3. Histogram of 2000 bootstrap correlation coefficients. Bivariate normal sampling model.

The empirical standard deviation of the 2000 $\hat{\theta}^*$ values is

$$\widehat{se}_{\text{boot}} = 0.112\,,$$

which is the normal-theory bootstrap estimate of standard error for $\hat{\theta}$. 2000 is ten times more than needed for a standard error, but we will need all 2000 later for the discussion of approximate confidence intervals.

## 5.2 The plug-in principle

The Fisher information and bootstrap standard error estimates, 0.107 and 0.112, are quite close to each other. This is no accident. Despite the fact that they look completely different, the two methods are doing very similar calculations. Both are using the 'plug-in principle' as a crucial step in getting the answer.

Here is a plug-in description of the two methods:

- Fisher Information: (i) Compute an (approximate) formula for the standard error of the sample correlation coefficient as a function of the unknown parameters $(\mu, \Sigma)$; (ii) Plug in estimates $(\hat{\mu}, \hat{\Sigma})$ for the unknown parameters $(\mu, \Sigma)$ in the formula.

- Bootstrap: (i) Plug in $(\hat{\mu}, \hat{\Sigma})$ for the unknown parameters $(\mu, \Sigma)$ in the mechanism generating the data; (ii) Compute the standard error of the sample correlation coefficient, for the plugged-in mechanism, by Monte Carlo simulation.

The two methods proceed in reverse order, 'compute and then plug in' versus 'plug in and then compute,' but this is a relatively minor technical difference. The crucial step in both methods, and the only statistical inference going on, is the substitution of the estimates $(\hat{\mu}, \hat{\Sigma})$ for the unknown parameters $(\mu, \Sigma)$, in other words the plug-in principle. Fisherian inference makes frequent use of the plug-in principle, and this is one of the main reasons that Fisher's methods are so convenient to use in practice. All possible inferential questions are answered by simply plugging in estimates, usually maximum likelihood estimates, for unknown parameters.

The Fisher information method involves cleverer mathematics than the bootstrap, but it has to because we enjoy a $10^7$ computational advantage over Fisher. A year's combined computational effort by all the statisticians of 1925 wouldn't equal a minute of modern computer time. The bootstrap exploits this advantage to numerically extend Fisher's calculations to situations where the mathematics becomes hopelessly complicated. One of the less attractive aspects of Fisherian statistics is its over-reliance on a small catalog of simple parametric models like the normal, understandable enough given the limitations of the mechanical calculators Fisher had to work with.

Modern computation has given us the opportunity to extend Fisher's methods to a much wider class of models, including nonparametric ones (the more usual arena of the bootstrap). We are beginning to see many such extensions, for example the extension of discriminant analysis to CART, and the extension of linear regression to Generalized Additive Models.

## 6. The Standard Intervals

I want to continue the CD4 example, but proceeding from standard errors to confidence intervals. The confidence interval story illustrates how computer-based inference can be used to extend Fisher's ideas in a more ambitious way.

The MLE and its estimated standard error were used by Fisher to form approximate confidence intervals, which I like to call the *standard intervals* because of their ubiquity in day-to-day practice,

$$(7) \qquad\qquad \hat{\theta} \pm 1.645 \, \widehat{se}.$$

The constant, 1.645, gives intervals of approximate 90% coverage for the unknown parameter $\theta$, with 5% non-coverage probabilities at each end of the interval. We could use 1.96 instead of 1.645 for 95% coverage, etc., but here I'll stick to 90%.

The standard intervals follow from Fisher's result that $\hat{\theta}$ is asymptotically normal, unbiased, and with standard error fixed by the sample size and the Fisher information,

$$(8) \qquad\qquad \hat{\theta} \to N\left(\theta, se^2\right),$$

as in (4). We recognize (8) as one of Fisher's ideal 'obvious' forms.

If usage determines importance then the standard intervals were Fisher's most important invention. Their popularity is due to a combination of optimality, or at least asymptotic optimality, with computation tractability. The standard intervals are

- *Accurate:* Their non-coverage probabilities, which are supposed to be 0.05 at each end of the interval, are actually

$$(9) \qquad\qquad 0.05 + c/\sqrt{n}$$

  where $c$ depends on the situation, so as the sample size $n$ gets large as we approach the nominal value 0.05 at rate $n^{-\frac{1}{2}}$.

- *Correct:* The estimated standard error based on the Fisher information is the minimum possible for any asymptotically unbiased estimate of $\theta$ so interval (7) doesn't waste any information nor is it misleadingly optimistic.

- *Automatic:* $\hat{\theta}$ and $\widehat{se}$ are computed from the same basic algorithm no matter how complicated the problem may be.

Despite these advantages, applied statisticians know that the standard intervals can be quite inaccurate in small samples. This is illustrated in the left panel of Figure 4 for the CD4 correlation example, where we see that the standard interval endpoints lie far to the right of the endpoints for the normal-theory exact 90% central confidence interval. In fact, we can see from the bootstrap histogram (reproduced from Figure 3) that in this case the asymptotic normality of the MLE hasn't taken hold at $n = 20$, so that there is every reason to doubt the standard interval. Being able to look at the histogram, which has a lot of information in it, is a luxury Fisher did not have.

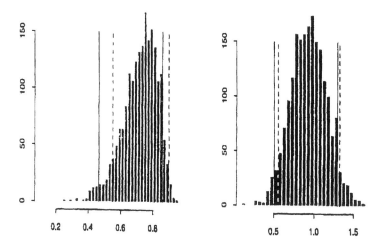

Figure 4. Left Panel: endpoints of exact 90% confidence interval for CD4 correlation coefficient (solid lines) are much different than standard interval endpoints (dashed lines), as suggested by the non-normality of the bootstrap histogram. Right Panel: Fisher's transformation normalizes the bootstrap histogram and makes the standard interval more accurate.

Fisher suggested a fix for this specific situation: transform this correlation coefficient to $\widehat{\phi} = \tanh^{-1}(\hat{\theta})$, that is to

$$(10) \qquad\qquad \widehat{\phi} = \frac{1}{2} \log \frac{1 + \hat{\theta}}{1 - \hat{\theta}} \,,$$

apply the standard method on this scale, and then transform the standard interval back to the $\theta$ scale. This was another one of Fisher's ingenious reduction methods. The $\tanh^{-1}$ transformation greatly accelerates convergence to normality, as we can see from the histogram of the 2000 values of $\hat{\theta}* = \tanh^{-1}(\hat{\theta}*)$ in the right panel of Figure 4, and makes the standard intervals far more accurate. However we have now lost the 'automatic' property of the standard intervals. The $\tanh^{-1}$ transformation works only for the normal correlation coefficient and not for most other problems.

The standard intervals take literally the large sample approximation $\hat{\theta} \sim N(\theta,\ se^2)$, which says that $\hat{\theta}$ is normally distributed, unbiased for $\theta$, and has a constant standard error. A more careful look at the asymptotics shows that each of these three assumptions can fail in a substantial way: the sampling distribution of $\hat{\theta}$ can be skewed, $\hat{\theta}$ can be biased as an estimate of $\theta$, and its standard error can change with $\theta$. Modern computation makes it practical to correct all three errors. I am going to mention two methods of doing so, the first using the bootstrap histogram, the second based on likelihood methods.

It turns out that there is enough information in the bootstrap histogram to correct all three errors of the standard intervals. The result is a system of approximate confidence intervals an order of magnitude more accurate, with non-coverage probabilities

$$0.05 + c/n$$

compared to (9), achieving what is called 'second order accuracy'. Table 1 demonstrates the practical advantages of second order accuracy. In most situations we would not have exact endpoints as a gold standard for comparison, but second order accuracy would still point to the superiority of the bootstrap intervals.

| | Exact | Bootstrap | Standard |
|---|---|---|---|
| .05: | .464 | .468 | .547 |
| .95: | .859 | .856 | .899 |

Table 1. Endpoints of exact and approximate 90% confidence intervals for the CD4 correlation coefficient assuming bivariate normality.

The bootstrap method, and also the likelihood-based methods of the next section, are *Transformation Invariant*. That is, they give the same interval for the correlation coefficient whether or not you go through the

$\tanh^{-1}$ transformation. In this sense they automate Fisher's wonderful
transformation trick.

I like this example because it shows how a basic Fisherian construction, the standard intervals, can be extended by modern computation. The
extension lets us deal easily with very complicated probability models, even
nonparametric ones, and also with complicated statistics such as a coefficient in a step-wise robust regression.

Moreover, the extension is not just to a wider set of applications. Some
progress in understanding the theoretical basis of approximate confidence
intervals is also made along the way. Other topics are springing up in the
same fashion. For example, Fisher's 1925 work on the information loss for
insufficient estimators has transmuted into our modern theories of the EM
algorithm and Gibbs sampling.

## 7. Conditional Inference, Ancillarity and the Magic Formula

Table 2 shows the occurrence of a very undesirable side effect in a randomized experiment that will be described more fully later. The treatment
produces a smaller ratio of these undesirable effects than does the control,
the sample log odds ratio being

$$\hat{\theta} = \log \left( \frac{1}{15} \Big/ \frac{13}{3} \right) = -4.2 .$$

|  | Yes | No |  |
|---|---|---|---|
| Treatment | 1 | 15 | 16 |
| Control | 13 | 3 | 16 |
|  | 14 | 18 |  |

Table 2. The occurrence of adverse events in a randomized experiment;

Fisher wondered how one might make appropriate inferences for $\theta$, the
true log odds ratio. The trouble here is nuisance parameters. A multinomial model for the $2 \times 2$ table has three free parameters, representing four
cell probabilities constrained to add up to 1, and in some sense two of the
three parameters have to be eliminated in order to get at $\theta$. To do this
Fisher came up with another device for reducing a complicated situation
to a simple form.

Fisher showed that if we condition on the marginals of the table, then the conditional density of $\hat{\theta}$ given the marginals depends only $\theta$. The nuisance parameters disappear. This conditioning is 'correct' he argued because the marginals are acting as what might be called *approximate ancillary statistics*. That is, they do not carry much direct information concerning the value of $\theta$, but they have something to say about how accurately $\hat{\theta}$ estimates $\theta$. Later Neyman gave a much more specific frequentist justification for conditioning on the marginals, through what is now called 'Neyman Structure'.

For the data in Table 2, the conditional distribution of $\hat{\theta}$ given the marginals yields [-6.3, -2.4] as a 90% confidence interval for $\theta$, ruling out the null hypothesis value $\theta = 0$ where Treatment equals Control. However, the conditional distribution is not easy to calculate, even in this simple case, and it becomes prohibitive in more complicated situations.

In his 1934 paper, which was the capstone of Fisher's work on efficient estimation, he solved the conditioning problem for translation families. Suppose that $\mathbf{x} = (x_1, x_2, \ldots, x_n)$ is a random sample from a Cauchy distribution (3), and that we wish to use $\mathbf{x}$ to make inferences about $\theta$, the unknown center point of the distribution. In this case there is a genuine ancillary statistic $\mathbf{A}$, the vector of spacings between the ordered values of $\mathbf{x}$. Again Fisher argued that correct inferences about $\theta$ should be based on $f_\theta(\hat{\theta}|\mathbf{A})$, the conditional density of the MLE $\hat{\theta}$ given the ancillary $\mathbf{A}$, not on the unconditional density $f_\theta(\hat{\theta})$.

Fisher also provided a wonderful trick for calculating $f_\theta(\hat{\theta}|\mathbf{A})$. Let $L(\theta)$ be the likelihood function: the unconditional density of the whole sample, considered as a function of $\theta$ with $\mathbf{x}$ fixed. Then it turns out that

$$(11) \qquad f_\theta\left(\hat{\theta} \,|\, \mathbf{A}\right) = c\,\frac{L(\theta)}{L(\hat{\theta})}$$

where $c$ is a constant. Formula (11) allows us to compute the conditional density $f_\theta(\hat{\theta}|\mathbf{A})$ from the likelihood, which is easy to calculate. It also hints at a deep connection between likelihood-based inference, a Fisherian trademark, and frequentist methods.

Despite this promising start, the promise went unfulfilled in the years following 1934. The trouble was that formula (11) applies only in very special circumstances, not including the $2 \times 2$ table example for instance. Recently though there has been a revival of interest in likelihood-based conditional inference. Durbin, Barndorff-Nielsen, Hinkley, and others have developed a wonderful generalization of (11) that applies to a wide variety of problems having approximate ancillaries, the so-called magic formula

$$(12) \qquad f_\theta\left(\hat\theta \mid \mathbf{A}\right) = c\,\frac{L(\theta)}{L(\hat\theta)}\left\{-\frac{d^2}{d\theta^2}\,\log L(\theta)\Big|_{\theta=\hat\theta}\right\}^{\frac{1}{2}}.$$

The bracketed factor is constant in the Cauchy situation, reducing (12) back to (11).

Likelihood-based conditional inference has been pushed forward in current work by Fraser, Cox and Reid, McCullagh, Barndorff-Nielson Pierce, DiCiccio, and many others. It represents a major effort to perfect and extend Fisher's goal of an inferential system based directly on likelihoods.

In particular the magic formula can be used to generate approximate confidence intervals that are more accurate than the standard intervals, at least second-order accurate. These intervals agree to second order with the bootstrap intervals. If this were not true then one or both of them would not be second order correct. Right now it looks like attempts to improve upon the standard intervals are converging from two directions: likelihood and bootstrap.

Results like (12) have enormous potential. Likelihood inference is the great unfulfilled promise of Fisherian statistics — the promise of a theory that directly interprets likelihood functions in a way that simultaneously satisfies Bayesian and frequentists. Fulfilling that promise, even partially, would greatly influence the shape of 21st Century statistics.

## 8. Fisher's Biggest Blunder

Now I'll start edging gingerly into the 21st Century by discussing some topics where Fisher's ideas have not been dominant, but where they might or might not be important in future developments. I am going to begin with the fiducial distribution, generally considered to be Fisher's biggest blunder. But in Arthur Koestler's words 'The history of ideas is filled with barren truths and fertile errors'. If fiducial inference is an error it certainly has been a fertile one.

In terms of Figure 1, the Bayesian-frequentist comparison chart, fiducial inference was Fisher's closest approach to the Bayesian side of the ledger. Fisher was trying to codify an objective Bayesianism in the Laplace tradition but without using Laplace's ad hoc uniform prior distributions. I believe that Fisher's continuing devotion to fiducial inference had two major influences, a negative reaction against Neyman's ideas and a positive attraction to Jeffreys' point of view.

The solid line in Figure 5 is the fiducial density for a binomial parameter $\theta$ having observed 3 successes in 10 trials,

$$s \sim \text{Binomial}\,(n,\theta), \qquad s = 3 \text{ and } n = 10\,.$$

Also shown is an approximate fiducial density that I will refer to later. Fisher's fiducial theory at its boldest treated the solid curve as a genuine a posteriori density for $\theta$ even though, or perhaps because, no prior assumptions had been made.

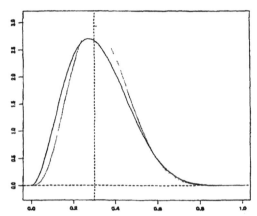

Figure 5. Fiducial density for a binomial parameter $\theta$ having observed 3 successes out of 10 trials. The dashed line is an approximation that is useful in complicated situations.

## 8.1 The confidence density

We could also call the fiducial distribution the 'confidence density' because this is an easy way to motivate the fiducial construction. As I said earlier, Fisher would have hated this name.

Suppose that for every value of $\alpha$ between 0 and 1 we have an upper 100 $\alpha$-th confidence limit $\hat{\theta}[\alpha]$ for $\theta$, so that by definition

$$\text{prob} \left\{ \theta < \hat{\theta}[\alpha] \right\} = \alpha .$$

We can interpret this as a probability distribution for $\theta$ given the data if we are willing to accept the classic *wrong* interpretation of confidence,

$\theta$ is in the interval $\left( \hat{\theta}[0.90], \ \hat{\theta}[0.91] \right)$ with probability 0.01, etc.

Going to the continuous limit gives the 'confidence density,' a name Neyman would have hated.

The confidence density *is* the fiducial distribution, at least in those cases where Fisher would have considered the confidence limits to be inferentially correct. The fiducial distribution in Figure 5 is the confidence density based on the usual confidence limits for $\theta$ (taking into account the discrete nature of the binomial distribution): $\hat{\theta}[\alpha]$ is the value of $\theta$ such that $S \sim$ Binomial $(10,\theta)$ satisfies

$$\text{prob}\{S > 3\} + \frac{1}{2}\,\text{prob}\{S = 3\} = \alpha.$$

Fisher was uncomfortable applying fiducial arguments to discrete distributions because of the ad hoc continuity corrections required, but the difficulties caused are more theoretical than practical.

The advantage of stating fiducial ideas in terms of the confidence density is that they then can be applied to a wider class of problems. We can use the approximate confidence intervals mentioned earlier, either the bootstrap or the likelihood ones, to get approximate fiducial distribution even in very complicated situations having lots of nuisance parameters. (The dashed curve in Figure 5 is the confidence density based on approximate bootstrap intervals.) And there are practical reasons why it would be very convenient to have good approximate fiducial distributions, reasons connected with our profession's 250-year search for a dependable objective Bayes theory.

## 8.2 Objective Bayes

By 'objective Bayes' I mean a Bayesian theory in which the subjective element is removed from the choice of prior distribution; in practical terms a universal recipe for applying Bayes theorem in the absence of prior information. A widely accepted objective Bayes theory, which fiducial inference was intended to be, would be of immense theoretical and practical importance.

I have in mind here dealing with messy, complicated problems where we are trying to combine information from disparate sources — doing a meta- analysis for example. Bayesian methods are particularly well-suited to such problems. This is particularly true now that techniques like the Gibbs sampler and Markov Chain Monte Carlo are available for integrating the nuisance parameters out of high-dimensional posterior distributions.

The trouble of course is that the statistician still has to choose a prior distribution in order to use Bayes theorem. An unthinking use of uniform priors is no better now than it was in Laplace's day. A lot of recent effort has been put into the development of uninformative or objective prior distributions, priors that eliminate nuisance parameters safely while remaining neutral with respect to the parameter of interest. Kass and Wasserman's 1996 *JASA* article reviews current developments by Berger, Bernardo, and many others, but the task of finding genuinely objective priors for high dimensional problems remains a daunting task.

Fiducial distributions, or confidence densities, offer a way to finesse this difficulty. A good argument can be made that the confidence density is the posterior density for the parameter of interest, after all of the nuisance parameters have been integrated out in an objective way. If this argument turns out to be valid then our progress in constructing approximate confidence intervals, and approximate confidence densities, could lead to an easier use of Bayesian thinking in practical problems.

This is all quite speculative, but here is a safe prediction for the 21st Century: statisticians will be asked to solve bigger and more complicated problems. I believe that there is a good chance that objective Bayes methods will be developed for such problems, and that something like fiducial inference will play an important role in this development. Maybe Fisher's biggest blunder will become a big hit in the 21st Century!

## 9. Model Selection

Model selection is another area of statistical research where important developments seem to be building up, but without a definitive breakthrough. The question asked here is how to select the model itself, not just the continuous parameters of a given model, from the observed data. $F$ tests, and $F$ stands for Fisher, help with this task, and are certainly the most widely used model selection techniques. However even in relatively simple problems things can get complicated fast, as anyone who has gotten lost in a tangle of forward and backward stepwise regression programs can testify.

The fact is that classic Fisherian estimation and testing theory are a good start, but not much more than that, on model selection. In particular, maximum likelihood estimation theory and model fitting do not account for the number of free parameters being fit, and that is why frequentist methods like Mallow's Cp, the Akaike Information Criterion, and Cross-Validation have evolved. Model selection seems to be moving away from its Fisherian roots.

Now statisticians are starting to see really complicated model selection problems, with thousands and even millions of data points and hundreds of candidate models. A thriving area called 'machine learning' has developed to handle such problems, in ways that are not yet very well connected to statistical theory.

Table 3, taken from Gail Gong's 1982 thesis, shows part of the data from a model selection problem that is only moderately complicated by today's standards, though hopelessly difficult from a pre-war viewpoint. A

'training set' of 155 chronic hepatitis patients were measured on 19 diagnostic prediction variables. The outcome variable $y$ was whether or not
the patient died from liver failure (122 lived, 33 died), the goal of the study
being to develop a prediction rule for $y$ in terms of the diagnostic variables.

| y | Cons-tant 1 | Age 2 | Sex 3 | Ster-oid 4 | Anti-viral 5 | Fatigue 6 | Mal-aise 7 | Anor-exia 8 | Liver Big 9 | Liver Firm 10 | Spleen Palp 11 | Spiders 12 | As-cites 13 | Varices 14 | Bili-rubin 15 | Alk Phos 16 | SGOT 17 | Albu-min 18 | Pro-tein 19 | Histo-logy 20 | • |
|---|---|---|---|---|---|---|---|---|---|---|---|---|---|---|---|---|---|---|---|---|---|
| 1 | 1 | 45 | 1 | 2 | 2 | 1 | 1 | 1 | 2 | 2 | 2 | 1 | 1 | 2 | 1.90 | -1 | 114 | 2.4 | -1 | -3 | 145 |
| 0 | 1 | 31 | 1 | 1 | 2 | 1 | 2 | 2 | 2 | 2 | 2 | 2 | 2 | 2 | 1.20 | 75 | 193 | 4.2 | 54 | 2 | 146 |
| 1 | 1 | 41 | 1 | 2 | 2 | 1 | 2 | 2 | 2 | 1 | 1 | 1 | 2 | 1 | 4.20 | 65 | 120 | 3.4 | -1 | -3 | 147 |
| 1 | 1 | 70 | 1 | 1 | 2 | 1 | 1 | 1 | -3 | -3 | -3 | -3 | -3 | -3 | 1.70 | 109 | 528 | 2.8 | 35 | 2 | 148 |
| 0 | 1 | 20 | 1 | 1 | 2 | 2 | 2 | 2 | 2 | -3 | 2 | 2 | 2 | 2 | .90 | 89 | 152 | 4.0 | -1 | 2 | 149 |
| 0 | 1 | 36 | 1 | 2 | 2 | 2 | 2 | 2 | 2 | 2 | 2 | 2 | 2 | 2 | .60 | 120 | 30 | 4.0 | -1 | 2 | 150 |
| 1 | 1 | 46 | 1 | 2 | 2 | 1 | 1 | 1 | 2 | 2 | 1 | 1 | 1 | 1 | 7.60 | -1 | 242 | 3.3 | 50 | -3 | 151 |
| 0 | 1 | 44 | 1 | 2 | 2 | 1 | 2 | 2 | 2 | 1 | 2 | 2 | 2 | 2 | .90 | 126 | 142 | 4.3 | -1 | 2 | 152 |
| 0 | 1 | 61 | 1 | 1 | 2 | 1 | 1 | 2 | 1 | 1 | 2 | 1 | 2 | 2 | .80 | 95 | 20 | 4.1 | -1 | 2 | 153 |
| 0 | 1 | 53 | 2 | 1 | 2 | 1 | 2 | 2 | 2 | 2 | 1 | 1 | 2 | 1 | 1.50 | 84 | 19 | 4.1 | 48 | -3 | 154 |
| 1 | 1 | 43 | 1 | 2 | 2 | 1 | 2 | 2 | 2 | 2 | 1 | 1 | 1 | 2 | 1.20 | 100 | 19 | 3.1 | 42 | 2 | 155 |

Table 3. 155 chronic hepatitis patients were measured on 19 diagnostic
variables; data shown for the last 11 patients; outcome $y = 0$ or 1 as patient
lived or died; negative numbers indicate missing data.

In order to predict the outcome a logistic regression model was built
up in 3 steps:

- Individual logistic regressions were run for each of the 19 predictors,
  yielding 13 that were significant at the 0.05 level.
- A forward stepwise logistic regression program, including only those
  patients with none of the 13 predictors missing, retained 5 of the 13
  predictors at significance level 0.10.
- A second forward stepwise logistic regression program, including those
  patients with none of the 5 predictors missing, retained 4 of the 5 at
  significance level 0.05.

These last four variables:

13. Ascites,    15. Bilirubin,    7. Malaise,    20. Histology,

were deemed the 'important predictors'. The logistic regression based on
them misclassified 16% of the 155 patients, with cross-validation suggesting
a true error rate of about 20%.

A crucial question concerns the validity of the selected model. Should
we take the four 'important predictors' very seriously in a medical sense?
The bootstrap answer seems to be 'probably not', even though it was natural for the medical investigator to do so given the impressive amount of
statistical machinery involved in their selection.

Gail Gong resampled the 155 patients, taking as a unit each patient's entire record of 19 predictors and response. For each bootstrap data set of 155 resampled records, she reran the three-stage logistic regression model, yielding a bootstrap set of 'important predictors'. This was done 500 times. Figure 6 shows the important predictors for the final 25 bootstrap data sets. The first of these is (13, 7, 20, 15), agreeing except for order with the set (13, 15, 7, 20) from the original data. This didn't happen in any of the other 499 bootstrap cases. In all 500 bootstrap replications only variable 20, histology, which appeared 295 times, was 'important' more than half of the time. These results certainly discourage confidence in the causal nature of the predictor variables (13, 15, 7, 20).

```
13    7   20   15
13   19    6
20   16   19
20   19
14   18    7   16    2
18   20    7   11
20   19   15
20
13   12   15    8   18    7   19
15   13   19
13    4
12   15    3
15   16    3
15   20    4
16   13    2   19
18   20    3
13   15   20
15   13
15   20    7
13
15
13   14
12   20   18
 2   20   15    7   19   12
13   20   15   19
```

Figure 6. The set of 'important predictors' selected in the last 25 of 500 bootstrap replications of the three-step logistic regression model selection program; original choices were (13, 15, 7, 20).

Or do they? It seems like we should be able to use the bootstrap results to quantitatively assess the validity of the various predictors. Perhaps they could also help in selecting a better prediction model. Questions like these are being asked these days, but the answers so far are more intriguing than conclusive.

It is not clear to me whether Fisherian methods will play much of a role in the further progress of model selection theory. Figure 6 makes model selection look like an exercise in discrete estimation, while Fisher's MLE theory was always aimed at continuous situations. Direct frequentist methods like cross-validation seen more promising right now, and there have been some recent developments in Bayesian model selection, but in fact our best efforts so far are inadequate for problems like the hepatitis data. We could badly use a clever Fisherian trick for reducing complicated model selection problems to simple obvious ones.

## 10. Empirical Bayes Methods

As a final example, I wanted to say a few words about empirical Bayes methods. Empirical Bayes seems like the wave of the future to me, but it seemed that way twenty-five years ago and the wave still hasn't washed in, despite the fact that it is an area of enormous potential importance. It is not a topic that has had much Fisherian input.

Table 4 shows the data for an empirical Bayes situation: independent clinical trials were run in 41 cities, comparing the occurrence of recurrent bleeding, an undesirable side effect, for two stomach ulcer surgical techniques, a new treatment and an older control. Each trial yielded an estimate of the true log odds ratio for recurrent bleeding, Treatment versus Control,

$$\theta_i = \text{log odds ratio in city } i, \qquad i = 1, 2, \ldots, 41.$$

| Experiment | a | b | c | d | $\hat{\theta}$ | $\widehat{SD}$ | Experiment | a | b | c | d | $\hat{\theta}$ | $\widehat{SD}$ |
|---|---|---|---|---|---|---|---|---|---|---|---|---|---|
| *1 | 7 | 8 | 11 | 2 | -1.84 | .86 | 21 | 6 | 34 | 13 | 8 | -2.22 | 6? |
| 2 | 8 | 11 | 8 | 8 | -.32 | .66 | 22 | 4 | 14 | 5 | 34 | .66 | .71 |
| *3 | 5 | 29 | 4 | 35 | .41 | .68 | 23 | 14 | 54 | 13 | 61 | .20 | .42 |
| 4 | 7 | 29 | 4 | 27 | .49 | .65 | 24 | 6 | 15 | 8 | 13 | -.43 | .64 |
| *5 | 3 | 9 | 0 | 12 | Inf | 1.57 | 25 | 0 | 6 | 6 | 0 | -Inf | 2.08 |
| *6 | 4 | 3 | 4 | 0 | -Inf | 1.65 | 26 | 1 | 9 | 5 | 10 | -1.50 | 1.02 |
| *7 | 4 | 13 | 13 | 11 | -1.35 | 68 | 27 | 5 | 12 | 5 | 10 | -.18 | .7 |
| *8 | 1 | 15 | 13 | 3 | -4.17 | 1.04 | 28 | 0 | 10 | 12 | 2 | -Inf | 1.68 |
| 9 | 3 | 11 | 7 | 15 | -.54 | .76 | 29 | 0 | 22 | 8 | 16 | -Inf | 1.48 |
| *10 | 2 | 36 | 12 | 20 | -2.38 | .75 | 30 | 2 | 16 | 10 | 11 | -1.98 | 86 |
| 11 | 6 | 6 | 8 | 0 | -Inf | 1.56 | 31 | 1 | 14 | 7 | 6 | -2.79 | 1.0" |
| *12 | 2 | 5 | 7 | 2 | -2.17 | 1.06 | 32 | 8 | 16 | 15 | 12 | -.92 | .5? |
| *13 | 9 | 12 | 7 | 17 | .60 | .61 | 33 | 6 | 6 | 7 | 2 | -1.25 | .9" |
| 14 | 7 | 14 | 5 | 20 | .69 | .66 | 34 | 0 | 20 | 5 | 18 | -Inf | 1.5" |
| 15 | 3 | 22 | 11 | 21 | -1.35 | .68 | 35 | 4 | 13 | 2 | 14 | .77 | .8? |
| 16 | 4 | 7 | 6 | 4 | -.97 | .86 | 36 | 10 | 30 | 12 | 8 | -1.50 | .5? |
| 17 | 2 | 8 | 8 | 2 | -2.77 | 1.02 | 37 | 3 | 13 | 2 | 14 | .48 | 9" |
| 18 | 1 | 30 | 4 | 23 | -1.65 | 98 | 38 | 4 | 30 | 5 | 14 | -.99 | 7" |
| 19 | 4 | 24 | 15 | 16 | -1.73 | .62 | 39 | 7 | 31 | 15 | 22 | -1.11 | .5? |
| 20 | 7 | 36 | 16 | 27 | -1.11 | .51 | *40 | 0 | 34 | 34 | 0 | -Inf | 2.01 |
|  |  |  |  |  |  |  | 41 | 0 | 9 | 0 | 16 | NA | 2.04 |

Table 4. Ulcer data. 41 independent experiments concerning the number of occurrences of recurrent bleeding following ulcer surgery; (a, b) = (# bleeding, # non-bleeding) for Treatment, a new surgical technique; (c, d) the same for Control, an older surgery $\hat{\theta}$; is the sample log odds ratio, with estimated standard deviation $\widehat{sd}$; stars indicate cases shown in Figure 7.

In city 8 for example we have the estimate seen earlier in Table 2,

$$\hat{\theta} = \log\left(\frac{1}{15} \Big/ \frac{13}{3}\right) = -4.2\,,$$

indicating that the new surgery was very effective in reducing recurrent bleeding, at least in city 8.

Figure 7 shows the likelihoods for $\theta_i$ in 10 of the 41 cities. These are conditional likelihoods, using Fisher's trick of conditioning on the marginals to get rid of the nuisance parameters in each city. It seems clear that the log odds ratios $\theta_i$ are not all the same. For instance, the likelihoods for cities 8 and 13 barely overlap. On the other hand, the $\theta_i$ values are not wildly discrepant, most of the 41 likelihood functions concentrating themselves on the range (-6, 3). (This is the kind of complicated inferential situation I was worrying about in the discussion of fiducial inference, confidence densities, and objective Bayes methods.)

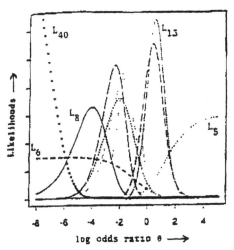

Figure 7. Individual likelihood functions for $\theta_i-$, for 10 of the 41 experiments in Table 4; $L_8$, the likelihood for the log odds ratio in city 8, lies to the left of most of the others.

Notice that $L_8$, the likelihood for $\theta_8$, lies to the left of most of the other curves. This would still be true if we could see all 41 curves instead of just 10 of them. In other words, $\theta_8$ appears to be more negative than the log odds ratios in most of the other cities.

What is a good estimate or confidence interval for $\theta_8$? Answering this question depends on how much the results in other cities influence our thinking about city 8. That is where empirical Bayes theory comes in, giving us a systematic framework for combining the direct information for $\theta_8$ from city 8's experiment with the indirect information from the experiments in the other 40 cities.

The ordinary 90% confidence interval for $\theta_8$, based only on the data (1, 15, 13, 3) from its own experiment, is

$$(13) \qquad\qquad \theta_8 \in [-6.3, \ -2.4] \ .$$

Empirical Bayes methods give a considerably different result. The empirical Bayes analysis uses the data in the other 40 cities to estimate a prior density for log odds ratios. This prior density can be combined with the likelihood $L_8$ for city 8, using Bayes theorem, to get a central 90% a posteriori interval for $\theta_8$,

$$(14) \qquad\qquad \theta_8 \in [-5.1, \ -1.8] \ .$$

The fact that most of the cities had less negatively-tending results than city 8 plays an important role in the empirical Bayes analysis. The Bayesian prior estimated from the other 40 cities says that $\theta_8$ is unlikely to be as negative as its own data by itself would indicate.

The empirical Bayes analysis implies that there is a lot of information in the other 40 cities' data for estimating $\theta_8$, as a matter of fact, just about as much as in city 8's own data. This kind of 'other' information does not have a clear Fisherian interpretation. The whole empirical Bayes analysis is heavily Bayesian, as if we had begun with a genuinely informative prior for $\theta_8$ and yet it still has some claims to frequentist objectivity.

Perhaps we are verging here on a new compromise between Bayesian and frequentist methods, one that is fundamentally different from Fisher's proposals. If so, the 21st Century could look a lot less Fisherian, at least for problems with parallel structure like the ulcer data. Right now there aren't many such problems. This could change quickly if the statistics community became more confident about analyzing empirical Bayes problems. There weren't many factorial design problems before Fisher provided an effective methodology for handling them. Scientists tend to bring us the problems we can solve. The current attention to meta- analysis and hierarchical models certainly suggests a growing interest in the empirical Bayes kind of situation.

## 11. The Statistical Triangle

The development of modern statistical theory has been a three-sided tug of war between the Bayesian, frequentist, and Fisherian viewpoints. What I have been trying to do with my examples is apportion the influence of the three philosophies on several topics of current interest: standard error estimation, approximate confidence intervals, conditional inference, objective Bayes theories and fiducial inference, model selection, and empirical Bayes techniques.

Figure 8, the statistical triangle, does this more concisely. It uses barycentric coordinates to indicate the influence of Bayesian, frequentist, and Fisherian thinking upon a variety of active research areas. The Fisherian pole of the triangle is located between the Bayesian and frequentist poles, as in Figure 1, but here I have allocated Fisherian philosophy its own dimension to take account of its distinctive operational features: reduction to 'obvious' types, the plug-in principle, an emphasis on inferential correctness, the direct interpretation of likelihoods, and the use of automatic computational algorithms.

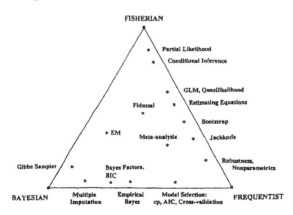

Figure 8. A barycentric picture of modern statistical research, showing the relative influence of the Bayesian, frequentist, and Fisherian philosophies upon various topics of current interest.

Of course, a picture like this cannot be more than roughly accurate, even if one accepts the author's prejudices, but many of the locations are difficult to argue with. I had no trouble placing conditional inference and partial likelihood near the Fisherian pole, robustness at the frequentist pole, and multiple imputation near the Bayesian pole. Empirical Bayes is clearly a mixture of Bayesian and frequentist ideas. Bootstrap methods combine the convenience of the plug-in principle with a strong frequentist desire for accurate operating characteristics, particularly for approximate confidence intervals, while the jackknife's development has been more purely frequentistic.

Some of the other locations in Figure 8 are more problematical. Fisher provided the original idea behind the EM algorithm, and in fact the self-consistency of maximum likelihood estimation (when missing data is filled in by the statistician) is a classic Fisherian correctness argument. On the other hand EM's modern development has had a strong Bayesian component, seen more clearly in the related topic of Gibbs sampling. Similarly, Fisher's method for combining independent $p$-values is an early form of meta-analysis, but the subject's recent growth has been strongly frequentist. The trouble here is that Fisher wasn't always a Fisherian, so it is easy to confuse parentage with development.

The most difficult and embarrassing case concerns what I have been calling 'objective Bayes' methods, among which I included fiducial inference. One definition of frequentism is the desire to do well, or at least not to do poorly, against every possible prior distribution. The Bayesian spirit, as epitomized by Savage and DeFinetti, is to do very well against one prior distribution, presumably the right one.

There have been a variety of objective Bayes compromises between these two poles. Working near the frequentist end of the spectrum, Welch and Peers showed how to calculate priors whose a posteriori credibility intervals coincide closely with standard confidence intervals. Jeffreys' work, which has led to vigorous modern development of Bayesian model selection, is less frequentistic. In a bivariate normal situation Jeffreys would recommend the same prior distribution for estimating the correlation coefficient or for the ratio of expectations, while the Welch-Peers theory would use two different priors in order to separately match each of the frequentist solutions.

Nevertheless Jeffreys' Bayesianism has an undeniable objectivist flavor. Erich Lehmann (personal communication) had this to say: "If one separates the two Bayesian concepts [Savage-DeFinetti and Jeffreys] and puts only the subjective version in your Bayesian corner, it seems to me that something interesting happens: the Jeffreys' concept moves to the right and winds up much closer to the frequency corner than to the Bayesian one. For example you contrasted Bayes as optimistic and risk-taking with frequentist as pessimistic and playing it safe. On both of these scales Jeffreys is much closer to the frequentist end of the spectrum. In fact, the concept of uninformative prior is philosophically close to Wald's least favorable distribution, and the two often coincide".

Lehmann's advice is followed a bit in Figure 8, where the Bayesian model selection (BIC) point, a direct legacy of Jeffreys' work, has been moved a little ways towards the frequentist pole. However, I have located fiducial inference, Fisher's form of objective Bayesianism, near the center of the triangle. There isn't much work in that area right now but there is a lot of demand coming from all three directions.

The point of my examples, and the main point of this talk, was to show that Fisherian statistics is not a dead language, and that it continues to inspire new research. I think this is clear in Figure 8, even allowing for its inaccuracies. But Fisher's language is not the only language in town, and it is not even the dominant language of our research journals. That prize would have to go to a rather casual frequentism, not usually as hard-edged as pure decision theory these days. We might ask what Figure 8 will look like 20 or 30 years from now, and whether there will be many points of active research interest lying near the Fisherian pole of the triangle.

## 12. R.A. Fisher in the 21st Century

Most talks about the future are really about the present, and this one has certainly been no exception. But here at the end of the talk, and nearly at the end of the 20th Century, we can peek cautiously ahead and speculate at least a little bit about the future of Fisherian statistics.

Of course Fisher's fundamental discoveries like sufficiency, Fisher information, the asymptotic efficiency of the MLE, experimental design, and randomization inference are not going to disappear. They might become less visible though. Right now we use those ideas almost exactly as Fisher coined them, but modern computing equipment could change that.

For example, maximum likelihood estimates can be badly biased in certain situations involving a great many nuisance parameters (as in the Neyman-Scott paradox.) A computer-modified version of the MLE that was less biased could become the default estimator of choice in applied problems. REML estimation of variance components offers a current example. Likewise, with the universal spread of high-powered computers statisticians might automatically use some form of the more accurate confidence intervals I mentioned earlier instead of the standard intervals.

Changes like these would conceal Fishers' influence, but not really diminish it. There are a couple of good reasons though that one might expect more dramatic changes in the statistical world, the first of these being the miraculous improvement in our computational equipment, by orders of magnitude every decade. Equipment is destiny in science, and statistics is no exception to that rule. Secondly, statisticians are being asked to solve bigger, harder, more complicated problems, under such names as pattern recognition, DNA screening, neural networks, imaging, and machine learning. New problems have always evoked new solutions in statistics, but this time the solutions might have to be quite radical ones.

Almost by definition it's hard to predict radical change, but I thought I would finish with a few speculative possibilities about a statistical future that might, or might not, be a good deal less Fisherian.

## 12.1 A Bayesian world

In 1974 Dennis Lindley predicted that the 21st Century would be Bayesian. (I notice that his recent *Statistical Science* interview now predicts the year 2020.) He could be right. Bayesian methods are attractive for complicated problems like the ones just mentioned, but unless the scientific world changes the way it thinks I can't imagine subjective Bayes methods taking over. What I called objective Bayes, the use of neutral or uninformative priors, seems a lot more promising, and is certainly in the air these days.

A successful objective Bayes theory would have to provide good frequentist properties in familiar situations, for instance reasonable coverage probabilities for whatever replaces confidence intervals. Such a Bayesian world might not seem much different than the current situation except for more straightforward analyses of complicated problems like multiple comparisons. One can imagine the statistician of the year 2020 hunched over his or her super-computer terminal, trying to get Proc Prior to run successfully, and we can only wish that future colleague 'good luck'.

## 12.2 Nonparametrics

As part of our Fisherian legacy we tend to overuse simple parametric models like the normal. A nonparametric world, where parametric models were a last resort instead of the first, would favor the frequentist vertex of the triangle picture.

## A new synthesis

The postwar years and especially the last few decades have been more notable for methodological progress than the development of fundamental new ideas in the theory of statistical inference. This doesn't mean that such developments are finished forever. Fisher's work came out of the blue in the 1920's, and maybe our field is due for another bolt of lightening.

It's easy for us to imagine that Fisher, Neyman, and the others were lucky to live in a time when all the good ideas hadn't been plucked from the trees. In fact, we are the ones living in the golden age of statistics — the time when computation has become fast and easy. In this sense we are overdue for a new statistical paradigm, to consolidate the methodological gains of the postwar period. The rubble is building up again, to use Joan Fisher Box's simile, and we could badly use a new Fisher to put our world in order.

My actual guess is that the old Fisher will have a very good 21st Century. The world of applied statistics seems to need an effective compromise between Bayesian and frequentist ideas, and right now there is no substitute in sight for the Fisherian synthesis. Moreover, Fisher's theories are well-suited to life in the computer age. Fisher seemed naturally to think in algorithmic terms. Maximum likelihood estimates, the standard intervals, ANOVA tables, permutation tests are all expressed algorithmically and are easy to extend with modern computation.

Let me say finally that Fisher was a genius of the first rank, who has a solid claim to being the most important applied mathematician of the 20th Century. His work has a unique quality of daring mathematical synthesis combined with the utmost practicality. The stamp of his work is very much upon our field and shows no sign of fading. It is the stamp of a great thinker, and statistics, and science in general, is much in his debt.

## References

### Section 1

[1] Savage, L.J.H. (1976), On Rereading R.A. Fisher. *Annals Stat*, **4**, 481–500. [Savage says that Fisher's work greatly influenced his seminal book on subjective Bayesianism. Fisher's great ideas are examined lovingly here, but not uncritically.]

[2] Yates, F. and Mather K. (1971) Ronald Aylmer Fisher, in: *Collected Papers of R.A. Fisher*, volume 1, 23–52. Univ. of Adelaide Press, ed. K. Mather. [Reprinted from a 1963 RSS memoir. Non-technical assessment of Fisher's ideas, personality, and attitudes toward science.]

[3] Box, J.F. (1978) *The Life of a Scientist*, Wiley, N.Y. [Both a personal and an intellectual biography by Fisher's daughter, a scientist in her own right and also an historian of science. Some unforgettable vignettes of precocious mathematical genius mixed with a difficulty in ordinary human interaction. The sparrow quote in Section 4 is put in context on page 130.]

### Section 2

[4] Fisher, R.A. (1925), Theory of Statistical Estimation. *Pro. Cambridge Phil. Soc.*, **22**, 200–225. [Reprinted in the Mather collection, and also in the 1950 Wiley Fisher Collection Contributions to Mathematical Statistics. My choice for the most important single paper

in statistical theory. A competitor might be Fisher's 1922 Philosophical Society paper, but as Fisher himself points out in the Wiley collection, the 1925 paper is more compact and businesslike than was possible in 1922 , and more sophisticated as well.]

[5] Efron B. (1995), The Statistical Century. *RSS News*, **22**, No. 5, 1–2. [Mostly about the post-war boom in statistical methodology, uses a different statistical triangle than Figure 8.]

*Section 3*

[6] Fisher, R.A. (1934), Two New Properties of Mathematical Likelihood. *Proc Royal Soc.* A, **144**, 285–307. [Concerns two situations when fully efficient estimation is possible in finite samples: One-parameter exponential families, where the MLE is a sufficient statistic, and location-scale families, where there are exhaustive ancillary statistics. Reprinted in the Mather and the Wiley Collections.]

*Section 4*

[7] Efron, B. (1978), Controversies in the Foundations of Statistics. *Amer. Math. Month.*, **85**, 231–246. [The Bayes-Frequentist-Fisherian argument in terms of what kinds of averages should the statistician take. Includes Fisher's famous circle example of ancillarity.]

[8] Efron, B. (1982), Maximum Likelihood and Decision Theory. *Annals Stat*, **10**, 340–356. [Examines five questions concerning maximum likelihood estimation: what kind of theory is it? How is it used in practice? How does it look from a frequentistic decision- theory point of view? What are its principal virtues and defects? What improvements have been suggested by decision theory?]

[9] Cifarelli, D. and Regazzini, E., (1996), DeFinettis contribution to probability and statistics. *Statistical Science*, **11**, 253–282. [The sec ond half of the quote in my Section 4, their Section 3.2.2., goes on to criticize the Neyman-Pearson school. DeFinetti is less kind to Fisher in the discussion following Savage's article.]

*Sections 5 and 6*

[10] DiCiccio, T. and Efron, B. (1991), Bootstrap Confidence Intervals. *Statistical Science*, **11**, 189–228. [Presents and discusses the CD4 data of Figure 2. The bootstrap confidence limits in Table 1 were obtained by the BCa method.]

*Section 7*

[11] Reid, N. (1995), The Roles of Conditioning in Inference. *Statistical Science*, **10**, 138–157. [A survey of the p* formula, what I called the magic formula following Ghosh's terminology, and many other topics in conditional inference; see also the discussion on pages 173-199, in

particular McCullagh's commentary. Gives an extensive bibliography.]

[12] Efron, B. and Hinkley, D. (1978), Assessing the Accuracy of the Maximum Likelihood Estimator: Observed versus Expected Fisher Information. *Biometrika*, **65**, 457–487. [Concerns ancillarity, approximate ancillarity, and the assessment of accuracy for a MLE.]

## *Section 8*

[13] Efron, B. (1993), Bayes and Likelihood Calculations from Confidence Intervals. *Biometrika*, **80**, 3–26. [Shows how approximate confidence intervals can be used to get good approximate confidence densities, even in complicated problems with a great many nuisance parameters.]

## *Section 9*

[14] Efron, B. and Gong, G. (1983), A Leisurely Look at the Bootstrap, the Jackknife, and Cross-validation. *The American Statistician*, **37**, 36–48. [The chronic hepatitis example is discussed in Section 10 of this bootstrap-jackknife survey article.]

[15] O'Hagan, A. (1995), Fractional Bayes Factors for Model Comparison. *Journal of Royal Statistical Society B*, **57**, 99–138. [This paper and the ensuing discussion, occasionally rather heated, give a nice sense of Bayesian model selection in the Jeffreys tradition.]

[16] Kass, R. and Raftery, A. (1995), Bayes Factors. *Journal of American Statistical Society*, **90**, 773–795. [This review of Bayesian model selection features five specific applications and an enormous bibliography.]

## *Section 10*

[17] Efron, B. (1996), Empirical Bayes Methods for Combining Likelihoods. *Journal of American Statistical Association*, **91**, 538–565.

## *Section 11*

[18] Kass, R. and Wasserman, L. (1996), The Selection of Prior Distributions by Formal Rules. *Journal of American Statistical Association*, **91**, 1343-1370. [Begins Subjectivism has become the dominant philosophical tradition for Bayesian inference. Yet in practice, most Bayesian analyses are performed with so-called noninformative priors]

## *Section 12*

[19] Lindley, D.V. (1974) The Future of Statistics — A Bayesian 21st Century, in: *Proceedings of the Conference on Directions for Mathematical Statistics*, University College, London.

[20] Smith, A. (1995), A Conversation with Dennis Lindley. *Statistical Science*, **10**, 305–319. [A nice view of Bayesians and Bayesianism. The 2020 prediction is attributed to DeFinetti.]

# Students Can Help Improve College Teaching: A Review and an Agenda for the Statistics Profession

John D. Emerson

Mathematics and Computer Science
Middlebury College
Middlebury, Vermont

Frederick Mosteller and Cleo Youtz

Department of Statistics
Harvard University
Cambridge, Massachusetts

**Abstract.** This report addresses two primary questions about common uses of student ratings. First, are student ratings of instruction valid measures of teaching effectiveness for evaluating college instructors? Second, are student ratings of instruction useful in improving the effectiveness of college teaching? In addressing these questions, this overview rests on detailed reviews of over 120 articles and books found by a computer search of the literature published between 1980 and 1997. The findings surveyed lead naturally to proposals for future research. Research is needed about rating instruments and other forms of student input that can support efforts to improve instruction. Research is also needed about how teachers and consultants work together to improve teaching. The types of needed empirical research are varied; they may range from carefully documented descriptive reports through controlled comparative experiments to research synthesis. The statistics profession should expand its role in aiding this important research effort.

## 1. Introduction

More than 14 million students enroll at over 3500 colleges and universities in the United States each year. Nearly all U.S. colleges and universities ask students to evaluate their instructors and courses. Both steady growth in the rate of college attendance and the fiscal fact that higher education is an expensive enterprise increase the demand for good teaching. This review explores the role student ratings can have in attaining this goal. It suggests ways in which the enormous mounds of data, already being assembled at considerable cost, can bring greater benefits to colleges, faculty members, and students. It also encourages statisticians to help design research to improve college teaching.

This review focuses on rating by students of college-level instruction. It addresses two primary questions about the common uses of student ratings.

1. Are student ratings of instruction reliable and valid measures of teaching effectiveness for evaluating college instructors?
2. Are student ratings of instruction useful in improving the effectiveness of college teaching?

The question about validity of ratings leads naturally to a possible relation between student ratings of teaching and measures of student learning. Have the students in classes that rate their teachers highly learned more on the average than other students who are taught by teachers with lower ratings?

The second question focuses on teaching improvement. If student ratings are to help instructors improve their teaching:

- results of the ratings need to be made known to the teacher in a form that is accurate and interpretable;
- the teacher needs to know how to apply the information in ways that improve student learning.

Some evidence about ratings and teaching improvement addresses how teachers can use feedback from student ratings to improve their teaching, but more information is needed, as we discuss in Section 4.

The two primary questions have been studied for more than seven decades. Section 2 of this report reviews briefly some key developments in this research program through the 1970s. The pace of research about student ratings has continued at a high level in the 1980s and 1990s. We estimate that by 1998 the number of articles and books on student ratings exceeded 2000.

Sections 3 and 4 review some of the research appearing between 1980 and 1997. We interpret the research on how ratings contribute to evaluating and improving teaching. Section 5 offers some practical guidance for using student ratings to improve instruction. Finally, in Section 6 we identify ways for statisticians to aid in empirical studies to improve college teaching.

We address this report to the statistics profession. Although the research literature on rating by students is vast, such publications in statistics journals are rare. The fields of education and psychology have led the research efforts, and we believe that the statistics profession also has much to contribute. Our discipline should help the world get additional payoff from higher education's investment.

## 2. Historical notes on student ratings

Student rating of university and college teaching made its first appearance in the 1920s. Students at Harvard University published student assessments of their courses and teachers in their student "Confidential Guide to Courses" in the early 1920s (Canelos, 1985, p. 187). Since 1925, students at the University of Washington have been completing questionnaires about their professors and courses. Cook (1989, who cites Darr, 1977) reports that the first published form for gathering student ratings of college teaching, the Purdue Rating Scale for Instructors, was released in 1926. Using such new data, researchers began empirical studies of student ratings of college teaching.

In an historical review, McKeachie (1990) summarizes empirical research programs on college teaching, which became well-established in the 1920s and 1930s. For example, research on class size showed that smaller classes favor better retention of learning and superior ability to use what is learned in problem solving. Other research showed that discussion teaching methods compare favorably to lecture teaching methods with respect to long-term retention and higher-level learning outcomes (as opposed to immediate factual recall). McKeachie reports that well-designed experiments addressed each of these areas. He further discusses a series of reports by Herman Remmers, the first completed in 1927, examining the validity and reliability of the Purdue Rating Scale for Instructors.

McKeachie concludes that questions about the validity and reliability of student ratings were "confronted and answered with a good deal of clarity" by Remmers's work, and that subsequent research has reinforced and extended those findings. He adds that, "Student ratings are the best validated of all practical sources of relevant data [about teaching effectiveness]." (McKeachie 1990, p. 195). (McKeachie 1990, pp. 194-195). Marsh and Roche (1997, pp: 1189-1190) compare several sources of teaching evaluation, including self-ratings by teachers, ratings by former students, ratings by colleagues and administrators following classroom visitations, and ratings by trained external observers; their own review of the research literature led Marsh and Roche to remark about the general lack of empirical support for other indicators of effective teaching than student ratings.

In the 1960s, student pressures for faculty accountability and for course improvements led to a greater role for student ratings in the evaluation of college and university teachers. By the early 1970s, the use of student ratings of college teaching had become widespread; it also remained a subject of controversy because some faculty members and administrators doubted the ability of students to evaluate teaching. Costin, Greenough, and Menges (1971) published a comprehensive review of research findings about student ratings. They concluded that "students' ratings can provide reliable and valid information on the quality of courses and instruction," and that such information "... may aid the individual instructor in improving his teaching effectiveness." (p. 530).

Much research in the 1970s focused on the validity of ratings. One concern was the possible impact of grades on ratings. Research findings fueled the lively debates about student ratings because they were often contradictory. The decade of the 1980s saw the emergence of meta-analysis as an important tool for sorting out disparate research findings about student ratings. Peter Cohen's 1981 formal meta-analysis of research findings addresses these issues and indicates that empirical studies, when combined, lend much support for ratings validity. Cashin (1994) summarizes the conclusions in the major reviews of the ratings literature, beginning with the report by Costin, Greenough, and Menges (1971) and extending through several papers published in 1987.

In late 1997, a series of articles (d'Apollonia and Abrami, 1997; Greenwald, 1997; Greenwald and Gillmore, 1997; Marsh and Roche, 1997; McKeachie, 1997) in <u>American Psychologist</u> reviewed the status of the debates about student ratings and their uses. These articles reveal that not all questions about the validity of student ratings have been settled. For example, do higher grades given in courses with better student ratings indicate that students reward grading leniency? Or do they reflect that students who are better taught learn more and (appropriately) earn higher grades? The resolution of this puzzle awaits future research, but is not likely to change the primary finding of the research on the validity of ratings. We believe there is compelling evidence for the validity of student ratings of college instruction.

We began our literature review with a computer search of several library data bases of research articles and books in the fields of education and psychology. The data bases included ERIC (Educational Resources Information Center), Psychology Abstracts, and the Harvard University Libraries Catalog. These searches located over 460 review articles and general articles containing references to student ratings of college instruction. From the abstracts and reference lists in the articles, we identified and read over 120 articles and chapters bearing on college teaching. These materials form the backbone for this review article.

## 3. Student ratings and student learning

Student ratings of college instruction are positively associated with student academic achievement. An impressive body of primary research and several research syntheses support this finding, as we now describe.

### 3.1 Evidence from multisection studies

Persuasive evidence for the validity of student ratings comes from large courses with many sections where the teacher **has full charge** of the lectures and instruction. (These are **not** courses taught by a single lecturer with a teaching assistant meeting the section occasionally.) The students take common examinations with uniform grading. The basic unit of information about ratings of the instructor, the course, or self-reported learning is the average rating over all the students in a course section. Similarly, the unit of information about achievement is the average examination score for all students in the section.

The **multisection study design** enables researchers to analyze the average rating paired with average student achievement across the sections of the course taught by different instructors. A key question is whether teachers whose sections give higher ratings for teaching performance tend to have greater student achievement. In Section 3.2, we see that they do.

The multisection study uses the same instructional rating form for all students in all sections of a course. The course uses a common grading scheme to ensure unbiased assessment. The multisection design enables many factors to be the same or nearly the same in all course sections: the course text, the assigned readings and homework, and even the syllabus.

Random assignment of students to the sections would avoid the obvious dangers of student self-selection. Unfortunately, the random assignment of college and university students to their course sections (and meeting times) is often neither feasible nor politically acceptable. Initial systematic differences among sections are thus likely. A common pretest or other measure of student ability levels can, through adjustment, help provide comparability of outcomes across the sections.

Multisection studies evaluate teaching in sections from a **single course**; this feature limits the generalizability of a study of the validity of student ratings. When a large number of multisection studies of many courses in diverse subject areas at various institutions give reasonably congruent results, then we are more confident about drawing general conclusions. Still, evidence from large multisection courses may not apply to smaller courses or to more advanced courses, and there seems to be no way around this.

## 3.2 Findings from research synthesis

Peter A. Cohen (1980a, 1981, 1982, 2986, 1987) conducted pioneering research syntheses of multisection validity studies.

Cohen reviewed results of many multisection courses and used various degrees of stringency in admitting these courses into evidence. A set of 41 courses he regarded as "well-designed studies" met 7 criteria. These studies must

1. provide data from actual college and university classes;
2. analyze the course section (and its instructor) rather than individual students;
3. use a multisection format and a common achievement measure to assess learning;
4. provide data within the course from which a correlation between student ratings and student achievement can be calculated;
5. offer evidence of equivalence of student ability across course sections, either through random assignment of students to sections or through statistical adjustment for initial student differences;
6. require that teachers being rated have primary responsibility for instruction; and
7. reduce possible grade/rating bias by having students rate instructors before they obtain their final course grades.

|  | Instructor Ratings | Course Ratings | Self-Learning Ratings |
|---|---|---|---|
| Number Combined | 35 | 16 | 11 |
| Correlation | .45 | .49 | .39 |

Table 1. Background Data on Cohen Research Synthesis of 41 Multisection Studies of Correlation, r, Between Overall Student Ratings of the Instructor, the Course, and the Students' Self-Learning, Each with the Students' Performance on a Common Examination.

*Source*: Adapted from P. A. Cohen (1987), "A Critical Analysis and Reanalysis of the Multisection Validity Meta-Analysis," Paper presented at the Annual Meeting of the American Educational Research Association, Washington, D.C., April 20-24, 1987.

Table 1 shows the average correlation between the student ratings given for instructors and the students' performance on the common test as .45; between the overall course rating and the students' common test results as .49; and between the students' self-ratings of learning and performance on the common test as .39. We believe that these findings, derived from the systematic integration of data from 41 courses representing many academic areas, offer strong evidence that student ratings of college teaching are positively correlated with their academic achievement. This conclusion strengthens the claim that student ratings are valid measures of teaching effectiveness.

After the publication of Cohen's initial research synthesis (1981) on the validity of ratings, other investigators carried out and published their own syntheses (Dowell and Neal, 1982 and 1983; Abrami, 1984; and McCallum, 1984). Their estimates of correlation and their conclusions about the validity of student ratings varied considerably and they differed from those of Cohen.

Cohen's 1987 report responded to this state of affairs. He benefited from the work of his colleagues (a) by revising and updating his own earlier work, and (b) by presenting multiple analyses using increasingly stringent criteria for including primary research studies. Further analyses by Abrami, Cohen, and d'Apollonia (1988) probed the sources of differences among six published research syntheses. They attributed some differences to major variation in the implementation of meta-analysis (for example, varying selection criteria) and others to varying interpretations of the research findings. Their success in providing a detailed accounting for differences among these research syntheses left Cohen's original finding that student ratings are positively correlated with student learning more strongly established than before.

d'Apollonia and Abrami (1997, p. 1202) conducted a meta- analysis of 43 multisection validity studies with a total of 741 instructors. They report a mean correlation between student ratings of general instructional skill and student learning of 0.33. After adjustment for the reduction (attenuation) of the correlation coefficient by unreliability of the instruments, d'Appolonia and Abrami report a correlation of 0.47. They conclude that "student ratings of General Instructional Skill are valid measures of instructor-mediated learning in students." (p. 1202). Remmers's finding for the validity of ratings seven decades earlier is supported once again.

### 3.3 Dimensions of Teaching

The form used for gathering student ratings may be designed by persons in the school or chosen from among several standard forms. The forms often contain more than just a few general items rating the instructor, the course, and the amount learned. Marsh (1987, pp. 379-388) reproduces five examples of computer- scored forms used to collect student ratings of teaching effectiveness. The numbers of rating items range from 11 to 43. Some forms also provide space for additional questions, and all but one form invite written comments by students.

The rating items ask students to assess several aspects of "teaching," such as: Did the instructor organize the course well? Were lectures well-prepared and clearly presented? Was instructor feedback to students useful and timely? Each question addresses one or more dimensions of teaching. We distinguish between specific items – for example, "Was the instructor dynamic?," and overall rating items – for example, "Was the instruction better than that in most courses?" Some forms use both types for each of three targets: instructor, course, and student learning.

Several research studies have asked whether student ratings of specific items about instruction add substantially to the information gained for summative evaluation from overall ratings of the instructor and the course. Kenneth Feldman (1989, 1997) published a synthesis of the data used by Cohen (1987) for his expanded analysis emphasizing specific components of teaching and their relation to student achievement. Feldman uses Cohen's data base to examine 28 dimensions of instruction (Cohen used 8) plus three more "overall" dimensions (for teachers, lectures, and courses).

Feldman's analysis shows that the dimensions of instruction having highest correlation with student achievement are:

Teacher Preparation and Organization of the Course: $r=.57$;
Clarity and Understandableness: $r=.56$; and
Perceived Outcome or Impact of Instruction: $r=.46$.

The first of these dimensions is an important component of Cohen's "Structure" dimension; the second contributes to what Cohen labels "Skill"; and the third is Feldman's description of students' rating of their own learning.

Student responses to specific rating items can be useful to instructors who seek to improve their teaching, as we discuss in Section 4. But we conclude, based on an extensive debate in the literature, that responses to a large number of questions may not be needed to assist in judging the overall quality of instruction. Student responses to open-ended questions may also give specific information useful to instructors who are trying to improve their teaching. But these responses may pose a risk for summative evaluation because one or two strongly worded assessments can weigh too heavily in the evaluation process. In sum, the choice of the type of ratings items should reflect their intended use.

## 4. Student ratings and teaching improvement

Another major purpose of student rating of instruction is to provide information to instructors that is useful for improving their teaching. There has been less research about using student ratings for teaching improvement than about their use in evaluating instruction.

We distinguish between teaching improvement and course improvement even though these two goals are related. Teaching refers to the activities of the instructor designed to facilitate student learning. Here, course improvement refers to broad choices in course content and to the general structure of the course.

After courses are over and student grades are determined, instructors usually receive summaries of the students' ratings for each item on the rating form. Comparative information about ratings for other teachers helps faculty members interpret their own ratings (Marsh and Roche 1997, p. 1194).

Researchers have asked whether feeding back student ratings to instructors leads to better teaching. They also ask whether future students learn more from a teacher who has received information about ratings, but there is little research about this question.

## 4.1 Sources of evidence

Most evidence about the relation between student rating and teaching improvement derives from studies of the short-term effects of feedback from student ratings to instructors. In these studies, instructors who teach the sections of a large multisection course usually are randomly assigned to one of two groups. Instructors in the experimental group receive midterm feedback from student ratings and those in the control group do not. Feedback is sometimes augmented by a meeting with a consultant who has relevant expertise in interpreting collections of ratings. The outcomes measuring the improvement assessed in these studies rely on ratings and student performance at the end of the course.

Comparing two groups of instructors

The data gathered at the end of the course for instructors getting midterm feedback are compared with data from those in the control group to give evidence about short-term teaching improvement. The information used for comparison often includes: student ratings of overall effectiveness of instruction; student ratings of specific components of teaching; student ratings of their own learning; student attitudes about the subject matter; and objective measures of student achievement.

## 4.2 Chief findings about feedback

End-of-term differences between average ratings by the sections for instructors getting midterm feedback and those for instructors in the control group do suggest that the feedback of midterm ratings can be effective.

Much of the research reported here includes syntheses of collections of studies, sometimes meta-analyses. An illustrative report of a single primary study is informative. Overall and Marsh (1979) carried out a large experiment to see whether midterm feedback to the instructor produced better learning at the end of the term. They reported that empirical research by others had not consistently supported the value of such feedback. Their study used a computer programming course for social scientists with 30 sections, each with its own instructor. Based on pretest and other information, the sections of the course, averaging 31 students, were nearly equal in ability and preparation. Instruction time was 6 hours per week, split equally between lectures and class meetings providing individual attention to students.

Overall and Marsh randomly chose 6 instructors (handling 12 sections) to give midterm feedback from student ratings, and they withheld the midterm information from instructors in other sections. In addition to giving the rating feedback, Overall and Marsh met with the 6 feedback instructors and discussed the meanings of the ratings and how these evaluations might be used to improve the effectiveness of their teaching. Overall and Marsh encouraged instructors to formulate their own approaches to improving the teaching. To summarize the results, at the end of the term the sections that received the midterm feedback and advice: (a) received more favorable teaching ratings from the students; (b) had better final examination scores; and (c) had more favorable affective outcomes than the no-feedback sections.

Overall and Marsh attribute some of the systematic success of the feedback measured on a number of different scales (rating components) to the inclusion of discussions with an external consultant (themselves). They note that McKeachie and Lin (1975) had reported a parallel study and success.

Conclusions about the favorable effects of feedback follow largely from two syntheses of the research about midterm feedback to instructors.

A research synthesis by Cohen (1980b) uses 22 comparisons of the effectiveness of midterm feedback of student ratings. He reports that the end-of-term rating for the group given midterm feedback had an effect size of .38 (the gain, in raw standard deviation units). The second synthesis, by L'Hommedieu, Menges, and Brinko (1990), integrates results from 28 studies not used in Cohen's synthesis; they report an effect size of .34 for mid- course feedback of student ratings. Both syntheses use primary research articles that incorporate the design features outlined in Section 3. Both investigations also analyze study characteristics that might modify findings, such as whether treatments were randomly assigned and whether the institution was a doctorate-granting university.

Cohen (1980b) examines the impact of nine study characteristics on his findings by calculating the correlations of each study characteristic with a measure of the overall rating. Only one of these correlations is significant and substantial: The correlation between use of augmentation of feedback and a summary measure of the overall rating of teaching is 0.64. Cohen also concludes that the augmentation of feedback, most often through consultation with an expert facilitator, contributes substantially to the efficacy of supplying information to teachers about ratings.

Marsh and Roche (1997, p. 1194) report that student-ratings feedback and consultation are "an effective means to improve teaching effectiveness," and that they provide a useful basis for consultation with faculty members about teaching improvement. Teachers like many in other settings need coaches to help them develop their skills.

### 4.3 Strengths and weaknesses of feedback design

The midterm feedback design has the usual strengths associated with randomized, controlled experiments. It makes comparisons that should be relatively free of bias. Still, the design poses some difficulties for interpreting its findings.

A research synthesis by L'Hommedieu, Menges, and Brinko (1990) focuses on the potential impact of 40 variables that they believe might moderate the effect of feeding back student ratings. They cite as a threat to the validity of their findings "diffusion or imitation of treatment." They note that, in at least eight of the primary studies they examined, student ratings were already a matter of policy before the courses began. These studies thus do not compare feedback of student ratings to a no- feedback control; rather they compare the additional effect of new midterm feedback with the effect of all feedback received by the control group before the present course began. This diffusion of the treatment may.substantially reduce the measured beneficial effects of feedback from student ratings.

Similarly, in other situations where methodological weaknesses may be present, L'Hommedieu, Menges, and Brinko (1990) suggest that these threats act to reduce feedback benefits to faculty members. They "expect that improved research will document effects that are more substantial and robust than those shown so far." (p. 240).

In a brief review of feedback studies, Herbert Marsh (1987, Chapter 7) remarks that the long-term effect of feedback is more important than its short-term effect. He found only two studies that compared midterm feedback with a control group for a period of more than one semester. Both studies found only weak effects for feedback, and each had design limitations because the groups of teachers being compared may not have been equivalent at the outset.

Marsh acknowledges the difficulties inherent in carrying out long term follow-up studies of the effects of feeding back student ratings. It is usually not appropriate to withhold the results of student ratings from randomly selected instructors for several semesters. Marsh and Roche (1993) conducted an experimental study in Australia in which they did just that; we review their study in Section 6.

## 5. From ratings to teaching improvement: practical guidance

Feeding back rating information to faculty members before or at the middle of the course may offer special opportunities. Students should be told that their thoughtful responses may lead directly to beneficial changes during the present course; they should also be told that the teacher values their opinions and hopes to use them to improve the course.

### 5.1 A Role for Consulting

One important role for a consultant is to help faculty members focus their efforts on a few specific aspects of their teaching. These areas should be those for which responses can be tangible and immediate. A faculty member whose personal style is serious and formal may not be able to introduce humor overnight. But a faculty member can convey useful information and facilitate organization, for example, by placing a brief 3-or-4 line summary of the day's agenda on the chalkboard at the start of each class.

Robert C. Wilson, an experienced faculty developer, gathered advice from top-rated teachers about what they did in their teaching that led students to give them high ratings. Wilson used "summary comments by eight to ten high-rated faculty members as a basis for giving other faculty members ideas about how to improve the lowest rated aspects of their teaching" (Wilson 1990, p. 273). When a faculty member decided to target a particular aspect of teaching for improvement, Wilson (1986, p. 198) shared with the client teacher several specific ideas used by successful teachers. Wilson reported that the actual use of these ideas by teachers resulted in higher ratings from their students.

### 5.2 The minute paper

Most uses of student ratings are applied the next time the instructor gives the course, but teachers and students can benefit from improvement in courses being presented now. Wilson (1986) cited a physics professor who knew if his class understood him because he used a device he called a "minute paper." At the end of some classes the professor reserved 2 or 3 minutes for students to write down and submit queries, or mention areas they would like to know more about. Mosteller (1989) tried this in every lecture of a beginning applied statistics course, inviting the students to tell (a) the key point in the lecture, (b) the muddiest point in the lecture, and (c) what they would like to know more about (about half the students replied per lecture). He prepared supplementary handouts for the next lecture in response to these stimuli. Although the minute paper adds to the teacher,s preparation time, it gives students a daily invited opportunity to provide guidance in the form of immediate feedback to the instructor. The students liked the procedure.

## 5.3 Practical advice for improvement

We report practical advice about the use of student feedback for teaching improvement.

1. Use student comments, where available, as an aid to interpreting the ratings data, and use the ratings data to help place the comments in context. The two sources of evidence, if used separately, sometimes give quite different messages. Cashin (1990) remarks that the comments alone may give a negative impression, whereas the numerical ratings may show relatively high numbers. Supplemental written comments give insights about students' judgments and may point toward specific steps for improving teaching.

2. Use a trained instructional consultant to help in making practical interpretations of ratings. Class visitation by the consultant may aid in doing this. For information about ratings to be useful for improving teaching, an instructor needs to generate diagnostic interpretations from the ratings (Arreola and Aleamoni, 1990; Braskamp and Ory, 1994, Chapter 12).

3. Use other available resources to aid in interpreting ratings: videotape classes; have informal exploratory conversations with supportive colleagues (perhaps even from other disciplines); and exchange classroom visits with other instructors who are motivated to improve their teaching.

4. Get some ratings early in the term, even during the first few weeks, in order to make adjustments that might benefit present students (McKeachie, 1997; McKeachie and Kaplan, 1996).

5. Collect short written feedback from students routinely (Braskamp and Ory 1994).

6. Target one course at a time for teaching improvement (Cashin, 1990). Identifying and implementing specific strategies for improvement needs focused energy.

7. Examine differences among student ratings of the various types of courses that a faculty member has taught. Murray, Rushton, and

Paunonen (1990) found that these differences are often substantial. Thus, faculty members may want to initiate conversations in their departments about the allocation of teaching assignments.

For additional practical advice, we recommend the collection of articles in Teaching and Learning in the College Classroom, edited by Kenneth A. Feldman and Michael B. Paulson (1994). We also recommend the monograph by John A. Centra (1993), Reflective Faculty Evaluation. Enhancing Teaching and Determining Faculty Effectiveness.

## 6. Helping teachers improve: what comes next?

Research findings reveal a promise that student ratings can improve college teaching. A concerted research initiative should now be directed toward uses of feedback from students about teaching.

The effective use of feedback from students can benefit from further research in two areas:

1. Gathering useful input
2. Applying input to improve teaching.

Many articles give advice about these areas for future research. We summarize a few important points and add our own suggestions.

### 6.1 Gathering feedback to support good teaching

Rating forms with items and questions designed explicitly for teaching improvement may be better suited to that purpose than forms developed for rating teachers and courses. One veteran researcher in this field, Wilbert McKeachie, recommends a research program on how to elicit more useful input from students. He asks "what kinds of structure for ratings, and what balance of ratings and open-ended questions" (1997, p. 1223) would stimulate more thought. Research about ratings does not yet respond to McKeachie,s question.

Global rating items, although useful in summative evaluation, seem less useful than more specific items that address components of teaching which can change (McKeachie and Kaplan, 1996; McKeachie, 1997). Two examples of such items are: "The instructor uses illustrations and examples effectively in conveying newly introduced material," and "The instructor provides adequate feedback to students so that they know how well they are meeting the expectations of the course."

Schmelkin, Spencer, and Gellman (1997) report empirical findings that faculty members view the usefulness of student feedback differentially; they find feedback on their interaction with students most useful, and they also benefit from feedback on their grading practices.

Established rating systems

L'Hommedieu, Menges, and Brinko (1990) recommend that research on using ratings for improvement adopt existing ratings systems with tested reliability and validity. Timpson and Andrew (1997) report on the development and adoption of two new ratings instruments, Teaching Feedback and Subject Evaluation, at the University of Queensland in Australia. They further indicate that a third instrument, Approaches to Studying, can assist with the interpretation of student ratings. The new instruments, developed and tested in Australia, seem well-suited to future research about using ratings for instructional improvement.

New modes of input

McKeachie envisions a new research program on how teachers can encourage students to be "more sophisticated evaluators as well as ways that the experience of filling out the rating form can become more educational for students." (1997, p. 1223). McKeachie and Kaplan illustrate the kind of item that might encourage more reflection: "Think about the conditions where you have learned well. Describe them." (1996, p.6).

Student interpretations of rating items

The effective use of ratings information for teaching improvement requires: (a) that teachers have an accurate understanding of what the responses mean; and (b) that the advice from students, if acted upon, can lead to better instruction. Empirical research by Benz and Blatt (1996) reveals that the meaning of a rating item to students may differ for different students, and that students' interpretations of an item may differ from what faculty members assume. For example, students who agree with the item "The instructor prepared well for classes" may be referring either to the efficient use of class time, or to the effective use of materials such as handouts, overheads, and lecture notes. But Benz and Blatt report that the strongest theme in student responses to this item was that instructors "knew their subject matter." (p. 422). Thus Benz and Blatt's (1996) exploration of students' interpretations of rating items identifies an area for further research. Both teachers and consultants need to interpret correctly what students mean when they give rating information; we must learn to avoid treating this information too casually.

Our own experiences with introductory statistics courses reveal that some students say that it is important to use illustrative examples when presenting new material. Benz and Blatt report that students often mention "personal examples" when considering two rating items, "The subject matter was clearly presented by the instructor" and "The instructor put material across in an interesting way." Yet even when every lecture of an introductory statistics course incorporates at least one illustrative example with social relevance, students sometimes appear to be looking for something else. Benz and Blatt report that "story telling" and "stories and personal examples from real life" are frequently advocated by students. We wonder whether the word "examples" holds the same meaning for teachers of statistics (or mathematics) as it does for many college students? Empirical research about interpretation can provide information to assist teachers who want to use ratings for improving teaching.

### Other sources of input from students

Not all student input about teaching, whether quantitative or descriptive, needs to come from rating forms. The minute paper, described in Section 5 is one alternative. A second approach uses a neutral assistant, perhaps a student or a teaching consultant, to actively solicit information and advice from students while they are taking the course. Both approaches offer the advantage of providing input continuously throughout a course to encourage timely and beneficial changes. Empirical studies are needed to identify effective ways of implementing each approach.

### 6.2 Using consultation with feedback to improve teaching

Professional consultation enhances the benefits of rating feedback, but we know little about how to use consulting effectively. Marsh and Roche (1997, p. 1194) report that there is insufficient information about effective models for using feedback, although they themselves carried out a large empirical study of feedback and consultation (Marsh and Roche 1993). They want additional empirical work to focus on how (and in what form) consultation can lead to improved teaching.

### Models for instructional consultation

Kathleen Brinko (1990) describes and contrasts several different models for teaching consultancy. She analyzes tapes of the work of 10 teaching consultants, and finds that all these consultants used both a prescriptive approach (diagnose and suggest, like a physician) and a collaborative approach (interact and question, like a peer colleague). Thus she finds no empirical evidence that teaching consultants conform to specific practices which differentiate successful consultation from less effective work; see also Marsh and Roche (1993, p. 223). Likely, effective consultants can adapt their style of interaction to the needs of the client teacher.

Our interpretation of the empirical research literature on instructional consultation leads us to consider instead the potential targets of the consultation. Efforts to aid teaching improvement may address either:

- the individual teacher, in a personal and private way,

or

- the instructional unit, usually a department or academic program.

Robert C. Wilson was a consultant who used the first approach, as described in Section 5.1. He worked with his individual client teachers in a discreet and private way, using a style that is low key, sympathetic, and collegial. Nira Hativa is a consultant who targeted the instructional program of an entire department. Next we summarize and compare two research efforts which implemented and reported on these two contrasting approaches.

### An experiment using private feedback and consultation

Marsh and Roche (1993) study Robert C. Wilson's (1986) model for feeding back student ratings and offering assistance by a trained professional. Although they acknowledge the appeal of Wilson's consultation strategy, Marsh and Roche recognize weaknesses in Wilson's research design: It was nonexperimental; its study participants were volunteers whereas the controls were teachers who did not volunteer; and the reported gains for targeted teaching items could have potentially resulted from regression to the mean. Marsh and Roche use a randomized controlled study to assess the effectiveness of students' evaluations of teaching when augmented by Wilson's intervention.

Marsh and Roche (1993, pp. 222–223) identify two important deficits in earlier research about student ratings feedback and consultation. One is the failure to compare systematically midterm feedback and intervention with end-of-term feedback and intervention. The second is the failure to provide empirical evidence that some practices make consultation more successful than others. Marsh and Roche provide experimental evidence that begins to respond to these challenges. Their research, done in Australia, is able to use an experimental design because the formal collection of student ratings with a standardized form had not already been introduced at the University of Western Sydney, Macarthur. This circumstance helps avoid ethical questions about withholding ratings information from teachers assigned at random to the control group until the end of the two-semester study.

Table 2, panel A, outlines Marsh and Roche's study design and research goals. The study implements Wilson's use of strategies, previously identified by distinguished teachers as effective in their own teaching. Panel B of Table 2 enumerates the support that the consultant provided to teachers in the treatment groups.

A. Research Goals and Stud Characteristics
- aimed at assessing feedback and consultation intervention model of Robert C. Wilson (1986), used across the university
- compares mid-course and end-of-course use of feedback and consultation with no intervention (control)
- an experiment that uses stratified random assignment (using previous overall teacher ratings by students) to one of three groups:
  a. feedback/consultation in middle of semester 1
  b. feedback/consultation at end of semester 1
  c. no feedback/consultation prior to end of study year (control)
- two-semester intervention
- n = 83 faculty participants provide complete data for assessment

B. Support Provided by Consultant to Client Teacher
- individual feedback and consultation sessions (varies with treatment group)
- Australian version of Students' Evaluation of Educational Quality (SEEQ) forms assessing 8 dimensions of teaching
- completed SEEQ forms, computerized summary sheet for ratings, normative data about ratings, and a guide to interpreting the ratings
- assistance in developing "importance ratings" for each SEEQ dimension
- assistance in developing teaching strategies in areas selected by the instructor
- summary ratings using weights determined by a faculty member's own ratings of the importance of various dimensions of teaching
- teaching idea packets (based on concrete strategies used by award-winning teachers) relevant for each of two or three teaching dimensions targeted for improvement
- assistance in plans to adapt chosen strategies to the teaching by the client instructor of the course
- a summary letter and subsequent telephone calls to the client teacher as follow-up on the stated plans for implementing teaching strategies

C. Principal Findings by Marsh and Roche
- feedback from ratings coupled with Wilson's consultation technique are effective in improving university teaching
- gains in ratings are larger for the initially least effective teachers
- gains are largest for the specific areas each teacher targeted as the focus of the intervention

- effects of end-of-term feedback are stronger than those for midterm ratings; effects are measured at end of the subsequent term
- effect size of intervention using end-of-term feedback is
- 50 for overall teacher ratings

Table 2. An Experiment on the Effectiveness of an Individually Structured Feedback/Consultation Intervention Approach to Improving Instruction (Marsh and Roche, 1993)

Marsh and Roche's experiment gives evidence that individually-structured feedback and consultation leads to improved instruction in the semester following the feedback. It concludes that end-of-course feedback is more effective for improving instruction than mid-course feedback. Although the investigators note that gains for the initially least-effective teachers are greatest and that improvements in targeted areas are largest, they caution that regression to the mean is also a plausible explanation for these effects. Still, this experiment shows that Wilson's use of feedback and consultation for individual teachers improves college instruction. Panel C of Table 2 summarizes the findings.

Marsh and Roche's two-semester study of teachers who had not previously received systematic feedback on standardized rating forms avoids some of the limitations of the feedback design described in Section 4.3. That student ratings were not already a matter of policy before the study began helps to reduce the possibility of "diffusion of treatment." That the study extends through two full semesters helps answer the concern that teachers may not have the time needed to address identified weaknesses. Indeed, Marsh and Roche's findings indicate that mid-course feedback (with consultation) is less effective than end-of-course feedback, when effectiveness is measured near the end of the following semester.

An observational study targeting an entire department

Nira Hativa (1995) studies feedback and consultation that focuses on an entire academic department. She uses intervention by a consultant that is comprehensive in addressing the teaching program of the department with a long-range, multi-faceted, and sustained approach. Hativa's two-year intervention and data collection responds to some of the limitations of the more traditional midterm feedback study.

This research at the University of Tel Aviv focuses on the teaching program of the physics department. Table 3, panels A and B, provide an outline of the study, and panel C of Table 3 presents the chief findings reported by Hativa.

## A. Research Goals and Study Characteristics
- aimed at improving teaching of the entire 60-member physics department, but restricted to the single discipline
- uses mid-course and end-of-course ratings with feedback
- uses continuous student-based feedback on teaching
- observational study, with partial controls provided by other physics teachers, physics teaching assistants, and by the departments of chemistry and mathematics
- four-semester intervention
- between 26 and 29 undergraduate courses given each semester, with ratings gathered in each course
- 20 faculty participants received intervention

## B. Support Provided by Consultant to Client Teachers
- background data, including some ratings information from three preceding years, retrospective ratings by departmental alumni
- initial meeting with teacher to identify specific difficulties and potential strategies to address them
- advice on initial planning of the course and syllabus
- validated rating form with one global item and six instructor-attribute items
- written analysis and interpretation of ratings information
- observation of several classes by consultants; discussion based on class observation
- weekly relaying of input from two student class monitors
- evaluation session with instructor at end of term

## C. Principal Findings by Hativa
- 11 of 20 teachers who got aid showed apparent improvement in ratings
- 5 of 20 teachers showed very substantial improvement (from below 3.1 to above 3.9 on a five-point Likert scale)
- second-semester ratings higher than first-semester ratings
- for entire department there was minor improvement in first year and distinct improvement in post-study year (TAs did not show comparable improvements)
- ratings by all physics majors showed statistically significant improvement from first to second year in
  (a) instruction in general
  (b) instruction by faculty members in particular; and
  (c) faculty attitudes toward students (TAs showed no such increases and sometimes showed decreases)

- physics faculty rated considerably higher than the TAs
- survey of faculty attitudes toward importance of teaching in promotions and the need to improve instruction were more positive than those for faculty in chemistry and mathematics who did not undergo treatment

Table 3. An Empirical Study of the Effectiveness of Teaching Consultants in a Department-Wide Approach to Improving Instruction in Physics (Hativa, 1993)

Hativa offers one form of control by collecting data on all physics professors, irrespective of whether they were among the 20 instructors with whom Hativa communicated about teaching during the two years of the study. The student ratings of the graduate teaching assistants (TAs) provide a second control; she compares these ratings to those of the entire faculty in the physics department. In principle, they might explain gains (or declines) in rating scores from the first to the fourth semester of the two-year intervention. Finally, gathering information about attitudes for all teachers in chemistry and in mathematics enables comparison of the results for physics teachers, on whom Hativa's intervention focuses, with those from similar departments.

A further strength of Hativa's research design n is that it avoids volunteering: All physics instructors who taught undergraduates participated in many of the improvement activities. Some highly-rated teachers sought assistance from the consultant and received all elements of the support program listed in Table 3, panel B. Her outcome assessment uses multiple instruments and several comparisons.

Hativa concludes that a teaching consultant can use student feedback to help teachers make improvements, and that a comprehensive intervention program can bring improvements to the instructional program of an entire department.

Lessons from the two studies

The contrasts between the study by Marsh and Roche (1993) and that by Hativa (1995) are sharp. The former study is an experiment and the latter is an observational study. The first study adopts a confidential and private approach to working with individual teachers, and the second study involves the immersion of a consultant in the culture of a department. Marsh and Roche examine a program that targets teachers in a wide variety of disciplines across a university, whereas Hativa focuses on the instructional program of a particular department and, presumably, on aspects of teaching that may be unique to the needs of the discipline. The first study offers concrete "solutions" to teachers seeking improvement in areas they have identified as needing attention; the second study provides nearly continuous feedback in the form of responses and reactions to what the teacher has just done. Marsh and Roche measure success by evaluating the instruction by a targeted teacher in a particular course; so does Hativa, and she also measures success by evaluating responses to the instructional program of an entire department.

Programs for improving instruction that address entire departments or other academic units over two or more semesters seem promising. L'Hommedieu, Menges, and Brinko (1990, p. 239) stress the importance of research across subject areas, types of institutions, and levels of student experiences. Carefully documented reports of a variety of such "department-holistic" case studies, modeled after Hativa's project, can provide valuable information about this approach to better teaching. They would provide an information base useful to statisticians and other researchers in seeking generalizations and synthesis.

The department-wide, and thus more "public", approach by consultants should encourage conversations among colleagues about issues that arise. For example, Murray, Rushton, and Paunonen (1990) report that good teachers are rarely strong at all levels: introductory, advanced, laboratory, and seminars. Consultants should encourage departments to reflect on this finding, and to report on their deliberations and conclusions about what this "diversity of teaching strengths" implies for the instructional programs of the departments. A collection of such reports could help identify the direction for future research and aid other departments in thinking about their own best course of action.

Programs that work privately with individual teachers in a succession of courses could mirror and extend the research by Marsh and Roche. Again, experience with a variety of types of institutions and with various styles of consulting will be valuable.

## 6.3 Proposed research

We take special note of Peter Cohen's plea (Cohen 1990). He remarked that little is known about the actual practice of using student ratings to improve teaching. Cohen urges that "future research efforts focus on conditions and settings affecting the practice of using student ratings." (p. 130) Studies like those by Marsh and Roche, and by Hativa, establish a valuable beginning for this next research phase.

We recommend further research on gathering information from students that can strengthen efforts for instructional improvement, as described in Section 6.1. This research would:

- identify specific ratings items that support teaching improvement;
- investigate new ratings items that promote deeper reflection and responses from student raters;
- explore uses of open-ended questions;
- investigate the ranges of meanings students attach to their ratings responses;
- identify alternative vehicles for student feedback.

We also recommend research on the work of consultants who aid faculty members in translating feedback from students into improved teaching, as described in Section 6.2. This empirical research would:

- document what takes place in the feedback and consultation process
- identify the specific practices that characterize successful consultation
- explore strengths of consultation that targets entire departments or other instructional units
- distinguish strategies for change that can be effective while a course is underway from those that need a longer time frame
- investigate the use by a consultant and teacher of both ratings feedback and independent information sources (e.g., classroom videos or teaching portfolios).

We offer a few additional suggestions beyond those listed for the two main areas.

New teaching modalities and new technology

McKeachie (1997, p. 1223) draws attention to new teaching modalities and the use of new technologies in teaching. For example, in today's classroom each student may have a computer connected to the internet, with access to supporting software and to the World-Wide Web. Teaching in this environment is far different from teaching at the chalkboard. Of course such changes affect teaching in the statistics profession as much as in others.

Many in the profession would agree that there are pressing needs for teaching improvement in statistics courses- especially the introductory courses. Most universities and colleges teach these courses, and the enrollments in them are often substantial. Changes in the course delivery are rapid; recent advances in technology have led to the introduction of highly interactive and visual software packages for data analysis and statistics, and to course materials (even entire courses) on CD-ROM and on the World-Wide Web. Emerson and Mosteller (1998a, 1998b) and Miech, Nave, and Mosteller (1997) review work in this area. Feedback based on student ratings can help us learn how to use such technologies more effectively. New research is needed to teach us how to interpret, and respond to, these ratings as we use technology for instructional improvement.

## Choosing outcome measures

Student rating items, both global and specific, are natural outcome measures in empirical research. Measures of affect deserve attention, in part because they are likely important indicators of the long-term impact of teaching on choices students make–choices of courses, majors, fields of graduate or professional study, and careers. Measures of "deeper" learning, such as conceptual insight and capacity for creative problem solving, may also benefit from more attention.

## Contributions from the statistics profession

The challenges and rewards of translating vast quantities of student feedback into improved teaching are great. Teachers and students demand flexibility in their teaching and learning, thus imposing limitations on the choice of research designs and analyses. Imaginative research efforts are needed to respond to the challenge. We urge the statistics profession to join efforts to improve college teaching and learning.

### Acknowledgments

The preparation of this material was supported in part by a grant from the Andrew W. Mellon Foundation to the American Academy of Arts and Sciences in support of the Center for Evaluation of the Academy's project "Initiatives for Children."

We thank Jason Sachs, who called our attention to a number of articles on student ratings, and Edward Miech for pointing out the 1995 Annual Meeting Program of the American Educational Research Association. We appreciated receiving materials from the following: Susan A. Basow; Peter A. Cohen; Kenneth A. Feldman; Anthony G. Greenwald; Larry H. Ludlow; Wilbert J. McKeachie; Joseph M. Ryan and Paul D. Harrison; and Karin J. Spencer, Liora Pedhazur Schmelkin, and Estelle S. Gellman.

Marjorie Olson helped by summarizing articles, making suggestions, and preparing the manuscript. We thank the following colleagues who gave us advice on early versions of the manuscript: John H. Buehler, Howard Frazier, Mark Glickman, Howard Hiatt, Nathan Keyfitz, Edward Knox, Russell Leng, Richard Light, John McCardell, Martin McIntosh, Wilbert McKeachie, John McWilliams, Edward Miech, Lincoln Moses, Bill Nave, Carol Rifelj, George Saul, and Richard Wolfson.

### References

[1] Abrami, P.C. (1984), Using Meta-Analytic Techniques to Review the Instructional Evaluation Literature. *Postsecondary Education Newsletter*, **6**, 8.

[2] Abrami, P.C., Cohen, P.A., and d'Apollonia, S. (1988), Implementation Problems in Meta-Analysis. *Review of Educational Research*, **58**, 151–179.

[3] Arreola, R.A. and Aleamoni, L.M. (1990) Practical Decisions in Developing and Operating a Faculty Evaluation System, in: *Student Ratings of Instruction: Issues for Improving Practice*, eds. M. Theall and J. Franklin, San Francisco, Jossey-Bass, 37–55.

[4] Benz, C.R. and Blatt, S.J. (1996), Meanings Underlying Student Ratings of Faculty. *The Review of Higher Education*, **19**, 411–433.

[5] Braskamp, L.A. and Ory, J.C. (1994) *Assessing Faculty Work: Enhancing Individual and Institutional Performance*, San Francisco, Jossey-Bass.

[6] Brinko, K.T. (1990), Instructional Consultation with Feedback in Higher Education. *Journal of Higher Education*, **61**, 65–83.

[7] Canelos, J. (1985), Teaching and Course Evaluation Procedures: A Literature Review of Current Research. *Journal of Instructional Psychology*, **12**, 187–195.

[8] Cashin, W.E. (1990) Student Ratings of Teaching: Recommendations for Use, IDEA Paper No. **22**, Center for Faculty Evaluation and Development, Kansas State University, Manhattan, Kansas.

[9] Cashin, W.E. (1994) Student Ratings of Teaching: A Summary of the Research, in: *Teaching and Learning in the College Classroom*, eds. K.A. Feldman and M.B. Paulson, Ginn Press, Needham Heights, MA.

[10] Centra, J.A. (1993) *Reflective Faculty Evaluation. Enhancing Teaching and Determining Faculty Effectiveness*, San Francisco, Jossey-Bass.

[11] Cohen, P.A. (1980a) A Meta-Analysis of the Relationship between Student Ratinas of Instruction and Student Achievement, Doctoral Dissertation, The University of Michigan, Ann Arbor, MI.

[12] Cohen, P.A. (1980b), Effectiveness of Student-Rating Feedback for Improving College Instruction: A Meta-Analysis of Findings. *Research in Higher Education*, **13**, 321–341.

[13] Cohen, P.A. (1981), Student Ratings of Instruction and Student Achievement: A Meta-analysis of Multisection Validity Studies. *Review of Educational Research*, **51**, 281–309.

[14] Cohen, P.A. (1982), Validity of Student Ratings in Psychology Courses: A Research Synthesis. *Teaching of Psychology*, **9**, 78–82.

[15] Cohen, P.A. (1986) An Updated and Expanded Meta-Analysis of Multisection Student Rating Validity Studies, Paper presented at the Annual Meeting of the American Educational Research Association,

San Francisco, April 16–20, 1986, (ERIC ED 270 471).

[16] Cohen, P.A. (1987) A Critical Analysis and Reanalysis of the Multi-section Validity Meta-Analysis, Paper presented at the Annual Meeting of the American Educational Research Association, Washington, D.C., April 20–24, 1987, (ERIC ED 283 876).

[17] Cohen, P.A. (1990) Bringing Research into Practice, in: *Student Ratings of Instruction: Issues for Improving Practice*, eds. M. Theall and J. Franklin, San Francisco, Jossey-Bass, 123–132.

[18] Cook, S.S. (1989), Improving the Quality of Student Ratings of Instruction: A Look at Two Strategies. *Research in Higher Education*, **30**, 31–45.

[19] Costin, F., Greenough, W.T., and Menges, R.J. (1971), Student Ratings of College Teaching: Reliability, Validity, and Usefulness. *Review of Educational Research*, **41**, 511–535.

[20] Darr, R.F., Jr. (1977) Evaluation of College Teaching: State of the Art. 1977, Paper presented at the Ohio Academy of Science, Psychology Division, Columbus, OH, (ERIC ED 162 559).

[21] d'Apollonia, S. and Abrami, P.C. (1997), Navigating Student Ratings of Instruction. *American Psychologist*, **52**, 1198–1208.

[22] Dowell, D.A. and Neal, J.A. (1982), A selective Review of the Validity of Student Ratings of Teaching. *Journal of Higher Education*, **53**, 51–62.

[23] Dowell, D.A. and Neal, J.A. (1983), The Validity and Accuracy of Student Ratings of Instruction: A Reply to Peter A. Cohen. *Journal of Higher Education*, **54**, 459–463.

[24] Emerson, J.D. and Mosteller, F. (1998a) Interactive Multimedia in College Teaching, Part I: A Ten-Year Review of Reviews, in: *Educational Media and Technology Yearbook 1998*, eds. R.M. Branch and M.A. Fitzgerald, Englewood, CO: Libraries Unlimited, Inc., **23**, 43–58.

[25] Emerson, J.D. and Mosteller, F. (1998b) Interactive Multimedia in College Teaching, Part II: Lessons from Research in the Sciences, in: *Educational Media and Technology Yearbook 1998*, eds. R.M. Branch and M.A. Fitzgerald, Englewood, CO: Libraries Unlimited, Inc., **23**, 59–75.

[26] Feldman, K.A. (1989), The Association Between Student Ratings of Specific Instructional Dimensions and Student Achievement: Refining and Extending the Synthesis of Data from Multisection Validity Studies. *Research in Higher Education*, **30**, 583–645.

[27] Feldman, K.A. (1997) Identifying Exemplary Teachers and Teaching: Evidence from Student Ratings, in: *Effective Teaching in Higher Ed-*

*ucation: Research and Practice*, eds. R.P. Perry and J.C. Smart, New York, Agathon Press, 368–395.

[28] Feldman, K.A. and Paulson, M.B. (eds) (1994) *Teaching and Learning in the College Classroom* Needham Heights, MA, Ginn Press.

[29] Greenwald, A.G. (1997), Validity Concerns and Usefulness of Student Ratings of Instruction. *American Psychologist*, **52**, 1182–1186.

[30] Greenwald, A.G. and Gillmore, G.M. (1997), Grading Leniency Is a Removable Contaminant of Student Ratings. *American Psychologist*, **52**, 1209–1217.

[31] Hativa, N. (1995), The Department-Wide Approach to Improving Faculty Instruction in Higher Education; A Qualitative Evaluation. *Research in Higher Education*, **36**, 377–413.

[32] L'Bommedieu, R., Menges, R.J., and Brinko, K. T. (1990), Methodological Explanations for the Modest Effects of Feedback from Student Ratings. *Journal of Educational Psychology*, **82**, 232–241.

[33] Marsh, H.W. (1987), Students' Evaluations of University Teaching: Research Findings, Methodological Issues, and Directions for Future Research. *International Journal of Educational Research*, **11**, 253–388 (entire issue).

[34] Marsh, H.W. and Roche, L.A. (1993), The Use of Students' Evaluations and an Individually Structured Intervention to Enhance University Teaching Effectiveness. *American Educational Research Journal*, **30**, 217–251.

[35] Marsh, H.W. and Roche, L.A. (1997), Making Students' Evaluations of Teaching Effectiveness Effective: The Critical Issues of Validity, Bias, and Utility. *American Psychologist*, **52**, 1187–1197.

[36] McCallum, L.W. (1984), A Meta-Analysis of Course Evaluation Data and Its Use in the Tenure Decision. *Research in Higher Education*, **21**, 150–158.

[37] McKeachie, W. J. (1990), Research on College Teaching. The Historical Background. *Journal of Educational Psychology*, **82**, 189–200.

[38] McKeachie, W.J. (1997), Student Ratings: The Validity of Use. *American Psychologist*, **52**, 1218–1225.

[39] McKeachie, W.J. and Kaplan, M. (1996), Persistent Problems in Evaluating College Teaching. *AAHE Bulletin*, **February 1996**, 5–8.

[40] McKeachie, W.J. and Lin, Y.G. (1975) *Study I: Using Student Ratings to Improve Teaching*, University of Michigan, Ann Arbor, MI, (ERIC ED 104 284).

[41] Miech, E.J., Nave, B., and Mosteller, F. (1997) On CALL: A Review of Computer-Assisted Language Learning in U.S. Colleges and Universities, in: *Educational Media and Technology Yearbook 1997*, eds.

R.M. Branch and B. B. Minor, Englewood, CO, Libraries Unlimited, Inc., **22**, 61–84.

[42] Mosteller, F. (1989), The 'Muddiest Point in the Lecture' as a Feedback Device On Teaching and Learning, 3. *The Journal of the Harvard Danforth Center for Teaching and Learning*, **April, 1989**, 10–21.

[43] Murray, H.G., Rushton, J.P., and Paunonen. S.V. (1990), Teacher Personality Traits and Student Instructional Ratings in Six Types of University Courses. *Journal of Educational Psychology*, **82**, 250–261.

[44] Overall, J.U. and Marsh, H.W. (1979), Midterm Feedback from Students: Its Relationship to Instructional Improvement and Students' Cognitive and Affective Outcomes. *Journal of Educational Psychology*, **71**, 856–865.

[45] Schmelkin, L.P, Spencer, K.J., and Gellman, E. S. (1997), Faculty Perspectives on Course and Teacher Evaluations. *Research in Higher Education*, **38**, 575–582.

[46] Timpson, W.W. and Andrew, D. (1997), Rethinking Instruments for Change at the University of Queensland. *Studies in Higher Education*, **22**, 55–65.

[47] Wilson, R.C. (1986), Improving Faculty Teaching: Effective Use of Student Evaluations and Consultants. *Journal of Higher Education*, **57**, 196–211.

[48] Wilson, R.C. (1990), Commentary: The Education of a Faculty Developer. *Journal of Educational Psychology*, **82**, 272–274.

# The Classical Extreme Value Model: Mathematical Results Versus Statistical Inference

Janos Galambos

Department of Mathematics
Temple University
Philadelphia, Pennsylvania

**Abstract.** The distribution of quantities associated with several practical problems such as high winds, life length, equipment failure due to fatigue, floods and others can be approximated by the classical extreme value model. The major conclusion of the classical extreme value model is that, apart from location and scale, a single one parameter family of distributions $H_c(z) = \exp\{-(1 + cz)^{-\frac{1}{c}}\}$, $1 + cz > 0$, approximates the linearly normalized maximum. The estimation of $c$ from data is crucial: if we decide that $c < 0$ we then have a finite upper bound for the random quantity that we approximate by $H_c(z)$. This can have serious scientific consequences. In the paper, apart from describing the classical extreme value model in some detail, the currently available estimations of $c$ are described. None of the three major estimation techniques is fully developed, hence the most important task for the next century is to further develop these methods. It is pointed out that , based on an earlier finding of Galambos and Macri, the threshold method is not to be used in its present form since more frequently than not it leads to finite upper bounds when the bound is definitely infinity.

## 1. Introduction

Several problems concerning the prevention of major damages from natural disasters such as floods and high winds, the prediction of the age of the longest living individual in a large society, the frequency of large claims at an insurance company, and the time to first failure of simple structures such as parallel or series systems (we shall return to general coherent systems in a subsequent section) require the study of the following statistical model. Let $Z = Z_n$ be a random quantity that can be expressed as the maximum (or minimum) of further random variables $X_1, X_2, \ldots X_n$. Assume that $n$ is sufficiently large and thus an asymptotic extreme value model well describes the distributional properties of $Z_n$. There are several extreme value models in the literature (see Chapter 3 in Galambos (1987)) in which the asymptotic distribution of the maximum of $n$ random variables, when linearly normalized, is the same as if the underlying random variables were independent and identically distributed. In the major part of the present paper we deal with problem, covering all those listed in the opening sentence above, in which the variables $X_j$, determining $Z_n$, satisfy the conditions of one of the cited models, entailing that, from the point of view of $Z_n$, it becomes irrelevant whether we use an exact model for the $X_j$ or if we assume that the $X_j$ are independent and identically distributed. We therefore assume that the observable quantity $Z_n = \max(X_1, X_2, \ldots, X_n)$ with $n$ large, and where the $X_j$ are distributed with the common distribution function $F(x)$. Furthermore, there are constant sequences $a_n$ and $b_n > 0$ suchthat $\lim F^n(a_n + b_n z) = H(z)$ exists and is nondegenerate and $\lim P(Z_n \leq a_n + b_n z) = H(z)$; in both limits $n \to +\infty$. These combined assumptions will be referred to as having a classical extreme value model for $Z_n$ and $H(z)$ is called a classical extreme value distribution function. Before proceeding let us remark that when $Z_n$ is in fact a minimum rather than maximum, neither the underlying mathematics nor the recommended methods for statistical inference would require a significant change, since

$$\max(X_1, X_2, \ldots, X_n) = -\min(-X_1, -X_2, \ldots, -X_n).$$

We therefore deal with the case of maximum only.

Now, it is known (see Galambos (1987), Chapter 2) that in a classical model for the maximum, apart from location and scale, $H(z)$ is a one-parameter family of distribution:

(1)         $H(z) = H_c(z) = \exp\left\{-(1 + cz)^{-1/c}\right\}, 1 + cz > 0,$

where $c = 0$ means the limit of $H_c(z)$ as $c \to 0$ with $c \neq 0$. We thus have $H_0(z) = \exp(-e^{-z})$, all $z$. The role of $c$ is crucial: the domain at (1) implies that

(2)    $Z = Z_n$ is bounded for $c < 0$ and unbounded if $c > 0$ or $c = 0$.

Further distinctions among the $H_c(z)$ include that the positive part of $Z$ has finite moments of all order if $c \leq 0$ but only a set of moments of the positive part of $Z_n$ is finite for $c > 0$. More details on the mathematical results for the classical model are given in Section 2.

In the mathematical theory, the $X_j$ and their distribution $F$ are used to determine the distribution $H_c$ of $Z = Z_n$. In practice the process is reversed: observations are taken on $Z$ only, and the $X_j$ and $F$ are put into the background and used perhaps if further theoretical statements are to be made. The $X_j$ may even be unobservable, or unpractical to observe, and $F$ is unknown in most situations. As an example, take the flood levels of a river at a given location. An instrument is placed into the river which marks the water level at the time the instrument is placed there. If the water level goes down, the mark stays in place, while if the water level goes up, a new mark is established. This way, the instrument measures the highest water level between two visits. Now, if we denote the water level at the $j$-th minute after a most recent reading of the instrument by $X_j$, and the instrument is read daily, then the recorded measurements $Z_{n,k}$ are the daily maxima of the minute by minute fluctuation $X_j$ of the water level. Here, $n = 24 \times 60 = 1440$, $X_j = X_{j,k}$ are identically distributed with unknown $F$, and the $X_j$ are not recorded on the instrument. The $X_j$ are not independent, but their interdependence weakens by the elapse of time, and this is sufficient for the approximation by the classical model. So, the theory tells us that the observations $Z_k = Z_{n,k}, 1 \leq k \leq N$, come from a distribution $H_c((x - A)/B)$, at least approximately. We want to estimate $A$, $B$, and $c$. The estimations of $A$ and $B$ are routine because of their close relation to the sample mean and standard deviation. But can we estimate $c$? If the estimated value of $c$ turns out to be -0.08, say, how confident can we be that $c$ is really negative? Recall (2) for the significance of such a decision. We shall discuss these problems, that is the estimation of $c$ and the decision on the sign of $c$, in Section 4. Section 3 is devoted to the listing of the currently available methods, and Section 4 is the critical demonstration of these methods.

## 2. The mathematical results on the classical model

We use the definition and the notation of the preceding section. We start by pointing out that we know quite much on the mathematical side of the classical model as Chapter 2 of Galambos (1987) and the historical overview Galambos (1994) demonstrate. We therefore restrict our attention to the most fundamental results here.

We have mentioned at (1) that $H_c$ and only $H_c$ of (1) is a classical extreme value distribution (note that our definition is for the maximum only). A population distribution $F$ generates $H_c$ via maxima if, and only if, one of the following conditions holds:

(i) $F(x) < 1$ for all $x$ and

$$\lim_{t \to +\infty} \frac{1 - F(tx)}{1 - F(t)} = x^{-1/c}, x > 0, c > 0;$$

(ii) There is a finite $\omega$ such that $F(x) < 1$ if $x < \omega$, $F(\omega) = 1$, $F(x)$ is continuous at $\omega$, and $F^*(x) = F(\omega - 1/x), x > 0$, satisfies (i);

(iii) Whatever $\omega$ below, $\int_0^{+\infty} (1 - F(x))dx < +\infty$, and there is a function $h(t) > 0$ such that, for all $x$,

$$\lim_{t \to \omega} \frac{1 - F(t + xh(t))}{1 - F(t)} = e^{-x},$$

where $\omega = \sup\{x : F(x) < 1\}$, which may be finite or infinity.

The statement that $F(x)$ generates $H_c(z)$ in the model means that, with some constants $a_n$ and $b_n > 0$, $H_c(z) = \lim F^n(a_n + b_n z)$ as $n \to +\infty$. We refer to this relation that $F(x)$ is in the domain of attraction of $H_c(z)$. In $H_c(z)$, $c > 0$ if $F(x)$ satisfies (i), $c < 0$ in the case of (ii), and $c = 0$ if (iii) is valid.

The theorem above on the domain of attraction is due to Gnedenko (1943). In order to eliminate the "guessing part" of finding $h(t)$ in part (iii), de Haan (1970) established that one can always choose $h(t) = R(t)$, where

$$(3) \qquad R(t) = E(X_1 - t | X_1 > t) = (1 - F(t))^{-1} \int_t^\omega (1 - F(y))dy.$$

The function $R(t)$ is known as the expected residual life (at time $t$). Note that the condition at (iii) for $F$ to be in the domain of attraction of $H_0(z)$ can be reformulated as saying that the asymptotic conditional distribution of $(X_1 - t)/R(t)$, given that $X_1 > t$, must be unit exponential.

In order to point out a strange, and yet somewhat expected, result for normal populations $F$, let us quote the general rule of finding the normalizing constants at (iii): one can choose $a_n = \inf\{x : 1 - F(x) \le 1/n\}$ and $b_n = R(a_n)$. Now, if $F(x)$ is standard normal, upon integrating by parts one obtains that $R(t) = 1/t + O(t^{-3})$ as $t \to +\infty$ (evidently, $\omega = +\infty$). Next, we check the condition at (iii) with $h(t) = 1/t$, an asymptotic value of $R(t)$, and indeed, l'Hospital's rule yields that (iii) applies. Finally, we compute the asymptotic values of $a_n$ and $b_n$ by the quoted rules above. Clearly, the main term of $a_n = (2 \log n)^{1/2}$, and thus $b_n = (2 \log n)^{-1/2}$.

This in turn implies that further terms in $a_n$ are to be computed until we reach a term whose magnitude is smaller than $(2 \log n)^{-1/2}$. We thus get

$$a_n = (2 \log n)^{1/2} - (1/2)(\log \log n + \log 4\pi)(2 \log n)^{-1/2}$$

and

$$b_n = (2 \log n)^{-1/2}.$$

An immediate consequence of these specific forms of $a_n$ and $b_n$ is that $Z_n /$ $[(2 \log n)^{1/2}]$ converges to one in probability. We also get, by some non-routine application of a Borel-Cantelli argument (an extension of the classical Borel-Cantelli lemmas is required), that (for standard normal populations)

(4) $\qquad Z_n - (2 \log n)^{1/2}$ converges to zero almost surely.

In other words, for large $n$, $Z_n$ is behaving like a constant. Since $Z_n$ is a bound on $n$ observations, and since $(2 \log 5,000)^{1/2} \approx 4.1$, samples of size not exceeding 5,000 behave like observations from a bounded population. See Section 4 for further implications of this confusing boundedness property of normal populations. For computations leading to (4), see pp. 265–266 in Galambos (1987).

The reader may be interested in alternative criteria for $F$ to belong to the domain of attraction of $H_0(z)$. We quote one that goes back to von Mises (1936) and another one due to Galambos and Xu (1990).

(iiia). (the von Mises condition) Assume that there is a real number $x_1$ such that for all $x$ between $x_1$ and $\omega$, the distribution function $F(x)$ has a differentiable positive density: $F'(x) = f(x) > 0$ and $f'(x)$ exists. Furthermore, if, as $x \to +\infty$,

$$\lim \frac{d}{dx} \left[ \frac{1 - F(x)}{f(x)} \right] = 0$$

then $F(x)$ is in the domain of attraction of $H_0(x)$.

(iiib). (Galambos and Xu) Let $F(x) < 1$ for all $x$. Assume that $R(t)$ is regularly varying, that is, $R(t) = t^b s(t)$, where $s(t)$ satisfies

$$\lim_{t \to +\infty} \frac{s(tx)}{s(t)} = 1 \text{ for all } x > 0$$

(such an $s(t)$ is called slowly varying). If $b < 1$, then $F(x)$ belongs to the domain of attraction of $H_0(x)$. The same conclusion about $F(x)$ is valid if $b = 1$ and $s(t) \to 0$ as $t \to +\infty$.

The fact that $b \leq 1$ in the domain of attraction of $H_0(x)$ follows from an observation of Gnedenko (1943) that in the cited domain $R(t)/t \to 0$ as $t \to +\infty$ (for $F$ with $\omega = +\infty$). The converse of this last statement is generally not true, but (iiib) states that $R(t)/t \to 0$ is both necessary and sufficient for $F$ with $\omega = +\infty$ and with regularly varying $R(t)$ to belong to the domain of attraction of $H_0(x)$. Note how attractive a condition (iiib) is: when we computed $R(t)$ for a normal population and found that $R(t) = t^{-1}s(t)$ with $s(t)$ converging to one, no further step is necessary to conclude that the normal distribution is in the domain of attraction of $H_0(x)$ ($s(t)$ is slowly varying because it has a finite limit, and thus, $b = -1 < 1$).

The von Mises condition may appear to be very restrictive because of the differentiability requirements in it. However, a second surprise of extreme value theory is associated with this condition: while a large variety of functions $F$ belongs to the domain of attraction of $H_0(x)$, to every distribution function $B(x)$ in this domain there is a function $F(x)$ that satisfies the von Mises condition and it is equivalent to $B(x)$ in the following sense:

$$(5) \qquad \text{as } x \to \omega \text{ of } B(x), \quad \lim \frac{1 - B(x)}{1 - F(x)} = 1.$$

It easily follows that the just defined equivalence entails that $F(x)$ is in the domain of attraction of $H_0(x)$ if, and only if, $B(x)$ is. Just write

$$F^n(x) = \left\{1 - \frac{1 - F(x)}{1 - B(x)}[1 - B(x)]\right\}^n = \left\{1 - \frac{1 - F(x)}{1 - B(x)}\frac{n[1 - B(x)]}{n}\right\}^n$$

and

$$B^n(x) = \exp[n \log B(x)] = \exp(n \log\{1 - [1 - B(x)]\}).$$

Now, as $x = x_n = a_n + b_n z \to \omega$ of $B(x)$, $1 - B(x) \to 0$, and thus, by Taylor's expansion, $\log\{1 - [1 - B(x)]\} \sim -[1 - B(x)]$. We thus have that if $B^n(x_n) \to H_0(z)$, then $n[1 - B(x_n)] \to -\log H_0(z) = e^{-z}$. Elementary calculus now yields that, under equivalence, $F^n(x_n) \to \exp(-e^{-z})$ as claimed. An explicit construction of $F(x)$ to a given $B(x)$ with the stated properties is given in Balkema and de Haan (1972).

We conclude the present section with some remarkable results on the frequency of times when a new maximum is reached, which we refer to as having reached a new record. That is, consider taking observations sequentially. We shall call the very first observation $X_1$ the first record. The second record is reached at $L(2) = t$ if $X_t > X_1$ but $X_j \leq X_1$ for all $1 < j < t$. Now, by induction, we say that $L(k) > L(k-1)$ is the $k^{th}$ record time if $X_{L(k)} > X_{L(k-1)}$ but $X_j \leq X_{L(k-1)}$ for all $L(k-1) < j < L(k)$. The value $X_{L(k)}$ is called the $k^{th}$ record in the sequence $X_j, j \geq 1$. In other words, $X_n$ is a record if $Z_n > Z_{n-1}$, and $X_n$ is not a record if $Z_n = Z_{n-1}$. The question is how long should we wait for a new record to occur, or equivalently, how large can $m$ be with $Z_n = Z_{n+1} = \ldots = Z_{n+m}$? The study of record times $L(k), k \geq 1$, is relatively simple because monotonic transformations preserve inequalities and therefore we can transform $X_j$ to $F(X_j)$, which is known to be uniform, without affecting $L(k)$. See Section 6.3 in Galambos (1987) for several results. We put down only one result: as $n \to +\infty$,

$$(6) \qquad \lim \frac{\log L(n)}{n} = \lim \frac{\log[L(n) - L(n-1)]}{n} = 1 \text{ almost surely.}$$

Although this theorem is very theoretical since it requires that the taking of observations never stops, it tells much to the practitioner as well. Vaguely translated, (6) tells that it takes exponentially long time $(\exp(n))$ to get to the $n^{th}$ record, and then a similarly long exponential time is required to reach the very next record. A less precise consequence of (6) is, however, that new records, and thus more devastating disasters, are sure to come which we must prepare for. Are we, the statisticians, prepared for advising engineers on disasters? The answer is, unfortunately, no. I wanted to show in this section that the mathematical theory of the classical model is very well developed, but how can we utilize these results for practical purposes? In the next section the methods for statistical inference are described, and only in Section 4 shall we see how far behind we are. That will set our goals for the next century.

## 3. Methods for statistical inference

In order to keep our notation of the Introduction, we denote by $Z$ the quantity to be observed, and we suppress the subscript $n$. Neither will it be necessary to refer to the implicit variables $X_1, X_2, \ldots, X_n$ whose maximum is $Z$. Their role has been utilized in justifying the assumption that the distribution of $Z$ is close to, but not exactly equal to, one member of the family $H_c[(z-A)/B]$. We take observations on $Z$, and we denote them by $Z_1, Z_2, \ldots, Z_N$. We rely on the ordered observations $Z_{1:N} < Z_{2:N} < \ldots < Z_{N:N}$, that is, the order statistics of the $Z_j$. It is well known (see David (1981), p. 255) that the quantities $Z_{r:N}, r = [pN], 0 < p < 1$ fixed, are asymptotically normally distributed whatever $c$ (in $H_c$ of $Z$). Consequently, only the top order statistics $Z_{N-k:N}, 0 \leq k \leq k_0(N), k_0(N)/N \to 0$ as $N \to +\infty$, are sensitive to $c$. (Those who fear that information is being lost by ignoring a large part of the sample may want to know that in extremal problems the top order statistics contain all information of the sample; see A. Janssen (1989).)

One of our basic tools is the empirical distribution function $F_N(x)$ and its graph on Gumbel probability paper. That is, $F_N(x) = t/N$, where $t$ is defined by $Z_{t:N} \le x \le Z_{t+1:N}$. Gumbel's probability paper chooses the scales on the $y$-axis in an $(x, y)$-Cartesian system in such a way that guarantees that $H_0(x) = \exp(-e^{-x})$, which is also known as Gumbel's distribution, becomes a straight line. That is, a function $F(x)$ is plotted at the scale $y = -log(-log(F(x))$ against the variable $x$. Note that $F_N(x)$ at $x = Z_{N:N}$ becomes infinity in our new $y$-scale. Since $Z_{N:N}$ is important for us, in some formulas that follow we redefine $F_N(x)$ as $(t - 0.5)/N$ in place of $t/N$. This has no effect on the underlying theoretical results on $F_N(x)$. We now start with the list of methods for statistical inference.

The Castillo-Galambos-Sarabia Curvature Method. Set $N_1 = N - N^{1/2}$ and $N_2 = N - 2N^{1/2}$. Fit two straight lines by the method of least squares to the tail points $Z_{r:N}$, once with $N_2 \le r < N_1$ and then with $N_1 \le r \le N$, of $F_N(x)$ drawn on Gumbel probability paper. Upon denoting the slopes of these two lines by $S_1$ and $S_2$, respectively, we use the statistic

$$(7) \qquad\qquad S = S_1/S_2$$

to decide whether $c > 0$, $c = 0$ or $c < 0$ in the approximate distribution $H_c((x - A)/B)$ of $Z$ on which the observations are taken.

The remarkable characteristics of the method are (i) $S$ is both location (A) and scale (B) invariant, (ii) $S$ relies on all upper extremes $Z_{r:N}$ with $N_2 \le r \le N$ (hence, all information of the sample is contained in $S$), (iii) no order statistic $Z_{r:N}$ with $r < N_2$ is used in any way in the decision of the method.

The drawback of the method is that the exact distribution of $S$ is not known, therefore errors of decision cannot directly be computed. However, because of property (i) of the last paragraph, one can generate tables by Monte Carlo methods for accepting or rejecting a specific type of $c$. With this in mind, we now choose two values $C_1 > 1$ and $0 < C_2 < 1$ and we accept $c > 0$ if $S > C_1$, $c = 0$ if $C_1 > S > C_2$, and $c < 0$ if $S < C_2$.

The underlying theory comes from a result of E. Castillo, J. Galambos and J.M. Sarabia (1989) who refined an earlier result of E. Castillo and J. Galambos (1986). E. Castillo (1988) gives Monte Carlo tables for actual computation of error probabilities. The heart of the method is the recognition that empirical distributions $F_N(x)$ for an arbitrary underlying distribution $F(x)$ from the domain of attraction of $H_c$ have the same geometric nature as $H_c$ itself when both are drawn on Gumbel probability paper. In our case, $F(x)$ is known to be close to $H_c$ from which we draw the advantage of requiring smaller sample sizes

for the similarity of $F_N(x)$ to the appropriate $H_c$ in their upper tail. It cannot be overemphasized that only the upper tail of $F_N(x)$ is relevant for identifying $H_c$. Now, since

$$y = -\log[-\log H_c(x)] = (1/c)\log(1 + cx), 1 + cx > 0$$

is convex for $c < 0$, straight line for $c = 0$ and concave for $c > 0$, the justification of the method of the previous paragraph follows.

A remarkable aspect of the curvature method is that, in several instances, a visual inspection suffices to conclude from the shape of the upper tail of $F_N(x)$ whether $c > 0$, $c = 0$ or $c < 0$.

A curvature method different from the one by Castillo, Galambos and Sarabia was first introduced by Gumbel (1965) and had been adopted in several applied oriented publications. Gumbel's method relies on the set of all observations, uses only a very few order statistics (mostly three altogether) which always include $Z_{1:N}$ and $Z_{N:N}$, a choice that is bound to lead to misleading conclusions, unless the underlying population distribution is exactly $H_c$. For exactly $H_c$-populations, see Tiago de Oliveira (1984) whose method is only implicitly a curvature method, and a fine analysis of Gumbel's method is found in Tiago de Oliveira and Gomes (1984). Pickands (1975) proposes a curvature method which is appropriately for the upper tail only, but uses three order statistics from the upper tail which can lead to incorrect conclusions.

Next we turn to another selection method whose origin is not in extreme value theory. Filliben (1975) introduced a method which we reformulate here for our model. For a set of data $Z_1, Z_2, \ldots, Z_N$ whose common distribution is known to be close to a member of the family of distributions $H_c$ in which $c$ is unknown, one would like to select that $H_c$ that is closest to the empirical distribution $F_N(x)$ in some sense. For this purpose, Filliben introduced a measure, termed by him probability plot correlation coefficient, by which it can be decided whether $H_{c(1)}$ or $H_{c(2)}$ is closer to $F_N(x)$. Therefore, if one chooses a sequence of $c$-values in the interval (-0.5, 0.5), say, and compares each $H_{c(j)}$ with $F_N(x)$, then a closest one can be selected. One can then refine the distance between $c$-s in the neighborhood of the closest $c$, and a better approximation may be found. It requires much computation, but if a computer program is already at hand then the method is very fast. The drawback of the method is that it was designed for nonextremal purposes, and therefore a study is not yet available for the effect on the choice by this method as we limit the upper extremes to be included in the distance formula.

The next method is an actual estimation of the value of $c$. We first assume that $c > 0$. From criterion (i) of Section 2 one can easily conclude (see p. 118 of Galambos (1987)) that

(see p. 118 of Galambos (1987))

$$(8) \qquad \lim_{t \to +\infty} \frac{1}{1 - F(t)} \int_t^{+\infty} \frac{1 - F(y)}{y} dy = c,$$

whenever $F(x)$ is in the domain of attraction of $H_c$. Upon replacing $F(y)$ by $F_N(y)$ and $t$ by $Z_{N-r:N}$, one gets

$$(9) \qquad \lim_{t \to +\infty} \frac{1}{r} \sum_{j=1}^r (\log Z_{N-j+1:N} - \log Z_{N-r:N}) = c$$

The fact that the above substitution preserves the limit (8), and thus yields (9), follows from known approximations of population distribution by empirical distributions. The limit (9) is due to Hill (1975). From (8) and (9) we have that $r = r(N) \to +\infty$ with $N$ and $r(N)/N \to 0$ as $N \to +\infty$, but otherwise the expression (without the limit sign) on the left hand side of (9) suggests itself to be an estimator of $c$, known in advance to be positive. Now, regardless of the sign of $c$, put

$$c_r(N) = \frac{1}{r} \sum_{j=1}^r (\log Z_{N-j+1:N} - \log Z_{N-r:N})$$

and

$$c_r^{(2)}(N) = \frac{1}{r} \sum_{j=1}^r (\log Z_{N-j+1:N} - \log Z_{N-r:N})^2.$$

A relatively routine calculation shows that $c_r(N) \to 0$ if $c = 0$ and

$$c_r^{(2)}(N)/c_r^2(N) \to 2 \text{ if } c \geq 0;$$

an extension of Hill's estimator to arbitrary $c$ is suggested by the formula

$$(10) \qquad C_r(N) = c_r(N) + 1 - \frac{1}{2} \left\{ 1 - c_r^2(N)/c_r^{(2)}(N) \right\}^{-1}$$

Indeed, it turns out (Dekkers et al (1989) and de Haan (1994)) that, once again assuming that $r \to +\infty$ with $N$ and $r/N \to 0$ as $N \to +\infty$, $C_r(N) \to c$ if $H_c$ is the proper approximation to the distribution of $Z$. Extension to confidence intervals on $c$, using asymptotic normality of normalized forms of $c_r(N)$, also is established in de Haan (1994) under further restrictions on $r = r(N)$. The usual choice of $r$ in $C_r(N)$ is a random procedure: a threshold $u$ is chosen in advance and then those $Z_{N-j+1:N}, 1 \leq j \leq r$, are chosen for estimation which exceed $u$. This is why estimating by $C_r(N)$ of (10) is referred to as the threshold method.

## 4. Facing some data and the goals for the next century

In the preceding section the three most powerful methods for evaluating data are formulated. In particular, the threshold method received quick acceptance which also produced the most dangerous conclusions. Statistical data analysis does not have the duty or the right to produce laws of science in such disciplines as biology or any other field. On the other hand, there are laws in every field which are statistical in nature for which urgent answers are needed. When can we expect the next big flood at a given location, can an atom reactor withstand the high winds at a certain location, how long can a well inspected aircraft fly without crashing due to fatigue, how old will be the very old in the not too distant future? These, and other questions need answers, and only extreme models can provide the correct answers. Note that we assume that engineers did their best when we speak of a well made and well inspected aircraft. That is, if we ask the inspector whether the aircraft will crash (soon), the answer is a definite no (with reference to mechanical problems), but behind such a no there is always the defensive thought of "very, very unlikely." but a crack might have already started on the body which is not visible yet, and thus the engineer cannot do anything, but statistical analysis can give an estimate for the time to first major problems due to fatigue. On the other hand, setting up rules and regulations from conclusions reached by purely statistical methods may require further analysis before making the rule stand. Take two examples both of which have recently been analyzed in the literature to some extent (see Galambos and Macri (1998) and (1999)): estimating the speed of high wind at a location and predicting the age of the longest living person in a society. By first grouping consecutive measurements and taking only the highest value from the group (thus replacing highest windspeed on a day to weekly highest values, say, and looking at each individual can be replaced by recording the longest living person of a city), we can achieve both independence and the fact that each observation is already a high value. Hence, we face the problem of finding the appropriate $c$-value for our model. If we get a negative value for $c$, this implies that there is a finite upper bound on the quantity being observed (wind speed or age). But mathematics tells us that if there is an upper bound then there is a least upper bound $K$ as well which can be reached arbitrarily closely but cannot be surpassed. For the wind speed it would imply that engineers face an easy task: if a long record is available then one observation has already come close to $K$, and it will never be higher, so building codes should be determined by the last highest wind. In the case of old age, if $K$ can be estimated then it means that there is a biological law which makes everyone die at age $K$. In spite of this obvious argument against finite upper bound, recent estimates by the threshold method tend to come with negative $c$, and $K$ has been estimated both for winds and for old age. See Simiu and Heckert (1996) for winds, and Aarssen and de Haan (1994) and R.-D. Reiss and M. Thomas (1997) for old age. All three works use the threshold method, and since the conclusions are unreasonable in these papers, Galambos and Macri (1998) and (1999) reevaluated these data sets by the curvature method of Castillo-Galambos-Sarabia and found that opposite conclusions are obtained by this latter method. Then to show that the arguments are not just personal views supported by two different statistical approaches, Galambos and Macri used independent normal variables and estimated $c$ by the threshold method. The estimated value did come up negative. Galambos and Macri suspected that this will be the case, based on the property expressed at (4) for the normal distribution. It turns out that because of the triple limit involved in the threshold method (including the limitations on the choice of $r$), the method is not sufficiently sensitive when the upper endpoint of the underlying distribution is slowly increasing to infinity with $n$ or $N$. But this seems to be the case for most practical problems, certainly for wind speed and age. Interestingly, Filliben's method by plot correlation supports the conclusion by the curvature method for winds; and thus let us formulate the major tasks ahead.

1. the curvature method of Castillo-Galambos-Sarabia should be made more acceptable by the statistical community by developing estimates (other than by Monte Carlo method) for decision errors. The method appears to be very reliable.

2. Filliben's method by plot correlation deserves a theoretical study for the case of extremal problems. In the method one should use a small part of the upper extremes only.

3. The current form of the threshold method should be put aside since it leads to wrong conclusions more often than not. This method should also be a warning sign for future development: a nice mathematical limit theorem is not necessarily useful in practice.

The papers of Galambos and Macri (1998) and (1999) are the first steps for future development. And our future in extreme value theory depends on our success in solving practical problems with accurate and reliable methods.

We conclude the paper by adding some further references both to the mathematical theory and to statistical inference. First, Gumbel (1958)'s classical book remains an important reference, although it is recommended as a supplement to Castillo (1988)'s book. In order to see where the theory and statistics of extremes stand at the moment, one may want to consult the three volume set edited by Galambos, Lechner and Simiu (1994). The relation of extremes to other order statistics, using an important metric, is discussed in Reiss (1989), the relation to rare events in general is developed in Falk,Hüsler and Reiss (1994); and the treatment of extremes by Bonferroni-type inequalities can be found in Galambos and Simonelli (1996). The comprehensive study of extremes by nonlinear normalization by Pancheva (1994) may turn out to be an alternative to the classical theory. Since parameters of a population distribution are estimated from data, one actually changes the underlying distribution. This way, limit theorems may have to be modified. A distinguished example for this is the penultimate form of approximation developed by Gomes (1978) and (1984). For results on dependent samples, see Chapter 3 in Galambos (1987).

### References

[1] Aarssen, K. and Haan, L. de (1994) *On the maximal life span of humans.* Reproduced in deHaan's contribution to Extreme Value Theory and Applications (eds.: J. Galambos et al), 117–121, Kluwer, Dordrecht.

[2] Balkema, A.A. and Haan, L. de (1972), On R. Von Mises's condition for the domain of attraction of $\exp(-e^{-x})$. *Ann. Math. Statist.*, **43**, 1352–1354.

[3] Castillo, E. (1988) *Extreme value theory in engineering*, Academic Press, New York.

[4] Castillo, E. and Galambos, J. (1986) *Determining the domain of attraction of an extreme value distribution*, Technical Report, Temple University.

[5] Castillo, E., Galambos, J. and Sarabia, J.M. (1989) The selection of the domain of attraction of an extreme value distribution from a set of data. In: *Extreme value theory* (eds.: J. Hüsler and R.-D. Reiss), Lecture Notes in Statistics, Vol. **51**, Springer, Heidelberg, 181–190.

[6] David, H.A. (1981) *Order Statistics,* 2nd ed., Wiley, New York.

[7] Dekkers, A.L.M., Einmahl, J.H.J. and Haan, L. de (1989), A moment estimator for the index of an extreme-value distribution. *Ann. Statist.*, **17**, 1833–1855.

[8] Filliben, J.J. (1975), The probability plot correlation test for normality. *Technometrics*, **17**, 111–117.

[9] Falk, M., Hüsler, J. and Reiss, R.-D. (1994) *Laws of small numbers: extremes and rare events*, DMV-Seminar, Vol **23**, Birkhauser, Basel.

[10] Galambos, J. (1987) *The asymptotic theory of extreme order statistics*, 2nd ed., Krieger, Malabar, Florida.

[11] Galambos, J. (1994), The development of the mathematical theory of extremes in the past half century. *Teoriya Veroyatnost. i Primenen.*, **39**, 273–293.

[12] Galambos, J. and Macri, N. (1998) The classical extreme value model and prediction of extreme winds. *J. Structural Engineering*, to appear.

[13] Galambos, J. and Macri, N. (1999) The life length of humans does not have a limit, to appear.

[14] Galambos, J. and Simonelli, I. (1996) *Bonferroni-type inequalities with applications,* Springer Verlag, New York.

[15] Galambos, J and Xu, Y. (1990), Regularly varying expected residual life and domains of attraction of extreme value distributions. *Ann. Univ. Sci. Budapest, Sectio Math.*, **33**, 105–108.

[16] Galambos, J., Lechner, J. and Simiu, E. (editors) (1994) Extreme value theory and applications, Vol. **I-III**. Vol. **I**: Kluwer, Dordrecht, Vol **II**: J. Special Research, NIST, Gaithersburg, MD, Vol **III**: Special Publications, NIST, Gaithersburg, MD.

[17] Gnedenko, B.V. (1943), Sur la distribution limite du terme maximum d'une serie aleatoire. *Ann. Math.*, **44**, 423–453.

[18] Gomes, M.I. (1978) *Some probabilistic and statistical problems in extreme value theory,* Thesis for Ph.D., University of Sheffield.

[19] Gomes, M.I. (1984), Penultimate limiting forms in extreme value theory. *Ann. Inst. Statist. Math.*, **36**, 71–85.

[20] Gumbel, E.J. (1958) *Statistics of extremes,* Columbia University Press, New York.

[21] Gumbel, E.J. (1965), A quick estimation of the parameters in Frechet's distribution. *Rev. ISI*, **33**, 349–363.

[22] Haan, L. de (1970) *On regular variation and its application to the weak convergence of sample extremes,* Math. Centre Tracts, Vol. **32**, Amsterdam.

[23] Haan, L. de (1994) Extreme value statistics, in: Extreme value theory and applications (eds.: J. Galambos et.al), Kluwer, Dordrecht, 93–122.

[24] Hill, B. (1975), A simple general approach to inference about the tail of a distribution. *Ann. Statist.*, **3**, 1163–1174.

[25] Jannsen, A. (1989) The role of extreme order statistics for exponential families, in: Extreme value theory (eds.: J. Husler and R.-D. Reiss), Lecture Notes in Statist. Vol **51**, Springer, Heidelberg, 204–221.

[26] Mises, R. von (1936) La distribution de la plus grande de n valeurs. Reprinted in: *Selected Papers* **II**, Amer. Math Soc., Providence, RI, 1954, 271–294

[27] Pancheva, E. (1994) Extreme value limit theory with nonlinear normalization, in: *Extreme value theory and applications* (eds.: J. Galambos et.al), Kluwer, Dordrecht, 305–318.

[28] Pickands, J. III (1975), Statistical inference using extreme order statistics. *Ann. Statist*, **3**, 119–131.

[29] Reiss, R.-D. (1989) *Approximate distributions of order statistics*, Springer, New York.

[30] Reiss, R.-D. and Thomas, M. (1997) *Statistical analysis of extreme values*, Birkhauser, Basel.

[31] Simiu, E. and N.A. Heckert (1996), Extreme wind distribution tails: A peaks over threshold approach. *J. Structural Engineering*, **122**, 539–547.

[32] Tiago de Oliveira, J. (1984) Univariate extremes; statistical choice, in: *Statistical extremes and applications* (ed.: J. Tiago de Oliveira), Reidel, Dordrecht, 91–107.

[33] Tiago de Oliveira, J. and Gomes, M.I. (1984) Two test statistics for choice of univariate extreme models, in: *Statistical extremes and applications* (ed.: J. Tiago de Oliveira), Reidel, Dordrecht, 651–668.

# The Analysis of Cross-Classified Data: Notes on a Century of Progress in Contingency Table Analysis, and Some Comments on Its Prehistory and Its Future

Leo A. Goodman

Department of Statistics
University of California
Berkeley, California

**Abstract.** In this article, attention is focused on the analysis of categorical data that is cross-classified; i.e., on the analysis of contingency tables. Measures, models, and graphical displays are considered here for the analysis of nonindependence (or association) in the contingency table. Some of the work done on this topic during the twentieth century is discussed, including work done very early in the century and related work done near the middle and near the end of the century. Brief comments are also included here on the prehistory of this topic (on related work on this topic done in the nineteenth century) and on its future in the twenty-first century. In addition, a new measure of association is described in this article, and some new methods pertaining to graphical displays are also considered. These new methods can serve, for example, as a necessary correction to methods commonly used now pertaining to graphical displays in correspondence analysis.

## 1. Introduction and summary

The problem of measuring the association (nonindependence or dependence) between, say, two dichotomous variables (two dichotomous classifications or two attributes) is a topic that was considered by eminent scholars beginning in the nineteenth century. Among the nineteenth-century scholars who considered this topic, we have, for example, C.S. Peirce, the great American philosopher and logician, and able scientist and mathematician (in addition to the recognition he has received for some of his other work, he is also sometimes referred to as the "founder of pragmatism"); Adolphe Quetelet, the Belgian astronomer, statistician, and sociologist (sometimes referred to as the "founder of sociology"); József Kőrösy, the pioneer Hungarian statistician and demographer; and M.H. Doolittle, the well-known American mathematician (of Doolittle's method for the inversion of matrices). Some of the nineteenth-century work on this topic was unknown to twentieth-century scholars (or was unacknowledged by them), and was reinvented by them. Brief comments will be included later herein on nineteenth-century scholarship on this topic. I begin this Introduction to my article now with a brief quotation from the nineteenth-century, which the late-twentieth-century reader might find quaint (and amusing). Here is M.H. Doolittle at a 1887 meeting of the Mathematical Section of the Philosophical Society of Washington, struggling at that time to describe verbally the problem of measuring the association in what we now call the 2 × 2 contingency table:

> "The general problem may be stated as follows: Having given the number of instances respectively in which things are both thus and so, in which they are thus but not so, in which they are so but not thus, and in which they are neither thus nor so, it is required to eliminate the general quantitative relativity inhering in the mere thingness of the things, and to determine the special quantitative relativity subsisting between the thusness and the soness of the things." (Doolittle, 1888; Goodman and Kruskal, 1959)

The twentieth century is a period in which great progress was made in the analysis of cross-classified data (contingency tables). I shall comment in this article on work done on this topic very early in the century and on related work done near the middle and near the end of the century. In the early part of the century, Karl Pearson and G. Udny Yule introduced two different approaches, two apparently incompatible approaches, to the measurement of the association in cross-classified data. The differences between their approaches led to intense debate. In this Introduction to my article, I include now some brief quotations from this early-twentieth-century debate, which the late-twentieth-century reader might find of interest:

> Pearson: "If Mr. Yule's views are accepted, irreparable damage will be done to the growth of modern statistical theory ... [Yule's Q] has never been and never will be used in any work

done under [my (Pearson's)] supervision ... We regret having to draw attention to the manner in which Mr. Yule has gone astray at every stage in his treatment of association ... " (Pearson and Heron, 1913)

Yule: The value of Pearson's method "depends entirely on the empirical truth of the assumptions made ... These assumptions were never adequately tested [by Pearson] ... and the few tests which I applied ... sufficed to show that they were, to say the least, of exceedingly doubtful validity ... At best the normal correlation can only be said to give us in cases like these a hypothetical correlation between supposititious variables. The introduction of needless and unverifiable hypotheses does not appear to me a desirable proceeding in scientific work." (Yule, 1912)

We shall return to the Pearson-Yule debate later in this article. With respect to the two different approaches of Pearson and Yule, it can not be said that one approach is correct and the other approach incorrect. Each approach has its merits. In some contexts, one approach will be more appropriate; and in other contexts, the other approach will be. In fact, it is possible to obtain, in a certain sense, a reconciliation and synthesis of some parts of the two approaches (see, e.g., Goodman, 1981a, 1996). Further details on these matters will be included later herein.

Lest the reader think that this article will consider only matters of historical interest, I hasten to note that a new index for measuring the association (nonindependence or dependence) in the 2 × 2 contingency table will be considered here, and some new methods pertaining to graphical displays will also be discussed. With respect to the new index for measuring association, since the general subject under consideration here has been considered for at least the past one hundred and sixty five years (see, e.g., Goodman and Kruskal, 1959, and Section 3 later herein), it comes as something of a surprise that it is still possible to produce something on this subject that is new and that may be of interest.

Our focus of attention in the main part of this article will be on work done during the twentieth century on the analysis of contingency tables; and later in the article, brief comments will also be included on the prehistory of this topic (on related work on this topic done in the nineteenth century) and on its future.

For the 2 × 2 contingency table, we shall consider two classic measures of the nonindependence between the row classification and the column classification in the contingency table, and two more recent measures of this nonindependence, and one new measure of this nonindependence; and we shall show how each of these five measures can be described as a special case within a single general framework. We then note that, for each of these five measures of nonindependence, there is a corresponding generalization that can be applied in the analysis of the $I \times J$ contingency table (for $I \geq 2$ and $J \geq 2$).

For the $I \times J$ contingency table, the five different generalizations provide five different sets of statistical methods for measuring the nonindependence between the row classification and the column classification in the contingency table; and each of these five different sets of statistical methods can be described as a special case within a single general framework. The five different sets of statistical methods are: (1) correspondence analysis (or the canonical correlation approach to the analysis of contingency tables), (2) association analysis (or the log-bilinear approach to the analysis of contingency tables), (3) concomitance analysis (or the "midway" approach to the analysis of contingency tables), (4) marginal-free correspondence analysis (or the canonical correlation approach to the analysis of standardized contingency tables), (5) marginal-free association analysis (or the log-bilinear approach to the analysis of standardized contingency tables). Because of space constraints, attention will be focused here mainly on only one of these five approaches and on the single general framework that yields each of the five approaches as a special case.

With each of the five different approaches listed above, a different set of graphical displays can be obtained. Each set of graphical displays can be viewed as a set of pictures that portray the nonindependence in the rows, columns, and cells of the contingency table. Our attention here will be focused mainly on the portrayal of the nonindependence in the *cells* of the contingency table. For each of the five different approaches listed above, the corresponding portrayal of the nonindependence in the cells of the contingency table can be viewed using any one of three different graphical displays; and each of these three displays can be "deconstructed" in order to see more clearly the relative magnitude of the nonindependence in the cells of the contingency table. For each of the five different approaches, the three graphical displays for the analysis of the nonindependence in the cells of the table are quite different from each other; but we shall see here that the three displays are, in a certain sense, equivalent to each other.

With each of the five different approaches listed above, we can obtain a set of three graphical displays to portray (1) the "row profiles" (i.e., the magnitude of the nonindependence in each of the $I$ rows of the $I \times J$ table), (2) the "column profiles" (i.e., the magnitude of the nonindependence in each of the $J$ columns of the $I \times J$ table), and (3) the "cell nonindependence" (i.e., the magnitude of the nonindependence in each of the

$I \times J$ cells of the $I \times J$ table). As noted in the preceding paragraph, for the "cell nonindependence" graphical display, any one of three different (but equivalent) graphical displays can be used. For the row-profiles display, the "inter-point distance" (viz., the "inter-row difference") is relevant; and for the column-profiles display, the "inter-point distance" (viz., the "inter-column difference") is relevant; but for the cell-nonindependence display, the "inter-vector angle" and vector lengths (viz., the "row-vector-by-column-vector angle" and row-vector and column-vector lengths) are relevant. The cell-nonindependence display can also be viewed as a "point-onto-vector projection" display, in which the projection of the row-point onto the column-vector and the length of the column-vector are relevant, or in which the projection of the column-point onto the row-vector and the length of the row-vector are relevant.

With respect to, say, the correspondence analysis approach listed above, the nonindependence in the rows of the contingency table can be viewed in one graphical display, the nonindependence in the columns of the contingency table can be viewed in another graphical display, and any one of three different (but equivalent) graphical displays can be used to view the nonindependence in the cells of the contingency table; and each of the three graphical displays of the nonindependence in the cells can be "deconstructed" for greater clarity. These new methods can serve to correct methods commonly used now in correspondence analysis, when the usual correspondence analysis simultaneous display of row points and column points is used to assess the nonindependence in the cells, with the nonindependence in each cell determined by the usual correspondence analysis placement of the corresponding row point and column point.

Our attention here will be focused on the analysis of the two-way contingency table; a corresponding analysis of the multiway table will not be included here, again because of space constraints. On this topic, the interested reader is referred to, for example, Goodman (1979a, 1981b, 1986), Clogg (1982), Agresti and Kezouh (1983), Gilula and Haberman (1988), Choulakian (1988), Becker (1989), Becker and Clogg (1989), Green (1989), and Goodman and Hout (1998).

There are many other topics in contingency table analysis that could be included in an article of this kind. Reference will be made to some of these other topics in the final section of the present article; but again because of space constraints, these other topics will not be discussed in any detail here. For some of these other topics, we refer the interested reader to, for example, recently published textbooks that are concerned with this subject (see, e.g., Agresti, 1990, 1996, and Andersen, 1990).

Interest in this subject continues to grow at a rapid pace, and work on this subject also continues to develop at a rapid pace. We can look forward with great interest to further developments in this field in the twenty-first century.

## 2. Five different views of nonindependence in the contingency table

### 2.1 Preliminaries

For the $I \times J$ table, let $P_{ij}$ denote the probability that an observation will fall in the $i$-th row and $j$-th column of the table ($i = 1, \dots, I$; $j = 1, \dots, J$). We begin this section with the usual null model of statistical independence between the row classification and the column classification in the contingency table. This model is usually expressed as

$$(2.1) \qquad P_{ij} = P_{i+}P_{+j},$$

where $P_{i+}$ and $P_{+j}$ are the row and column marginal distributions,

$$(2.2) \qquad P_{i+} = \sum_{j=1}^{J} P_{ij}, \; P_{+j} = \sum_{i=1}^{I} P_{ij}.$$

We shall consider here the case where the null model (2.1) may not hold true.

For expository purposes, we shall consider first the $2 \times 2$ table, and the analysis of nonindependence between the row classification and the column classification in this table. To simplify notation in this special case, we let

$$a = P_{11}, \; b = P_{12}, \; c = P_{21}, \; d = P_{22},$$

and                                                                          (2.3)

$$A = P_{1+}P_{+1}, \; B = P_{1+}P_{+2}, \; C = P_{2+}P_{+1}, \; D = P_{2+}P_{+2}.$$

We shall next consider five different measures of nonindependences in the $2 \times 2$ table.

### 2.2 The analysis of nonindependence in the $2 \times 2$ contingency table

The five measures of nonindependence that will be considered in this section appear to be quite different from each other, but we shall show here that each of these five measures can be viewed as a special case within a single general framework. The five measures can be described as follows:

## 2.2a The coefficient of correlation in the $2 \times 2$ table

I shall use the symbol $\rho$ here to denote the usual (Pearsonian) correlation coefficient in the $2 \times 2$ table:

$$(2.4a) \qquad \rho = (ad - bc)/(P_{1+}P_{2+}P_{+1}P_{+2})^{1/2}.$$

By assigning scores to the two row categories (say, $X_1 = -1$ and $X_2 = +1$ for row categories 1 and 2, respectively) and scores to the two column categories (say, $Y_1 = -1$ and $Y_2 = +1$ for column categories 1 and 2, respectively), the usual correlation coefficient between the row scores and the column scores will be equal to the coefficient (2.4a).

## 2.2b The coefficient of concomitance in the $2 \times 2$ table

Coefficient (2.4.b) is a new coefficient:

$$(2.4b) \qquad \tau = 2[\sqrt{aD} + \sqrt{dA} - \sqrt{bC} - \sqrt{cB}].$$

Comparing (2.4b) with (2.4a), we see that (2.4b) is related, to some extent, to the numerator of (2.4a). We also note that the factor $(P_{1+}P_{2+}P_{+1}P_{+2})^{1/2}$, which is needed as the *divisor* in (2.4a) (and which will be needed as the *multiplier* in the coefficient (2.4c) that will be considered next), is not needed in (2.4b). Justification for the use of coefficient (2.4b) will become clear later herein when coefficients (2.4a), (2.4b), and (2.4c) are considered as special cases within a single general framework.

## 2.2c The coefficient of weighted association in the $2 \times 2$ table

The odds-ratio (or cross-product ratio) in the $2 \times 2$ table is defined as $(ad)/(bc)$ (i.e., as $(a/b)/(c/d)$ or as $(a/c)/(b/d)$), and the coefficient (2.4c) is based, in part, on the corresponding log-odds-ratio:

$$(2.4c) \qquad \tilde{\phi} = \log[(ad)/(bc)](P_{1+}P_{2+}P_{+1}P_{+2})^{1/2},$$

where log denotes the natural logarithm.

For a sample of size $N$ from the $2 \times 2$ contingency table, the coefficients $\rho$, $\tau$, and $\tilde{\phi}$ in (2.4a), (2.4b), and (2.4c), respectively, can be estimated by replacing the parameters in (2.4a), (2.4b), and (2.4c) by the corresponding sample estimates. And the distribution of each of the estimated coefficients multiplied by $\sqrt{N}$ will be approximately normal with mean zero and variance one, under the usual null hypothesis (2.1) that the row classification and the column classification are statistically independent of each other.

**2.2d The coefficient of unweighted association in the $2 \times 2$ table**

The weighted association coefficient (2.4c), and also the preceding coefficients (2.4a) and (2.4b), are *marginal-dependent*. In contexts in which it is desirable to use *marginal-free* coefficients, we can replace (2.4c) by the following coefficient:

$$(2.4d) \qquad \phi^\dagger = \frac{1}{4} \log[(ad)/(bc)].$$

Coefficient (2.4d) is obtained directly from (2.4c) simply by replacing each of the marginals (viz., $P_{1+}$, $P_{2+}$, $P_{+1}$, and $P_{+2}$) in (2.4c) by $1/2$. Instead of referring to coefficients (2.4c) and (2.4d) as the weighted association and unweighted association, respectively, we could also have referred to them as the unstandardized association and standardized association, respectively.

**2.2e The coefficient of colligation in the $2 \times 2$ table**

In addition to coefficient (2.4d), here is another coefficient that is *marginal-free*:

$$(2.4e) \qquad \phi^* = (\sqrt{ad} - \sqrt{bc})/(\sqrt{ad} + \sqrt{bc}).$$

This is Yule's coefficient of colligation.

Both coefficients (2.4d) and (2.4e) remain unchanged when the $P_{ij}$ in the $2 \times 2$ table are replaced by $P_{ij}^* = P_{ij}\gamma_i\delta_j$, where $\gamma_i$ and $\delta_j$ are positive constants. When the $P_{ij}^*$ replacement for the $P_{ij}$ has uniform marginals (i.e., $P_{1+}^* = P_{2+}^* = .5$, and $P_{+1}^* = P_{+2}^* = .5$), the coefficient $\phi^\dagger$ in (2.4d) is obtained when (2.4c) is applied to the corresponding $P_{ij}^*$, and the coefficient $\phi^*$ in (2.4e) is obtained when (2.4a) is applied to the corresponding $P_{ij}^*$.

**2.2f A single general coefficient of nonindependence in the $2 \times 2$ table**

With the quantities $a, b, c, d$ and $A, B, C, D$ defined as in (2.3), I start this section with the ratios

$$(2.5) \qquad \psi_a = a/A, \ \psi_b = b/B, \ \psi_c = c/C, \ \psi_d = d/D,$$

which I shall call the "Pearson ratios". Next consider a simple function of the Pearson ratios, viz.,

$$(2.6) \qquad R_a = R[\psi_a], \ R_b = R[\psi_b], \ R_c = R[\psi_c], \ R_d = R[\psi_d],$$

where $R[x]$ is a power function of $x$ (for $x > 0$)), viz.,

$$(2.7) \qquad R[x] = x^\gamma/\gamma,$$

and where $\gamma$ is a positive constant. The general coefficient of nonindependence is defined as

$$(2.8) \qquad \Delta = \{R_a - R_b - R_c + R_d\}(P_{1+}P_{2+}P_{+1}P_{+2})^{1/2}.$$

Let us now consider the following three simple power functions; viz.,

$$(2.9a) \qquad\qquad R[x] = x,$$

the identity function (i.e., $\gamma = 1$ in (2.7));

$$(2.9b) \qquad\qquad R[x] = 2\sqrt{x},$$

the square-root function multiplied by 2 (i.e., $\gamma = 1/2$ in (2.7)); and

$$(2.9c) \qquad\qquad R[x] = \log x,$$

the logarithmic function (i.e., $\gamma = 0$ in a modified form for $R[x]$ in (2.7), viz., $R[x] = (x^\gamma - 1)/\gamma$; see, e.g., Tukey, 1957; Box and Cox, 1964). (The logarithm in (2.9c) and earlier in (2.4c) and (2.4d) is the natural logarithm.) When (2.9a) is used to define the function $R[x]$ in (2.6) and (2.7), the general coefficient (2.8) becomes (2.4a). Similarly, when (2.9b) is used to define the function $R[x]$, the general coefficient (2.8) becomes (2.4b); and when (2.9c) is used, the general coefficient (2.8) becomes (2.4c).

We noted earlier herein that, for a sample of size $N$ from the $2 \times 2$ table, the distribution of the estimated coefficients obtained from (2.4a), (2.4b), and (2.4c) multiplied by $\sqrt{N}$ will be approximately normal with mean zero and variance one, under the null hypothesis (2.1); and this is also the case more generally for the estimated coefficient obtained from (2.8) when the function used for $R[x]$ in (2.6) and (2.8) is any power function of the form (2.7).

Since the power functions (2.9a), (2.9b), and (2.9c) are obtained from (2.7) with $\gamma = 1$, $\gamma = 1/2$, and $\gamma = 0$, respectively (see, e.g., Tukey, 1957), the power function (2.9b) can be viewed as being, in a certain sense, "midway" between the power functions (2.9a) and (2.9c); and so we can also view the corresponding coefficient (2.4b) as being, in a certain sense, "midway" between the coefficients (2.4a) and (2.4c). The concomitance coefficient (2.4b) is obtained from a view of nonindependence that is midway between the view of nonindependence that yields the usual (Pearsonian) correlation coefficient (2.4a) and the view of nonindependence that yields the coefficient (2.4c) of weighted association.

The coefficients (2.4d) and (2.4e) can also be obtained directly from the general coefficient (2.8), when (2.9c) and (2.9a) are used to define the function $R[x]$ in (2.6) and (2.7), and the $P_{ij}$ in the $2 \times 2$ table are replaced by the corresponding $P_{ij}^*$ defined earlier herein (with $P_{1+}^* = P_{2+}^* = .5$, and $P_{+1}^* = P_{+2}^* = .5$).

We have now indicated how each of the five coefficients (2.4a)–(2.4e) considered in this section can be viewed as a special case of the general coefficient (2.8). For illustrative purposes we present in Table 1 the measures of nonindependence obtained when each of the five coefficients is applied to three examples of $2 \times 2$ tables.

The reader might get the impression from Table 1 that $\tau$ is always between $\rho$ and $\tilde{\phi}$ for $2 \times 2$ tables; but this need not be the case. Consider, for example, a $2 \times 2$ table similar to the examples of $2 \times 2$ tables in Table 1, but with entries 9, 5, 5, and 25. For this table, we find that $\rho = .476$, $\tau = .468$, $\tilde{\phi} = .477$, $\phi^\dagger = .549$, and $\phi^* = .500$.

| Example | 1 | | 2 | | 3 | |
|---|---|---|---|---|---|---|
| Table | 3 | 1 | 3 | 2 | 1 | 10 |
|       | 1 | 3 | 2 | 12 | 10 | 900 |
| Correlation $\rho$ | .500 | | .457 | | .080 | |
| Concomitance $\tau$ | .518 | | .432 | | .043 | |
| Weighted association $\tilde{\phi}$ | .549 | | .426 | | .026 | |
| | | | | | | |
| Unweighted association $\phi^\dagger$ | .549 | | .549 | | .549 | |
| Yule's colligation $\phi^*$ | .500 | | .500 | | .500 | |

Table 1. Measures of nonindependence, applied to three examples of $2 \times 2$ cross-classifications

The coefficient of colligation (2.4e) was introduced by Yule (1912), as noted earlier herein; and the coefficient of unweighted association (2.4d) might also be referred, in a broad sense, back to Yule.[1] The usual (Pearsonian) coefficient of correlation (2.4a) in the $2 \times 2$ table is attributed here to

---

[1] Although Yule did not introduce the coefficient of unweighted association (2.4d), this coefficient is based on the odds-ratio, and so are Yule's coefficient of colligation (2.4e) and another measure of association introduced by him (viz., Yule's Q). On the other hand, in comparing Yule's coefficients with the coefficient of unweighted association (2.4d), we see that the latter coefficient is based more directly on the *log*-odds-ratio than are Yule's coefficients. (With respect to Yule's Q, see, e.g., footnote 6

Pearson in part because the usual (product-moment) formula for calculating the correlation between two quantitative variables, when applied to the row variable and the column variable of the $2 \times 2$ table, will yield the coefficient (2.4a), as noted in the sentence immediately following (2.4a).[2] (A different reason for this attribution could be that $\rho$ in (2.4a) can be viewed as the basis of Pearson's "mean squared contingency" in the $2 \times 2$ table — in particular, as the square root of this "mean squared contingency".[3] ) On the other hand, it should be noted that it was Yule who saw that (2.4a) might be used as a coefficient of association (see, e.g., Yule 1912), whereas Pearson objected to this use except "in the rare cases in which the frequency is collected into points – absolute homogeneity of category contents ... " (see, e.g., Pearson 1913).[4]

The general coefficient of nonindependence (2.8), and each of the five coefficients (2.4a)–(2.4e), were presented earlier herein for the analysis of the $2 \times 2$ table. For each of these coefficients, there is a corresponding generalization that can be applied in the analysis of the $I \times J$ table (for $I \geq 2$ and $J \geq 2$). Each of the coefficients for the analysis of the $2 \times 2$ table can be viewed as a special case of a corresponding more general set of coefficients obtained in the analysis of the nonindependence between the row classification and the column classification in the $I \times J$ table.[5] This generalization is considered next.

---

in Section 3 later herein; and, with respect to the unweighted association (2.4d), see also footnote 5 below.)

[2] The usual product-moment formula for calculating the correlation between quantitative variables is usually referred to as "Pearson's" product-moment estimate. (For more on the history of the product-moment formula and on the history of correlation and related matters, see, e.g., Stigler, 1986.)

[3] Note also that, for a sample of size $N$ from the $2 \times 2$ contingency table, the square of the estimated coefficient obtained from (2.4a), viz., $\hat{\rho}^2$, multiplied by $N$ is equal to *Pearson's* goodness-of-fit (chi-square) statistic for testing the null hypothesis (2.1) in the $2 \times 2$ table. (See also related comment in the first paragraph after (2.4c) above and in the first paragraph after (2.19) later herein.)

[4] For an alternative interpretation of the usual (Pearsonian) coefficient of correlation (2.4a) in the $2 \times 2$ table, and an alternative interpretation of Yule's coefficient of colligation (2.4e), see, e.g., Goodman and Kruskal (1954, 1959).

[5] The coefficients of association, (2.4c) and (2.4d), were introduced as a

## 2.3 The analysis of nonindependence in the $I \times J$ contingency table (for $I \geq 2$ and $J \geq 2$)

In the preceding section on the $2 \times 2$ table, there was a separate subsection pertaining to each of the five coefficients and a separate subsection pertaining to the general coefficient. In the present section on the $I \times J$ table, because of space constraints, we shall illustrate how each of five coefficients for the $2 \times 2$ table can be generalized by considering how one of the coefficients (viz., coefficient (2.4a)) is generalized and how the general coefficient (2.8) for the analysis of the $2 \times 2$ table is generalized for the analysis of the $I \times J$ table.

### 2.3a Correspondence analysis or the canonical correlation approach to the analysis of contingency tables

For the $I \times J$ table, with the definition of $P_{ij}$, $P_{i+}$, and $P_{+j}$ presented earlier herein (Section 2.1), we take note again of the usual null model (2.1) of statistical independence between the row classification and the column classification in the contingency table, and we next consider the following generalization of (2.1):

$$(2.10) \qquad P_{ij} = P_{i+}P_{+j}\Big(1 + \sum_{m=1}^{M} \rho_m x_{im} y_{jm}\Big),$$

where $M = \min(I, J) - 1$, and where the row scores $x_{im}$ $(m = 1, \ldots, M)$ and column scores $y_{jm}$ $(m = 1, \ldots, M)$ satisfy the following conditions:

$$(2.11) \qquad
\begin{array}{l}
\sum_{i=1}^{I} x_{im} P_{i+} = 0,\ \sum_{j=1}^{J} y_{jm} P_{+j} = 0, \\
\sum_{i=1}^{I} x_{im}^2 P_{i+} = 1,\ \sum_{j=1}^{J} y_{jm}^2 P_{+j} = 1, \\
\sum_{i=1}^{I} x_{im} x_{im'} P_{i+} = 0,\ \sum_{j=1}^{J} y_{jm} y_{jm'} P_{+j} = 0,
\end{array}$$

for $m \neq m'$. The parameters $x_{im}$ and $y_{jm}$ in (2.10)–(2.11) are standardized scores for the row categories $(i = 1, \ldots, I)$ and column categories $(j = 1, \ldots, J)$, respectively, pertaining to the $m$-th component

---

by-product of the association approach to the analysis of nonindependence in the $I \times J$ table (see, e.g., Goodman, 1985, 1991a); and the concomitance coefficient (2.4b) is a by-product of the midway view of nonindependence (Goodman, 1997). Also, the general coefficient of nonindependence (2.8) was introduced as a by-product of the generalized nonindependence approach to the analysis of the $I \times J$ table (see, e.g., Goodman, 1993, 1996, 1997).

$(m = 1, \ldots, M)$ on the right side of (2.10); and, for $m \neq m'$, the corresponding row scores $x_{im}$ and $x_{im'}$ are uncorrelated with each other, and the corresponding column scores $y_{jm}$ and $y_{jm'}$ are uncorrelated with each other. The parameter $\rho_m$ in (2.10) is a measure of the correlation between the row score $x_{im}$ and the column score $y_{jm}$, since

$$(2.12) \qquad \sum_{i=1}^{I} \sum_{j=1}^{J} x_{im} y_{jm} P_{ij} = \rho_m,$$

for $m = 1, \ldots, M$, under (2.10) - (2.11). Without loss of generality, the correlation parameters can be ordered so that $1 \geq \rho_1 \geq \rho_2 \geq \ldots \geq \rho_M \geq 0$.

The row and column scores, $x_{i1}$ and $y_{j1}$, in (2.10)–(2.11) are the standardized scores that maximize the correlation $\rho_1$; the $x_{i2}$ and $y_{j2}$ are the standardized scores that maximize the correlation $\rho_2$ subject to the constraint that the $x_{i2}$ and $x_{i1}$ are uncorrelated and that the $y_{j2}$ and $y_{j1}$ are uncorrelated (see (2.11)); et cetera.

I now rewrite (2.12) as follows:

$$(2.13) \qquad \sum_{i=1}^{I} \sum_{j=1}^{J} (x_{im} P_{i+})(y_{jm} P_{+j})[P_{ij}/(P_{i+} P_{+j})] = \rho_m,$$

for $m = 1, \ldots, M$. From (2.11) - (2.13), we see that the correlation $\rho_m$ in (2.13) can be viewed as a row $\times$ column contrast in the $P_{ij}/(P_{i+} P_{+j})$. Thus, the row and column scores, $x_{i1}$ and $y_{j1}$, in (2.13) are the standardized scores that maximize the above row $\times$ column contrast; the $x_{i2}$ and $y_{j2}$ are the standardized scores that maximize the above row $\times$ column contrast subject to the constraint that the $x_{i2}$ and $x_{i1}$ are uncorrelated and that the $y_{j2}$ and $y_{j1}$ are uncorrelated; et cetera.

Another way to view the parameters in model (2.10) is as follows: The coefficient $\rho_1$ is the first canonical correlation between the linear functions $\sum_i x_{i1} X_i$ and $\sum_j y_{j1} Y_j$, where $x_{i1}$ and $y_{j1}$ are the coefficients in the linear functions and where $X_i$ and $Y_j$ are dummy (indicator) variables corresponding to the $i$-th row category ($i = 1, \ldots, I$) and $j$-th column category ($j = 1, \ldots, J$), respectively; and, similarly, the coefficient $\rho_m$ is the $m$-th canonical correlation between the linear functions $\sum_i x_{im} X_i$ and $\sum_j y_{jm} Y_j$.

Let us now consider the measure of nonindependence

$$(2.14) \qquad \lambda_{ij} = (P_{ij} - P_{i+} P_{+j})/(P_{i+} P_{+j}),$$

which I shall call the "Pearson contingency". Note that the $\lambda_{ij}$ satisfy the following conditions:

$$(2.15) \quad \sum_{i=1}^{I} \lambda_{ij} P_{i+} = 0 \text{ (for } j = 1, \ldots, J), \quad \sum_{j=1}^{J} \lambda_{ij} P_{ij} = 0 \text{ (for } i = 1, \ldots, I).$$

We can now rewrite model (2.10) as

$$(2.16) \qquad\qquad \lambda_{ij} = \sum_{m=1}^{M} \rho_m x_{im} y_{jm},$$

where the row scores $x_{im}$ $(m = 1, \ldots, M)$ and column scores $y_{jm}$ $(m = 1, \ldots, M)$ satisfy conditions (2.11). (We obtain (2.10)–(2.11), or the corresponding (2.16) and (2.11), by a singular value decomposition of the $\lambda_{ij}$, and by making use of the fact that the $\lambda_{ij}$ satisfy conditions (2.15); see, e.g., Goodman, 1991a.)

In the usual correspondence analysis, the row scores $x_{im}$ and column scores $y_{jm}$ are reparameterized to obtain the following related scores:

$$(2.17) \qquad\qquad x'_{im} = \rho_m x_{im}, \quad y'_{jm} = \rho_m y_{jm};$$

and the $x'_{im}$ and $y'_{jm}$ are the row and column scores that are used in the usual correspondence analysis graphical display. We shall comment later in Section 2.4 on this usage of the $x'_{im}$ and $y'_{jm}$ in graphical displays.

With the Pearson contingency defined by (2.14), we next consider Pearson's "mean squared contingency" for the $I \times J$ table:

$$(2.18) \qquad\qquad \lambda^2 = \sum_{i=1}^{I} \sum_{j=1}^{J} \lambda_{ij}^2 P_{i+} P_{+j}.$$

From (2.10)–(2.11), or the corresponding (2.16) and (2.11), we obtain

$$(2.19) \qquad \lambda^2 = \sum_{m=1}^{M} \rho_m^2 = \sum_{m=1}^{M} \sum_{i=1}^{I} x'^2_{im} P_{i+} = \sum_{m=1}^{M} \sum_{j=1}^{J} y'^2_{jm} P_{+j};$$

with the $x'_{im}$ and $y'_{jm}$ defined by (2.17). Formula (2.19) partitions $\lambda^2$ into $M$ components, $\rho_m^2$ (for $m = 1, \ldots, M$), with the further partitioning of each component into subcomponents pertaining to the additive contribution of the $i$-th row (for $i = 1, \ldots, I$) or the additive contribution of the $j$-th column (for $j = 1, \ldots, J$).

For a sample of size $N$ from the $I \times J$ contingency table, the $\lambda_{ij}$ defined by (2.14) and the $\lambda^2$ defined by (2.18) can be estimated by replacing the parameters in (2.14) and (2.18) by the corresponding sample estimates. The estimated $\lambda^2$ thus obtained multiplied by $N$ is Pearson's goodness-of-fit statistic; and the distribution of this statistic will be approximately chi-square with $(I - 1)(J - 1)$ degrees of freedom, under the usual null hypothesis (2.1). For Pearson's goodness-of-fit statistic, the corresponding components and subcomponents can be obtained from (2.19), using the estimated $\rho_m^2$, $x_{im}'^2$, and $y_{jm}'^2$, multiplied by $N$.

Both the correspondence analysis approach and the canonical correlation approach to the analysis of contingency tables rest, either explicitly or implicitly, on model (2.10); and so we refer to (2.10) as the canonical correlation model or the correspondence analysis model. We shall return to this model later herein.

We present next some brief historical comments on the development of the correspondence analysis approach and the canonical correlation approach to the analysis of contingency tables:

Benzécri (1973) is usually cited as the founder of correspondence analysis, and Fisher (1940) is usually cited in connection with the application of canonical correlation to the analysis of contingency tables. As noted earlier (Goodman, 1981), reference on this topic should also be made to other early contributions, e.g., Hotelling (1933, 1936), Hirschfeld (1935), and Maung (1941).

The contributions of Hotelling, Hirschfeld (a.k.a. Hartley), Fisher, and Maung can be described briefly as follows: Hotelling introduced canonical correlation for the analysis of sets of continuous variates; Hirschfeld introduced a special set of scores for the rows and a special set of scores for the columns in the contingency table, and he then calculated the correlations between the corresponding row scores and column scores using related characteristic polynomial equations; Fisher, using his own theory of discriminant functions, applied the (Hotelling) canonical correlation approach to contingency tables by introducing a set of variates pertaining to the row categories and a set of variates pertaining to the column categories, and then calculating what Hotelling had called canonical components; and Maung developed further Fisher's approach using the canonical correlations as defined by Hotelling. The canonical correlation approach to the analysis of contingency tables could be called the Hirschfeld–Fisher–Maung approach (or the Hartley–Fisher–Maung approach). Additional reference might also be made to related early work by Guttman (1941, 1950) and others. (For additional references, see, e.g., Goodman, 1991b.)

Before closing this section, let us return for a moment to the $2 \times 2$ table. For this special case, we note that $M = 1$ in the canonical correlation or correspondence analysis model (2.10), and that the correlation $\rho_1$ in (2.10) and (2.12) is equal to the absolute value of the coefficient (2.4a) considered earlier herein.

The next section will present a single general approach to the analysis of nonindependence in the $I \times J$ table that will include as special cases correspondence analysis (or the canonical correlation approach), association analysis (or the log-bilinear approach), and three other approaches that provide generalizations of the corresponding approaches considered earlier in Section 2.2. Other special cases of the general approach presented next could also be considered; and other generalizations of correspondence analysis could be considered too. Again, due to space constraints, these topics will not be included here. Instead we refer the interested reader to, e.g., Rao (1995) for the development of a general theory of canonical coordinates (that includes correspondence analysis as a special case) and for an alternative to correspondence analysis that replaces the chi-square-type measure of distance used in correspondence analysis (as in, e.g., (2.42)–(2.43) later herein) by the Hellinger distance.

## 2.3b A single general approach to the analysis of contingency tables

For the $I \times J$ table, with the definition of $P_{ij}$, $P_{i+}$, and $P_{+j}$ presented earlier herein, we take note again of the usual null model (2.1) of statistical independence between the row classification and the column classification in the contingency table, and we next consider the following ratios in the case where the null model (2.1) may not hold true:

$$(2.20) \qquad \psi_{ij} = P_{ij}/(P_{i+}P_{+j}).$$

As earlier herein with the ratios (2.5), I shall call the $\psi_{ij}$ in (2.20) the "Pearson ratios"; and, as with (2.14), I call the quantities $\psi_{ij} - 1$ the "Pearson contingencies". (Under the null model of statistical independence, $\psi_{ij} = 1$, for $i = 1, \dots, I$ and $j = 1, \dots, J$.) Let $R[x]$ be any monotonically increasing function of $x$ (for $x > 0$), and let

$$(2.21) \qquad R_{ij} = R[\psi_{ij}].$$

I shall call $R[x]$ the "interaction link function" or the "nonindependence link function". We now define the interaction

$$(2.22) \qquad \tilde{\lambda}_{ij} = R_{ij} - \tilde{R}_{i\cdot} - \tilde{R}_{\cdot j} + \tilde{R}_{\cdot\cdot},$$

where

$$(2.23) \qquad \begin{array}{l} \tilde{R}_{i\cdot} = \sum_{j=1}^{J} R_{ij} P_{+j}, \quad \tilde{R}_{\cdot j} = \sum_{i=1}^{I} R_{ij} P_{i+}, \\ \tilde{R}_{\cdot\cdot} = \sum_{i=1}^{I}\sum_{j=1}^{J} R_{ij} P_{i+} P_{+j} = \sum_{i=1}^{I} \tilde{R}_{i\cdot} P_{i+} = \sum_{j=1}^{J} \tilde{R}_{\cdot j} P_{+j}. \end{array}$$

The interaction $\tilde{\lambda}_{ij}$ in (2.22)–(2.23) is somewhat analogous to the usual two-factor (row-by-column) interaction defined in the usual analysis of variance for a two-way array with continuous data. (In the present context, the interaction is applied to the $R_{ij}$ calculated from the $\psi_{ij}$.) The definition of the $\tilde{\lambda}_{ij}$ in (2.22)–(2.23) uses as weights the $P_{i+}$ and $P_{+j}$, the row and column marginal distributions. This definition is directly analogous to a definition of two-factor (row-by-column) interaction in the analysis of variance (for a two-way array with continuous data), which is different from the usual definition, and which uses a set of row weights and column weights in defining the general mean, main effects, and interactions (see, e.g., Scheffé, 1959). We can also view the $\tilde{\lambda}_{ij}$ in (2.22)–(2.23) as "generalized Pearson contingencies".

In defining the weighted interaction $\tilde{\lambda}_{ij}$ in (2.22)–(2.23), we use as weights the $P_{i+}$ and $P_{+j}$ for expository purposes and to facilitate the comparisons that will be made here; but we could use, more generally, any specified set of positive weights.

From (2.22) and (2.23), we see that the $\tilde{\lambda}_{ij}$ satisfy the following conditions:

(2.24)
$$\textstyle\sum_{i=1}^{I} \tilde{\lambda}_{ij} P_{i+} = 0 \ (\text{for } j = 1, \ldots, J), \ \sum_{j=1}^{J} \tilde{\lambda}_{ij} P_{+j} = 0 \ (\text{for } i = 1, \ldots, I).$$

Let us consider for a moment the special case where $R[x]$ is the identity function, with $R[x] = x$. In this case, we see from (2.22)–(2.23) that

(2.25)
$$\tilde{R}_{i.} = \tilde{R}_{.j} = \tilde{R}_{..} = 1,$$

and the $\tilde{\lambda}_{ij}$ in this case are simply the $\psi_{ij} - 1$. In other words, in this special case, the $\tilde{\lambda}_{ij}$ in (2.22)–(2.23) are equal to the "Pearson contingencies", which appeared in the preceding section as (2.14).

The interaction $\tilde{\lambda}_{ij}$ defined by (2.22)–(2.23) can now be rewritten as

(2.26)
$$\tilde{\lambda}_{ij} = \textstyle\sum_{m=1}^{M} \tilde{\phi}_m \tilde{\mu}_{im} \tilde{\nu}_{jm},$$

where $M = \min(I, J) - 1$, and where the row scores $\tilde{\mu}_{im}$ $(m = 1, \ldots, M)$ and column scores $\tilde{\nu}_{jm}$ $(m = 1, \ldots, M)$ satisfy the following conditions:

(2.27)
$$\textstyle\sum_{i=1}^{I} \tilde{\mu}_{im} P_{i+} = 0, \ \sum_{j=1}^{J} \tilde{\nu}_{jm} P_{+j} = 0,$$
$$\textstyle\sum_{i=1}^{I} \tilde{\mu}_{im}^2 P_{i+} = 1, \ \sum_{j=1}^{J} \tilde{\nu}_{jm}^2 P_{+j} = 1,$$
$$\textstyle\sum_{i=1}^{I} \tilde{\mu}_{im} \tilde{\mu}_{im'} P_{i+} = 0, \ \sum_{j=1}^{J} \tilde{\nu}_{jm} \tilde{\nu}_{jm'} P_{+j} = 0,$$

for $m \neq m'$. (We obtain (2.26)–(2.27) by a singular value decomposition of the $\tilde{\lambda}_{ij}$, and by making use of the fact that the $\tilde{\lambda}_{ij}$ satisfy conditions (2.24);

see, e.g., comment in the preceding section in the parenthetical sentence immediately following (2.16).) The parameters $\tilde{\mu}_{im}$ and $\tilde{\nu}_{jm}$ in (2.26)–(2.27) are standardized scores for the rows and columns, respectively, pertaining to the $m$-th component $(m = 1, \ldots, M)$ on the right side of (2.26); and, for $m \neq m'$, the corresponding row scores $\tilde{\mu}_{im}$ and $\tilde{\mu}_{im'}$ are uncorrelated with each other, and the corresponding column scores $\tilde{\nu}_{jm}$ and $\tilde{\nu}_{jm'}$ are uncorrelated with each other. For the $I \times J$ contingency table, the parameter $\tilde{\phi}_m$ in (2.26) can be viewed as a row-by-column interaction in the $R_{ij}$ (i.e., a row-by-column contrast in the $R_{ij}$), since

$$(2.28) \qquad \sum_{i=1}^{I} \sum_{j=1}^{J} (\tilde{\mu}_{im} P_{i+})(\tilde{\nu}_{jm} P_{+j}) R_{ij} = \tilde{\phi}_m,$$

for $m = 1, \ldots, M$, under (2.26)–(2.27). Without loss of generality, the parameters $\tilde{\phi}_m$ can be ordered so that $\tilde{\phi}_1 \geq \tilde{\phi}_2 \geq \ldots \geq \tilde{\phi}_M \geq 0$.

From (2.22) and (2.26), we find that

$$(2.29) \qquad R_{ij} = \bar{R}_{i\cdot} + \bar{R}_{\cdot j} - \bar{R}_{\cdot\cdot} + \sum_{m=1}^{M} \tilde{\phi}_m \tilde{\mu}_{im} \tilde{\nu}_{jm};$$

and we see from (2.20) and (2.21) that

$$(2.30) \qquad P_{ij} = P_{i+} P_{+j} R^{-1}[\bar{R}_{i\cdot} + \bar{R}_{\cdot j} - \bar{R}_{\cdot\cdot} + \sum_{m=1}^{M} \tilde{\phi}_m \tilde{\mu}_{im} \tilde{\nu}_{jm}],$$

where $R^{-1}$ denotes the inverse function of $R[x]$.

With the "generalized Pearson contingency" defined by (2.22)–(2.23), we next consider the corresponding generalization of Pearson's mean squared contingency:

$$(2.31) \qquad \tilde{\lambda}^2 = \sum_{i=1}^{I} \sum_{j=1}^{J} \tilde{\lambda}_{ij}^2 P_{i+} P_{+j}.$$

And from (2.26)–(2.27) we obtain

$$(2.32) \qquad \tilde{\lambda}^2 = \sum_{m=1}^{M} \tilde{\phi}_m^2 = \sum_{m=1}^{M} \sum_{i=1}^{J} \tilde{\mu}_{im}^{\prime 2} P_{i+} = \sum_{m=1}^{M} \sum_{j=1}^{J} \tilde{\nu}_{jm}^{\prime 2} P_{+j},$$

where the $\tilde{\mu}_{im}'$ and $\tilde{\nu}_{jm}'$ are the following reparameterized row scores and column scores:

$$(2.33) \qquad \tilde{\mu}_{im}' = \tilde{\phi}_m \tilde{\mu}_{im}, \quad \tilde{\nu}_{jm}' = \tilde{\phi}_m \tilde{\nu}_{jm}.$$

Formula (2.32) partitions the generalized mean squared contingency $\bar{\lambda}^2$ into $M$ components, $\bar{\phi}_m^2$ (for $m = 1, \ldots, M$), with the further partitioning of each component into its subcomponents.

We noted in the preceding section that, for a sample of size $N$ from the $I \times J$ table, the distribution of the estimated mean squared contingency obtained from (2.18) multiplied by $N$ will be approximately chi-square with $(I - 1)(J - 1)$ degrees of freedom, under the null hypothesis (2.1); and this is also the case more generally for the estimated coefficient obtained from (2.31) when the function used for $R[x]$ in (2.21)–(2.23) is any power function of the form (2.7) (see, e.g., Goodman, 1997). Also, for the corresponding generalized goodness-of-fit statistic, viz., the estimated coefficient obtained from (2.31) multiplied by $N$, the corresponding components and subcomponents can be obtained from (2.32), using the estimated $\bar{\phi}_m^2$, $\bar{\mu}_{im}'^2$, and $\bar{\nu}_{jm}'^2$, multiplied by $N$.

In the special case where $R[x]$ is the identity function (with $R[x] = x$), each of the more general formulas presented in the present section can be replaced by the corresponding less general formula presented in the preceding section. For example, in this special case we see from (2.25) and (2.30) that

$$(2.34) \qquad P_{ij} = P_{i+} P_{+j} \left(1 + \sum_{m=1}^{M} \rho_m x_{im} y_{jm}\right),$$

where the $\rho_m$, $x_{im}$, and $y_{jm}$ in (2.34) denote the corresponding $\bar{\phi}_m$, $\bar{\mu}_{im}$, and $\bar{\nu}_{jm}$ in (2.30) in this case. Thus, we obtain in this special case the same formula (2.10) presented in the preceding section. As we noted in the preceding section, this formula (viz., (2.10) or (2.34)) can be viewed as the basic formula used in the canonical correlation approach to contingency table analysis and in the correspondence analysis approach.

In the special case where $R[x]$ is the logarithmic function (with $R[x] = \log x$), we see from (2.30) that

$$(2.35) \qquad P_{ij} = \bar{\alpha}_i \bar{\beta}_j \exp\left(\sum_{m=1}^{M} \bar{\phi}_m \bar{\mu}_{im} \bar{\nu}_{jm}\right),$$

where $\bar{\alpha}_i$ and $\bar{\beta}_j$ are parameters that are related to the $P_{i+}$ and $P_{+j}$. The $\bar{\alpha}_i$ and $\bar{\beta}_j$ are such that the row and column marginals,

$$\bar{\alpha}_i \sum_j \bar{\beta}_j \exp\left[\sum_m \bar{\phi}_m \bar{\mu}_{im} \bar{\nu}_{jm}\right] \text{ and } \bar{\beta}_j \sum_i \bar{\alpha}_i \exp\left[\sum_m \bar{\phi}_m \bar{\mu}_{im} \bar{\nu}_{jm}\right],$$

are equal to the corresponding $P_{i+}$ and $P_{+j}$, respectively. (From (2.29) we also see that the $\bar{\alpha}_i \bar{\beta}_j$ in (2.35) can be rewritten as $\exp[\bar{G}_i. + \bar{G}_{.j} - \bar{G}..]$, where the $\tilde{G}_{i\cdot}$, $\tilde{G}_{\cdot j}$, and $\tilde{G}_{\cdot\cdot}$ are calculated as in (2.23) with the $R_{ij}$ there replaced by the corresponding $G_{ij} = \log P_{ij}$.) Formula (2.35) can be viewed as the basic formula used in weighted association analysis (see, e.g., Goodman, 1985, 1986, 1987).

When $R[x]$ is the identity function (2.9a), we have now seen that the general approach presented in the present section yields the usual canonical correlation approach to the analysis of the contingency table and the correspondence analysis approach; and for the corresponding $2 \times 2$ table special case, the coefficient (2.4a) is obtained. Similarly, in the special case when $R[x]$ is the logarithmic function (2.9c), the general approach presented in the present section yields the usual association analysis approach (or the log-bilinear approach to the analysis of the contingency table); and for the corresponding $2 \times 2$ table special case, the coefficient (2.4c) is obtained. In addition, in the special case when $R[x]$ is the square-root function multiplied by 2 (see (2.9b)), the general approach yields the concomitance analysis approach (or the "midway" approach to the analysis of the contingency table); and for the corresponding $2 \times 2$ table special case, the coefficient (2.4b) is obtained. (For more details pertaining to the "midway" approach or the concomitance analysis approach, see, e.g., Goodman, 1996, 1997.)

When the $P_{ij}$ in the contingency table are replaced by the corresponding standardized values, $P_{ij}^*$ (with $P_{i+}^* = 1/I$, for $i = 1,\ldots,I$, and $P_{+j}^* = 1/J$, for $j = 1,\ldots,J$), the general approach presented in this section can yield the marginal-free association model approach (or the log-bilinear approach to the analysis of the standardized table) and/or a marginal-free correspondence analysis approach (or a canonical correlation approach to the analysis of the standardized table). And, for the corresponding $2 \times 2$ table special case, the coefficients (2.4d) and/or (2.4e) are obtained, respectively.

It may also be worth noting here that the interaction $\tilde{\lambda}_{ij}$ in (2.22)–(2.23) can be rewritten as follows in the special case of the $2 \times 2$ table:

$$(2.36) \qquad \tilde{\lambda}_{ij} = (R_{11} - R_{12} - R_{21} + R_{22})(-1)^{i+j}(1 - P_{i+})(1 - P_{+j}).$$

For this special case, we note that $M = 1$ in (2.26), and that the interaction parameter $\tilde{\phi}_1$ in (2.26) and (2.28) is equal to the absolute value of the coefficient (2.8) considered earlier herein. Thus, when the general approach presented in the present section is applied to the special case of the $2 \times 2$ table, this approach yields the same general coefficient of nonindependence (2.8) presented earlier herein (Section 2.2f) for the analysis of the $2 \times 2$ table.

The general approach presented in the present section for the analysis of the $I \times J$ contingency table yields as special cases (1) the usual canonical correlation approach (or the correspondence analysis approach) to the analysis of the $I \times J$ table, (2) the usual association analysis approach (or the log-bilinear approach) to the analysis of the $I \times J$ table, and the three other approaches noted above. The usual canonical correlation approach can be viewed as an approach based on the analysis of correlation coefficients, and the usual association analysis approach can be viewed as an approach based on the analysis of odds-ratios or log-odds-ratios. The difference between these two approaches brings to mind the difference between Pearson's approach and Yule's approach in the Pearson-Yule debate referred to in the Introduction to the present article. We shall now return for a moment to this debate.

At the time of the Pearson-Yule debate on the analysis of cross-classified data, Pearson's approach was based primarily on the bivariate normal. He assumed that the row and column classifications in the contingency table arise from underlying continuous random variables having a bivariate normal distribution, so that the sample contingency table comes from a discretized bivariate normal; and he then was concerned with the estimation of the correlation coefficient for the underlying bivariate normal. On the other hand, Yule felt that, for many kinds of contingency tables, it was not desirable in scientific work to introduce assumptions about an underlying bivariate normal in the analysis of these tables; and for such tables, he used, to a large extent, coefficients based on the odds-ratios (e.g., his coefficient of colligation (2.4e)), coefficients that did not require any assumptions about underlying distributions. The Pearson approach and the Yule approach appear to be wholly different from each other; but a kind of reconciliation of the two perspectives was obtained in Goodman (1981a). In that article, association analysis (which, as I noted in the preceding paragraph, can be viewed as an analysis of odds-ratios or log-odds-ratios) was used to obtain association models for (1) examining whether the assumption of an underlying bivariate normal distribution is congruent with the observed contingency table, and (2) estimating the correlation coefficient for the underlying bivariate normal in the case where this distribution is found to be congruent with the observed data. It may also be worthwhile to note here that the use of association analysis (based on an analysis of odds-ratios or log-odds-ratios) to examine Pearson's bivariate normal assumption turns out to yield more parsimonious results, in a certain sense, than a corresponding use of the approach that is based on an analysis of correlation coefficients (viz., Fisher's canonical correlation approach – i.e., the Hartley-Fisher-Maung approach – or the correspondence analysis approach); see, e.g., Goodman (1985, 1991a).

## 2.4 Graphical displays for the analysis of nonindependence in the $I \times J$ contingency table

For expository purposes, I shall consider in this section graphical displays for the analysis of nonindependence between the row classification and the column classification in the contingency table using correspondence analysis and the canonical correlation approach in Section 2.3a herein. Corresponding results could also be presented pertaining to graphical displays for the analysis of this nonindependence using the more general approach presented in Section 2.3b, and/or using the various special cases (of the more general approach) noted in that section.

In Section 2.3a, we noted that, in the usual correspondence analysis, the row scores $x_{im}$ and column scores $y_{jm}$ (in (2.10)–(2.11)) are reparameterized to obtain the corresponding scores $x'_{im}$ and $y'_{jm}$ defined by (2.17), and the $x'_{im}$ and $y'_{jm}$ are the row and column scores that are used in the usual graphical display in correspondence analysis. We next consider a different reparameterization with the row scores $x^{\dagger}_{im}$ and column scores $y^{\dagger}_{jm}$ defined as follows:

$$(2.37) \qquad x^{\dagger}_{im} = \sqrt{\rho_m}\, x_{im} = x'_{im}/\sqrt{\rho_m},\ y^{\dagger}_{jm} = \sqrt{\rho_m}\, y_{jm} = y'_{jm}/\sqrt{\rho_m}.$$

The reparameterization (2.37) is a special case of the more general reparameterization in which the row scores $x^{*}_{im}$ and column scores $y^{*}_{jm}$ are defined as follows:

$$(2.38) \qquad x^{*}_{im} = x_{im}\rho_m^{\gamma} = x'_{im}/\rho_m^{\delta},\ y^{*}_{jm} = y_{jm}\rho_m^{\delta} = y'_{jm}/\rho_m^{\gamma},$$

for $\gamma + \delta = 1$. As special cases of (2.38), when $\gamma = 0$ and $\delta = 1$, we have $x^{*}_{im} = x_{im}$ and $y^{*}_{jm} = y'_{jm}$; when $\gamma = 1$ and $\delta = 0$, we have $x^{*}_{im} = x'_{im}$ and $y^{*}_{jm} = y_{jm}$; and when $\gamma = \delta = \frac{1}{2}$, we have $x^{*}_{im} = x^{\dagger}_{im}$ and $y^{*}_{jm} = y^{\dagger}_{jm}$. From (2.10) and (2.38) we see that

$$(2.39) \qquad\qquad P_{ij} = P_{i+}P_{+j}\Big(1 + \sum_{m=1}^{M} x^{*}_{im} y^{*}_{jm}\Big).$$

Formula (2.39) simplifies (2.10) and (2.16). From (2.39) we see that

$$(2.40) \qquad (P_{ij} - P_{i+}P_{+j})/(P_{i+}P_{+j}) = \sum_{m=1}^{M} x^{*}_{im} y^{*}_{jm} = |\mathbf{x}^{*}_i|\,|\mathbf{y}^{*}_j|\cos(\mathbf{x}^{*}_i, \mathbf{y}^{*}_j)$$

$$= \{|\mathbf{x}^{*}_i|^2 + |\mathbf{y}^{*}_j|^2 - |\mathbf{x}^{*}_i - \mathbf{y}^{*}_j|^2\}/2,$$

where $\mathbf{x}_i^* = (x_{i1}^*, \ldots, x_{iM}^*)$ and $\mathbf{y}_j^* = (y_{j1}^*, \ldots, y_{jM}^*)$, and where $\cos(\mathbf{x}_i^*, \mathbf{y}_j^*)$ is the cosine of the angle subtended at the origin by the points $\mathbf{x}_i^*$ and $\mathbf{y}_j^*$. Formula (2.40) provides a straightforward geometric interpretation for the comparison of a row point with a column point using the simultaneous display of the $\mathbf{x}_i^*$ and $\mathbf{y}_j^*$ in the same graph. On the right side of (2.40), the $|\mathbf{x}_i^*|$ is the distance of the point $\mathbf{x}_i^*$ from the origin, the $|\mathbf{y}_j^*|$ is the distance of the point $\mathbf{y}_j^*$ from the origin, and $\cos(\mathbf{x}_i^*, \mathbf{y}_i^*)$ is the angle between the vector formed by $\mathbf{x}_i^*$ and the vector formed by $\mathbf{y}_j^*$ (where the vector formed by a point is simply the directed line segment from the origin to the point). The right side of (2.40) can also be expressed as the product of $|\mathbf{x}_i^*|$ and the length of the projection of the point $\mathbf{y}_j^*$ onto the vector formed by $\mathbf{x}_i^*$, and as the product of $|\mathbf{y}_j^*|$ and the length of the projection of the point $\mathbf{x}_i^*$ onto the vector formed by $\mathbf{y}_j^*$; see, e.g., Goodman (1986, 1991a) and related literature cited there.

From formula (2.40), we also see that the larger the relative difference $(P_{ij} - P_{i+}P_{+j})/(P_{i+}P_{+j})$, the larger will be the difference between $(|\mathbf{x}_i^*|^2 + |\mathbf{y}_j^*|^2)$ and $|\mathbf{x}_i^* - \mathbf{y}_j^*|^2$, and the larger will be the product $|\mathbf{x}_i^*||\mathbf{y}_j^*|\cos(\mathbf{x}_i^*, \mathbf{y}_j^*)$. Note that $|\mathbf{x}_i^* - \mathbf{y}_i^*|^2$ is simply the squared distance between the points $\mathbf{x}_i^*$ and $\mathbf{y}_j^*$, and the sum of $|\mathbf{x}_i^*|^2$ and $|\mathbf{y}_j^*|^2$ can be interpreted simply as the squared distance between the points $\mathbf{x}_i^*$ and $\mathbf{y}_j^*$ when the vectors formed by $\mathbf{x}_i^*$ and $\mathbf{y}_j^*$ are made orthogonal (perpendicular) to each other.

In the special cases when $\gamma = \delta = \frac{1}{2}$ in (2.38), when $\gamma = 0$ and $\delta = 1$, and when $\gamma = 1$ and $\delta = 0$, the simultaneous displays obtained with the $\mathbf{x}_i^*$ and $\mathbf{y}_j^*$ are particularly easy to interpret for the comparison of a row point with a column point. The special case when $\gamma = \delta = \frac{1}{2}$ has the advantage that the $\mathbf{x}_i^*$ and $\mathbf{y}_j^*$ thus obtained are on the same scale (see (2.37)), and the other two special cases have special features that we will take note of later in this section.

Benzécri (1973, 1992) has emphasized that "a distinguishing feature of correspondence analysis is the perfect symmetry of the roles assigned to the [rows and columns] in correspondence", and that "this permits the simultaneous representation of the [row points and column points] on the same axes". (See, e.g., Benzécri, 1992, pp. 2–3.) In his correspondence analysis, he uses the row points $\mathbf{x}_i'$ and column points $\mathbf{y}_j'$ in the same graph (i.e., on the same axes), where $\mathbf{x}_i' = (x_{i1}', \ldots, x_{iM}')$ and $\mathbf{y}_j' = (y_{j1}', \ldots, y_{jM}')$, and where the row scores $x_{im}'$ and column scores $y_{jm}'$ are equal to the quantities defined by (2.17) earlier herein. (In practice, graphical displays in correspondence analysis will usually include only the first two components of the row points and column points – supplementary displays may sometimes in-

clude the first and third components and the second and third components.)
The simultaneous display of the $\mathbf{x}'_i$ and $\mathbf{y}'_j$ in the same graph is different
from the simultaneous display of the $\mathbf{x}^*_i$ and $\mathbf{y}^*_j$ (and it is also different from
the simultaneous display of the $\mathbf{x}_i$ and $\mathbf{y}_j$, where $\mathbf{x}_i = (x_{i1}, \ldots, x_{iM})$ and
$\mathbf{y}_j = (y_{j1}, \ldots, y_{jM})$). Justification for the simultaneous display of the $\mathbf{x}^*_i$
and $\mathbf{y}^*_j$ in the same graph is obtained using formula (2.40) above; whereas
the simultaneous display of the row points $\mathbf{x}'_i$ and column points $\mathbf{y}'_j$ in the
same graph can *not* be justified by the usual correspondence analysis for-
mulae, although this simultaneous display is part of routine practice in the
usual correspondence analysis.

Although the simultaneous display of the $\mathbf{x}'_i$ and $\mathbf{y}'_j$ in a single graph
can not be justified, we can use (2.40) to justify the simultaneous display of
the $\mathbf{x}_i$ and $\mathbf{y}'_j$ (with $\gamma = 0$ and $\delta = 1$ in (2.38)), and we can also use (2.40)
to justify the simultaneous display of the $\mathbf{x}'_i$ and $\mathbf{y}_j$ (with $\gamma = 1$ and $\delta = 0$
in (2.38)). Additional justification for the simultaneous display of the $\mathbf{x}_i$
and $\mathbf{y}'_j$ and the simultaneous display of the $\mathbf{x}'_i$ and $\mathbf{y}_j$ is obtained from the
following results, which can be derived from (2.10)–(2.11) and (2.17):

(2.41)
$$\sum_{i=1}^{I} x_{im} P_{ij}/P_{+j} = \rho_m y_{jm} = y'_{jm},$$
$$\sum_{j=1}^{J} y_{jm} P_{ij}/P_{i+} = \rho_m x_{im} = x'_{im}.$$

As we noted earlier, in the situation in which the comparison of a row
point with a column point is of interest, the simultaneous display of $\mathbf{x}^*_i$
and $\mathbf{y}^*_j$ using $\gamma = \delta = \frac{1}{2}$ in (2.38) has the advantage that the $\mathbf{x}^*_i$ and $\mathbf{y}^*_j$
thus obtained (viz., the $\mathbf{x}^\dagger_i$ and $\mathbf{y}^\dagger_j$ in (2.37)) are on the same scale in this
portrait of the nonindependence in the cells in the contingency table; and
the simultaneous display of $\mathbf{x}^*_i$ and $\mathbf{y}^*_j$ using $\gamma = 0$ and $\delta = 1$, or using
$\gamma = 1$ and $\delta = 0$, has the advantage that the $\mathbf{x}^*_i$ and $\mathbf{y}^*_j$ thus obtained can
be justified using (2.41).

The preceding comments in the present section were concerned with
the simultaneous representation of row points and column points in the
same graph for the purpose of comparing row points with column points.
On the other hand, in the situation in which only the comparison of the
rows is of interest, a separate graphical display that presents the $\mathbf{x}'_i$ ($i =
1, \ldots, I$), or a display that presents the $\mathbf{x}'_i$ and $\mathbf{y}_j$, should be used; and in
the situation in which only the comparison of the columns is of interest, a
separate graphical display that presents the $\mathbf{y}'_j$ ($j = 1, \ldots, J$), or a display
that presents the $\mathbf{y}'_j$ and $\mathbf{x}_i$, should be used.

The justification for a separate graphical display that presents the $x_i'$ $(i = 1, \ldots, I)$ and/or for a separate graphical display that presents the $y_j'$ $(j = 1, \ldots, J)$ is obtained from the following results, which can be derived from (2.10) - (2.11) and (2.17): Using the measure of nonindependence $\lambda_{ij}$ defined by (2.14), we find that

(2.42)

$$
\sum_{j=1}^{J} \lambda_{ij}^2 P_{+j} = \sum_{j=1}^{J} [(P_{ij}/P_{i+}) - P_{+j}]^2/P_{+j} = \sum_{j=1}^{J} P_{+j} [\sum_{m=1}^{M} \rho_m x_{im} y_{jm}]^2
$$

$$
= \sum_{m=1}^{M} \rho_m^2 x_{im}^2 = \sum_{m=1}^{M} x_{im'}^2 = |\mathbf{x}_i'|^2,
$$

and

$$
\sum_{j=1}^{J} (\lambda_{ij} - \lambda_{i'j})^2 P_{+j} = \sum_{j=1}^{J} [(P_{ij}/P_{i+}) - (P_{i'j}/P_{i'+})]^2/P_{+j}
$$

$$
= \sum_{j=1}^{J} P_{+j} \left[ \sum_{m=1}^{M} \rho_m (x_{im} - x_{i'm}) y_{jm} \right]^2
$$

(2.43)

$$
= \sum_{m=1}^{M} \rho_m^2 (x_{im} - x_{i'm})^2
$$

$$
= \sum_{m=1}^{M} (x_{im}' - x_{i'm}')^2 = |\mathbf{x}_i' - \mathbf{x}_{i'}'|^2;
$$

and similar formulae can be obtained for $|\mathbf{y}_j'|^2$ and $|\mathbf{y}_j' - \mathbf{y}_{j'}'|^2$. Thus, we see from (2.42) - (2.43) that the squared distance, $|\mathbf{x}_i'|^2$, of the row point $x_i'$ from the origin is, in a sense, the "average squared nonindependence" in the $i$-th row; and the squared distance, $|\mathbf{x}_i' - \mathbf{x}_{i'}'|^2$, of the row point $x_i'$ from the row point $x_{i'}'$ is, in a sense, the "average squared difference in nonindependence" comparing the corresponding nonindependence in rows $i$ and $i'$; and a similar kind of interpretation can be obtained for the $|\mathbf{y}_j'|^2$ and the $|\mathbf{y}_j' - \mathbf{y}_{j'}'|^2$.

Using again the measure of nonindependence $\lambda_{ij}$ defined by (2.14), we see from (2.40) that

$$
\lambda_{ij} P_{+j} = [(P_{ij}/P_{i+}) - P_{+j}] = P_{+j} \sum_{m=1}^{M} x_{im}^* y_{jm}^* = P_{+j} |\mathbf{x}_i^*| |\mathbf{y}_j^*| \cos(\mathbf{x}_i^*, \mathbf{y}_j^*)
$$

(2.44)

$$
= |\mathbf{x}_i^*| |P_{+j} \mathbf{y}_j^*| \cos(\mathbf{x}_i^*, P_{+j} \mathbf{y}_j^*) = |\mathbf{x}_i^*| |P_{+j} \mathbf{y}_j^*| \cos(\mathbf{x}_i^*, \mathbf{y}_j^*);
$$

and a similar formula can be obtained for $\lambda_{ij}P_{i+} = [(P_{ij}/P_{+j}) - P_{i+}]$. Thus, we obtain again justification for the simultaneous display of $\mathbf{x}_i^*$ and $\mathbf{y}_j^*$ even in the asymmetric context in which $\lambda_{ij}P_{+j}$ and/or $\lambda_{ij}P_{i+}$ is of interest (as well as in the symmetric context in which $\lambda_{ij}$ is of interest), and also justification for the corresponding simultaneous display of $\mathbf{x}_i^*$ and $P_{+j}\mathbf{y}_j^*$ and/or the corresponding simultaneous display of $P_{i+}\mathbf{x}_i^*$ and $\mathbf{y}_j^*$.

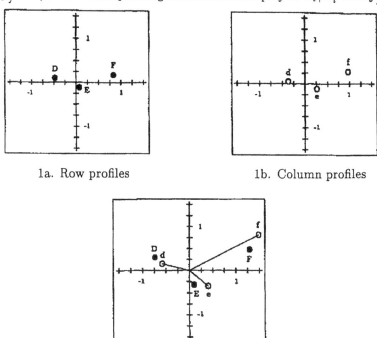

1a. Row profiles             1b. Column profiles

1c. Cell nonindependence

Figure 1.    Nonindependence in the rows, columns, and cells in the example in Table 2, using the correspondence analysis or canonical correlation approach. The inter-point distance displays in Figs. 1a and 1b, and the inter-vector angle display (or point-onto-vector projection display) is in Fig. 1c. Each cell in the two-way table is represented by its row and column in Fig. 1c.

To illustrate the use of some of the graphical displays described above, we shall consider next the example presented in Table 2. This example is a $3 \times 3$ table. For this example, Table 2 presents the intrinsic correlation $\rho_m$ (for $m = 1, 2$), the row scores $x_{im}$ and column scores $y_{jm}$ obtained from (2.10) - (2.11), the reparameterized scores $x'_{im}$ and $y'_{jm}$ (see (2.17)), and the reparameterized scores $x_{im}^{\dagger}$ and $y_{jm}^{\dagger}$ (see (2.37)). Fig. 1a displays the $\mathbf{x}_i' = (x_{i1}', x_{i2}')$, Fig. 1b displays the $\mathbf{y}_j' = (y_{j1}', y_{j2}')$, and Fig. 1c displays the $\mathbf{x}_i^{\dagger} = (x_{i1}^{\dagger}, x_{i2}^{\dagger})$ and $\mathbf{y}_j^{\dagger} = (y_{j1}^{\dagger}, y_{j2}^{\dagger})$.

| Table | | | 60 20 1 60 63.25 10 6 20 10 | | | | |
|---|---|---|---|---|---|---|---|

| Intrinsic correlation | | | $\rho_1 = .42$ | | $\rho_2 = .12$ | | |
|---|---|---|---|---|---|---|---|
| Row coordinate | $x_{i1}$ | $x_{i2}$ | $x_{i1}'$ | $x_{i2}'$ | $x_{i1}^{\dagger}$ | $x_{i2}^{\dagger}$ |
| 1 | −1.14 | 0.89 | −0.48 | 0.10 | −0.74 | 0.30 |
| 2 | 0.16 | −0.92 | 0.07 | −0.11 | 0.10 | −0.32 |
| 3 | 1.99 | 1.42 | 0.83 | 0.17 | 1.29 | 0.48 |
| Column coordinate | $y_{j1}$ | $y_{j2}$ | $y_{j1}'$ | $y_{j2}'$ | $y_{j1}^{\dagger}$ | $y_{j2}^{\dagger}$ |
| 1 | −0.89 | 0.44 | −0.37 | 0.05 | −0.58 | 0.15 |
| 2 | 0.62 | −1.02 | 0.26 | −0.12 | 0.40 | −0.35 |
| 3 | 2.30 | 2.37 | 0.97 | 0.28 | 1.49 | 0.81 |

Table 2. Correspondence analysis applied to an example of a $3 \times 3$ cross-classification. Coefficients of intrinsic correlation, and the corresponding row scores and column scores.

Using D, E, and F to denote rows 1, 2, and 3, respectively, in Fig. 1a, this figure can be used to describe the magnitude of the noninde-pendence (see (2.42)) in rows 1, 2, and 3 of the $3 \times 3$ table, and the magnitude of the difference in nonindependence (see (2.43)) comparing rows 1 and 2, rows 1 and 3, and rows 2 and 3. Similarly, using d, e, and f to denote columns 1, 2, and 3, respectively, in Fig. 1b, this figure can be used to describe the corresponding magnitudes pertaining to the columns of the $3 \times 3$ table. Using D, E, F, and d, e, f to denote rows 1, 2, 3, and columns 1, 2, 3, respectively, in Fig. 1c, this figure can be used to describe the magnitude of the nonindependence (see (2.40)) in the cells of the $3 \times 3$ table. For example, using Fig. 1c, the length of the projection of row points D, E, and F onto the vector formed by col-umn point f can describe the relative magnitude of the nonindependence in cells $(1, 3)$, $(2, 3)$, and $(3, 3)$, respectively. (Note, for example, that non-independence is negative in cell $(1, 3)$, slightly negative in cell $(2, 3)$, and positive in cell $(3, 3)$, using Fig. 1c. This will be discussed further later herein.) Two dimensions are needed (in Fig. 1) to describe these mag-nitudes with the usual correspondence analysis or canonical correlation approach.

Fig. 1a can be viewed as an "inter-point distance display" (viz., as an "inter-row difference display"); Fig. 1b can also be viewed as an "inter-point distance display" (viz., as an "inter-column difference display"); and Fig. 1c can be viewed as an "inter-vector angle display" or as a "point-onto-vector projection display" (viz., as a "row-vector-by-column-vector angle display" or as a "row-point-onto-column-vector projection display" or a "column-point-onto-row-vector projection display").

For the sake of simplicity, we have included in the display for Fig.1c only the row points and the column vectors; but instead of this display, we could have presented the corresponding display that includes only the row vectors and the column points, or the display that includes the row vectors and the column vectors, or the display that includes the row-point projections onto the column vectors (see, e.g., Fig. 2 later herein), or the display that includes the column-point projections onto the row vectors. In addition, instead of the display presented here for Fig. 1c (or the displays noted above as possible alternatives to the display presented for Fig. 1c), we could also have presented the corresponding display that includes only the row points $x_i^\dagger$ and the column points $y_j^\dagger$, with the understanding that, for each cell in the two-way table under consideration, nonindependence is portrayed in the display in terms of the corresponding row-column pair of vectors and the cosine of the angle between these two vectors (see (2.40)), or in terms of the corresponding row-point-onto-column-vector projection, or in terms of the corresponding column-point-onto-row-vector projection. (It is important to note that, in the display presented here for Fig. 1c, and in the displays noted above as the various possible alternatives to the display presented for Fig. 1c, the points and vectors included in the display are all based upon the $x_i^\dagger$ and $y_j^\dagger$.)

From Fig. 1a, we see that nonindependence in row E is small in magnitude; and we also see that, with respect to the primary axis, nonindependence in the rows can be ordered from the nonindependence in row D to the nonindependence in row E to the nonindependence in row F; and, with respect to the secondary axis, the nonindependence in row E can be contrasted with the nonindependence in row D and the nonindependence in row F. (To gain additional insight into the possible ways of interpreting the portrait of nonindependence as presented in Fig. 1a, consider also the nonindependence that can be calculated pertaining to each of the rows, D, E, and F, of the two-way table presented in Table 2, and the corresponding conditional distribution that can be calculated pertaining to each of the rows of this two-way table.) Comparing Fig. 1b with Fig. 1a, we see that a somewhat similar kind of interpretation applies to nonindependence in columns d, e, and f as applied to nonindependence in rows D, E, and F.

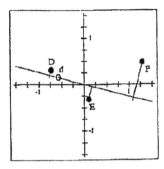 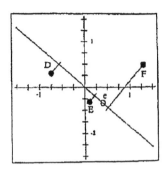

2a. Cell nonindependence
in column d

2b. Cell nonindependence
in column e

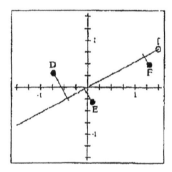

2c. Cell nonindependence in column f

Figure 2.    Nonindependence in the cells in the first column (column d), second column (column e), and third column (column f) in the example in Table 2, using the correspondence analysis or canonical correlation approach. The row-points-onto-column-vector projection displays, for column-vectors d, e, and f. Each cell in the two-way table is represented by its row-point and column-vector.

To gain additional insight into the Fig. 1c portrait of nonindependence, we introduce here Fig. 2, a "deconstruction" of the corresponding display presented earlier as Fig. 1c. Three separate displays are included in Fig. 2, with a separate display corresponding to each of the three column vectors included in Fig. 1c. The first, second, and third displays in Fig. 2 present the row points projected onto the axis formed using the column vectors d, e, and f, respectively. By applying (2.40) (and the comments following

immediately after (2.40)) in turn to the first, second, and third displays in
Fig. 2, we obtain a portrait of the nonindependence in the cells in columns
d, e, and f, respectively.

From the first display in Fig. 2, we see that nonindependence is positive in cell (1,1), slightly negative in cell (2,1), and more negative in cell
(3,1); from the second display, we see that nonindependence is negative in
cell (1,2), somewhat positive in cell (2,2), and more positive in cell (3,2);
and from the third display, we see that nonindependence is negative in cell
(1,3), slightly negative in cell (2,3), and positive in cell (3,3). From each
display, we can also assess the relative magnitude of the nonindependence
in the cells in the column pertaining to that particular display; and, in
addition, we can also compare the magnitude of the nonindependence in
the cells in different columns, using the corresponding displays and taking
into account the length of the corresponding vectors that serve to form the
axes (onto which the projections were made) in the displays (see comments
following (2.40)).

While Fig. 2 presents a portrait of the nonindependence in the cells
in each of the columns in the two-way table in Table 2, a corresponding
set of graphical displays could also have been presented to portray the
nonindependence in the cells in each of the rows in the two-way table.

As we noted earlier, Fig. 2 can be viewed as a "deconstruction" of
Fig. 1c. The above discussion of Fig. 2 is intended to provide additional
insight into the kind of information contained in Fig. 2, in the graphical
display presented for Fig. 1c, and in the displays mentioned earlier herein
as the various possible alternatives to the display presented for Fig. 1c.

In Fig. 1c and in the corresponding deconstructions in Fig. 2, the
displays use row points $x_i^\dagger$ and column points $y_j^\dagger$ in the same graph. As we
noted earlier (see (2.37) - (2.38)), the simultaneous display of the $x_i^\dagger$ and $y_j^\dagger$
is a special case of the simultaneous display of the $x_i^*$ and $y_j^*$ in (2.38), and
two other such special cases are presented in Fig. 3; viz., the simultaneous
display of $x_i'$ and $y_j$ in Fig. 3a, and the simultaneous display of $x_i$ and $y_j'$ in
Fig. 3b. Justification for the simultaneous displays in Fig. 1c, Fig. 3a, and
Fig. 3b is obtained using (2.34); and, in this sense, the displays in the three
figures are equivalent. Although Figs. 1c, 3a, and 3b look very different
from each other, each of these figures includes the same information about
the nonindependence in each cell in the contingency table. As we noted
earlier, Fig. 1c has the advantage that row points and column points are
on the same scale in this portrait of the nonindependence in the cells in
the contingency table; whereas, with respect to Figs. 3a and 3b, additional
justification for these two displays is obtained using (2.41).

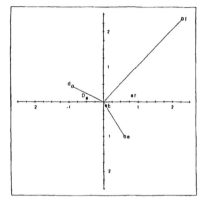
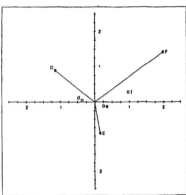

3a. Row point as a weighted
average of column scores

3b. Column point as a weighted
average of row scores

Figure 3. Nonindependence in the cells in the example in Table 2, using the correspondence analysis or canonical correlation approach, with each row point viewed as a weighted average of column scores in Fig. 3a, and each column point viewed as a weighted average of row scores in Fig. 3b. The inter-vector angle display (or point-onto-vector projection display) compares row points with column points (or vectors) in Fig. 3a, and it compares column points with row points (or vectors) in Fig. 3b. Each cell in the two-way table is represented by its row and column in Figs. 3a and 3b.

Figs. 3a and 3b can be deconstructed in a similar way to the deconstruction presented here for Fig. 1c, obtaining for Fig. 3a one set of graphical displays (deconstructions), and for Fig. 3b another set of graphical displays (deconstructions); and each set of displays would be similar in form to the set of graphical displays presented in Fig. 2. From the deconstruction of Fig. 3a, we obtain additional insight into the Fig. 3a portrait of nonindependence, focusing attention on the nonindependence in the cells in each of the columns in the two-way table; and from the deconstruction of Fig. 3b, we obtain additional insight into the Fig. 3b portrait of nonindependence focusing attention now on the nonindependence in the cells in each of the rows in the table. Since Fig. 1c has the advantage (when compared with Figs. 3a and 3b) that the row points and column points are on the same scale, the corresponding deconstruction of Fig. 1c (viz., the graphical displays in Fig. 2) has the same advantage when compared with the corresponding deconstruction of Fig. 3a and/or the corresponding deconstruction of Fig. 3b.

Before closing this section, we note that the kinds of displays presented here (e.g., Figs. 1a, 1b, 1c) are different from the usual correspondence analysis graphical display. In the usual correspondence analysis, the row points $(\mathbf{x}_i')$ in Fig. 1a and the column points $(\mathbf{y}_j')$ in Fig. 1b would have been included together in a single graphical display (rather than in the two separate displays); and the row points $(\mathbf{x}_i^{\dagger})$ and column points $(\mathbf{y}_j^{\dagger})$ in Fig. 1c would not have been used. Also, in the usual correspondence analysis, the distinction between an "inter-point distance display" and an "inter-vector angle display" (or a "point-onto-vector projection display") is usually not made.

The usual correspondence analysis graphical display can be misleading, as noted in, e.g., Goodman (1986, 1991a). One of the ways in which the usual correspondence analysis display can be misleading pertains to the interpretation of the distance between a row point and a column point in that display. In some cases, this distance is interpreted in ways that are not warranted. An illustration of this kind of incorrect interpretation is considered in Goodman (1993), and some counter-examples are introduced there that show that this kind of interpretation pertaining to the usual correspondence analysis display is incorrect. When correspondence analysis is used in the analysis of contingency tables, the usual correspondence analysis display should be replaced by the kinds of displays presented in the present article.

As noted at the beginning of this section, we presented in this section graphical displays using correspondence analysis for expository purposes; and corresponding results could also have been presented pertaining to graphical displays using the more general approach presented in Section 2.3b, and/or using the various special cases (of the more general approach) noted in that section. Instead of using correspondence analysis to analyze the example presented in Table 2 (as we have done in the present section) we could have used, say, association analysis (and/or concomitance analysis) to analyze this example. With the correspondence analysis approach applied to the example (the $3 \times 3$ table) in Table 2, in order to portray the nonindependence in the table, the row points (and vectors) and column points (and vectors) in the graphical displays were all two dimensional (see graphical displays in Figs. 1, 2, and 3); whereas, if we had analyzed this table using instead the association analysis approach, we would have found that, in order to portray the nonindependence in the table, only *one* dimension (rather than two dimensions) is needed to describe the corresponding row points and column points in the graphical displays. (For further details, see, e.g., Goodman, 1996, 1997.) It might also be worthwhile to note here that, if the example in Table 2 had been replaced by an $I \times J$ table in which the joint distribution of the entries could be described by, say, a discretized bivariate normal (or, more generally, by a discretized generalized bivariate normal with arbitrary marginals), again only *one* dimension would be needed to describe the corresponding row points and column points in the graphical displays that portray the nonindependence in the table, when the association analysis approach is applied to analyze the table; whereas, several dimensions would be needed when the correspondence analysis approach is applied in this case (see, e.g., Goodman, 1981a, 1985, 1991a).

As noted at the end of Section 2.3a, in addition to the generalization of correspondence analysis and association analysis presented in Section 2.3b, other generalizations of correspondence analysis and other alternatives to correspondence analysis could have been presented; and we could also have presented other graphical displays using these generalizations and/or these alternatives. Here again we refer the interested reader to, e.g., Rao (1995).

### 3. Brief comments on the prehistory of contingency table analysis and on its future in the twenty-first century

All of the work cited in the preceding section was published in the twentieth century. In the present section, I shall include some brief comments on related work on contingency table analysis done in the nineteenth century, and also a brief comment on the future of this topic in the twenty-first century.

As noted in the Introduction to the present article, the problem of measuring the association between, say, two dichotomous variables (two dichotomous classifications or two attributes) is a topic that was considered by eminent scholars beginning in the nineteenth century. The question of priority in the use of simple measures of association for the $2 \times 2$ table scarcely seems very important, but it may be of historical interest now to the late-twentieth-century reader to take note here of this nineteenth century scholarship. Our brief comments here supplement related comments in Goodman and Kruskal (1954, 1959) and Stigler (1986).

In the Introduction to the present article, I began with a quotation from a paper presented by the American mathematician M.H. Doolittle in 1887. A few years earlier, at a 1884 meeting of the Mathematical Section of the Philosophical Society of Washington, Doolittle introduced the square of what is now called the coefficient of correlation (2.4a) for the $2 \times 2$ table; and he considered this index further in 1887 at subsequent meetings of the Mathematical Section (Doolittle, 1885, 1888). He called his index the discriminant association ratio; and the index is now referred to as Pearson's mean squared contingency for the $2 \times 2$ table. (Doolittle's late-nineteenth-century work in the U.S. does not appear to have been known to (or cited by) Pearson and/or Yule in early-twentieth-century England.)

The coefficient of correlation (2.4a) and the corresponding mean squared contingency for the $2 \times 2$ table are symmetric coefficients (coefficients that remain unchanged when the row variable and column variable are interchanged). Some related asymmetric coefficients were introduced prior to the introduction of these symmetric coefficients. For the asymmetric context (e.g., when the row dichotomy is, say, predict tornado and predict no tornado, and the column dichotomy is tornado occurrence and no tornado occurrence), the American philosopher-logician-scientist-mathematician C.S. Peirce, in 1884, proposed the coefficient

$$
\begin{aligned}
\theta &= (ad - bc)/(P_{+1}P_{+2}) = (a - P_{1+}P_{+1})/(P_{+1}P_{+2}) \\
&= [(a/P_{+1}) - (b/P_{+2})].
\end{aligned}
$$
(3.1)

A somewhat different (but related) symmetric coefficient was proposed earlier by the Belgian astronomer-statistician-sociologist A. Quetelet, in 1832; viz.,

$$
\begin{aligned}
\theta^* &= (a - P_{1+}P_{+1})/(P_{1+}P_{+1}) = (ad - bc)/(P_{1+}P_{+1}) \\
&= [(a/P_{+1}) - P_{1+}]/P_{1+}.
\end{aligned}
$$
(3.2)

If the two categories of the dichotomous column variable are interchanged, the absolute value of the coefficient $\theta$ will remain unchanged (the sign of the coefficient $\theta$ will change); but the absolute value of the coefficient $\theta^*$ will change (except in the special case when $P_{+1} = P_{+2} = 1/2$), and the sign of the coefficient will also change.

In the special context in which the row marginal distribution and the column marginal distribution are equal to each other, the German meteorologist W. Köppen proposed the following index in 1870:

$$
(3.3) \qquad \theta^{**} = (P_{1+}P_{+2} - b)/(P_{1+}P_{+2}) = (ad - bc)/(P_{1+}P_{+2}).
$$

In this special case when the two marginal distributions are equal to each other, we then find that $\theta^{**} = \theta$; and we see that the denominators of $\theta^*$ and $\theta^{**}$ are different from each other (except in the special case when $P_{+1} = P_{+2} = 1/2$). We also find that the coefficients $\theta$ and $\theta^{**}$ are both equal to the coefficient of correlation (2.4a) in the special case when the marginal distributions are equal to each other.

Note that the coefficients $\theta^*$ and $\theta^{**}$ described above have the same general form. The two coefficients are special cases of the following coefficient for the $2 \times 2$ table:

$$
(3.4) \qquad \bar{\theta}_{ij} = \begin{cases} [P_{ij}/(P_{i+}P_{+j})] - 1 = (ad - bc)/(P_{i+}P_{+j}), & \text{for } i = j, \\ 1 - [P_{ij}/(P_{i+}P_{+j})] = (ad - bc)/(P_{i+}P_{+j}), & \text{for } i \neq j, \end{cases}
$$

for $i = 1, 2$, and $j = 1, 2$. We also find that

$$(3.5) \qquad \bar{\theta}_{11}\bar{\theta}_{22} = \bar{\theta}_{12}\bar{\theta}_{21},$$

and each of the above products is equal to the square of what is now called the coefficient of correlation (2.4a) for the $2 \times 2$ table.

In the Introduction to the present article, I also referred to the Hungarian statistician-demographer J. Körösy. We find that he proposed and used the odds-ratio, $(ad)/(bc)$, as a "coefficient of relative intensity" in the $2 \times 2$ table in 1887, and he also considered the coefficient $a/(P_{1+}P_{+1})$ (see, e.g., Körösy, 1887; Jordan, 1927). With the latter coefficient viewed as a special case of the coefficient

$$(3.6) \qquad \kappa_{ij} = P_{ij}/(P_{i+}P_{+j}) = (P_{ij}/P_{i+})/P_{+j} = (P_{ij}/P_{+j})/P_{i+},$$

we see that the coefficient $\bar{\theta}_{ij}$ considered in the preceding paragraph is related as follows to the $\kappa_{ij}$:

$$(3.7) \qquad \bar{\theta}_{ij} = \begin{cases} \kappa_{ij} - 1, \text{ for } i = j, \\ 1 - \kappa_{ij}, \text{ for } i \neq j, \end{cases}$$

for $i = 1, 2$, and $j = 1, 2$. We also find that

$$(3.8) \qquad (\kappa_{11} - 1)(\kappa_{22} - 1) = (\kappa_{12} - 1)(\kappa_{21} - 1),$$

and each of the above products is also equal to the square of the coefficient of correlation (2.4a) for the $2 \times 2$ table.

As we noted earlier in this section, the square of the coefficient of correlation (2.4a) for the $2 \times 2$ table is now referred to as Pearson's mean squared contingency for the $2 \times 2$ table. And so, comments in the preceding two paragraphs indicate how the $\bar{\theta}_{ij}$ (with Quetelet's $\theta^*$ and Köppen's $\theta^{**}$ as special cases) and the $\kappa_{ij}$ (with Körösy's coefficient as a special case) are related to Pearson's mean squared contingency. A similar kind of result can be obtained with Peirce's asymmetric coefficient $\theta$ described earlier in this section and the corresponding coefficient $\theta'$ obtained when the row variable and column variable are interchanged (with $b$ and $c$ interchanged, and with the replacement of $P_{+1}$ and $P_{+2}$ by $P_{1+}$ and $P_{2+}$ in the formula (3.1) for $\theta$). The product of Peirce's $\theta$ and the corresponding $\theta'$ is equal to Pearson's mean squared contingency. (Doolittle noted in 1884 that the product of $\theta$ and $\theta'$ was equal to his "discriminant association ratio" — i.e., to what is now called Pearson's mean squared contingency.)

In view of the fact that (1) Kőrösy proposed and used the odds-ratio, $(ad)/(bc)$, in 1887; (2) Yule's Q, a well-known measure of association introduced by Yule in 1900, was based on the odds-ratio;[6] and (3) Kőrösy's coefficient $a/(P_{1+}P_{+1})$ and the corresponding $\kappa_{ij}$ defined by (3.6) are related to Pearson's mean squared contingency (as indicated above), the Hungarian statistician C. Jordan (1927) asserted priority (in some sense) for Kőrösy's work in the following terms: "Le mérite de Kőrösy consiste a avóir introduit et utilisè on 1887, c.á-d. avant l'avènement de la Statistique Mathématique, des grandeurs, mesurant l'association, en bon accord avec les coefficients ... de Yule et ... de Pearson, utilisés aujourd'hui". Now, with respect to the work of the other nineteenth-century scholars considered in the present section, taking into account the results presented earlier in this section (see, e.g., (3.4)–(3.5)), we might also state here that (1) Quetelet's coefficient $\theta^*$ proposed in 1832, (2) Köppen's coefficient $\theta^{**}$ proposed in 1870, and (3) Peirce's coefficient $\theta$ proposed in 1884 are all "en bon accord avec le coefficient de Pearson utilisé aujourd'hui;" and, as we noted earlier, Doolittle's coefficient proposed in 1884 is actually the same as "le coefficient de Pearson utilisé aujourd'hui" (i.e., Pearson's mean squared contingency) for the $2 \times 2$ table.

As we noted earlier herein, the general subject under consideration in the present article has been considered for at least the past one hundred and sixty five years. (We cited above, e.g., Quetelet's work in 1832.) The general subject has been of interest to statisticians, physical scientists, social scientists, biological scientists, mathematicians, philosophers, and others. One reason for this broad interest is that the problem of measuring the association (nonindependence or dependence) between two qualitative/categorical variables [or the association (nonindependence or dependence) among more than two such variables] arises naturally in many different fields of study, and many important substantive questions can be expressed in terms of the measurement of this association. These kinds of important substantive questions arose in many different fields of study in the nineteenth and twentieth centuries, and I am certain that such questions will continue to arise in many different fields of study in the twenty-first century. At the end of the nineteenth century, we would not have been able to see clearly what forms such questions would take in the twentieth century and what kinds of developments would take place stimulated, in part, by such questions. And now near the end of the twentieth century, we

---

[6] The formula for Yule's Q is the same as the formula for Yule's coefficient of colligation (2.4e) except that the square-root signs appearing in (2.4e) are deleted in the corresponding formula for Yule's Q.

would again not be able to see clearly what forms such questions will take
in the twenty-first century and what kinds of developments will take place
stimulated, in part, by such questions. Considering the great progress that
was made in the twentieth century in the development of methods for the
analysis of cross-classified data, we can look forward with great interest to
further developments in this field in the twenty-first century.

## 4. Some additional contingency table topics

We noted earlier herein that (2.10) - (2.11) was the basic formula used
in the canonical correlation approach to contingency table analysis and in
the usual correspondence analysis, and that (2.35) with (2.27) provided
the basic formula used in association analysis. Also, we see that (2.26) -
(2.27) provides the basic formula for the general approach to the analysis
of contingency tables presented in Section 2.3b. These basic formulas can
be viewed as providing saturated models for the $P_{ij}$; and the formulas can
be modified in a straightforward way in order to obtain a corresponding set
of unsaturated models. For each of the saturated models (see, e.g., (2.10),
(2.30), (2.35)), there are $M$ components that describe the nonindependence
in the contingency table, with $M = min(I, J) - 1$; and a corresponding set
of unsaturated models can be obtained simply by replacing the $M$ in these
formulas by $M^*$, where $1 \leq M^* < M$. [When there are no components
to describe the nonindependence in the contingency tables (say, $M^* = 0$),
we then have the usual model (2.1) of statistical independence between
the row classification and column classification in the contingency table;
and when $M^* = 1$, we have models in which the row scores and column
scores are one dimensional.] Because of space constraints, this topic will not
be considered further here. The reader interested in this topic is referred
to, e.g., Goodman (1979a, 1985, 1986), Gilula and Haberman (1986), and
Becker (1990).

The analysis of nonindependence between the row classification and
the column classification in the contingency table has been viewed in the
present paper primarily from a symmetric perspective, with the row classification and column classification treated in a symmetric way; but many of
the results presented here can also be applied in an asymmetric context in
which interest is focused on the possible dependence of one classification in
the contingency table on the other classification in the table. The change
from a symmetric perspective to an asymmetric one is quite straightforward with many of the results presented here; and their application and
interpretation in the asymmetric context is also straightforward. Again, because of space constraints, this topic was not considered here. The reader
interested in this topic is referred to, e.g., Goodman (1983, 1986, 1987) and

Rao (1995) for related material.

In the analysis of the nonindependence (association) between the row classification and the column classification in the contingency table, the nonindependence is usually analyzed taking into account the observed row marginal distribution and the observed column marginal distribution. Models of the nonindependence (association) between the row classification and the column classification will usually fit both the observed row marginal and the observed column marginal; while models of the dependence of the row classification on the column classification will usually fit the observed column marginal but need not fit the observed row marginal, and a corresponding remark applies also to models of the dependence of the column classification on the row classification. Models for the joint distribution of the row and column classifications need not fit the observed row marginal nor the observed column marginal. We have focused our attention in the present paper on the analysis of nonindependence (association), rather than on the analysis of dependence or on the analysis of the joint distribution. Some references that pertain to the analysis of dependence were mentioned at the end of the preceding paragraph; and for the analysis of the joint distribution in the contingency table, the interested reader is referred to, e.g., Goodman (1979b, 1981b), Lang and Agresti (1994), and Sobel, Becker, and Minick (1998).

As was noted in the Introduction to the present article, there are many topics on contingency table analysis that are not included in the article because of space constraints – topics that could have been included (if not for space constraints) in an article of this kind. For some of these topics, the interested reader was referred, in the Introduction, to some recently published textbooks that are concerned with this subject.

## References

[1] Agresti, A. (1990) *Categorical Data Analysis*, New York, Wiley.

[2] Agresti, A. (1996) *An Introduction to Categorical Data Analysis*, New York, Wiley.

[3] Agresti, A. and Kezouh, A. (1983), Association Models for Multi-dimensional Cross-Classifications of Ordinal Variables. *Communications in Statistics, Part A – Theory and Methods*, **12**, 1261–1276.

[4] Andersen, E.G. (1990) *The Statistical Analysis of Categorical Data*, Berlin, Springer-Verlag.

[5] Becker, M.P. (1989), Models for the Analysis of Association in Multivariate Contingency Tables. *Journal of the American Statistical Association*, **84**, 1014–1019.

[6] Becker, M.P. (1990), Maximum-Likelihood Estimation of the RC(M) Association Model. *Applied Statistics*, **39**, 152–167.

[7] Becker, M.P., and Clogg, C.C. (1989), Analysis of Sets of Two-Way Contingency Tables Using Association Models. *Journal of the American Statistical Association*, **84**, 142–151.

[8] Benzécri, J.-P. (1973) *L'Analyse des Données: 1. La Taxonomie: 2. L'Analyse des Correspondances*, Paris, Dunod (2nd ed., 1976).

[9] Benzécri, J.-P. (1991a)Discussion of article by L.A. Goodman, Measures, Models, and Graphical Displays in the Analysis of Cross-Classified Data. *Journal of the American Statistical Association*, **86**, 1112–1115.

[10] Benzécri, J.-P. (1991b), Mémoire recu: Measures, Models, and Graphical Displays in the Analysis of Cross-Classified Data. Point de vue de J.-P. Benzécri sur un exposé de L.A. Goodman. *Les Cahiers de l'Analyse des Données*, **16**, 119–126.

[11] Benzécri, J.-P. (1992) *Correspondence Analysis Handbook*, New York, Marcel Dekker.

[12] Box, G.E.P., and Cox, D.R. (1964), An Analysis of Transformations. *Journal of the Royal Statistical Society, Series B*, **26**, 211–243.

[13] Choulakian, V. (1988), Exploratory Analysis of Contingency Tables by Loglinear Formulation and Generalization of Correspondence Analysis. *Psychometrika*, **53**, 235–250.

[14] Clogg, C.C. (1982), Some Models for the Analysis of Association in Multiway Cross-Classifications Having Ordered Categories. *Journal of the American Statistical Association*, **77**, 803–815.

[15] Doolittle, M.H. (1885), The Verification of Predictions (Abstract). *Bulletin of the Philosophical Society of Washington*, **7**, 122–127. A summary of this article was published in 1885-86 in *The American Meteorological Journal*, **2**, 327–329.

[16] Doolittle, M.H. (1888), Association Ratios (Abstract). *Bulletin of the Philosophical Society of Washington*, **10**, 83–87 and 94–96.

[17] Fisher, R.A. (1940), The Precision of Discriminant Functions. *Annals of Eugenics, London*, **10**, 422–429.

[18] Gabriel, K.R. (1971), The Biplot Graphic Display of Matrices with Application to Principal Component Analysis. *Biometrika*, **54**, 453–467.

[19] Gilula, Z., and Haberman, S.J. (1986), Canonical Analysis of Two-Way Contingency Tables by Maximum Likelihood. *Journal of the American Statistical Association*, **81**, 780–788.

[20] Gilula, Z., and Haberman, S.J. (1988), The Analysis of Multivariate Contingency Tables by Restricted Canonical and Restricted Association Models. *Journal of the American Statistical Association*, **83**, 760–771.

[21] Goodman, L.A. (1979a), Simple Models for the Analysis of Association in Cross-Classifications Having Ordered Categories. *Journal of the American Statistical Association*, **74**, 537–552.

[22] Goodman, L.A. (1979b), Multiplicative Models for Square Contingency Tables with Ordered Categories. *Biometrika*, **66**, 413–418.

[23] Goodman, L.A. (1981a), Association Models and the Bivariate Normal for Contingency Tables With Ordered Categories. *Biometrika*, **68**, 347–355.

[24] Goodman, L.A. (1981b) Three Elementary Views of Log-Linear Models for the Analysis of Cross-Classifications Having Ordered Categories, in: *Sociological Methodology, 1981*, ed. S. Leinhardt, San Francisco: Jossey-Bass, 193–239.

[25] Goodman, L.A. (1983), The Analysis of Dependence in Cross-Classifications Having Ordered Categories Using Log-Linear Models for Frequencies and Log-Linear Models for Odds. *Biometrics*, **39**, 149–160.

[26] Goodman, L.A. (1984) *The Analysis of Cross-Classified Data Having Ordered Categories,* Cambridge, MA, Harvard University Press.

[27] Goodman, L.A. (1985), The Analysis of Cross-Classified Data Having Ordered and/or Unordered Categories: Association Models, Correlation Models, and Asymmetry Models for Contingency Tables With or Without Missing Entries. *The Annals of Statistics*, **13**, 10–69.

[28] Goodman, L.A. (1986), Some Useful Extensions of the Usual Correspondence Analysis Approach and the Usual Log-Linear Models Approach in the Analysis of Contingency Tables; with discussion. *International Statistical Review*, **54**, 243–309.

[29] Goodman, L.A. (1987), New Methods for Analyzing the Intrinsic Character of Qualitative Variables Using Cross-Classified Data. *American Journal of Sociology*, **94**, 529–583.

[30] Goodman, L.A. (1991a), Measures, Models, and Graphical Displays in the Analysis of Cross-Classified Data; with discussion. *Journal of the American Statistical Association*, **86**, 1085–1138.

[31] Goodman, L.A. (1991b), New Methods in the Age of Progress in Contingency Table Analysis Using Measures, Models, and Graphical Displays. Reply to discussion of "Measures, Models, and Graphical Displays in the Analysis of Cross-Classified Data". *Journal of the American Statistical Association*, **86**, 1124–1138.

[32] Goodman, L.A. (1993) Correspondence Analysis, Association Analysis, and Generalized Nonindependence Analysis of Contingency Tables: Saturated and Unsaturated Models and Appropriate Graphical Displays, in: *Multivariate Analysis: Future Directions 2*, eds. C.M.

Cuadras and C.R. Rao, Amsterdam, North Holland, 265–294.

[33] Goodman, L.A. (1996), A Single General Method for the Analysis of Cross-Classified Data: Reconciliation and Synthesis of Some Methods of Pearson, Yule, and Fisher, and Also Some Methods of Correspondence Analysis and Association Analysis. *Journal of the American Statistical Association*, **91**, 408–428.

[34] Goodman, L.A. (1997) Statistical Methods, Graphical Displays, and Tukey's Ladder of Re-expression in the Analysis of Nonindependence in Contingency Tables: Correspondence Analysis, Association Analysis, and the Midway View of Nonindependence, in: *The Practice of Data Analysis: Essays in Honor of John W. Tukey*, eds. D. Brillinger, L.T. Fernholz, and S. Morgenthaler, Princeton, New Jersey, Princeton University Press, 101–132.

[35] Goodman, L.A., and Hout, M. (1998) Statistical Methods and Graphical Displays for Analyzing How the Association Between Two Qualitative Variables Differs Among Countries, Among Groups, or Over Time: A Modified Regression-Type Approach; with discussion, in: *Sociological Methodology, 1998*, **28**, ed. A.E. Raftery, Cambridge, Massachusetts, Blackwell Publishers, 175–261.

[36] Goodman, L.A., and Kruskal, W.H. (1954), Measures of Association for Cross-Classifications. *Journal of the American Statistical Association*, **49**, 732–784.

[37] Goodman, L.A., and Kruskal, W.H. (1959), Measures of Association for Cross-Classifications. II: Further Discussion and References. *Journal of the American Statistical Association*, **54**, 123–163.

[38] Green, M. (1989) Generalizations of the Goodman Association Model for the Analysis of Multidimensional Contingency Tables, in: *Statistical Modeling* (Lecture Notes in Statistics **57**), eds. A Decarli, B.J. Francis, R. Gilchrist, and G.V.H. Seber, New York, Springer-Verlag, 165–171.

[39] Greenacre, M.J. (1993), Biplots in Correspondence Analysis. *Journal of Applied Statistics*, **20**, 251–269.

[40] Guttman, L. (1941) The Quantification of a Class of Attributes, in: *The Prediction of Personal Adjustment*, P. Horst, with collaboration of P. Wallin and L. Guttman, prepared for the Social Science Research Council Committee on Social Adjustment under the direction of the Subcommittee on Prediction of Social Adjustment, New York, Social Science Research Council, 319–348.

[41] Guttman, L. (1950) The Principal Components of Scale Analysis, in: *Measurement and Prediction. Studies in Social Psychology in World War II (Vol. 4)*, S.A. Stouffer, L. Guttman, E.A. Suchman,

P.F. Lazarsfeld, S.A. Star, and J.A. Clausen, Princeton, Princeton University Press, 312–361.

[42] Hartley, H.O. (1935). See Hirschfeld (1935).

[43] Hirschfeld, H.O. (1935), A Connection Between Correlation and Contingency. *Proceedings of the Cambridge Philosophical Society*, **31**, 520–524.

[44] Hotelling, H. (1933), Analysis of a Complex of Statistical Variables into Principal Components. *Journal of Educational Psychology*, **24**, 417–441, 498–520.

[45] Hotelling, H. (1936), Relations Between Two Sets of Variables. *Biometrika*, **28**, 321–377.

[46] Jordan, C. (1927), Les coefficients d'intensité relative de Kőrösy. *Revue de la Société Hongroise de Statistique (Magyar Statisztikai Társaság, Revue)*, **5**, 332–345.

[47] Köppen, W. (1870-1), Die Aufeinanderfolge der unperiodischen Witterungserscheinungen nach den Grundsätzen der Wahrscheinlichkeitsrechnung. *Repertorium für Meteorologie, Akademiia Nauk, Petrograd*, **2**, 189–238.

[48] Kőrösy (Kőrösi), J. (1887), Kritik der Vaccinations - Statistik und Neue Belträge zur Frage des Impfschutzes" ("Critical Review of Vaccinational Statistics, with New Contributions Relating to Its Protective Power"). *Transactions of the Ninth International Medical Congress, Washington, D.C.*, **1**, 238–418; also published in 1889 in Berlin, Puttkamer & Mühlbrecht..

[49] Lang, J.B., and Agresti, A. (1994), Simultaneously Modeling Joint and Marginal Distributions of Multivariate Categorical Responses. *Journal of the American Statistical Association*, **89**, 625–632.

[50] Maung, K. (1941), Measurement of Association in a Contingency Table With Special Reference to the Pigmentation of Hair and Eye Color in Scottish School Children. *Annals of Eugenics, London*, **11**, 189–223.

[51] Pearson, K. (1913), On the Measurement of the Influence of 'Broad Categories' on Correlation. *Biometrika*, **9**, 116–139.

[52] Pearson, K., and Heron, D. (1913), On Theories of Association. *Biometrika*, **9**, 159–315.

[53] Peirce, C.S. (1884), The Numerical Measure of the Success of Predictions. *Science*, **4**, 453–454.

[54] Quetelet, A. (1832), Sur la possibilité de mesurer l'influence des causes qui modifient les éléments sociaux. Lettre à M. de Dr. Villermé. *Correspondance mathématique et physique*, **7**, 321–346.

[55] Rao, C.R. (1995), A Review of Canonical Coordinates and an Alter-

native to Correspondence Analysis Using Hellinger Distance. *Qüestiió*, **19**, 23–63.

[56] Scheffé, H. (1959) *The Analysis of Variance,* New York, Wiley.

[57] Sobel, M.E., Becker, M.P., and Minick, S.M. (1998) Origins, Destinations, and Association in Occupational Mobility, *American Journal of Sociology* **104**

[58] Stigler, S.M. (1986) *The History of Statistics. The Measurement of Uncertainty before 1900,* Cambridge, Massachusetts, Harvard University Press.

[59] Tukey, J.W. (1957), On the Comparative Anatomy of Transformations. *The Annals of Mathematical Statistics*, **28**, 602–632.

[60] Yule, G.U. (1900), On the Association of Attributes in Statistics. *Philosophical Transactions of the Royal Society,*Series A, **194**, 257–319.

[61] Yule, G.U. (1912), On the Methods of Measuring Association Between Two Attributes. *Journal of the Royal Statistical Society*, **75**, 579–642.

# Diffusion Models for Neural Activity

Satish Iyengar[1]

Department of Statistics
University of Pittsburgh
Pittsburgh, Pennsylvania

**Abstract.** Stochastic models of neural activity are a well developed application in biology. Diffusion models hold a prominent place because of the many synaptic inputs to a neuron, and because these models arise out of noisy versions of differential equations for the neural membrane's electrical properties. While the probabilistic aspects of such models have been well studied, inferential procedures for them are not. In this paper, we first outline the physiological background, and then give an account of the deterministic and stochastic models of spike generation. We also describe the statistical inference problems that arise, especially for models which include a finite membrane time constant and reversal potentials. In the course of that description, we point to directions for further research.

## 1. Introduction

The brain and the central nervous system have a large number of specialized cells, broadly categorized as neurons and glia. Neurons, which comprise only about ten percent of these cells, sense changes in the environment, communicate these changes to other neurons, and help to control the body's responses to them. The glial cells support neural function in many ways, but beyond that their role is not well understood. Despite their smaller numbers, neurons are of great interest because of their functions.

---

[1] Supported by NIMH grant MH-55123.

They communicate by sending electrical impulses, or spikes, to other neurons or muscles; a sequence of spikes from a single neuron is called a spike train. Much of the current work in neuroscience is based on the hypothesis that spike trains encode information, and that perception is the result of the neural computation, or processing of these spike trains [33].

Neural spikes were first observed long ago: in the eighteenth century, Luigi Galvani observed the contraction of a frog's nerve and muscle tissue due to electrical discharge [25]. It was not until 1939, however, that Hodgkin and Huxley [11] reported on the first successful recording within the cell of a neuron. Using a squid's neuron, they recorded both its electrical state leading to a spike and the spike itself. This and later Nobel prize winning work on nerve conduction led to the Hodgkin-Huxley equations, a system of differential equations that describe how the impulse travels along a nerve fiber called the axon. Around the same time, several experiments on the muscle spindles of frogs revealed that there is variability in the times between successive spikes, or interspike interval (ISI) [3]. It soon became clear that this variability was not a result of measurement error, and that noise is a prominent feature of neural activity generally. The Hodgkin-Huxley equations and its variants are deterministic in nature, so they do not capture this variability. Thus, several stochastic models for spike generation were proposed, starting with the Brownian motion model of Gerstein and Mandelbrot [9].

There are several accounts of stochastic models for neural activity. They include the books by Holden [14], Ricciardi [32], Sampath and Srinivasan [34], and Tuckwell [39, 40], and the articles by Fienberg [8], and Lánský, Smith, and Ricciardi [28]. Most of this work has an applied mathematical or probabilistic flavor to it, with relatively little attention paid to statistical inference for the models. The main reason for this is that standard inferential procedures for the physiologically more realistic models are analytically intractable, and easily computed approximations for them are not available. The paper by Yang and Chen [43] is one of the few devoted to statistical methods in this area.

An unfortunate consequence of this lapse is that while many models of spike generation have been proposed, the fitting of these models to experimental data and the comparison of models has not been systematically done. Furthermore, without knowing the variability (through standard errors) of parameter estimates, it is difficult to assess the effect of varying experimental conditions. Some investigators have even suggested that the more realistic models should not be considered at all because it is hard to separately estimate the various physiological parameters contained in them: see, for instance, Levine [30].

D.R. Cox [6] has said that "[s]tatistical subjects can be characterized qualitatively by their statistical analysis to stochastic model ratio." One of the aims of this paper is to call attention to a subject that has a somewhat low value of this Cox ratio, and to describe some avenues of research that will raise it. In Section 2, we provide the physiological background needed to understand the models that have been proposed. In Section 3, we review the deterministic models that attempt to represent neural mechanisms, and which give rise to the stochastic models of spike generation. In Section 4, we describe these stochastic models, in particular random walks and diffusions, and point to directions for future research.

## 2. Physiological background

We begin with a brief account of the physiology of the neuron. More complete treatments intended for mathematical and statistical audiences are given in the books by Tuckwell [39] and Cronin [7], which in turn rely on neurophysiology texts such as the ones by Kandel and Schwartz [21], Katz [25], and Shepherd [36].

Neurons are generally classified into three groups. Sensory (or afferent) neurons transmit information for perception and motor coordination; motor neurons carry commands to muscles and glands; and interneurons relay signals from one neuron to another, but at present appear to have no specific sensory or motor purpose. All of these neurons share some basic characteristics: that is, they consist of the soma, the dendrites, the axon, and the synapses (Figure 1).

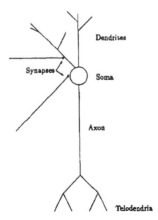

Figure 1. Schematic drawing of a typical neuron

The soma, or cell body, consists of a conducting electrolyte gel surrounded by a permeable membrane that selectively allows sodium and potassium ions to flow through its channels. Thus, the potential difference that exists across the membrane acts as a battery or source of stored energy, the release of which is a rapid electrical signal called a spike or action potential. The typical soma diameter is around 50 to 100 $\mu$m.

The dendrites are a tree-like structure consisting of processes that gather signals and transmit them to the soma for integration. The dendritic tree often branches to a considerable extent, with its surface area often being ten to twenty times that of the soma. In extreme cases, such as Purkinje cells in the cerebellum, the dendritic surface area is about a hundred times that of the soma, with about 100,000 input sites called synapses.

The signals are received at the dendrites and the soma surface itself through a synapse or synaptic juncture, which is a point of contact between a neuron and other cells. The cell transmitting (receiving) the signal is called the presynaptic (postsynaptic) cell. These two cells do not touch each other; rather, they are separated by a space called the synaptic cleft, through which electrochemical signals are transmitted. Synapses are generally classified as excitatory or inhibitory. The excitatory (inhibitory) ones depolarize (hyperpolarize) the postsynaptic cell membrane, which facilitates (discourages) spike generation. An excitatory (inhibitory) signal is called an excitatory (inhibitory) postsynaptic potential, and is abbreviated EPSP (IPSP). The numbers of excitatory and inhibitory synapses and their distribution over the postsynaptic dendritic tree and soma vary with the type of neuron.

When a neuron has sufficient excitatory input, it transmits a spike along its axon, which is a long tubular process that grows out of a specialized region of the neuron called the axon hillock. When the axon nears the target cells, it branches out to form telodendria, at the ends of which are the synapses. While neurons typically have just one axon, they can reach up to a thousand other neurons through the telodendria. The axons are quite thin (from 0.2 to 20 $\mu$m), and long (from 1 mm to 2 m!). To ensure high-speed conduction of the action potentials, large axons are often covered by a fatty insulating material called the myelin sheath.

The processes leading to spike generation are undoubtedly complicated, for they can involve each of the many parts of the neuron. Because mathematical models that account for all or most of these processes would be intractable, the ones in use attempt to reproduce the characteristics of observed spike trains by considering only a small number of physiological features. An important quantity in all of these models is the membrane potential, $V(x, t)$, which is measured in millivolts (mV) at time $t$ and position $x$ on the membrane. The simplest models ignore the spatial variability, or fix $x$ at the place on the neural membrane where the measuring electrode is implanted, and derive differential equations for the potential $V(t)$ using electrical circuit representations of the membrane and its surroundings.

We now describe the key physiological features that are incorporated in the models of spike generation below. The membrane potential varies because of inputs to the neuron, and the membrane's tendency to exponentially decay towards an equilibrium called its resting potential, $V_r$. Ordinarily, the membrane potential varies within a range $[V_I, V_E]$, but values outside that range can be reached by applying an electrical field. The effect of an input upon the neuron depends upon its membrane potential. That is, as $V(t)$ moves closer to $V_E$ ($V_I$), the effect of an EPSP (IPSP) input is blunted, and it even reverses: it becomes inhibitory (excitatory) when the membrane potential goes beyond $V_E$ ($V_I$). Hence, $V_I$ and $V_E$ are called the inhibitory and excitatory reversal potentials, respectively. According to the simplest models, when $V(t)$ reaches a constant threshold $V_f$, the neuron discharges an action potential or spike. Such models are called integrate-and-fire models; examples with a stochastic component include those based on (5), (7), (9)–(11), and (12) below. The spike itself has a characteristic shape, and is the object of other studies; however, its duration is short enough that it is often considered a point event. After the spike, there is a short refractory period, which consists of two parts: an absolute refractory period, when the neuron cannot emit a spike, and a relative refractory period, when it can, but with a different threshold. In more complicated models, the threshold is a nonconstant function $V_f(t)$; for instance, setting $V_f(t) = \infty$ for $0 \le t \le t_R$ leads to an absolute refractory period of length $t_R$.

Given data from a neuron, a basic problem is to learn about its anatomical and physiological properties, and its inputs. Of course, this rather broad statement must be narrowed before much progress can be made. The data come in many forms. First, there are anatomical studies that describe the physical structure of a neuron, such as its location in the nervous system, and the sizes and relative positions of its constituent parts. Next, there are physiological studies that describe the electrical and chemical activity of the neuron or parts thereof, and its connections to other neurons. In this paper, we will focus on two types of physiological data: the neuron's spike train, and its subthreshold membrane potential. The spike train is the more common of the two because it is obtained extracellularly, for example by placing a microelectrode just outside the axon. A record of the membrane potential itself requires intracellular methods or the penetration of the membrane, which is more difficult in small cells.

## 3. Deterministic models

The stochastic models that we consider here arise out of deterministic models of spike generation, which are typically expressed as ordinary or partial differential equations that describe the subthreshold potential at one or more locations on the neural membrane. In some models a firing threshold must be imposed, while in others the threshold is a natural property of the equations.

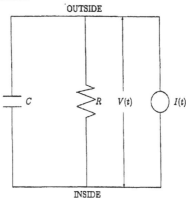

Figure 2. Electrical circuit for the Lapicque model of subthreshold potential.

The simplest such approach, due to Lapicque in 1907 (see [39]), represents the neuron as a single entity, or lumped circuit, instead of separately considering its parts (Figure 2). In this model, $V(t)$ is the potential difference across the membrane, which consists of a resistor and capacitor in parallel, and an input current $I(t)$. When the membrane resistance $(R)$ and capacitance $(C)$ are constant, the current through the resistor is $V/R$ by Ohm's law, and the current through the capacitance is $C dV/dt$, so that conservation of current implies that

$$(1) \qquad\qquad C\frac{dV}{dt} + \frac{V}{R} = I,$$

with an initial condition depending upon the details of the experimental preparation. This model has no natural threshold, so a firing threshold $V_f(t)$ is imposed. Thus, the time to firing is the first time that $V(t) = V_f(t)$. Using standard (Green's function, Laplace transforms, numerical) methods, equation (1) is sufficiently tractable to investigate the neural response to various inputs. Despite the simplicity of this model, it has provided insights into experimental observations, such as a neuron's response to sinusoidal inputs and phase locking. We shall see below that the Lapicque model is also the basis of the stochastic model of Stein which is well approximated by the Ornstein-Uhlenbeck process.

The next approach to modeling the subthreshold behavior, due to Hodgkin and Rushton [13], considers the propagation of signals through the dendritic tree and axon, which are regarded as nerve cylinders. Assuming axial symmetry, the voltage now depends on time $t$ and distance $x$ along the cylinder, both relative to some origin. For a homogeneous segment (that is, one for which the internal and membrane resistances and membrane capacitance are constant), the equation in its dimensionless form is

$$(2) \qquad \frac{\partial V}{\partial t} - \frac{\partial^2 V}{\partial x^2} + V = I,$$

with appropriate boundary and initial conditions. This time, the input $I(t, x)$ is also a function of both arguments. The derivation of this equation is based on circuit diagrams representing a small segment of the nerve cylinder. Equation (2) is known as the cable equation because it arose out of Kelvin's investigations into the propagation of signals along the first transatlantic cable. Rall [31] and Tuckwell [39] give a thorough treatment of the nature of steady state and transient solutions to (2). As with the Lapicque model, a firing threshold must be additionally specified for cable models; typically, this specification is at the neuron's soma near its trigger zone.

The cable equation is appropriate for a single dendrite or nerve cylinder. However, a model that is physiologically more accurate must account for the complicated dendritic tree structure. Under certain circumstances, using boundary conditions at the tree's branch points and terminals that assure the continuity of the voltage and the conservation of current, the Laplace transform of the voltage over the entire tree is tractable, but that transform is hard to invert. To address this difficulty, two approaches have been proposed. The first is Rall's notion of an equivalent cylinder [31], which gives conditions for the analysis of a dendritic tree to be effectively reduced to that of a related single cylinder to which the cable equation then applies. Tuckwell and Walsh [41] have proved a theorem which gives a precise statement of these conditions. Rall has also suggested a lumped circuit model for the soma together with the equivalent cylinder reduction for the dendrites. The second approach, also due to Rall, is called compartmental modeling. Here, each small region of the neuron is regarded as a homogeneous compartment for which a Lapicque-type lumped circuit model applies. For each small region, the circuit satisfies an ordinary differential equation that is more easily solved than a partial differential equation. In practice, the model can consist of thousands of compartments, the implementation of which requires specialized software. See Segev, Fleshman, and Burke [35] for recent developments in compartmental modeling.

Next, we discuss the approach of Hodgkin and Huxley [12]; see also Cronin [7] for an outline of the experimental methods and a detailed description of the mathematical aspects of the Hodgkin-Huxley equations. Hodgkin and Huxley studied the propagation of the action potential down the giant axon of a squid neuron (the term giant here refers to the size of the axon, about 0.5 mm, not to the squid itself). Through careful experiments, they first divided the current through the membrane into components consisting of three ion species: potassium ($K$), sodium ($Na$), and a heterogeneous mix of other ions such as chloride called the leakage ($l$) current. Using pharmacological probes, they showed that the $K$ and $Na$ currents were independent of each other. Treating the leakage current as independent of the other two, they used an electrical circuit model with membrane capacitance $C$ in parallel with the resistances $R_K$, $R_{Na}$, and $R_l$, and with corresponding driving forces $V - V_K$, $V - V_{Na}$, and $V - V_l$, where $V_K$, $V_{Na}$, and $V_l$ are the equilibrium potentials for the three ion species. Using methods similar to the the cable model, they derived the following equation for the voltage, expressed in standardized form:

$$(3) \quad \frac{\partial V}{\partial t} = \frac{\partial^2 V}{\partial x^2} + \bar{g}_K n^4 (V_K - V) + \bar{g}_{Na} m^3 h (V_{Na} - V) + g_l (V_l - V) + I,$$

where $\bar{g}_K$ and $\bar{g}_{Na}$ are constants, and $g_l$ is the leakage conductance. The variables $n$, $m$, and $h$ are dimensionless quantities representing the potassium activation, sodium activation, and sodium inactivation. The exponents of $n$, $m$, and $h$ are the result of curve fitting to data from separate experiments. These three variables satisfy the following differential equations:

$$(4) \quad \begin{aligned} \frac{\partial n}{dt} &= \alpha_n (1 - n) - \beta_n n, \\ \frac{\partial m}{dt} &= \alpha_m (1 - m) - \beta_m m, \\ \frac{\partial h}{dt} &= \alpha_h (1 - h) - \beta_h h, \end{aligned}$$

where the coefficients $\alpha$ and $\beta$ are functions of voltage that are once again derived from fitting curves to data from auxiliary experiments.

It should be clear from this account that the Hodgkin-Huxley equations are empirical descriptions. They are not derived purely from first principles such as from Kirchhoff's laws. There are several alternatives to these equations, such as those of Frankenhaeuser and Huxley which are more appropriate for myelinated axons. Neither set of equations admits of an analytic solution, so they have been studied by numerical methods; there are also simpler variants, such as the Fitzhugh-Nagumo equations [39]. Despite these difficulties, the Hodgkin-Huxley equations are important because they are close to the physiology. They also contain the firing threshold as a natural property of the equations: that is, the solutions to these equations resemble action potentials, and the traveling wave solutions to these equations model the propagation of the action potential along the axon. They have also spawned a considerable amount of research in partial differential equations.

## 4. Diffusion models

The diffusion models for the membrane potential and the resulting spikes generally arise out of the deterministic models above, together with noisy input. The resulting stochastic differential equations involve white noise; see Kloeden and Platen [26] and the two books by Karlin and Taylor [23, 24] for a good account of these topics and background on stochastic processes in general.

The inputs from the synapses are modeled as point processes, typically Poisson processes that are independent of each other. The membrane potential then executes a random walk with discrete jumps when the random inputs arrive, and in between varies according to the differential equation describing its dynamics. When the number of inputs is large, weak convergence methods lead to white noise inputs and diffusion approximations to the random walks. Ricciardi [32] gives a good heuristic description of the limiting process, and there are several rigorous statements and proofs of such weak convergence results for various cases: see for instance, Kallianpur [22] and Lánský [27].

These stochastic inputs are intended to model the spontaneous activity of neurons. The term spontaneous is a bit ambiguous. It roughly means that there is no intended input with structure, such as periodic excitatory impulses, or known inputs from other neurons. Understanding the neuron's spontaneous activity is a first step in understanding its response to signals with structure.

Gerstein and Mandelbrot [9] proposed the earliest integrate-and-fire diffusion model for neural activity. They ignored the boundaries due to reversal potentials and the decay of the membrane potential $V(t)$ due to leakage. Their model can be derived using the following argument. Let $\{N_i^E(t) : i = 1, \dots, n_E\}$ and $\{N_j^I(t) : j = 1, \dots, n_I\}$ represent the excitatory and inhibitory input Poisson processes, respectively. Suppose that $N_i^E$ and $N_j^I$ have rates $\lambda_i^E$ and $\lambda_j^I$, and magnitudes $\alpha_i^E$ and $\alpha_j^I$, respectively. Then the stochastic differential equation for $V(t)$ is

$$
(5) \qquad dV = \sum_{i=1}^{n_E} \alpha_i^E dN_i^E - \sum_{j=1}^{n_I} \alpha_j^I dN_j^I .
$$

When $\lambda_i^E$ and $\lambda_i^I$ both tend to infinity and $\alpha_i^E$ and $\alpha_i^I$ tend to zero so that

$$
\sum_{i=1}^{n_E} \alpha_i^E \lambda_i^E - \sum_{j=1}^{n_I} \alpha_j^I \lambda_j^I \to \mu \quad \text{and} \quad \sum_{i=1}^{n_E} \left[\alpha_i^E\right]^2 \lambda_i^E + \sum_{j=1}^{n_I} \left[\alpha_j^I\right]^2 \lambda_j^I \to \sigma^2,
$$

then (5) is well approximated by

$$
(6) \qquad dV = \mu dt + \sigma dW,
$$

where $dW$ is white noise.

Thus, $V(t) = V_0 + \mu t + \sigma W(t)$ is a Brownian motion starting at $V_0$ with drift $\mu$ and diffusion or variance $\sigma^2$. Gerstein and Mandelbrot considered a constant firing threshold $V_f$, so that the random time to firing is $T = \inf\{t > 0 : V(t) = V_f\}$. When the net input is excitatory, $\mu > 0$ and $T$ is a proper random variable with the inverse Gaussian density

$$
f(t; d, \nu) = \frac{d}{\sqrt{2\pi t^3}} \exp\left[-\frac{(d - \nu t)^2}{2t}\right],
$$

where $\nu = \mu/\sigma$ is the standardized drift and $d = (V_f - V_0)/\sigma$ is the standardized distance from the initial and firing potentials. Thus, if the membrane potential resets to the same initial value upon firing, only the two parameters $\nu$ and $d$ are identifiable from the resulting ISI renewal process, rather than the four parameters $\mu$, $\sigma$, $V_0$, and $V_f$ of the underlying model.

The inverse Gaussian is a two-parameter exponential family that has been studied for over fifty years [5]. Standard inferential procedures, such as computing the maximum likelihood estimates and their standard errors, and assessing the model's goodness of fit are well understood and quite simple for this model. Until recently, it was regarded as the only diffusion model for which inference based on ISI data is tractable. Gerstein and Mandelbrot found that this model adequately fit several units from the cochlear nucleus of cats. Despite this and other empirical evidence, this model has been considered only a rough approximation because of its "poor

physiological foundation." ([39, p. 144]) We will see below that the inverse
Gaussian model also arises from models with better physiological basis.

Soon after Gerstein and Mandelbrot's paper appeared, Stein [38] and
Calvin and Stevens [4] proposed models that also considered the membrane
potential decay. Under appropriate conditions, the diffusion approximation
to these models is the Ornstein-Uhlenbeck process whose stochastic differential equation is

$$(7) \qquad dV = \left( \mu - \frac{V}{\tau} \right) dt + \sigma dW,$$

where the new parameter $\tau$ is the time constant for the membrane potential's exponential decay. In the simplest case, the firing threshold $V_f$ is
assumed to be constant, and the spike train a renewal process. The $-V/\tau$
term in (7) provides a mean reversion in this model which makes the firing
time $T$ almost surely finite. Notice that when $\tau = \infty$, this model reduces
to that of Gerstein and Mandelbrot in (6); and when the white noise input
$\sigma dW$ is replaced by a deterministic input $I = I(t)$, it reduces to a slight
variant of the Lapicque model in (1).

Fitting the Ornstein-Uhlenbeck model to the spike train is difficult
because a tractable form of the density of $T$ is not available. Instead, the
Laplace transform is known:

$$E(e^{-\lambda T}) = \frac{He_{-\lambda \tau} \left( -\sqrt{2} \frac{V_0 - \mu \tau}{\sigma \sqrt{\tau}} \right)}{He_{-\lambda \tau} \left( -\sqrt{2} \frac{V_f - \mu \tau}{\sigma \sqrt{\tau}} \right)},$$

where $He_\lambda$ is the parabolic cylinder function

$$He_\nu(x) = 2^{-(1+\nu/2)} \sum_{m=0}^{\infty} \frac{\Gamma \left( \frac{m-\nu}{2} \right)}{\Gamma(-\nu)} \frac{(-x\sqrt{2})^m}{m!}.$$

While the function $He_\nu(x)$ may appear unfamiliar in this form, it is connected to well known quantities in statistics: it reduces to a Hermite polynomial with respect to $e^{-x^2/2}$ when $\nu$ is a positive integer, and is related
to derivatives of Mills' ratio when $\nu$ is a negative integer. Using the known
properties of parabolic cylinder functions and about diffusions in general,
Ricciardi and his colleagues have developed a rather large literature on approximations to the moments and the density of $T$, even for nonconstant
thresholds: see [28] for an entry into that literature.

The first serious attempt towards inference for the Ornstein-Uhlenbeck
process based on ISI data is due to Inoue, Sato, and Ricciardi [15]. They

assumed that the neuron's intrinsic parameters $V_f$, $V_0$, and $\tau$ were known. They then used moment methods — in particular the mean and coefficient of variation, which are commonly reported indices of ISIs — to estimate $\mu$ and $\sigma$. They then did an extensive numerical study to assess the sensitivity of these estimates to changes in the assumed values of $V_f$, $V_0$, and $\tau$. They did not, however, provide any standard errors for these estimates. They illustrated their method on spontaneous firings of neurons in a region called the mesencephalic reticular formation of anesthetized cats during different stages of sleep, and while they were awake.

More recently, Iyengar [16, 17] considered the estimation of all parameters in (7) along with $V_0$ and $V_f$ based on ISI data. A brief summary of the findings follows. As with the case of Brownian motion, not all five parameters of the underlying process are identifiable; rather, the three functions $\tau$, $(V_f - \mu\tau)/\sigma\sqrt{\tau}$ and $(V_0 - \mu\tau)/\sigma\sqrt{\tau}$ of them are. Furthermore, through a detailed study of the Laplace transform, the standard asymptotics of the maximum likelihood estimates (MLEs) of these parameters is valid: that is, the MLEs are asymptotically normal and efficient. This result provides an alternative to the less efficient approach using the empirical Laplace transform. There is also a program to compute the estimates and their standard errors, the latter by inverting the observed Fisher information matrix.

The next level of complexity involves reversal potentials. Recall from Section 2 that the depolarizing (hyperpolarizing) effect of an EPSP (IPSP) depends upon the membrane potential. The closer that $V(t)$ is to the excitatory (inhibitory) reversal potential $V_E$ ($V_I$), the less the effect of an excitatory (inhibitory) input. Many models that incorporate this phenomenon have been proposed. One starting point is the following stochastic differential equation:

$$(8) \quad dV = \left(\mu_0 - \frac{V}{\tau}\right)dt + \sum_{i=1}^{n_E} \alpha_i^E (V_E - V)dN_i^E + \sum_{j=1}^{n_I} \alpha_j^I (V_I - V)dN_j^I,$$

where $\alpha_i^E$ and $\alpha_j^I$ are positive. Passing to the limit, this leads to

$$(9) \qquad\qquad dV = \mu(V)dt + \sigma(V)dW,$$

where

$$(10) \qquad\qquad \mu(v) = \mu_0 - \frac{v}{\tau} + \mu_1(V_E - v) + \mu_2(V_I - v)$$

and

$$(11) \qquad\qquad \sigma^2(v) = \sigma_1^2(V_E - v)^2 + \sigma_2^2(V_I - v)^2$$

with constants $\mu_0$, $\mu_1$, $\mu_2$, $\sigma_1 > 0$, and $\sigma_2 > 0$. Thus, the mean coefficient is linear in the voltage, and the variance coefficient is quadratic. Lánský, Sacerdote, and Tomassetti [29] have suggested that as a good approximation the excitatory reversal potential is negligible and used a different starting point to get

$$(12) \qquad dV = \left( \mu_0 - \frac{V}{\tau} \right) dt + \sigma \sqrt{V - V_I} dW,$$

so that both the mean and variance coefficients are linear in the voltage. While there does not appear to be an analytically tractable density for the ISI based on (12), Iyengar and Liao [18, 19] have shown that the generalized inverse Gaussian family can serve as ISI densities for the model given by (9)-(11).

The generalized inverse Gaussian distribution [20] has density

$$(13) \qquad h(y|\psi, \chi, \lambda) = \frac{(\psi/\chi)^{\lambda/2}}{2K_\lambda(\sqrt{\psi\chi})} y^{\lambda-1} \exp\left[ -\frac{1}{2}(\psi y + \frac{\chi}{y}) \right],$$

where $K_\lambda$ is the modified Bessel function of the third kind with index $\lambda$ (see Abramowitz and Stegun [1]),

$$K_\lambda(u) = \frac{1}{2} \int_0^\infty x^{\lambda-1} \exp\left[ -\frac{u}{2}(x + \frac{1}{x}) \right] dx.$$

Let $\theta = (\psi, \chi, \lambda)$, and let $\Theta$ be the parameter space. Then for $-\infty < \lambda < \infty$, $\Theta = \bigcup_\lambda \Theta_\lambda$, where

$$\Theta_\lambda = \begin{cases} \{(\psi, \chi) : \psi > 0, \chi \geq 0\} & \text{if } \lambda > 0 \\ \{(\psi, \chi) : \psi > 0, \chi > 0\} & \text{if } \lambda = 0 \\ \{(\psi, \chi) : \psi \geq 0, \chi > 0\} & \text{if } \lambda < 0. \end{cases}$$

Barndorff-Nielsen, Blaesild, and Halgreen [2] have shown that when $\lambda \leq 0$, there exists a first passage time for which the the inverse Gaussian arises as a first passage time through a constant boundary. Specifically, suppose that $\theta = (\psi, \chi, \lambda) \in \Theta$, and $\lambda \leq 0$. For $u > 0$, let $\sigma^2(u)$ be a strictly positive differentiable function, and suppose that

$$\gamma(v) = \int_0^v \frac{du}{\sigma(u)}$$

is finite for all $v > 0$, and that $\gamma(v) \to \infty$ as $v \to \infty$. Also, let

$$\mu(v) = \begin{cases} -\sigma(v)\left[\frac{2\lambda-1}{2\gamma(v)} + \sqrt{\psi}\frac{K_{\lambda-1}(\gamma(v)\sqrt{\psi})}{K_\lambda(\gamma(v)\sqrt{\psi})}\right] + \frac{1}{4}\frac{d}{dv}\sigma^2(v) & \text{for } \psi > 0, \lambda \leq 0, \\ \sigma(v)\frac{2\lambda+1}{2\gamma(v)} + \frac{1}{4}\frac{d}{dv}\sigma^2(v) & \text{for } \psi = 0, \lambda < 0. \end{cases}$$

Now let $\{V(t) : t \geq 0\}$ be a time-homogeneous diffusion process with state space $[0, \infty)$, infinitesimal variance $\sigma^2(v)$, infinitesimal mean $\mu(v)$ given above, and initial value $v_0 > 0$ satisfying $\gamma^2(v_0) = \chi$. If $\lambda \leq 0$, then the first passage time $T = \inf\{t > 0 : X(t) = 0\}$ has the generalized inverse Gaussian distribution with parameter $(\psi, \chi, \lambda)$. The diffusion $V(t)$ is not unique, in part because there is considerable freedom in choosing the starting point, $\sigma(u)$ in (14).

Iyengar and Liao [18] showed that the inverse Gaussian can be used to model reversal potentials. In particular, starting with a Brownian motion, and then applying the transformation $v \to e^{-v}$, Ito's lemma yields a diffusion with linear mean and quadratic variance, as required by the model (9)-(11). This may well explain why this simple model has fit several experimental data sets adequately despite its apparent lack of sound physiological basis. Iyengar and Liao have also suggested the use of mixtures of generalized inverse Gaussian densities to fit multimodal ISI histograms. The details of that suggestion are currently under investigation.

We conclude with a brief sketch of stochastic cable models and Hodgkin-Huxley equations, for a detailed description is beyond the scope of this paper. Both stochastic versions are gotten by adding Poisson noise is in (5), which for high rates is approximated by white noise input. As before, our main interest is in the time to firing.

For cable models, the stochastic process representing the voltage along the cable is quite complicated: for Poisson noise input, it is an infinite sum of independent mean-reverting discontinuous Markov processes, and for white noise input it is an infinite sum of independent Ornstein-Uhlenbeck processes. While approximations to the first two moments of the process have been derived for various special cases, the time to firing itself has not been well studied analytically. There are many choices for firing threshold conditions. Tuckwell, Wan and Wong [42] used simulations to study a single-point trigger zone: if the voltage at time $t$ and location $x$ along a cable of length $L$ is $V(t, x)$, and the trigger point is $x_\theta$, the time to firing is

$$T_\theta = \inf\{t : V(t, x_\theta) \geq \theta\}.$$

Other models for firing include an extended trigger zone $[x_1, x_2]$, or that the potential exceed the threshold over the entire interval.

Next, Tuckwell [40] has sketched some results for the Hodgkin-Huxley and Fitzhugh-Nagumo equations with white noise input. It turns out that one case, known as a space-clamped system, is tractable because the corresponding Kolmogorov backward equations are linear even though the stochastic Hodgkin-Huxley equations are nonlinear. However, many questions such as the existence of approximately periodic solutions corresponding to spike trains have yet to be answered. Recent investigations of such complicated models attempt to explain data from experiments using simulations, and they rely on rather basic indices such as the coefficient of variation $(CV)$ of the ISIs. For instance, Softky and Koch [37] studied the highly irregular $(CV \simeq 1)$ firing of cortical neurons. Using simulations, they compared an integrate-and-fire model with a large compartment model. They concluded that the integrate-and-fire model consistently predicted low values of the $CV$, while the compartment model with Hodgkin-Huxley type currents on certain dendrites matched the cortical neurons better. The work of Softky and Koch gave rise to a number of other works. For instance, Gutkin and Ermentrout [10] proposed a one-dimensional dynamical model with white noise input for membrane potential that is capable of high values of the $CV$. They suggest their model neuron as an alternative to the more conventional integrate-and-fire neuron model in neural networks. As with the cable models, the properties of these proposals, and a comparison of their predictions with experimental data are interesting problems that await investigation.

## Acknowledgment

This work was done in part at the Centre for Mathematics and its Applications at the Australian National University, Canberra.

## References

[1] Abramowitz, M., Stegun, I. A. (1965) *Handbook of Mathematical Functions*, New York, Dover.

[2] Barndorff-Nielsen, O., Blaesild, P., Halgreen, C. (1978), First hitting time models for the generalized inverse Gaussian distribution. *Stoch. Proc. Appl.*, **7**, 49–54.

[3] Brink, F., Bronk, D., Larrabee, M. (1946), Chemical excitation of nerve. *Ann. New York Acad. Sci.*, **47**, 457–485.

[4] Calvin, W., Stevens, C. (1965), A Markov process model for neuron behavior in the interspike interval. *Proc. Ann. Conf. Engrg. Med. Biol.*, **7**, 118 (Abstract).

[5] Chhikara, R., Folks, L. (1989) *The Inverse Gaussian Distribution: Theory, Methodology and Applications*, New York, Marcel Dekker.

[6]  Cox, D. (1974), Discussion. *J. Roy. Statist. Soc. Ser. B.*, **36**, 225.

[7]  Cronin, J. (1987) *Mathematical Aspects of Hodgkin-Huxley Theory,* New York, Cambridge University Press.

[8]  Fienberg, S. (1974), Stochastic models for single neuron firing trains: a survey. *Biometrics*, **30**, 399–427.

[9]  Gerstein, G., Mandelbrot, B. (1964), Random walk models for the spike activity of a single neuron. *Biophys. J.*, **4**, 41–68.

[10] Gutkin, B., Ermentrout, G.B. (1997) *Type I membrane excitability as a mechanism for cortical spike train statistics,* Technical Report, Mathematics Department, University of Pittsburgh.

[11] Hodgkin, A.L., Huxley, A.F. (1939), Action potentials recorded from inside a nerve fibre. *Nature*, **144**, 710–711.

[12] Hodgkin, A.L., Huxley, A.F. (1952), A quantitative description of membrane current and its application to conduction and excitation in nerve. *J. Physiol.*, **117**, 500–544.

[13] Hodgkin, A.L., Rushton, W.A.H. (1946), The electrical constants of a crustacean nerve fibre. *Proc. Roy. Soc., London, Ser. B*, **133**, 444–479.

[14] Holden, A.V. (1976) *Models for the Stochastic Activity of Neurones,* New York, Springer-Verlag.

[15] Inoue, J., Sato, S., Ricciardi, L. (1995), On the parameter estimation for diffusion models of single neurons. *Biol. Cybernet.*, **73**, 209–221.

[16] Iyengar, S. (1996) Parameter estimation for a diffusion approximation to Stein's model, in: *Computation in Cellular and Molecular Biological Systems.* Singapore, World Scientific, 265–277.

[17] Iyengar, S. (1998) *Maximum likelihood estimation for the Ornstein-Uhlenbeck process,* Technical Report, Statistics Department, University of Pittsburgh.

[18] Iyengar, S., Liao, Q. (1997), Modeling neural activity using the generalized inverse Gaussian distribution. *Biol. Cybernet.*, **77**, 289–295.

[19] Iyengar, S., Liao, Q. (1998) *Model selection for neural spike train data,* Technical Report, Statistics Department, University of Pittsburgh.

[20] Jorgensen, B. (1982) *Statistical Properties of the Generalized Inverse Gaussian Distribution,* New York, Springer-Verlag.

[21] Kandel, E.R., Schwartz, J.H. (1985) *Principles of Neural Science,* Second edition, Amsterdam, Elsevier

[22] Kallianpur, G. (1983) On a diffusion approximation to a discontinuous model for a single neuron, in: *Contributions to Statistics.* Amsterdam: North Holland.

[23] Karlin, S., Taylor, H.M. (1975) *A First Course in Stochastic Processes,* Second edition, New York, Academic Press.

[24] Karlin, S., Taylor, H.M. (1981) *A Second Course in Stochastic Processes,* New York, Academic Press.

[25] Katz, B. (1966) *Nerve, Muscle, and Synapse,* New York, McGraw-Hill.

[26] Kloeden, P.E., Platen, E. (1992) *Numerical Solution of Stochastic Differential Equation,* New York, Springer-Verlag.

[27] Lánský, P. (1984), On approximations of Stein's neuronal model. *J. Theor. Biol.,* **107**, 631–647.

[28] Lánský, P., Smith, C.E., Ricciardi, L. (1990), One-dimensional stochastic diffusion models of neuronal activity and related first passage time problems. *Trends Biol. Cybernet.,* **1**, 153–162.

[29] Lánský, P., Sacerdote, L., Tomassetti, F. (1995), On the comparison of Feller and Ornstein-Uhlenbeck models for neural activity. *Biol. Cybernet.,* **73**, 457–465.

[30] Levine, M.W. (1991), The distribution of intervals between neural impulses in the maintained discharges of retinal ganglion cells. *Biol. Cybernet.,* **65**, 459–467.

[31] Rall, W. (1989) Cable theory, in: *Methods in Neuronal Modeling: From Synapses to Networks,* Cambridge, MIT Press, 9–62.

[32] Ricciardi, L. (1977) *Diffusion Processes and Related Topics in Biology,* Berlin, Springer-Verlag.

[33] Rieke, F. Warland, D., de Ruyter van Steveninck, R.R., Bialek, W. (1997) *Spikes,* Cambridge, MIT Press.

[34] Sampath, G., Srinivasan, S.K. (1977) *Stochastic Models for Spike Trains of Single Neurones,* Berlin, Springer-Verlag

[35] Segev, I., Fleshman, J.W., Burke, R. (1989) Compartmental models of complex neurons, in: *Methods in Neuronal Modeling: From Synapses to Networks,* Cambridge, MIT Press, 63–96.

[36] Shepherd, G.M. (1994) *Neurobiology,* New York, Oxford University Press.

[37] Softky, W.R., Koch, C. (1993), The highly irregular firing of cortical cells is inconsistent with temporal integration of random EPSPs. *J. Neurosci.,* **13**(1), 334–349.

[38] Stein, R.B. (1965), A theoretical analysis of neuronal variability. *Biophys. J.,* **5**, 173–194.

[39] Tuckwell, H.C. (1988) *Introduction to Theoretical Neurobiology,* Volumes 1 and 2 New York, Cambridge University Press.

[40] Tuckwell, H.C. (1989) *Stochastic Processes in the Neurosciences,* Philadelphia, SIAM.

[41] Tuckwell, H.C., Walsh, J.B. (1983), Random currents through nerve membranes. I. Uniform Poisson or white noise current in one-dimensional cables. *Biol. Cybernet.*, **49**, 99–110.

[42] Tuckwell, H.C., Wan, F.Y.M., Wong,Y.S. (1984), The interspike interval of a cable model neuron with white noise input. *Biol. Cybernet.*, **49**, 155–167.

[43] Yang, G.L., Chen, T.C. (1978), On statistical methods in spike train analysis. *Math. Biosci.*, **38**, 1-34.

# Estimation with Quadratic Loss

Lucien Le Cam

Department of Statistics
University of California
Berkeley, California

**Abstract.** The paper revolves around inequalities for the risk when estimating a function of the parameter. If the function takes its values in a Hilbert space, the loss function used is the square of the Hilbert norm. For estimating real valued functions this reduces to the square difference $|T - \gamma(\theta)|^2$.

Section 1 gives a short history and describes the problem. Section 2 and Section 3 are about estimating real valued functions. In Section 2, the estimates are required to be unbiased. One striking result is that in this case, the minimax risk is the supremum of risks obtainable on *one dimensional* sub-problems. The unbiasedness restriction is not imposed in Section 3. One obtains then a similar result where the supremum is taken over *two dimensional* problems. Various inequalities such as those of Cramér-Rao, Barankin and Kholevo are discussed as they occur.

Section 4 considers the estimation of Hilbert valued functions. Section 5 gives a list of related problems that may receive solutions early in the next Century.

## 1. Introduction

Our story begins with Laplace, [16], who introduced the use of loss functions. Laplace's favorite was the absolute value $|T - \theta|$. This was replaced in 1809, [12], and 1821, [13], by Gauss who proposed $|T - \theta|^2$, saying that Laplace's choice was "no less arbitrary". Other loss functions were studied, in particular by Edgeworth, [10], (1883), see Stigler, [29], (1980).

251

However quadratic loss functions have endured and are still the most commonly used ones. They give rise to one of the important contributions of this century, namely, the Cramér-Rao inequality, [25], (1945).

The Cramér-Rao inequality is still very much alive, with many variants, some of them Bayesian. See for instance Brown and Gajek, [6], (1990), or Borovkov. See also the proof by Simons and Woodroofe, [26], (1983), that the Cramér-Rao inequality holds almost everywhere.

The problem of finding *achievable* lower bounds was considered some years after Rao's paper by Barankin, [2], (1949). Barankin considers a statistical experiment $\mathcal{E} = \{P_\theta; \theta \in \Theta\}$ where the $P_\theta$ are measures on a $\sigma$-field $\mathcal{A}$. One is given a real-valued function $\gamma : \theta \rightsquigarrow \gamma(\theta)$. One looks at estimates $T$ defined on $\mathcal{E}$ such that $\int T dP_\theta = \gamma(\theta)$, for all $\theta$, (Unbiasedness). Barankin considers only variances (or moments of order larger than one) for *one probability measure*. We shall describe in Section 2 what his framework can support, with emphasis on the fact that it introduces "moduli of continuity" and bounds computed on pairs of measures. This is for 'unbiased' estimates. Section 3 considers general estimates and shows that one is again led to bounds computed using two or three probability measures. That is for the estimation of real valued functions. Section 4 considers the estimation of Hilbert space valued functions and recalls an upper bound of Le Cam, [17], (1975), reproduced in Le Cam and Yang, [19], (1990), pages 130-135. Section 5 considers related problems.

## 2. Unbiased Estimation

Consider an experiment $\mathcal{E}$ as described above. Introduce the linear space $S(\mathcal{E})$, linear span of the family $\mathcal{E}$ , that is, the space of all finite linear combinations of the $P_\theta$ and a set $C$, subset of the space $S'$ of real-valued measurable functions $f$ such that $\int f dP_\theta$ be finite for every $\theta$. These functions will be used as quality standard. For instance $C$ could be the set of functions $f$ such that $\int |f - \gamma(\theta)|^2 dP_\theta \leq 1$ for a particular $\theta$, or for a selected set of values of $\theta$. One measures how difficult it is to estimate $\gamma$ by taking the infimum of the numbers $r$ for which there is in $rC$ an unbiased estimate of $\gamma$. Assume $C$ convex, containing zero and closed for the weak topology induced by $S(\mathcal{E})$. Let $\mathcal{M}$ be the space of finite signed measures with finite support on $\Theta$. Map $\mathcal{M}$ to $S(\mathcal{E})$ by $\mu \rightsquigarrow P_\mu = \int P_\theta \, \mu(d\theta)$. (The notation $P_\mu$ is a bit confusing, but understandable. In that notation $P_\theta$ should really be $P_{\delta_\theta}$ for the probability measure $\delta_\theta$ that gives mass 1 to $\theta$ We shall avoid a surfeit of indices). Take a number $r \in [0, \infty)$. Then:

**Theorem 1.** *Assume that $C$ is convex, closed and contains the origin. There exist a $T \in rC$ such that $\int T dP_\theta = \gamma(\theta)$ identically, if and only if $\int \gamma d\mu \leq r$ for all $\mu$ such that $P_\mu \in C^0$, (polar of $C$ in $S$ ).*

(The polar of $C$ is the set of $P \in S$ such that $\int f dP \leq 1$ for all $f \in C$).

*Proof:* The proof is a simple application of the polar theorem: In the duality between $S(\mathcal{E})$ and the space of integrable functions, if $C$ is closed, convex and contains the origin then $C = C^{00}$. This Theorem is itself a consequence of the Hahn-Banach theorem.

This can be put in a different form. Assume that $C$ is symmetric around zero and that there is a probability measure $\lambda \in \mathcal{M}$ such that $\int f dP_\lambda = 0$, all $f \in C$. Then

$$\|\mu\| = \inf\{r : |\int f dP_\mu| \leq r, \text{for all} f \in C\}$$

is a semi-norm on $\mathcal{M}$. The statement of the theorem becomes: There is a $T \in rC$ that is unbiased for $\gamma$ if and only if

$$\sup_{\mu,\nu} \frac{|\langle \gamma, \mu \rangle - \langle \gamma, \nu \rangle|}{\|\mu - \nu\|} \leq r.$$

with $\langle \gamma, \mu \rangle = \int \gamma(\theta) \mu(d\theta)$. There is a smallest $r$ satisfying the inequality for all pairs $(\mu, \nu)$, and an estimate in the corresponding $rC$. Note that this presents a peculiar feature: It involves only two probability measures, $\mu, \nu$, elements of $\mathcal{M}$, at a time, and the 'quality' of the estimate is given by a 'modulus of continuity for $\gamma$.

The case considered by Barankin involves a particular $\lambda$ and a set $C$ of functions $f$ such that $\int f dP_\lambda = 0$ and $\int f^2 dP_\lambda \leq 1$. One seeks a $T$ such that $\int T dP_\theta \equiv \gamma(\theta)$ and such that $\int |T|^2 dP_\lambda$ be as small as possible. Introduce the ratio:

$$R(\lambda, \mu, \nu) = \frac{|\langle \gamma, \mu \rangle - \langle \gamma, \nu \rangle|^2}{\int (dP_\mu - dP_\nu)^2 / dP_\lambda}$$

The bound becomes:

$$Var_\lambda T \geq \sup_{\mu,\nu} R(\lambda, \mu, \nu),$$

where the sup is taken over all pairs of probability measures $\mu, \nu$ elements of $\mathcal{M}$.

The same bound was given by Kiefer, [15], (1952). It is true that Kiefer uses probability measures that are more general than our finite support ones, but that does not change the result because of the $\sup_{\mu,\nu}$. Our reason for using measures with finite support is just to avoid mention of integrability restrictions.

Let us take a closer look at the ratio $R(\lambda, \mu, \nu)$. *This is precisely the minimum variance* for $P_\lambda$ of estimates $T$ that satisfy the unbiasedness conditions $\int T dP_\lambda = 0$ and $\int T(dP_\mu - dP_\nu) = \langle \gamma, \mu - \nu \rangle$ *irrespective of what they might do at other measures* $P_\theta$. Then, to get Barankin's bound, one takes a supremum over the pairs $(\mu, \nu)$. If looking for a minimax risk, one would take a supremum of $R(\lambda, \mu, \nu)$ over the triplets $(\lambda, \mu, \nu)$. But then one can replace $\frac{\int (dP_\mu - dP_\nu)^2}{dP_\lambda}$ by $4k^2(P_\mu, P_\nu)$ where for two probability measures $P$ and $Q$, the number $k^2(P, Q)$ is the square distance

$$k^2(P, Q) = (1/2) \int \frac{(dP - dQ)^2}{dP + dQ}.$$

This same distance occurs in an inequality of Le Cam and Yang, [19], (1990),page 127.

$$Var_P(T) + Var_Q(T) \geq (1/2)\Delta^2(1 - k^2)/k^2,$$

where $k = k(P, Q)$, and $\Delta = E_Q(T) - E_P(T)$. It is to be noted that this bound is *always* attained by a suitable estimate.

The Hellinger distance $h$ defined by $h^2(P, Q) = (1/2) \int (\sqrt{dP} - \sqrt{dQ})^2$, satisfies

$$0 \leq h^2 \leq k^2 \leq h^2(2 - h^2) \leq 1.$$

Substituting in the Le Cam-Yang expression, one obtains Kholevo's (1973) inequality, (See[14], and also Pitman, [24], 1979):

$$Var_P(T) + Var_Q(T) \geq (1/2)\Delta^2 \rho^2/(1 - \rho^2)$$

with $\rho = 1 - h^2$ and $\Delta$ equal to the difference of expectations. Pitman, [24], and also Simons and Woodroofe, [26], (1983) pass to the limit in such inequalities as $k(P, Q)$ tends to zero to obtain their versions of the Cramér-Rao inequality. If $\Theta$ is an interval of the line, Pitman argues that the Fisher Information should be defined as the limit, if it exists, of $8\frac{h^2(P_{\theta+t}, P_\theta)}{t^2}$. This definition is not available here since $\Theta$ has no structure. However, one should note that as $k^2$ tends to zero the ratio of the Kholevo bound to that of Le Cam& Yang is approximated by $\frac{k^2}{2h^2}$, which can tend to any limit in $[1/2, 1]$ and will tend to 1 only under suitable equi-integrabiliy conditions. Thus, it may not always be desirable to use Kholevo's inequality in place of that of Le Cam& Yang. It might result in too low a bound.

There are other possible interpretations of the Fisher information. One of them has to do with local approximation of the given experiment by a Gaussian one. For instance, in the case of a real $\Theta$ as mentioned above, one often wishes to approximate $\log \frac{dP_{\theta+t}}{dP_\theta}$ by a linear plus quadratic $tV - \frac{1}{2}\sigma^2 t^2$ where $V$ is random but $\sigma^2$ is not and plays the role of Fisher Information. This type of approximation could be extended to an abstract $\Theta$.

Barankin considers also the case where one wants estimates $T$ that satisfy $\int T dP_\lambda = 0$ and where one wants to minimize $\int |T|^\alpha dP_\lambda$ for an $\alpha > 1$. This works similarly. The argument given above can be used in cases other than those considered by Barankin, [2]. For instance, instead of fixing a particular measure $\lambda$, one can take for $C$ the class of estimates that have a variance less than or equal to one for all the $P_\theta, \theta \in \Theta$, thus obtaining a bound of a minimax nature.

One of the virtues of Barankin's formulation is that his theorem asserts the *existence* of an estimate $T \in rC$, as does our Theorem 1. Thus it automatically provides the best possible bound. This existence is due to the assumption that $C$ is *closed* in the weak topology described. For instance, in the case of the particular set $C$ of functions such that $\int f dP_\lambda = 0$ and $\int f^2 dP_\lambda \leq 1$, to insure that $C$ is closed, Barankin assumes that the $P_\theta$ are absolutely continuous with respect to $P_\lambda$ and satisfy the square integrability condition $\int [\frac{dP_\theta}{dP_\lambda}]^2 dP_\lambda < \infty$. This same assumption will also insure that the set used for the minimax variance problem is closed. If $C$ is not closed, or if Barankin's conditions are not satisfied, the lower bound is still valid and, occasionally, there may still exist estimates reaching it, as is the case in Kiefer's examples. The role of Barankin's absolute continuity and square integrability condition can be seen as follows. Take an experiment that consists of just two measures, say $(P, Q)$. Look for estimates $T$ such that $\int T dP = 0$ and $\int T dQ = 1$. Let us consider the smallest possible variance $v_0$ of such a $T$ under $P$. If $Q$ is not dominated by $P$, then $v_0 = 0$ and this is attained. If $Q$ is dominated by $P$ but $\int [\frac{dQ}{dP}]^2 dP = \infty$, then $v_0 < \epsilon$ for any positive $\epsilon$, but there are no $T$ achieving zero variance.

## 3. General estimates

All of this has to do with "Unbiased" estimates. What about estimates in general? Here a basic formula is for the Bayes risk for a *prior* probability measure $\mu$ with finite support on $\Theta$. It is stated here for a function $\gamma$ and estimates $T$ with values in a Hilbert space with norm $\|.\|$. This is because it will be used in that form in Section 4 and because the Hilbert case formula differs from the one in the real case only by the substitution of $\|.\|$ for the absolute value $|.|$.

Let $R_\mu = \inf_T \int E_\theta \|T - \gamma(\theta)\|^2 \mu(d\theta)$ be the Bayes risk for $\mu$. Also, let $A(s,t) = \int \frac{dP_s \, dP_t}{dP_\mu}$.

Then

(1)  $$R_\mu = 1/2 \int \|\gamma(s) - \gamma(t)\|^2 A(s,t)\mu(ds)\mu(dt)$$

(2)  $$= V(\gamma, \mu) - \int \left\| \int \gamma(s)\frac{dP_s}{dP_\mu}\mu(ds) \right\|^2 dP_\mu.$$

In these expressions, one can and will assume $\int \gamma(s)\mu(ds) = 0$. That is that the expectation of $\gamma$ for $\mu$ is zero. The value called $V$ is then the "prior variance" $V = \int \|\gamma\|^2 \mu(ds)$. In the real case, the same expressions give rise to a "three-point" formula as follows. Assuming the expectation of $\gamma$ for $\mu$ equal to 0, let $c = c(\gamma, \mu) = \int |\gamma(s)|\mu(ds)$. Introduce the probability measures $\pi$ and $\nu$ by $d\pi = \frac{2}{c}\gamma^+ d\mu$ and similarly for $\nu$ using $\gamma^-$. (Here $\gamma^+$ and $\gamma^-$ are respectively the positive and negative parts of $\gamma$). Then:

(3)  $$R_\mu = V^2 - c^2 + c^2 \left[ 1 - \frac{1}{4} \int \frac{(dP_\pi - dP_\nu)^2}{dP_\mu} \right].$$

Those are exact expressions. To get a minimax value, one takes the supremum over $\mu$. This is not too simple since $\pi$ and $\nu$ are linked to $\mu$. One obtains an *upper bound* for $R_\mu$ by replacing the square bracket in the preceding expression by $\|P_\pi \wedge P_\nu\|(2 - \|P_\pi \wedge P_\nu\|)$, where $P_\pi \wedge P_\nu$ is the minimum of the two probability measures involved. It would be interesting to compare the bounds obtainable in this manner with those given by a different method by Donoho and Liu, [8], (1991). Their procedure will occur again in Section 4.

For real valued functions $\gamma$, the fact that the risk occurs as a sup over two or three point formulas hides the role of dimension or other measures of complexity, *but this is in the convex hull* of the family $\mathcal{E}$. An easy way of passing to convex hulls will probably wait for the next century.

Let us note that the kernel called $A(s,t)$ above should really be written $A(s,t;\mu)$ since it depends on the *prior* measure $\mu$. It is an interesting fact that the values of such kernels for the triplets $(s,t;\mu)$ already determines the type of the experiment, that is all the risk functions for totally arbitrary bounded loss functions. One could attempt to replace these kernels by the related Hellinger affinities $\int [dP_s dP_t]^{\frac{1}{2}}$. Those are easier to handle. They do not determine the type of the experiment. Indeed, for any Poisson experiment, there is a Gaussian one with the same affinities.

## 4. Estimating Hilbert valued functions

Except for a brief mention leading to formula (1) of Section 3, the preceding sections have dealt mostly with the estimation of a real valued

function $\gamma$ and emphasized the occurrence of suprema taken over problems
that involve only two or three measures at a time. If one lets $\gamma$ be Hilbert
valued, one obtains several different known results. For instance, results
similar to Fano's Lemma of 1954, (see [11]) or Assouad's Lemma, [1], (1983)
are obtainable by specialization of our basic formula. This is quite visible
for Assouad's lemma. Indeed, it was abstracted from previous results that
used a Hilbert space situation. It is not so clear for Fano's lemma. However
one can note the following. Take a finite set $\{\theta_j; j = 1, ..., k\}$ and assume
that the $\gamma(\theta_j)$ are orthonormal in the Hilbert space. Write $f_s$ for $\frac{dP_s}{dP_\mu}$ and
$w_s$ for the mass assigned to $s$ by $\mu$. Formula (2) of Section 3 becomes

$$R_\mu = 1 - \int \left[ \sum_s f_s{}^2 w_s{}^2 \right] dP_\mu$$

Since $\sum_s f_s w_s = 1$, this gives

$$R_\mu \geq 1 - \int [\max_s f_s w_s] dP_\mu.$$

In the case where all the $w_s$ are equal to, say, $\frac{1}{n}$, this gives a bound $R_\mu \geq 1 - \frac{1}{n} \| \max_s P_s \|$ from which one can obtain Fano's formula in terms of the
Kullback-Leibler numbers.

One obtains also crude lower bounds. Take for instance two disjoint
sets $A$ and $B$. Suppose that the shortest distance between values taken by
$\gamma$ in $A$ and values taken in $B$ is at least $b$. Take a probability measure $\lambda$
carried by $A$ and a $\nu$ carried by $B$. Then, for $\mu = \frac{1}{2}(\lambda + \nu)$,

$$R_\mu \geq \frac{b^2}{4}[1 - k^2(P_\lambda, P_\nu)]$$

This is not too appealing but will be contrasted to an upper bound given
below.

In view of the difficulty of passing to convex hulls, one may follow
Le Cam , [17],(1975),or [18], (1986), pages 481-508, and cover the fam-
ily $\{P_\theta; \theta \in \Theta\}$ by convex sets, such as balls $\{Q; h(Q, P) \leq a\}$ or balls
$\{Q; k(Q, P) \leq a\}$ in the distance $k$. One obtains then *upper bounds* of the
following type, for which see Le Cam & Yang, [19], pages 131-133. Let $\Theta$ be
metrized (or pseudo-metrized) by a distance $d$. Take decreasing sequences
of positive numbers $a_j$ and $b_j$ with $a_j < b_j$. For each $j$, cover $\Theta$ by sets of
diameter $\leq a_j$. Call a pair of such sets "$b_j$-distant" if they include points at
distance $\geq b_j$. Construct "confidence sets" step by step as follows. If a set
$S_{j-1}$ has been constructed, carry out testing procedures pairwise between

the sets of the $j^{th}$ cover that are $b_j$ distant and intersect $S_{j-1}$, retaining only those that pass *all* the tests. Intersect the union of those that pass with the previous $S_{j-1}$. This gives $S_j$. Stop the process at some integer $m$ and select a value $T = \hat{\theta} \in S_m$ for an estimate. Then:

$$(4) \qquad E_\theta g[d(T, \theta)] \leq g(b_m) + 2 \sum_{j=1}^{m} g(b_{m-j}) \beta(m - j + 1) C(m - j + 1).$$

In this expression $g$ is an increasing function. The distance used is $d$. The numbers $\beta(j)$ are probabilities of errors for tests of "$b_j$-distant" pairs and $C(j)$ is the number of sets of diameter $a_j$ needed to cover a set of diameter $2a_j + b_j$. To apply the procedure, one must select the numbers $a_j$ and $b_j$. It is often convenient to take $b_j = A a_j$ for some fixed $A > 1$ and let the $a_j$ decrease in geometric progression. This is the system used, for instance, in Le Cam, [17], (1975), but may not be the "optimal" system.

Often the first two terms on the right side of (4) are dominant. Then this *upper bound* differs from our previous *lower bound* mostly by the factor $2C(m)$, whose *log* (base 2 ) is proportional to the "dimension" of $\Theta$ at the level $a_m$, as defined in Le Cam, [18], page 497. In the cases considered by Le Cam, [18], the probabilities of error decrease exponentially as the distances increase and a suitable selection of the stopping size $b_m$ makes the bound proportional to the "dimension" itself.

The procedure has been applied with some success by Le Cam, [17], (1975), for the case of independent observations taking for square distance the sum of the square Hellinger distances on the components. A drastic improvement was provided by an inequality of L. Birgé, [5], (1983) for the case of non identically distributed observations. A variation on the "stepwise" method of selecting successive confidence sets is given in Le Cam, [18], pages 500-503. Birgé himself used a different method of construction of the estimates. He covers the entire parameter set at once by sets of a small, well chosen diameter. Birgé also considers the case of Markov Chains. See Birgé, [4], (1983).

Choosing, in the case of independent observations, as square distance the sum of square Hellinger distances computed on components , or in the case of infinitely divisible experiments, a square distance $- \log \int [dP_s dP_t]^{\frac{1}{2}}$, plunges the experiment into a particular Hilbert space in which the formulas (1) and (2) of Section 3 become applicable. We do not know at present of a firm argument linking these formulas (1) or (2) to the bounds just discussed. The "stepwise" procedure has also been adapted by Donoho and Liu, [8], (1991) for estimating real valued functions. It has been shown to be effective in providing there the best asymptotic rates of estimation, but a link to, say, formula (2), Section 3 is still missing. Similarly the bounds of this section seem to provide the best rates for estimating densities in very many nonparametric cases. The constants involved seem to be in need of improvement. This is not surprising considering how crude the arguments are. A better use of formula (1), Section 3 may be necessary.

## 5. Related problems

The arguments of Sections 3 and 4 raise a number of problems:

1) *Convex hulls.*
In the formulas of Section 2 and in the formula (3) of Section 3, a minimax risk is obtained as a supremum over risks of problems involving just 2 or 3 measures, that is 2 or 3 points in the convex hull of the family $\mathcal{E}$. As already indicated there, we do not have simple effective procedures to select the appropriate points. The subject has been with us for a long time in the form of finding "least favorable" distributions, (or sequences least favorable in the limit). In the general decision theory, this is largely done by guessing. Still, there must exist algorithms that would lead to least favorable sequences, perhaps even in such a way that mathematical properties would be ascertainable.

At one time we thought we could obtain such algorithms by a method akin to the "Tâtonnement de Walras" as presented by H. Uzawa, [30], (1959). This has not been done in a satisfactory way so far.

The fact that convex hulls can be tricky is exemplified by the "signal + noise" problem: $Y(dt) = f(t)dt + W(dt)$, where $dt$ is the Lebesgue measure on $[0, 1]$, $W$ is the associated Gaussian white noise and $f$ is an unknown function such that $\|f\|^2 = \int f^2(t)dt < \infty$. The observable is the process $Y$.

Suppose that it is desired to estimate $\|f\|$. It has probably been known for a long time, but was pointed out to this author by H. von Weiszäcker, that no good estimate can be found because the Gaussian distribution $G_0$ corresponding to $f = 0$ is in the closed convex hull of the distributions $G_f$ for $\|f\| = a$ and this for every $a > 0$. Thus the norm of $f$, a real valued function of $f$, cannot be estimated. Formula (1) of Section 3 must say so, but it is not an obvious statement.

2) *Gaussian analogues.*
With the work of C. Stein, [27], [28], (1956), of Pinsker, [23], (1980) and of Donoho and his colleagues, [9], (1990) one has achieved a definite insight for the problems of estimation arising in Gaussian-shift experiments. However, this is mostly for the cases where the centers of the Gaussian distributions form a convex set or satisfy other restrictive conditions.

The upper bounds on minimax or Bayes risks given in Le Cam, [18], pages 495-520 also have a special feature: For the case of infinitely divisible experiments, they depend *only* on the Hellinger affinities of the probability measures forming $\mathcal{E}$. For the case of independent observations, they depend *only* on sums $\sum_j h_j^2(s, t)$ of square Hellinger distances computed on the individual observations, that is on Hellinger affinities for the accompanying Poisson experiments.

In both cases there are well defined Gaussian-shift experiments with *exactly* the same affinities. Thus, one is led to wonder whether some of the results, or the intuition, available in the Gaussian case can be transferred to other experiments with the same affinities.

Of course, the affinities themselves cannot tell us whether an experiment $\mathcal{E}$ is close to a Gaussian one $\mathcal{G}$ in the sense of the distance $\Delta(\mathcal{E}, \mathcal{G})$ described by Le Cam, [18]. Something more is needed, but exactly what is unclear. If $\Delta(\mathcal{E}, \mathcal{G})$ is small, many things transfer 'approximately',between $\mathcal{E}$ and $\mathcal{G}$ but can one hope for more ?

3) *The risk of Bayes procedures at a point.*
Given an experiment $\mathcal{E}$ and a prior measure $\mu$, formula (1) or (2) of Section 3 gives a formula for the *Bayes risk* for $\mu$ of the Bayes estimate for $\mu$. It does not say at all what the value of the risk function could be at a specified $\theta \in \Theta$. This could be an important piece of information. An attempt to get upper bounds for such pointwise risks was made in Le Cam, [18], pages 509-515. It cannot be called entirely successful. Perhaps some other approach is needed.

4) *The quilt-patch problems.*
It often happens that, by crude procedures, An experiment $\mathcal{E}$ provides confidence sets, say confidence balls. It may also happen that if such balls had been selected in a nonrandom way, one would have for the experiment restricted to them good tractable approximations leading to good 'local' estimates. The question is: How to put the two things together ?

The problem was encountered by Le Cam, [18], pages 76-80 and 266-267, as well as by many other authors. It has been solved reasonably under 'dimension' restrictions, or as in the case of Nussbaum, [20], (1996), by using one part of the sample to construct the confidence balls and an independent part for the 'local' argumentation. A general study is lacking.

Of course, a Bayes approach does this almost automatically for the construction of estimates, (but not for the approximation of experiments). This is, however, poor consolation in non-parametric cases, where Bayes estimates are typically not very well behaved. See Diaconis and Freedman, [7], (1986).

5) *Relations with semiparametrics.*

The estimation of just one real valued function of $\theta$ when $\Theta$ can be very large comes under the label of "semiparametric" problems. these problems have been considered by Pfanzagl, [21], (1982), [22], (1990) and by Bickel et al.[3], (1993).

The formulas of Section 3 apply to that situation and suggest the following approach. Assume that the function $\gamma$ to be estimated takes only integer values. (By discretization and change of scale, this is not a major assumption). For each integer $r$, positive or negative, let $K_r$ be the closed convex hull of the set of $P_\theta$ for which $\gamma(\theta) = r$. Assume for simplicity that the sets $K_r$ are two by two disjoint.

Then the choice of a prior probability $\mu$ amounts to two choices: First, one selects the probabilities $\beta_r$ that $\gamma(\theta) = r$. Second, one selects an element, say, $Q_r$ in the set $K_r$.

To lead to a high value of the risk $R_\mu$, one should select the $Q_r$ close to each other in some sense. One should also select the $\beta_r$ in some most unfavorable fashion.

This geometric picture of "most unfavorable paths" seems to deserve investigation.

6) *What does i.i.d. mean?*

In the common acceptance of the term, the i.i.d. case is one where one has a sequence $X_1, X_2, ..., X_n$ of observations, all distributed according to a measure $p_\theta$ independent of $n$. then, one often lets $n$ tend to infinity. One obtains a much richer structure if one keeps the independence assumption but let the underlying measure be $p_{\theta,n}$ , changing as $n$ changes. The structure becomes even richer if, as practitioners often do, one lets the parameter set become $\Theta_n$, varying with $n$. Some portions of Le Cam, [18], are written in that spirit, but more needs to be done. Note for instance that when $\Theta_n$ depends on $n$ , the highly praised "consistency" of estimates can lose meaning entirely. Statements of almost sure convergence as $n$ tends to infinity typically are meaningless,since the underlying measure spaces vary with $n$ and are not linked in a natural way.

Many of the standard statements should be reviewed in the more general framework. This would lead to a lot of work that does not require much brain power, but some does and may be interesting.

**Acknowledgements**

I am greatly indebted to Professor Grace Lo Yang who read a draft of the paper and suggested many improvements. I also owe thanks to Professor David Freedman for comments on Section 2.

### References

[1] Assouad, P. (1983), Deux remarques sur l'estimation. *C.R. Acad.Sci. Paris Ser. I Math.*, **296**, 1021–1024.

[2] Barankin, E.W. (1949), Locally best unbiased estimates. *Ann. Math. Stat.*, **20**, 477–501.

[3] Bickel, P.J., Klaassen, C.A.J., Ritov, Y. and Wellner, J.A. (1993) *Efficient and adaptive estimation for semiparametric models,* Johns Hopkins Series in the Mathematical Sciences.

[4] Birgé, L. (1983a), Approximation dans les espaces métriques et théorie de l'estimation. *Z. Wahrsch. verw. Gebiete*, **65**, 181–237.

[5] Birgé, L. (1983b), Robust testing for independent non identically distributed variables and Markov chains. *Lecture notes in Statistics*, **16**, Springer Verlag.

[6] Brown, L.D. and Gajek, L. (1990), Information inequalities for the Bayes risk. *Ann. Statist.*, **18**, 1578–1594.

[7] Diaconis, P. and Freedman, D.A. (1986), On the consistency of Bayes estimates. *Ann. Statist.*, **14**, 1–67.

[8] Donoho, D.L. and Liu, R. (1991), Geometrizing rates of convergence. *Ann. Statist.*, **19**, part I, 633-667, Part II, 668-701.

[9] Donoho, D.L., Liu, R.C. and MacGibbon, B. (1990), Minimax risk over hyperrectangles and implications. *Ann. Statist.*, **18**, 1416–1437.

[10] Edgeworth, F.Y. (1883), The law of error. *Philosophical Magazine*, **16**, 300–309.

[11] Fano, R.M. (1954) *Statistical Theory of Communication*, M.I.T. unpublished Notes.

[12] Gauss, C.F. (1809) *The heavenly bodies moving around the sun in conic sections,* Reprint, Dover Pub. N.Y. 1963.

[13] Gauss, C.F. (1821) Theoria combinationis observationum erroris minimis obnoxiae, *Commentationes soc. reg. serien. Gottingenses.*

[14] Kholevo, A.S. (1973), A generalization of the Cramér-Rao inequality. *Theory of Probab., Appl.*, **18**, 359–362.

[15] Kiefer, J.C. (1952), On minimum variance estimators. *Ann. Math. Stat.*, **23**, 627–629.

[16] Laplace, P.S. (1814) *Théorie Analytique des Probabilités,* Second édition. Paris.

[17] Le Cam, L. (1975) On local and global properties in the theory of asymptotic normality of experiments, in: *Stochastic Processes and related topics* 13–54, M. L. Puri *ed*, Academic Press.

[18] Le Cam, L. (1986) *Asymptotic Methods in Statistical Decision Theory*, Springer Verlag.

[19] Le Cam, L. and Yang, G.L. (1990) *Asymptotics in Statistics*, Springer Verlag.

[20] Nussbaum, M. (1996), Asymptotic equivalence of density estimation and Gaussian white noise. *Ann. Statist.*, **24**, 2399–2430.

[21] Pfanzagl, J. and Wefelmeyer, W. (1982) *Contributions to a General Asymptotic Statistical Theory*, Lecture Notes in Statistics **13**, Springer Verlag.

[22] Pfanzagl, J. (1990) *Estimation in Semiparametric Models*, Lecture Notes in Statistics **63**, Springer Verlag.

[23] Pinsker, M.S. (1980), Optimal filtering of square integrable signals in Gaussian white noise. *Problems Inform. Transmission*, **16**, 120–133.

[24] Pitman, E.J.G. (1979) *Some basic theory for Statistical Inference*, Chapman and Hall.

[25] Rao, C.R. (1945), Information and accuracy attainable in the estimation of statistical parameters. *Bull. Calcutta Math. Soc.*, **37**, 81–91.

[26] Simons, G. and Woodroofe, M. (1983) The Cramér-Rao inequality holds almost everywhere, in: *Papers in the honor of Herman Chernoff*, 69–93, Rizvi, Rustagi and Siegmund eds, Academic Press.

[27] Stein, C. (1956), Efficient non parametric testing and estimation. *Proc. Third Berkeley Symp. Math. Stat. Probab.*, **1**, 187-195.

[28] Stein, C. (1956), Inadmissibility of the usual estimator for the mean of a multivariate Normal distribution. *Proc. Third Berkeley Symp. Math. Stat. Probab.*, **1**, 197-206.

[29] Stigler, S.M. (1980), An Edgeworth curiosum. *Ann. Statist.*, **8**, 931–934.

[30] Uzawa, H. (1959) Walras' tâtonnement and the existence of an equilibrium for a competitive economy, *Technical report*, Stanford University.

# Econometrics in the 21st Century

## G.S. Maddala

Department of Economics
Ohio State University
Columbus, Ohio

**Abstract.** The paper surveys the major developments in econometrics since the founding of the Econometric Society in 1931 until 1998.It points out the major currents the limitations of the existing work and the problems that need to be pursued in the 21$^{st}$ century.

> The Gods love the obscure and hate the obvious
> Brihadaranyaka Upanishad

## 1. Introduction

I shall review some broad trends in econometrics during the last 70 years and comment on what problems need to be tackled in the 21$^{st}$ Century. The actual problems analyzed might differ substantially from what I suggest because some existing trends will continue even if the marginal benefit from work in these areas is close to zero. I am reminded of a story told by Milton Friedman. He saw a man searching intensely for something under a street lamp. "What are you looking for?" asked Friedman. "My ring, I lost it" replied the man. "Where did you lose it?" asked Friedman. "Over there" replied the man. "If you lost it over there, why are you looking here?" asked Friedman. "It is dark over there, there is light here, so I am looking here," replied the man.

To make matters simple, there are no equations in the paper. It is written in plain English.

## 2. The Founding of the Econometric Society and Econometrica

The Econometric Society and the journal *Econometrica*, started around 1931. (See Christ, 1983). Econometrics (unlike biometrics and psychometrics) was not defined as the application of statistical methods to economics. Ragnar Frisch defined it as the application of statistical and mathematical methods in economics. He reiterated this definition in Frisch (1936). As a consequence mathematical economics also comes under the umbrella of econometrics. This has produced strange results. In recent years the issues of *Econometrica* have had only a couple of papers on econometrics (statistical methods in economics) and the rest are all on game theory and mathematical economics. If you look at the list of fellows of the Econometric Society, you find one or two econometricians and the rest are game theorists and mathematical economists. Econometricians have lost control of both the econometric society and the journal *Econometrica*. When Frisch defined econometrics as including mathematical economics, it was appropriate to do so, because there was no other avenue for mathematical economists. But now there are journals in mathematical economics and game theory: there is *Economic Theory*, *Journal of Mathematical Economics* and journals in game theory. It is not fair that *Econometrica* and the Econometric Society are dominated by these groups.

In the 21st century, econometricians should work to regain control of *Econometrica*, or split it into two sections (as was done with the *Annals of Mathematical Statistics*), *Annals of Econometrics* and *Annals of Mathematical Economics*. Something has to be done to at least put a quota—that the number of econometricians elected as fellows of the Econometric Society should be 50% of the total number elected. Currently each year of the 20 who are elected, about 2 or 3 are econometricians: this is crazy.

## 3. Frisch's Early Work

Ragnar Frisch worked on errors in variables (EIV) models, (See Frisch 1934). So did Koopmans, (See 1937). Subsequently, the emphasis in econometrics shifted to the "errors in equations" models. This is what Rudolf Kalman complains about in his tirade on econometrics under the title "Crisis in Econometrics" (he goes around lecturing all over the world). I heard from Ted Anderson that he has been lecturing on "crisis in statistics" too. Although errors in variables are all pervasive, it is true that econometricians have ignored them. There was a resurrection of interest in these models in the 1970's in the papers by Goldberger, Griliches and Jöreskog, which are reprinted in Aigner and Goldberger, (1977). But this trend has died. Recently there was the Ragnar Frisch symposium Strom and Holly,

(1998) on the occasion of the 100th anniversary of Frisch's birthday, but apart from my paper, there were no papers on EIV models.

In the statistical literature, there is the book by Fuller, (1987) but the problems encountered in econometrics in EIV models are different. The models have correlated errors as discussed and analyzed in Ashenfelter and Krueger (1994).

The problems of identification in EIV models have been discussed in Kalman, (1982), Becker et.al., (1985), and Leamer, (1987) and estimation of regression models with all variables measured with error in Klepper and Leamer, (1984).

In the 21$^{\text{st}}$ Century, more attention needs to be paid to EIV models. There is a forthcoming book by Kapteyn and Wansbeek in this area that reviews the problems solved and the problems that remain to be solved but the book may not come out until the 21$^{\text{st}}$ Century (year 2000?).

## 4. Econometrics in the 1950's

Haavelmo's paper (1944) provided the impetus for the work on errors in equations models in econometrics. (Kalman calls it the "Haavelmo" prejudice in econometrics). Later Anderson and Rubin suggested limited information maximum likelihood (LIML) for simultaneous equations models. The estimation procedure was cumbersome and Theil in 1956 suggested the two stage least squares (2SLS) method. The 2SLS method had the robustness property of all least squares methods compared with the LIML and it became very popular because it was much easier to compute. Strictly speaking, it is not a simultaneous equations method, as LIML is. If the equation is over identified then which way you normalize the equation matters (whereas LIML is invariant to normalizations) and this is contrary to the concept of simultaneity. This problem was solved much later when Hiller, (1990) suggested the symmetrically normalized two stage least squares (SN2SLS) method. As with 2SLS, instrumented variable methods are also not invariant to normalization and we consider the symmetrically normalized analogues (SNIV).

There has been a lot of work on the existence of moments of 2SLS and LIML estimators, their small sample properties and so on. What remains of all this for the 21$^{\text{st}}$ century? In the 21$^{\text{st}}$ Century there will be more work on symmetrically normalized methods and more use of these methods. They have been found more useful in the presence of weak instruments. Also the Bayesian approach to simultaneous equations will be an active area in view of the recent developments in simulation methods.

## 5. Distributed Lags of the 1960's

The 1960's witnessed a lot of work on distributed lags. This is summarized in Griliches, (1967) and Nerlove, (1972). This has now died. As remarked by Nerlove, (1972, p. 246), "current research on lags is not "good" because neither is the empirical research soundly based on economic theory nor is the theoretical research strongly oriented." Later the VAR models (due to Sims, 1960), and the error-correction models (ECM) (due to Sargan, 1964) became popular in modelling lags and dynamic behavior. Sargan also introduced (and David Hendry later followed it up and preached in several of his papers) the general to specific approach to dynamic econometric modelling, that is: start with a highly parameterized model and then simplify it by progressive testing. (This is also known as the "tops down" approach in contrast to "bottoms-up" approach–that is progressively complicating a simple model).

There are several areas in econometrics where Monte Carlo studies have shown that the tops down approach is the preferable one. One instance is the choice of the lag length in ADF tests for unit roots. Another is in tests for seasonal unit roots.

What is left of all this work to be pursued in the $21^{st}$ Century?

Not much except that more evidence needs to be gathered on the two approaches to econometric methodology: the tops down approach vs. the bottoms up approach. Another area to pursue is the estimation of nonlinear dynamic models. This can lead to artificial neutral networks and chaos theory. Also, as discussed in item 16 later, in dynamic models, the unconditional ML methods need to be used.

## 6. The Diagnostic Revolution of the 1970's Score Tests, Hausman Test

In the 1950's a big headache was estimating a given model with the given data. With the progress in computer technology, this is not a big problem. So attention shifted to the question: what could be wrong with my model? Several diagnostic tests have been devised. For instance, if a regression is estimated with time series data, you test for heteroscedasticity and autocorrelation of the residuals, test for parameter stability, test for location of structural breaks (CUMSUM tests), test for ARCH effects in the residuals and so on. Hendry popularized this by his slogan: "test, test, test."

As Johnston, (1991) remarks, the number of tests performed is often higher than the total number of observations. Friedman and Schwartz, (1991) also complain that it is not clear how many degrees of freedom are used in all this testing and how to evaluate the significance of the final results.

All these tests are now built into several computer packages: MICROFIT, PC-GIVE, SHAZAM and so on. You run a regression and you get the results of all these diagnostic tests in the output. Often these are just reported and no further action is taken.

Many of these diagnostic tests are score tests. (Rao, 1947) which are popularly known in the econometric literature as LM (Lugrangian Multiplier) tests following the work of Silvey more than a decade later. The LM terminology is appealing to econometricians because they are used to shadow prices and Lagrangian multipliers. Surveys on the score test can be found in Godfrey, (1988) and Bera and Ullah, (1991).

Most of these diagnostic tests have a specific alternative. You know what action to take if the test rejects the null. The information matrix test of White, (1982) is a diagnostic test that does not specify an alternative. All it can tell is that your model is wrong. (There have been attempts to remedy this situation and there is a discussion on some plausible alternatives). Anyway, it reminds me of the conversation between Alice and the cat in Alice in Wonderland. "Which way do I go from here?" asks Alice. "It depends on where you want to go," says the cat. "I don't care," says Alice. "Then it doesn't matter which way you go," replies the cat. The information matrix test just says "go". It doesn't matter which way. This takes the econometrician to Alice's Wonderland. Which is a great feeling. I always wondered why so many are excited about this useless test. They must all be a set of bored creatures and what better thing is there than being in Alice's Wonderland?

Another popular test in econometrics is the Hausman, (1978) test, which is based on a theorem in Rao, (1973, p. 317). Tests asymptotically equivalent to the Hausman test can be computed as score tests. Maddala, (1995) reviews other specification tests like conditional moment tests and relates them to score tests. There is also a discussion of extensions of the Hausman test. The paper concludes that score principle remains by far the most commonly used approach to specification testing.

Not much is known about the small sample performance of the score tests, Hausman test and the conditional moment tests. It is well known that the small sample performance of the information matrix test is very poor. there are substantial site distortions (the actual significance level could be 50 percent to 70 percent when the nominal level is 5 percent.

In the 21ˢᵗ Century, the small sample performance of all these diagnostic and specification tests need to be studied. One other area, which is a big headache, is to investigate the consequences of this "test, test, test" business. What do you know about the statistical properties of the final outcome?

## 7. The Czarina's Violet of Statistics

Prince Bismark, the "iron chancellor" of Germany went to Russia on a state visit. Since he was a very important person the Czar put him up at a guest house across from his palace. Bismark couldn't get to sleep all night because there was a guard going round and round, pounding his boots noisily. In the morning Bismark complained about this to the Czar and the Czar assured him that he would ask that guard not to make noise. But Bismark wouldn't let go. He wanted an answer as to why that guard was there going round and round. The Czar asked all his ministers and no one knew. Bismark got angry and said he was going back if he did not get an answer within 24 hours. The Czar got scared. He issued a proclamation that anyone who could tell what that guard was doing would get a thousand gold coins. But on one came forward until the last minute. Finally an old lady in her late nineties showed up. She said she knew why the guard was there.

When the Czarina, the grand mother of the present Czar was pregnant (with the father of the current Czar, one morning she saw a beautiful violet through her window. She asked her husband to protect it so she could look at it every morning. So, the Czar appointed a guard to go round the violet and see that no one trampled it. The violet was gone, no more violets grew there because of the guard's pounding of the ground with his boots. The original guard died, but his son got the job, and then his son and so on.

The story behind the 5% and 1% levels of significance in statistics like the story behind the czarina's violet, where did they come from? They came from R.A. Fisher the father of modern statistics, who suggested them in a paper in 1926 in the Journal of Ministry of Agriculture in Great Britain. His prescriptions have been followed religiously ever since.

There have been many arguments in the literature as to why the significance level should be changed with the sample size. Leamer, (1978) discusses this issue citing the earlier literature. He discusses how the model selection criteria suggest significance levels increasing with sample size.

Moreover, most statistical tests in econometrics are a prelude to further changes in the estimation method. The DW test is a prelude to GLS estimation. So are tests for heteorskedasticity. The unit root tests are a prelude to cointegrations analysis. Thus most tests are pre-tests. What significance levels should be used for pre-tests? There is no clear-cut answer but often a significance level of 0.35 or 0.4 is considered more appropriate than the conventional 0.05 significance level. (This is the case with the DW test).

Most tables of significance levels, for instance of unit root tests are given at the 1% and 5% levels. these are all useless.

In the 21$^{\text{st}}$ Century more attention should be paid to this problem and guidelines prepared for this pre-testing problem—as to what to do about it. All developments in deriving the (asymptotic) distributions of the test statistics are useless, if you do not know at what significance level you should be applying the tests.

## 8. Limited Dependent Variable Models

This has been an intense area of research starting with the work of Heckman in the early 70's. The book by Maddala, (1983) reviews the work of the late 1970's. Since then a lot of progress has been made on the estimation of a variety of models which are extensions of those discussed in Maddala, (1983).

In this area, the work that needs to be done is estimation of panel data limited dependent variable models. Some simulation methods of estimation are discussed in the papers by Keane and Hajivassiliou in Maddala, Rao and Vinod eds, (1993). Some other methods for models with lagged dependent variables are discussed in Honore, (1993) and Kyriazidou (1997).

In the 21$^{\text{st}}$ Century several types of panel data limited dependent variable models will be studied and estimation methods devised. There are now many panel data sets available and these methods are very important from the practical point of view. The problem of serial correlation and lagged dependent variables in logit, probit, tobit and other limited dependent variable models discussed in Maddala, (1983) need to be pursued in the 21$^{\text{st}}$ Century.

## 9. The Time Series Revolutions of the 1980's and 1990's

Before 1980 the commonly used methods in time series econometrics were the Box-Jenkins methods. There was work on distributed lag models as well.

Around 1980 three papers appeared: the paper by Rob Engle on ARCH models, the paper by Nelson and Plosser on the importance of unit roots in the study of macroeconomic fluctuations, and the paper by Clive Granger on cointegration. That changed everything. Each of the ideas took off. There was the ARCH revolution which excited people working in finance, and there were the unit root revolution and cointegration revolution that excited macroeconomists and also econometricians who found there a gold mine for the use of complicated and impressive mathematical techniques.

Why all this happened is a mystery to me. ARCH models have no firm basis in economic theory. Regarding unit roots, Harvey, (1997, p. 196) rightly remarks, "Testing for unit roots has become almost mandatory in applied economics. This is despite the fact that much of the time, it is either unnecessary or misleading or both." As for cointegration he says (p. 199), "I cannot think of one article that has come up with a cointegration relationship that we did not know about already from economic theory."

Granger's first paper on cointegration was presented at a conference I had organized in 1980. I thought it was an interesting paper but never thought it would create a revolution. Ten years later I woke up and felt like Rip Van Winkle. I started learning about it, the outcome of which is my book, Maddala, (1998).

What is the future for all this in the $21^{st}$ Century?

Regarding ARCH the bandwagon will keep rolling but more useful work will come from stochastic volatility models. Regarding unit roots, more and more unit root tests will come out (stochastic unit roots and so on) but the returns from all this unit root mania will be marginal. As for cointegration, again the bandwagon will keep rolling. What are we learning from all this, if anything, is a separate question.

All these will be areas of intense research activity in the $21^{st}$ Century.

## 10. Computer-Intensive Methods

Given the advances in computer technology, it makes sense to use computer-intensive methods in econometrics. There have been two methods that have become popular in econometrics: bootstrap methods and simulation methods.

Bootstrap Methods

Regarding bootstrap methods quite a few modifications need to be made when considering time-series models. These are reviewed with reference to econometrics in Li and Maddala, (1996).

Since then, there has been a lot of work on bootstrap methods for impulse response functions, where it has been found necessary to make bias corrections to the autoregressive parameters before bootstrapping. (This is different from the bias-corrected bootstrap methods suggested by Efron). Bootstrap methods have also been used in the finance literature but here there have been many defects in the methods used. These are reviewed in Maddala and Li, (1996). There is, however, a novel use of bootstrap methods. This is the use of trading rules with bootstrap data for model selection. (This is discussed in the paper by Maddala and Li).

Many tests in use in econometrics are derived using asymptotic theory (the diagnostic tests reviewed earlier, unit root tests, cointegration tests and so on). These tests all exhibit substantial size distortions in samples of sizes often encountered in econometric work. The use of bootstrap methods has been found to be useful in correcting these size distortions.

In the 21$^{\text{st}}$ Century bootstrap methods will be increasingly used in a variety of econometric applications. Particularly hypothesis testing where bootstrap generated critical values will be used rather than the critical values based on asymptotic theory (except in finance where samples are very large). There are numerous developments in the statistical literature (not all useful in econometrics) and these have to be incorporated in econometrics or suitably modified.

## Simulation-Based Inference

There were two papers in the late 1980's both in <u>Econometrica</u>. The first by Kloek and Van Dijk on "Integration by Monte Carlo" and the other by McFadden on a "Method of Simulated Moments." The first was on how multidimensional posterior distributions can be numerically integrated to get marginal distributions. This cleared a major hurdle in the development of Bayesian methods based on arbitrary priors. There have been further developments in this area since the paper by Kloek and Van Dijk. They are all surveyed in Geweke, (1996).

The simulation methods discussed in McFadden are also designed to evaluate multiple integrals. The estimation of limited dependent and qualitative response models in the presence of serially correlated errors and other dynamics involves very high-dimensional integrals and was given up as a hopeless computational problem until the advent of the simulation-based estimation methods. There are several surveys of this area now. There are two papers by Keane and Hajjivassiliou in <u>Handbook of Statistics</u>, Vol. 11, (1993), a paper by Hajjivassiliou and Ruud in <u>Handbook of Econometrics</u>, Vol. 4, (1994), the paper by McFadden and Ruud, (1994), and most recently the volume edited by Mariano et.al., (1998) that contains several papers on simulation based inference. The original simulation of moments method has been extended to the <u>method of simulated scores</u>.

In the 21$^{\text{st}}$ Century there will be a lot of work both in the Bayesian front using Monte Carlo integration methods and on the classical methods of estimation of complicated limited dependent variable models using simulation methods. There is not yet much empirical work in this area. In the 21$^{\text{st}}$ Century this gap between theory and practice will be filled.

## 11. Semiparametric and Non-parametric Methods

There has been an enormous amount of work at the theoretical level on semiparametric and non-parametric methods in econometrics. Some recent literature in this area is Powell, (1994), Manski, (1987), Honore, (1992), Hardle, (1990) and Ahn and Powell, (1993). But not much of it has been used in empirical work. During recent years, these techniques have spread into the area of finance as well, and have been used with financial data. See the papers in Maddala and Rao, (1996).

In the $21^{st}$ Century, the gap between theory and practice in this area should close and more wide spread use of these methods should take place.

## 12. Financial Econometrics

This is an area that is catching up. Many departments in economics are offering courses in financial econometrics, and the economics and finance departments are merging at many of the smaller universities. The finance area provides large data sets and much of econometrics based on asymptotic theory can be applied there with no worry about small sample sizes, and small sample corrections.

Until a few years ago, financial econometrics merely meant "ARCH" models. But ARCH models do not have any economic justification and are ad hoc. The stochastic volatility models reviewed in Ghysels, et. al., (1996) are an important alternative to ARCH.

Financial data do not satisfy the normality assumption, they have fatter tails. The paper by McCulloch, (1996) presents methods of analysis based on stable Paretian distributions. The papers in Maddala and Rao, (1996) deal with different problems in financial econometrics: tests of asset pricing models, IV estimation, semi-parametric estimation, bootstrap methods, different probability models for the error distributions, models of option pricing and so on.

In the $21^{st}$ Century, econometrics of finance will be an active area of research. The availability of large data sets, the invalidity of the normality assumption so commonly used in econometric models, both present opportunities and challenges.

## 13. The Frustration of Macroeconomists with Econometrics: Calibration

Macroeconomists have been frustrated with econometricians that the methods developed by econometricians have not been of much help in resolving major disputes in macroeconomics. There have been many attacks on traditional econometrics. The earlier attacks were by Lucas, (1976) and

Sims, (1980). Lucas said that econometric models estimated with historical data are useless for policy evaluation because economic policies will change the way economic agents behave. This means that the parameter estimates change. Hence the model estimated with historical data is useless. This critique received a lot more attention than it should have, and Ericsson, (1995) has documented that there is no empirical evidence in its favor.

The response to this criticism has been the Hansen-Sargent, (1980) approach to econometric modeling, whose aim is to estimate the parameters of the decision making process. The argument is that only by doing so can we estimate the effects of changes in the policy variables, and that conventional econometric modeling is useless in this respect. Though this is a valid argument, the practical usefulness of this approach is questionable. One needs extremely simplifying assumptions about the decision problem (quadratic objective functions and only one decision variable) to derive an estimable equation. On top of that we have to assume that this maximization problem formulated at the individual level is valid at the aggregate economy-wide level. Though quite a bit of conventional econometric work depends on this assumption, it is much stronger in the Hansen-Sargent approach because it imposes far more restrictions on the parameters of the model than the conventional approach. The Hansen- Sargent approach involves very intricate algebra but has not produced any useful results. What all the Lucas critique says is that we should test for parameter constancy at the aggregate level and make them functions of policy variables at the estimation stage.

There is no more future for this in the 21$^{st}$ Century. This does not mean that more work won't be done in this area. But it will all be useless.

Sims' critique is that the classification of endogenous and exogenous variables in econometric models used to achieve identification is quite arbitrary, and hence we should just dump all these and estimate vector autoregression (VAR) models. These models suffered from over-parametrization, (Zellner says VAR stands for "very awful regressions"), and further developments have been based on imposing some restrictions on the parameters through Bayesian priors. The Bayesian vector autoregressions (BVAR) seemed to perform well in forecasting exercises. All this came to be known as atheoretical econometrics. The literature on cointegration produced further restrictions on the VAR models and has led to Bayesian vector error correction models (BVECM) which have been found to be superior to the BVAR models. (LeSage, 1990).

The VAR and VECM models are used as starting points in discussions of cointegration and tests for cointegration. But much of cointegration is also "althoretical econometrics."

One other attack on the use of traditional econometric methods in dynamic macroeconomic models has been by Prescott (See Kydland and Prescott, 1982 and Prescott, 1986). This approach has been called "calibration." In this approach one asks whether data from a real economy share certain characteristics with data generated by the artificial economy described by an economic model. It is not claimed that the model explains all the characteristics of the data. What is required is that it explain the characteristics of interest. The results from this approach are easier to interpret than the results from the standard econometric approach because the economic model is not complicated by random elements added for statistical convenience. But the inferences have no statistical basis because there are no errors in the model and no specified probability distributions.

The idea behind calibration is simply to reduce uncertainty about a model's predictions, and hence to strengthen tests by assigning parameter values using point estimators from related studies. Gregory and Smith, (1990) study mixed estimators in which some parameters are pre-set (calibrated) and others estimated, as in Kydland and Prescott, (1982). Many models contain parameters which cannot be set with reference to other studies. In such cases calibration often proceeds by setting parameters to match a population moment from the model with the sample moment from the data. There have been some studies analyzing the statistical aspects of calibration (Gregory and Smith, 1993 and Watson, 1993). There have also been papers discussing "which moments to match" in the context of asset pricing models.

One other area of frustration for macroeconomists has been the area of unit roots. But here the problem is with the formulation of the question itself. As Harvey, (1997, p. 197) remarks, "trying to draw conclusion about whether the world is Keynesian or Neoclassical on the basis of the presence or otherwise of a unit root is complete nonsense." If you ask meaningless questions, of course you'll get meaningless answers.

The work in the $21^{st}$ Century is to examine the statistical problems with calibration closely and bridge the gap between calibration and traditional econometric methods. In summary, for the $21^{st}$ Century, the only thing left out of all these frustrations of macroeconmists, is to bring calibration closer to traditional econometric practice.

## 14. The Bayesian Approach to Econometrics

Arnold Zellner wrote his book on Bayesian inference in econometrics in 1971. He has been advocating Bayesian methods in econometrics for the last 30 years. But until recently, Bayesian econometrics did not receive widespread attention. Bayesian econometrics has always been hampered by the large computational costs associated with evaluating posterior distributions. Priors were chosen so that you could integrate the joint posterior to get marginal distributions of each of the parameters. A decade ago, Arnold Zellner was visiting Stanford. I asked Amemiya what he thought of Prof. Zellner. His reply was: "Zellner knows how to integrate."

Simulation methods and the rapidly improving computer technology knock down this computational hurdle. See Geweke, (1996) and several papers in Mariano et.al., (1998).

In the 21$^{\text{st}}$ Century. simulation based inference will be increasingly used by both the classical and Bayesian econometricians in a variety of problems–estimation of dynamic limited dependent variable models, nonlinear time-series models and stochastic volatility models Shephard, (1993), nonlinear errors in variables models (Newey, 1993) and so on. The gap between the Bayesian and classical econometricians is reducing.

Whatever your ideology, it is a good idea to learn the Bayesian approach to every econometric problem, although it is not good to get too deeply attached to Bayesianism. It has been said that "Bayesians are like Hare Krishna people. If you get too close to them, you will become one of them."

## 15. Likelihood Inference in Dynamic Models

Likelihood methods have been in wide use in econometrics. However, in dynamic models, in small samples it makes a lot of difference whether we consider the conditional likelihood function (conditional on initial observation) or the unconditional likelihood function taking account of the distribution of the initial observation. The wide differences in the estimates have been illustrated in Nerlove, (1998). This problem is also relevant in the context of unit root tests although the problems of taking into account the distribution of the initial observation is tricky. One choice would be to take it as a sample from the distribution assuming stationarity. In the 21$^{\text{st}}$ Century the problems with unconditional ML estimation in dynamic econometric models should receive more attention.

## 16. Problems of Weak Instruments

Instrumental variable (IV) estimation is a commonly used method in econometrics. The 2SLS estimator is an IV estimator. In errors in variables models, in dynamic models with serial correlation and a number of cases where the errors and explanatory variables are correlated, the IV methods are used. Sargan introduced the generalized IV estimation methods (GIVE) and the generalized methods of moments (GMM) estimator is a related development.

During recent years, there has been a lot of work on the problem of "weak instruments" (See Dufour, 1997 and the references in it). The questions of how to test whether an IV used is a valid IV ("instrument relevance"), how to choose among the different IV's available, and problems of estimation in the presence of weak instruments have received attention. Many of these problems have remained unsolved. In the $21^{st}$ Century, the instrumental variable estimation method with weak instruments will an important area of research.

## 17. What Will be Done?

In the preceding sections, I outlined some trends in econometrics and some possible topics to work on in the $21^{st}$ Century. These will not necessarily be the areas of active research. The momentum is now in time-series and those working in that area agree that the major thrust of activity in the next decade will be in non-linear time series—nonlinear error correction, nonlinear cointegration, nonlinear structural change, etc. Some references here are: Granger, (1995), Granger and Swanson, (1996, 1997) Granger and Terasvirta, (1993) and Granger et.al., (1997). This presumes that all the work on unit roots, cointegration, and structural change in the linear models is done. As I remarked earlier, I am skeptical about the usefulness of the work done until now. The work is voluminous, the mathematical developments have been imposing, but nothing useful and significant from the economic point of view has come out of all this. It's all much ado about a little something if not nothing. Jumping on non-linear extensions of work that in linear models is of questionable use is not necessarily a top priority topic. However, my opinion is a minority view. The concepts of unit roots and cointegration will be extended to non-linear time series models as well.

## 18. Conclusion

Econometrics is a very useful subject. In fact researchers in every branch of economics use econometric methods. But the gap between theory and practice has been widening. The theoretical research is growing more and more abstract and mathematically intricate papers are being written about problems that have no practical use or problems whose solutions are obvious. The quotation at the beginning of this paper from Brihadaranyaka Upanishad essentially summarizes my perception of what is going on. The econometrics profession needs to heed the admonition of the queen to Polonius in Shakespeare's <u>Hamlet</u>: "More matter, less art."

In December of 1996 there was an article in the New Yorker by John Cassidy titled, "The Decline of Economics." I agree with what the author was saying that economists have stopped working on economic problems of practical interest and are writing papers on abstract esoteric topics of no relevance to any economic problem. They live in a world of their own.

If the present trends continue in the 21$^{st}$ Century, there will also be an article on "The Decline of Econometrics."

## References

[1] Ahn, H. and J.L. Powell (1993), Semiparametric Estimation of Censored Selection Models With a Nonparametric Selection Mechanism. *Journal of Econometrics*, **58**, 3–29.

[2] Aigner, D.J. and A.S. Goldberger (eds) (1977) *Latent Variables in Socio-Economic Models*, North Holland, Amsterdam.

[3] Bekker, P.A. Kapteyn and T. Wansbeek (1985), Errors in Variables in Econometrics: New Developments and Recurrent Themes. *Statistica Nederlandica*, **39**, 129–141.

[4] Ashenfelter, O. and A. Krueger (1994), Estimates of the Economic Return to Schooling From a New Sample of Twins. *American Economic Review*, **84**, 1157–1173.

[5] Cassidy, J. (1996), The Decline of Economics. *New Yorker, (Dec. 2)*, , 52–58.

[6] Bera, A.K. and A. Ullah (1991), Rao's Score Test in Econometrics. *Journal of Quantitative Economics*, **7**, 189–220.

[7] Christ, C. (1983), The Founding of the Econometric Society and Econometrica. *Econometrica*, **51**, 3–6.

[8] Dufour, J.M. (1997), Some Impossibility Theorems in Econometrics with Applications to Instrumental Variables, Dynamic Models and Cointegration. *Econometrica*, **65**, 1365–1387.

[9] Friedman, M. and A.J. Schwartz (1991), Alternative Approaches to Analyzing Economic Data. *American Economic Review*, **81**, 39–49.

[10] Frisch, R. (1934) Statistical Confluence Analysis by Means of Complete Regression Systems, (Oslo, University Institute of Economics).

[11] Frisch, R. (1936), Note on the Term 'Econometrics'. *Econometrica*, **4**, 95.

[12] Fuller, W.A. (1987) *Measurement Error Models*, New York, Wiley.

[13] Geweke, J. (1996) Monte Carlo Simulation and Numerical Integration, in: H. Aman, D. A. Kendrick and J. Rust (eds) *Handbook of Computational Economics*, New York, Elsevier, 7310-800.

[14] Godfrey, L.G. (1988) *Misspecification Tests in Econometrics: the Lagrangian Multiplier Principle and Other Approaches*, Cambridge, Cambridge University Press.

[15] Granger, C.W.J. (1995), Modeling Nonlinear Relationships Between Extend Memory Variables. *Econometrica*, **63**, 265–279.

[16] Granger, C.W.J. and N.R. Swanson (1996), Future Developments in the Study of Cointegrated Variables. *Oxford Bulletin of Economics and Statistics*, **58**, 537–553.

[17] Granger, C.W.J. and N.R. Swanson (1997), An Introduction to Stochastic Unit Root Processes. *Journal of Econometrics*, **80**, 35–62.

[18] Granger, C.W.J. and T. Terasvirta (1993) *Modeling Non-Linear Economic Time Series,* Oxford, Oxford University Press.

[19] Granger, C.W.J., T. Inoue, and N. Norin (1997) Nonlinear Stochastic Trends, *Journal of Econometrics*, Forthcoming.

[20] Gregory, A.W. and G.W. Smith (1990), Calibration as Estimation. *Econometric Reviews*, **9**, 57–89.

[21] Gregory, A.W. and G.W. Smith (1993) Statistical Aspects of Calibration in Macroeconomics, in: G. S. Maddala, C. R. Rao and H. D. Vinod (eds) *Handbook of Statistics*, Vol. **11**: Econometrics, Amsterdam, Elsevier, 703–719.

[22] Griliches, Z. (1967), Distributed Lags: a Survey. *Econometrica*, **35**, 16–49.

[24] Ghyscls, E., A.C. Harvey and E. Renault (1996) Stochastic Volatility, in: G.S. Maddala and C. R. Rao (eds) *Handbook of Statistics*, Vol. **14**, 119–191.

[25] Haavelmo, T. (1944), The Probability Approach to Econometrics. *Econometrica*, **12**, 1–115.

[26] Hansen, L.P. and T.J. Sargent (1980), Formulating and Estimating Dynamic Linear Rational Expectations Models. *Journal of Economic Dynamics and Control*, **2**, 7–46.

[27] Hausman, J.A. (1978), Specification Tests in Econometrics. *Econometrica*, **46**, 1251–1271.

[28] Hardle, W. (1990) *Applied Nonparametric Regression,* Cambridge, Cambridge University Press.

[29] Harvey, A.C. (1997), Trends, Cycles and Autoregressions. *Economic Journal*, **107**, 109–201.

[30] Hillier, G.H. (1990), On the Normalization of Structural Equations: Properties of Direction Estimators. *Econometrica*, **56**, 1371–1395.

[31] Honore, B.E. (1992), Trimmed LAD and Least Squares Estimation of Truncated and Censored Regression Models With Fixed Effects. *Econometrica*, **60**, 533–565.

[32] Honore, B.E. (1993), Orthogonality Conditions for Tobit Models With Fixed Effects and Lagged Dependent Variables. *Journal of Econometrics*, **59**, 35–61.

[33] Johnston, J. (1991), Econometrics: Retrospect and Prospect. *Economic Journal*, **101**, 51–56.

[34] Kalman, R.E, (1982) System Identification From Noisy Data, in: A. Bednarek and L. Cesari (eds) *Dynamical System*, **II**, New York, Academic Press.

[35] Kyriazidou, E. (1997), Estimation of a Panel Data Sample Selection Model. *Econometrica*, **65**, 1335–1364.

[36] Klepper, S. and E.E. Leamer (1984), Consistent Sets of Estimates for Regression With Errors in All Variables. *Econometrica*, **55**, 163–184.

[37] Koopmans, T.C. (1937) *Linear Regression Analysis of Economic Time Series*, Haarlem, Netherlands Economic Institute.

[38] Kydland, F. and E.C. Prescott (1982), Time to Build and Economic fluctuation. *Econometrica*, **50**, 1345–1370.

[39] Leamer, E.E. (1978) *Specification Searches, AD-Hoc Inference With Non-Experimental Data*, New York, Wiley.

[40] Leamer, E.E. (1987), Errors in Variables in Linear Systems. *Econometrica*, **55**, 893–909.

[41] LeSage, J.P. (1990), A Comparison of the forecasting Ability of ECM and VAR Models. *The Review of Economics and Statistics*, **72**, 664–671.

[42] Li, H. and G.S. Maddala (1991), Bootstrapping Time Series Models (with discussion). *Econometric Reviews*, **15**, 115–195.

[43] Lucas, R.E. (1976) Econometric Policy Evaluation: A Critique, in: K. Brunner (ed) *Supplement to the Journal of Monetary Economics*, **1**, 19–64.

[44] Maddala, G.S. (1983) *Limited Dependent and Qualitative Variables in Econometrics*, Cambridge, Cambridge University Press.

[45] Maddala, G.S. (1995) Specification Tests in Limited Dependent Variable Models, in: G. S. Maddala, P. C. B. Phillips and T. N. Srinirasan (eds) *Advances in Econometrics and Quantitative Economics: Essays in Honor of C. R. Rao*, Blackwell, Oxford.

[46] Maddala, G.S. and H. Li (1996) Bootstrap Based Tests in Financial Models, in: *Handbook of Statistics*, Vol. **14**, 463–488.

[47] Maddala, G.S. and C.R. Rao (eds) (1996) *Handbook of Statistics, Vol. 14: Statistical Methods in Finance*, Amsterdam, Elsevier Science.

[48] Maddala, G.S. and I.M. Kim (1998) *Unit Roots, Cointegration and Structural Change*, Cambridge, Cambridge University Press.

[49] Manski, C. (1987), Semiparametric Analysis of Random Effects Linear Models From Binary Panel Data. *Econometrica*, **55**, 357–362.

[50] Mariano, R.S., M. Weeks and T. Schuermann (Eds) (1998) *Simulation Based Inference in Econometrics: Methods and Applications*, Cambridge, Cambridge University Press.

[51] McColluch, J.H. (1996) Financial Applications of Stable Distributions, in: G. S. Maddala and C. R. Rao (eds) *Handbook of Statistics*, Vol. **14**, 393–425.

[52] McFadden, D. and P.A. Ruud (1994), Estimation by Simulation. *Review of Economics and Statistics*, **76**, 591–608.

[53] Newey, W.K. (1985), Maximum Likelihood Specification Testing and Conditional Moment Tests. *Econometrica*, **53**, 1047–1073.

[54] Newey, W.K. (1993) Flexible Simulated Moments Estimation of Nonlinear Errors in Variables Models, MIT Dept. of Economics, Working Paper #93-18, Nov. 1993.

[55] Nerlove, M. (1972), Lags in Economic Behavior. *Econometrica*, **40**, 221–251.

[56] Nerlove, M. (1998) Properties of Alternative Estimators of Dynamic Panel Data Models, forthcoming in: C. Hsiao, K. Lahiri, L.F. Lee and M.H. Pesaran (eds) *Analysis of Panels and Limited Dependent Variable Models: Essays in honor of G.S. Maddala*, Cambridge University Press.

[57] Powell, J.L. (1994) Estimation of Semiparametric Models, *Handbook of Econometrics*, Vol. 4, 2444–2521.

[58] Prescott, E.C. (1986), Theory Ahead of Business Cycle Measurement. *Carnegie Rochester Conf. Series and Publ. Policy*, **25**, 11–44.

[59] Rao, C.R. (1947), Large Sample Tests of Statistical hypotheses Concerning Several Parameters With Applications to Problems of Estimation. *Proceedings of the Cambridge Philosophical Society*, **44**, 50–57.

[60] Rao, C.R. (1973) *Linear Statistical Inference and Its Applications*, New York, Wiley.

[61] Ruud, P.A. (1984), Tests of Specification in Econometrics. *Econometric Reviews*, **3**, 211–242.

[62] Sargan, J.D. (1964) Wages and Prices in the United Kingdom: A Study in Econometric Methodology, in: P.E. Hart (et al.) Eds. *Econometric Analysis for National Economic Planning*, Colston Papers #16 (London, Butterworth), 25-54.

[63] Shephard, N. (1993), Fitting Nonlinear Time Series Models With Applications to Stochastic Variance Models. *Journal of Applied Econometrics*, **8**, S135–S152.

[64] Sims, C. (1980), Macroeconomics and Reality. *Econometrica*, **48**, 1–48.

[65] Strom, S. and A. Holly (Eds) (1998) *Econometrics and Economic Theory in the 20th Century: The Ragnar Frisch Centennial Symposium*, Cambridge, Cambridge University Press, Forthcoming.

[66] Watson, M.W. (1993), Measures of Fit for Calibrate Models. *Journal of Political Economy*, **101**, 1011–1041.
[67] White, H. (1982), Maximum Likelihood Estimation of Misspecified Models. *Econometrica*, **50**, 1–25.

# On Stability of a Large (Ecological) System

V. Mandrekar

Department of Statistics and Probability
Michigan State University
East Lansing, Michigan

**Abstract.** We discuss the stabilization of a finite ecological system using results on stability in the finite dimensional stochastic differential equations.We then present results on stability for stochastic PDE's and propose for 21st century the problem of stabilizing infinite dimensional (large) ecological systems.

## 1. Introduction

When one starts to think about any problem for the twenty-first century, the major question one has to answer is whether there will be the twenty-first century. With the way the environment and ecological system is being abused, can the Nature preserve itself under such abuse. Obviously, a good example of the nature healing itself under some abuses is the human body. Since we are looking at Problems in Probability and Statistics, it is natural to assume that the nature acts in a random way. Mathematically, the problem can be stated by saying that if a system is unstable (deterministically) can one add "noise" to the system to stabilize the system a.s. This problem when the "noise" is deterministic was used in Physical problems by N. N. Bogolybov [3]. Stabilizing a system by adding "white noise" was first considered by R. Z. Khasminski [6] in analyzing questions arising in Physics. His method was very general and was used by L. Arnold and the collaborators [2] to study the stabilization problem in the finite-dimensional case. Our purpose here is to reinterpret this result in ecological context and propose the study in the infinite-dimensional case as ecological system is large dimensional. In this context, we give the results of the author on the stability of the infinite-dimensional system and propose methods to solve the stabilization problem. Finally, we propose new problems.

## 2. Preliminaries and notations

Let us begin with an example. Suppose $X(t)$ denotes the size of the population at time $t$ with growth rate $r$ then with starting population size $X_0$,

$$X(t) = X_0 e^{rt}.$$

Then so called Lyapunov exponent (LE), namely,

$$\lim_{t \to 0} \log X(t)/r.$$

If $r > 0$ the population explodes. It is tempting to observe that if we add noise of intensity $\sigma$ then solution of the Ito equation with initial value $X_0$,

$$dX(t) = rX(t)dt + \sigma dW(t)$$

with $W(t)$ being a Wiener process, is

$$X(t) = X_0 \exp \left[ (r - \frac{1}{2}\sigma^2)t + \sigma W(t) \right]$$

(See [13]) and using the law of iterated logarithm for $W(t)$, the LE of this random process is $(r - \frac{1}{2}\sigma^2)$. So it seems that if we take large enough intensity then a population of a single species in a fixed environment which is exploding can become extinct $(r - \frac{1}{2}\sigma^2 < 0)$. This is counter intuitive. The fallacy in the argument is that Ito stochastic differential equation (SDE) is not correct model for idealized continuous model which comes from a limit of discrete model. The correct SDE involves so-called Stratonovitch integral ([6], p.173) in the last term

$$dX(t) = rX(t)dt + \sigma X(t) \circ dW(t)$$

which can be written as Ito SDE

$$dX(t) = \left( r + \frac{1}{2}\sigma^2 \right) X(t)dt + \sigma X(t) \circ dW(t)$$

whose solution is

$$X(t) = X_0 \exp \left[ (r - \frac{1}{2}\sigma^2)t + \sigma W(t) \right].$$

The LE is r.

However if we have several species or several environments then suppose $\underline{X}(t)$ satisfies

$$d\underline{X}(t) = A\underline{X}(t)dt$$

With initial condition $\underline{X}_0$ the LE is

$$\lim_{t \to} \frac{\log |e^{At}\underline{X}_0|}{t} = \lambda_i$$

if $\underline{X}_0$ lies in the eigensubspace corresponding to eigenvalue $\lambda_i$ of $A$. The system is said to be stable if $\lambda_{\max} < 0$. Suppose the deterministic system is not stable, then the following result due to L. Arnold [1] (See also [2]) gives the appropriate result on the stabilization by introducing "white noise".

**Theorem A.** *A linear system $\dot{\underline{X}} = A\underline{X}$ can be stabilized by white noise iff* trace$A < 0$.

The proof extends the technique of Khasminski [6] from two to finite-dimensions. The major result is to relate $\lambda_{\max}$, the maximum LE of the random system to trace$A$ through the study of ergodic behavior of the markov process $X(t)/\|X(t)\|$ on the sphere.

As stated in the introduction the ecological systems have large number of species and/or large environments. Nature's ability to adapt the species to new environment thus creates even more variety of the species. Our purpose henceforth is to describe the results achieved in stability of infinite-dimensional systems in terms of Lyapunov function. These can be connected to stability in probability which, for a linear system is stability a.s. ([6], Thm 7.1, p.186 Thm 4.1, p.208 Thm 8.1, p.190). One can use ideas of Mao [12], to relate for general system Lyapunov function to LE. This will be done in future work.

## 3. Stability of system in infinite-dimensional case

We first start by looking at the analogue of the deterministic

$$d\underline{X}(t) = A\underline{X}(t)$$

in the finite-dimensional case. In order to facilitate the notation, we consider $u(t)$ for the infinite case instead of $X(t)$ or $\underline{X}(t)$. The problem arises from an ecological system $u(x, t)$ where $X$ is the location. (earth, three dimensional sphere !). The special relation is described by a differential operator in $x$. Thus the problem can be reformulated as solution of the Cauchy Problem

$$(2.1) \qquad \frac{d}{dt}u(t) = Au(t) \qquad t \geq 0 \quad \text{with} \quad u(0) = x \in H$$

($H$, a Hilbert space). The major difficulty is $A$ being a differential operator, it is not bounded on $H$. The solution can be represented as

$$u(t, x) = S(t)x$$

if $A$ generates a $C_0$-semigroup of bounded linear operators. $S(t)$ (i.e. $S(0) = I$, $S(t + s) = S(t)S(s)$ and $t \longrightarrow S(t)x$ is continuous on $[0, \infty)$). A solution is said to be (exponentially) stable if $||S(t)x|| \leq c||x||e^{-rt}$ for some $r > 0$. ( or equivalently $||S(t) \leq ce^{-rt}$). Unfortunately $\sup\{\lambda : \lambda \in \operatorname{spectrum}(A)\} < 0$ is not a necessary and sufficient condition for stability ([11]). In the finite-dimensional case Lyapunov showed that stability above is equivalent to the existence of a positive-definite matrix $R$ satisfying $A^*R + RA = -I$ with $c_1||x||^2 \leq (Rx, x) \leq c_2||x||^2$. In fact $(Rx, x) = \int_0^\infty ||u^x(t)||^2 dt$ where $u^x(t)$ is the solution with initial value $x$. The function $(Rx, x)$ is called Lyapunov function. The analogue of this result holds this result holds ([11]).

**Theorem 1.** *A necessary and sufficient condition that a $C_0$-semigroup on a real Hilbert space $H$ satisfies $||S(t)|| \leq ce^{-rt}$ (r > 0 and $1 \leq c < \infty$) is the existence of a Hermitian endomorphism $R$ on $H$ so that $2(RAx, x) = -||x||^2$ for all $x \in \mathcal{D}(A)$.*

Another approach to solving Cauchy problem is "variational". For this, we need the idea of coercivity. A linear operator $A : \mathcal{D} \subseteq \mathcal{H} \longrightarrow \mathcal{H}$ is coercive if

  i) There exists a Hilbert space $V$ densely embedded in $H$ and a continuous bilinear form $a : V \times V \longrightarrow \mathbb{R}^1$. $\ni$ . for $\alpha, \lambda_0 > 0$

$$-2a(v, v) \geq \alpha||v||_V^2 - \lambda_0||v||_H^2 \qquad \text{for all} \quad v \in V$$

  ii) $\mathcal{D}(A) = \{u \in V : a(u, \cdot) \text{ is continuous in the topology of } H\}$
  iii) $a(u, v) = <Au, v>$ for all $u \in \mathcal{D}(A)$ and $v \in V$.

**Theorem 2.** ([11]) *The solution of the Cauchy problem (2.1) is stable if there exists a continuous bilinear form $\Lambda$ on $H$ satisfying $c_1||x||_H^2 \leq \Lambda(x) \leq c_2||x||_H^2$ and $(\Lambda'(x), Ax) \leq -c_3\Lambda(x)$ where $c_1, c_2, c_3$ are positive, finite constants. In particular, $2(\Lambda'(x), Ax)_H = -||x||_V^2$, then we get the second part of the above condition.*

Let $(\Omega, \mathcal{F}, \mathcal{P})$ be a probability space with filtration $(\mathcal{F}_t)_{t \geq 0}$ (increasing family of) sub $\sigma$-fields of $\mathcal{F}$ such that $t \to \mathcal{F}_t$ is right continuous. Let $K$ be a real separable Hilbert space and $\{W(t), t \geq 0\}$ be a $K$-valued $\{\mathcal{F}_t\}$-brownian motion; i.e. $\{W(t), \mathcal{F}_t, t \geq 0$ is a martingale and $E(W(t), x)_K(W(s), y)_K = (t \wedge s)(Qx, y)_H$ with $Q$ a trace class operator. $Q$ is called covariance operator of $W_t$. First we consider stability of solutions of semi-linear equations, where methods used are with semigroup.

Let $H$ be a real separable Hilbert space and consider the semilinear Ito stochastic evolution equation ([5],[4])

$$(2.2) \qquad du(t) = (Au + F(u))dt + B(u)dW(t), \qquad u(0) = x \in H,$$

where $A$ is an infinitesimal generator of $C_0$-semigroup $\{S(t), t \geq 0\}$ on $H$ satisfying $\|S(t)\| \leq e^{wt}$ and $F$ and $B$ are in general non-linear mappings from $H$ to $H$ and $H \longrightarrow L(K, H)$ satisfying Lipschitz condition and growth condition

$$\|F(y) - F(z)\|_H + \|B(y) - B(z)\|_{L(K,H)} \leq d\|y - z\|_H$$

$$\|F(y)\|_H + \|B(y)\|_{L(K,H)} \leq d\|1 + y\|_H$$

for a constant $d$ and all $y, z \in H$. We introduce two kinds of solutions following [5]

*Definition 2.1.* A stochastic process $\{u(t), t \in [0, T]\}$ is a *strong solution* of (2.2) if
(i) $u(t)$ is $\mathcal{F}_t$-measurable,
(ii) $u(t)$ is is continuous in $t$ a.s.,
(iii) $u(t) \in \mathcal{D}(A)$ a.s. $[0, T) \times \Omega$ and $\int_0^t \|Au(t)\|_H$ is finite a.s.
(ii) $u(t) = x + \int_0^t Au(s)ds + \int_0^t F(u(s))ds + \int_o^t B(u(s))dW(s)$.

*Definition 2.2.* A stochastic process $\{u(t), t \in [0, T]\}$ is a *mild solution* of (2.2) if
(i) $u(t)$ is $\mathcal{F}_t$-measurable,
(ii) $\int_0^T \|u(t)\|_H^2 dt$ is finite a.s. and
(ii) $u(t) = S(t)x + \int_0^t S(t - s)F(u(s))ds + \int_0^t S(t - s)B(u(s))dW(s)$ for all $t \in [0, T]$.

Existence and uniqueness of solutions of equation (2.2) in $C([0, T]$, $L_p(\Omega, \mathcal{F}, \mathcal{P}))$ $(p \geq 2)$ is given by Ichikawa [5].

In case $F \equiv 0$ and $B(u) = Bu$ we call the equation (2.2) linear.

We consider slightly more general concept than (exponential) stability.

*Definition 2.3.* The solution $u^x(t)$ of (2.2) is exponentially ultimately bounded (UB) in m.s.s. (mean square sense) if there exist positive constants $c$, $r$, $M$ such that

$$E\|u^x(t)\|_H^2 \leq ce^{-rx}\|x\|^2 + M$$

If $M = 0$ we get (exponentially) stable in m.s.s. In the linear case, which is relevant to our problem in ecology, we get

**Theorem 3.** ([9]) *The mild solution* $\{u^x(t)\}$ *of (2.2) exponentially UB iff there exists a function* $\Lambda : H \hookrightarrow \mathbb{R}$, *twice Frechét differentiable with bounded derivatives satisfying conditions: with* $c_1$, $c_2$ *positive constants and* $k_1$, $k_2$, $k_3$ *real constants*

(i)
$$c_1||x||_H^2 - k_1 \leq \Lambda(x) \leq c_2||x||_H^2 - k_2$$

(ii)
$$\mathcal{L}\Lambda(x) \leq -c_2\Lambda(x) + k_2$$

*Here* $\mathcal{L}\Lambda(x) = <\Lambda'(x), Ax> + \frac{1}{2}tr(\Lambda''(x)BxQB^*(x))$ *with prime denoting Frechét derivatives.*

In case $M = 0$, $k_1 = k_2 = 0$, we get necessary and sufficient condition for stability in terms of Lyapunov function $\Lambda$ in the linear case. From this, one can derive stability in probability at zero (See Theorem 2.5 in [9]); i.e., for each $\epsilon > 0$,

$$\lim_{|x| \to 0} P\{\sup_t ||u^x(t)||_H > \epsilon\} = 0.$$

One can study the analogous for non-linear case by using first order linear approximation [9].

We now consider the analogue of variational technique for studying stochastic stability. For this we consider $V \subseteq H$ two real separable Hilbert space so that $V$ is dense in $H$, and injection $V \hookrightarrow H$ is continuous in $H$. We identify $H$ with its (continuous) dual space and $V^*$ be dual of $V$. Then, we have

$$V \subseteq H \subseteq V^*$$

Let $<, >$ denote the duality product between $V$ and $V^*$ and $(, )$ inner product on $H$. We assume that

$$< v, v^* > = (v, v^*) \qquad v^* \in H.$$

Let $M^2([0,T], V)$ denote the space of $V$-valued measurable processes satisfying:
- (i) $u(t, \cdot)$ is $\mathcal{F}_t$-measurable for each $t$ and
- (ii) $E \int_0^T ||u(t, w)||_V^2 dt$ is finite.

We consider now the following equation

(2.3)
$$u \in M^2([0,T], V)$$
$$du(t) = A(u(t))dt + B(u(t))dW(t)$$
$$u(0) = x$$

where $x \in H$, $A : V \longrightarrow V^*$ is a (non-linear) operator with $||A(u)||_V \leq c_1||u||_V$, $B(u) \in L(K, H)$ and $||B(u)||_{L(K,H)} \leq d_1||u||_V$ for $u \in V$ ($c_1$, $d_1$, positive constants). Under the following conditions:

*Coercivity*: $\exists \alpha > 0$ and $\lambda$ and $\beta$ such that for all $v \in V$,

(2.4)        $2 < v, A(v) + \text{tr}(B(v)QB^*(v)) \leq \lambda||v||_H^2 - \alpha||v||_V^2 + \beta$

and *monotonicity*: $\forall u, v \in V$,
(2.5)
$2 < u - v, A(u) - A(v) > +\text{tr}[(B(u) - B(v))Q(B(u) - B(v))^*] \leq \delta||u - v||_H^2$

we get that there exists a unique markovian solution with the corresponding semigroup Feller (See [14], [15], Ch.3).

The equation (2.3) is called linear if $A(u) = Au$ and $B(u) = Bu$ where $A$ is a linear operator on $V \longrightarrow V^*$. As before we are interested in the linear case.

**Theorem 4.** ([8]) *In the linear case of equation (2.3) satisfying (2.4) and (2.5), the solution $\{u^x(t), t \geq 0\}$ is exponentially UB in m.s.s. iff there exists a function $\Lambda : H \longrightarrow \mathbb{R}$ satisfying for all $x \in H$*

$$c_2||x||_H^2 - k_1 \leq \Lambda(x) \leq c_3||x||_H^2 + k_2$$

*and $\mathcal{L}\Lambda(x) \leq -c_2\Lambda(x) + k_2$ for all $x \in V$. Here $c_1, c_2, c_3 > 0$ and $k_1$, $k_2$, $k_3$ real constants and $\mathcal{L}\Lambda(x) = < \Lambda'(x), Ax > +\frac{1}{2}tr(\Lambda''(x)BxQ(Bx)^*)$.*

Again as before if $M = 0$ in Definition 2.3 then we get exponential stability and we get stability in probability at zero as was shown in [7].

Thus combining methods form Khasminski [6], with the above result we can establish a.s. path stability. In view of Example 2.1 ([8]) small noise does not disturb the stability behavior.

## Open Problems:

1. Extend the technique of Mao for relating Lyapunov function to Lyapunov Exponent.
2. Extension of Khasminski and Arnold et. al. work on the stabilization of infinite system. This requires existence of invariant measure ([9]).
3. Numerically check on existing data to see that it fits randomly perturbed model better than deterministic model. For stochastic PDE, numerical solution with single noise can be computed ([10]).
4. Stabilizing by noise other than "white noise" and choice of noise for a specific system.

## References

[1] Arnold, L. A (1979), A new example of an unstable system being stabilized by random parametric noise. *Inform. Comm. Math. Chem.*, 133–140.

[2] Arnold, L. and Crauel, H. and Wihstutz, V. (1983), Stabilization of linear system by noise. *SIAM J. Control and Optimization*, **21**, 451–461.

[3] Bogolybov, N. N. (1950), Perturbation theory in non-linear mechanics. *Sb. Stroit. Mekh. Akad. Nauk Ukr. SSR.*, **14**, 9-34 (Russian).

[4] Da Parto, G. and Zabczyk, J. (1992) *Stochastic equations in infinite dimensions,* Cambridge University Press, Cambridge, England.

[5] Ichikawa, A. (1892), Stability of semilinear stochastic evolution equations. *J. Math. Anal. Appl.*, **90**, 12-14.

[6] Khasminski, R. Z. (1981) *Stochastic stability of differential equations,* Sijthoff and Noordhoff, Netherlands.

[7] Khasminski, R. Z. and Mandrekar, V. (1994) On stability of solutions of stochastic evolution equations, *The Dynkin Festschrift* (Ed. M. Freidlin) Birkhauser, Boston.

[8] Liu, R. and Mandrekar, V. (1996), Ultimate boundedness and invariant measures of stochastic evolution equations. *Stochastics and Stochastic Reports*, **56**, 75–101.

[9] Liu, R. and Mandrekar, V. (1997), Stochastic semilinear evolution equations: Lyapunov function, stability and ultimate boundedness. *J. Math. Anal. and Appl.*, **212**, 537–553.

[10] Liu, R. and Mandrekar, V. Numerical solutions of stochastic heat equations perturbed by time white noise. Preprint.

[11] Mandrekar, V. (1994) On Lyapunov stability theorems for stochastic (deterministic) evolution equations, *Stochastic Analysis and Applications in Physics*, (Ed. Cardoso, A. I. et al), Kluwer, Netherlands.

[12] Mao Xuerong (1991) Lyapunov functions and almost sure exponentially stability. *Lyapunov Exponents*, (ed L. Arnold et al) Lecture notes in Math. **1486**, Springer-Verlag, New York.

[13] Oxendal, B. (1991) *Stochastic differential equations,* Springer-Verlag, Berlin.

[14] Pardoux, E. (1979), Stochastic partial differential equations and filtering of diffusion processes. *Stochastics*, **3**, 127–167.

[15] Rozovski, B. L. (1990) *Stochastic evolution systems,* Kluwer, Netherlands

# Multi-scale Statistical Approach to Critical-area Analysis and Modeling of Watersheds and Landscapes

G. P. Patil, G. D. Johnson and C. Taillie
Center for Statistical Ecology and Environmental Statistics
Department of Statistics
Pennsylvania State University
University Park, Pennsylvania

W. L. Myers
School of Forest Resources and
Environmental Resources Research Institute
Pennsylvania State University
University Park, Pennsylvania

**Abstract.** Environmental and ecological statistics is poised for dramatic growth both for reasons of societal challenge and information technology. It is becoming clear that environmental and ecological statistics is demanding more and more of non-traditional statistical approaches. This is partly because environmental and ecological studies involve space, time and relationships between many variables, and require innovative and cost-effective monitoring, sampling and assessment. Also, environmental and ecological statistics methodology must satisfy environmental policy needs

Prepared with partial support from the National Science Foundation Cooperative Agreement Number DEB-9524722. The contents have not been subjected to Agency review and therefore do not necessarily reflect the views of the Agency and no official endorsement should be inferred.

in addition to disciplinary and interdisciplinary environmental and ecological research imperatives. And all of this is true of research and policy issues involving water and watersheds for disciplinary and cross disciplinary research, such as: (i) a need to develop an enhanced predictive understanding of the processes and mechanisms that govern the dynamics and properties of surface and subsurface water and watershed ecosystems; (ii) a need to identify and research indicator variables, analytical methods, and other tools for determining waters and watersheds at risk and reducing the uncertainties of extrapolating information across broad spatial and temporal scales; and (iii) a need to interpret relationships between populations and communities of organisms and the quality and quantity of water, particularly as these relate to ecosystem processes, land use patterns, and landscape structure.

**Keywords**: Adaptive cluster sampling, Composite sampling, Environmental sampling, Hierarchical modeling, Landscapes, Multiscale assessment, Ranked set sampling, Spatial modeling, Watersheds.

## 1. Motivation, Issues and Background

### 1.1 Motivation

Many motivations arise for assessing the relationships between abiotic and biotic characteristics of a watershed or landscape, the concern normally being about biological effects of environmental changes. All inferences about such relationships are not only relative to specified spatial and temporal scales, but also depend on measurement scales used to obtain the necessary data for drawing inferential information.

Most ecological studies have been performed at very localized scales, either for academic research or for highly detailed site investigations. In recent times, there has been a rapid expansion of large regional ecological assessments. This is a result of increased awareness of regional environmental impacts, as well as development of regional measurements from remote sensing technologies and growing ability to manage spatial information through geographic information systems.

A thorough ecological assessment for a given region at a given time would encompass multiple scales. On the one hand, small-scale local investigations may reveal healthy communities, while a regional perspective reveals that habitat is sufficiently fragmented that the localized communities are threatened from lack of migratory routes and insufficient genetic mixing. On the other hand, what may appear to be only subtle changes from a large scale regional perspective can translate to potentially dramatic changes at smaller scale locales. For example, Graham, et al.(1991) show through spatial stochastic modeling of the Adirondack region that even though overall forest cover may not change much regionally, small areas of altered forest can have a dramatic effect on water quality in headwater watersheds.

There is a long standing awareness of regional impacts from airborne pollutants such as acidic deposition and ozone (Graham, et al.1991). Furthermore, the loss of species diversity through regional habitat modification is considered a top rated risk in the United States (Kiester, et al. 1993). Following a catastrophic event, large-scale regional assessments may need to be made rapidly in order to locate smaller scale areas that are most impacted. Furthermore, nationwide assessments of ecological and environmental resources are under way through the Environmental Monitoring and Assessment Program (EMAP) of the EPA and the Gap Analysis Program of the U. S. Geological Survey's Biological Resources Division. All of these motivations are directed toward an initially broad scale assessment that is used to identify local critical areas for more detailed higher resolution studies. It soon becomes clear that both data sources and inference objectives involve multiple scales, and therefore the relationships among scales must be understood if we are to reliably combine data from different measurement scales as well as make predictions at scales other than those for which we have sufficient data.

*Issues*: Environmental and ecological data are available at a variety of spatial scales, arising from different sources which range from satellite imagery to field plots. Meanwhile, we need to infer about characteristics and processes at other scales, such as may be delineated by watershed boundaries, political subdivisions, or a predefined unit of fixed size and shape like an EMAP 635 $km^2$ hexagon.

For example, various indicators may need to be computed for assessing ecological risk relative to a large watershed, followed by assessment for tributary sub watersheds, then for sub tributary sub-sub watersheds, and so on. For the purpose of resource allocation by a federal or state agency, we need summary assessments of large watersheds that correspond to higher order rivers. Within the domain of county planning, assessments may be needed for smaller sub watersheds corresponding to tributaries of the larger rivers. Still, ultimate management and mitigation efforts are applied at a very local scale which requires assessments of fine scale watersheds corresponding to first or second order streams.

A question then arises: Since we simply can not afford to perform a detailed ecological risk assessment for every watershed of every size, to what extent can we infer about the quality of smaller watersheds from composite information about their 'parent' watersheds and, vice versa, how reliably can we infer about a parent watershed from information obtained from a sample of its 'children' watersheds? As an example, if we are concerned with acid deposition impacts within a large watershed, all first order streams may be considered vulnerable due to low annual flows. However, the 'local' spatial distribution of limestone may be so heterogeneous that one headwater stream may have a high acid buffering capacity associated with high alkalinity while a neighboring watershed has very little buffering capacity. To what extent can we predict local variation and decide which first order watersheds should have field survey? Meanwhile, with what accuracy and precision can we compute estimates for larger watersheds based on a sample of their component watersheds?

Much of the source data are not only obtained at different measurement scales, but are also not strictly hierarchical, in that smaller scale sample units may not necessarily be nested exactly within larger scale units. In other words, the boundaries of sample units among the different measurement scales may be non-aligned. Also, the shapes of primary sample units may differ among various measurement scales.

Besides having to address both continuous and discrete response variables, some measurements may be additive as measurement scale increases, such as "number of trees" or "acreage in wetland", while other measurements may be non-additive, such as "species richness".

Of all the metrics that may be used for an ecological assessment, we desire those which not only contain the most information for a given scale of measurement, but also extrapolate with the least uncertainty to other scales. Conventional metrics should be tested and new metrics should be considered. The need for research into new metrics is especially relevant for characterizing landscape structure and pattern in such a way that allows sound, defensible statistical comparisons among landscape units such as watersheds or within a landscape unit over time.

## 1.2 Background

The issue of measurement scale was addressed early by agronomists, as Smith (1938) derived a regression relationship between the variance of crop yield and the size of quadrats used to sample the crops. This provided an 'empirical' law for relating variability to measurement scale which Smith claimed could be used to decrease the need for pilot studies.

Plant ecologists have also long recognized that measurement scale affects one's observations of species dispersion. In response, Grieg-Smith (1952) introduced an agglomerative approach to analyze contiguous quadrat data using nested analysis of variance. Measurement scale challenges mining engineers and soil scientists also. A need for representative sampling of heterogeneous discrete material prompted Gy (1982) to write a treatise on the subject. Pitard (1989) shows that it is a matter of maintaining a minimum mass or volume of each primary sample unit for a given amount of internal heterogeneity.

Although it seems to be often overlooked, the effect of measurement scale on statistical inference with ecological studies has been rigorously questioned (Schneider, 1994; Weins, 1989). Turner (1994) highlights the importance of defining the scale at which an ecological researcher desires to make inference. As stated by Turner, Dale and Gardner (1989), the primary question now appears to be "What, if any, are the rules of extrapolating across scales?".

When measurement scales are regular and strictly hierarchical, geostatisticians refer to changing scale as the change of support (Isaaks and Srivastava, 1989; Cressie, 1991). For additive variables, when only fine scale data are available for drawing inference based on larger scales, corrections need to be made for the change of support. Isaaks and Srivastava (1989) show two methods for adjusting an estimated distribution to account for the support effect, whereby the mean is unchanged and the variance is adjusted.

Progress has been made through fractal-like interpretations of predictability as a function of sampling resolution (Costanza and Maxwell, 1994). Through analysis of rasterized maps, these authors evaluated the spatial *auto-predictability* which is the reduction in uncertainty about the state of a pixel in a scene, given the states of adjacent pixels in that scene, and spatial *cross-predictability* which is the reduction in uncertainty about the state of a pixel in a scene, given the state of corresponding pixels in other scenes. While *auto-predictability* measures the ability of local data to predict, *cross-predictability* measures the ability of a model to predict. Their results clearly showed that *auto-predictability* increases while *cross-predictability* decreases as the resolution (number of pixels/unit area) increases. Since the predictability changed in a regular way, reflecting self-similarity regardless of scale, a "fractal dimension" was calculable that permits easy conversion of predictability taken at one resolution to other resolutions. The authors conjecture that other spatial measures may exhibit such self-similarity, perhaps leading to a generalized theory of scaling.

Scaling through a fractal dimension can also be used for estimating the spatial measure of an object, such as the length of a geographic boundary or the area of a geographic region. If the relationship between the log of estimated total length or area and the log of measurement scale is linearly decreasing, then an estimate of the slope of this linear trend is an estimate of the fractal dimension of the object in question where fractal dimension increases with spatial complexity (Sugihara and May, 1990; Maurer, 1994). With such an estimate in hand, the spatial measure of an object can be predicted at scales other than the scale of actual measurement.

When modeling is the objective, Rastetter *et al.*(1992) provide several methods for aggregating fine-scale data for coarser-scale model predictions.

So far, results appear to have been obtained for additive variables where smaller-scale measurement/inference units are regular and hierarchically nested within larger scale units. Irregular, non-aligned multi-scale units present the greatest challenge; however, irregular hierarchically nested multi-scale units, such as the structure of landscape units, have promise of being a tractable challenge.

## 2. Multi-scale Critical Area Assessment of Watersheds and Landscapes

### 2.1 Overview

For both hydrological and ecological aspects of the environment, differing sensitivity across landscapes is the norm rather than the exception. Certain localities of watersheds may be highly erodable due to localized factors and/or factor combinations such as steep slopes, soil type, sparse vegetative cover, etc. Particular areas contribute inordinately to nonpoint pollution due to highly generative land-use, impermeable soils, sparse cover, and so on. Wetlands, ephemeral ponds, old growth, corridors, and the like are especially important determinants of habitat suitability in a landscape context. Determination of degree and extent of risk therefore requires spatially specific reconnaissance to assess the need for remedial/preventative programs. When overall programmatic needs for remediation/conservation have been determined, spatially specific evaluation is again required to allocate resources and assistance in cost-effective manner. Spatial coincidence of risk and hazard are crucial in both respects.

Contemporary spatial information technologies such as remote sensing and geographic information systems provide coverage on several determinants of ecosystem function at different scales/resolutions with commensurate costs. Broad-area coverage with limited resolution is now available for most of the more important factors, but acquisition of highly detailed information is much more costly and must necessarily be limited to particular areas. As funding constraints for environmental management efforts become increasingly evident, it becomes correspondingly more important to develop efficient strategies for augmenting coarse scale information with finer-scale data in the more crucial areas. Each level of information must be exploited to predict more accurately the distribution of critical areas for

further attention at the next level of detail. Statistical inference provides
the only proven approach to such problems of sequential allocation, but has
been underutilized for spatial problems and particularly so for situations
involving polygonal mappings of categorized environmental variables.

Models are central to modern hydrologic environmental assessment,
and may take several forms according to specific purposes (Myers, 1994).
Although somewhat less well developed than those for hydrology, models
are also of growing importance for ecological environmental assessment.
Finite-difference flow and transport models use a rectangular tessellation
of space, whereas finite-element models use a triangular network. In either case, tessellations impose parameter lumping over space in accordance
with size of spatial partitions (Novotny and Chester, 1989). Coarse tessellations involve greater lumping and produce more generalized views of the
modeled process over space. Thus configuration of spatial partitions and
consequent spatial generalization of output constitutes a modeling version
of scale. Hydrologic models are computationally intensive and data requirements increase exponentially with degree of scale refinement. Practicality
thus dictates that scaling of models be commensurate with investigative
intent. The more sophisticated models make provisions for varying intensity of partitioning over space. Configuration of the spatial framework for a
modeling effort therefore substantially determines its cost and effectiveness.

Whereas the scientific modeling community has been most concerned
with the mathematical/mechanistic aspects of models, cost-effective environmental application of models requires that the spatial logistics of modeling be accorded greater attention. The latter is largely the province of
statistics. It may be interesting and worthwhile to develop generic multi-
phase (multi-scale) statistical strategies for model-based environmental assessment whereby progressively more spatially distributed (and perhaps
more mathematically sophisticated) models are exercised at each stage to
generate expectations of locally more variable and/or extreme model output at the next stage. Spatial configuration for the next stage of modeling
will then be formulated in accordance with expectations from the prior
stage. It should be noted that previous modeling experience over a region
also becomes an important source of expectation for subsequent modeling
work. We therefore have in view a statistical learning strategy for environmental critical-area assessment in a particular region.

### 2.2 Sources of Geographic Data Coverages

Pennsylvania provides an ideal setting for conducting this kind of research by virtue of the level of maturity in its synoptic spatial data resources
and the prior hydrologic modeling work which has been conducted in the
context of Chesapeake Bay watershed concerns. The Office for Remote

Sensing of Earth Resources (ORSER) in the Environmental Resources Research Institute (ERRI) houses an extensive library of physiographic spatial coverages. ORSER serves as a statewide spatial information repository and analysis center for the Pennsylvania Department of Environmental Protection. Nonpoint-source pollution potential has been modeled for major watersheds in Pennsylvania in conjunction with the multi-state Chesapeake Bay nutrient reduction program (Petersen *et al.*1991). Pennsylvania likewise shares in the MAHA (Mid-Atlantic Highlands Assessment) multiagency water quality thrust. The Pennsylvania Gap Analysis work is providing synoptic biotic and land-cover information to complement the existing physiographic information base, along with extensive satellite level digital information. The Pennsylvania Gap Analysis effort has likewise generated a new echelon theory of regionalization for spatial variables of ordinal and interval strength (Myers, Patil and Taillie, July 1997).

## 2.3 Types of Variables

When searching for finer-scale hotspots within larger-scale spatial units, mathematical properties of the variable of concern must be considered. Whether we are measuring discrete or continuous variables, we must consider whether the variable is additive or non-additive. Variables like "acreage of wetland" or "number of individuals in a population" are additive; whereas, variables like "species richness" are non-additive. In the case of additive variables, spatial correlation at one scale can be related to the correlation at both finer and coarser scales because of the bilinearity of the covariance operator. Establishing such relationships is more difficult for non-additive variables and, in general, has to be done empirically. However, there are particular classes of non-additive variables, such as species richness and presence/absence indicators, that obey specific subadditive rules under spatial aggregation; for these, it may be possible to relate spatial dependence at different scales.

## 2.4 Types of Scaling Hierarchies

Nested regular tessellations occur when smaller-scale units constitute a regular tessellation (systematic grid) whose elements are nested within larger-scale units. Such hierarchies may occur when a grid structure is imposed on a region for the purpose of modeling a spatial process such as sediment transport. Nested irregular tessellations occur when smaller-scale units have irregular shaped boundaries, but are still nested within larger-scale units, which may also have irregular boundaries. Such hierarchies arise naturally for watersheds, and may also occur when spatial units have political boundaries such as for townships within counties. Non-aligned hierarchical tessellations occur when small-scale units are not necessarily nested within larger-scale units but the tessellations can be hierarchically ranked according to scale despite imperfect boundary alignment across scales.

## 2.5 Prospective Approach

An extensive statewide raster database such as that developed by Petersen *et al.*(1991) at 100-meter resolution, provides a point of departure. Basic parameter information includes factors such as slope and soil which were used in weighted indexing models to generate potential loadings of sediment and nutrients.

A prospective research approach might consist of four major components.

Component A. A multi-scale assessment of spatial dependence structures and relationships among parameters. This is an important empirical component since its outcome determines the appropriate depth of regionalized data augmentation which is anticipated to be parameter-specific. Regionalized variograms and cross-variograms would be the primary assessment vehicle; however, marginal and hierarchical cumulative distribution functions would also be compared at differential scales of resolution for their assessment potential.

Finer scale data acquisition is needed in areas of high local variability, but this is expected to correlate with large values of model parameters (i.e., homogeneous coefficient of variation rather than homogeneous variance). Data augmentation would be then directed toward the extremes (typically, the large values) of process intensity and would therefore be a form of biased sampling. Accordingly, this component would also examine relationships between (i) the extremes of large-scale process intensity and (ii) smaller-scale spatial dependence and cross-dependence.

Component B. Development of large-scale to small-scale prediction algorithms for model parameters: While the predictions are of potential interest as model inputs, the primary application here envisioned is to drive the search algorithms for promising regions of data augmentation. Linear least squares methods would be employed in conjunction with the empirical results of Component A as well as available covariate information. Algorithms of this type have already been developed by Aragon, Patil, and Taillie (1994) in the context of two-way compositing but with two important differences. For two-way compositing, the correlation structure could be derived mathematically instead of obtained empirically. Secondly, the database size was comparatively small. Efficient and parsimonious data handling techniques are therefore needed in relation to the spatial referencing of the data.

Component C. Development of search strategies for localized data augmentation: "Optimal" strategies are considered to be impractical on both computational and theoretical grounds. Corresponding to each of the primary definitions of critical areas, the investigation might identify promising strategies based on the predictions of Component B. These strategies would be evaluated and compared with both real and simulated data acquisition and modeling protocols pertinent to water and watersheds. The evaluation would require the determination of realistic cost functions for information acquisition. An important question that would be addressed is the cost/benefit tradeoff corresponding to enhanced data resolution versus diminishing information returns.

Experience within the context of hotspot identification for composite sample measurements (Gore and Patil, 1994; Gore, Patil, and Taillie, 1995a,b) might offer some insights into developing such strategies.

Component D. Measurement error consequences: Model parameters often are estimated numbers that differ from true values. Adjustments for such errors are available at the prediction stage (non-interpolatory kriging, e.g., Vecchia, 1992). More important is the inflationary impact of error upon the assessment of local spatial variability due to the presence of this additional and short range component of variation. The resulting inflated assessment of variability would suggest that data be collected at finer scale and higher cost than is actually necessary. Sensitivity analyses could ascertain and document the magnitude of this effect. If necessary, simulated errors can be added to model inputs.

The methodology is intended to be applicable to broad classes of models and input variables. Effects of nonlinear models and of nonadditive input variables both need to be examined. Nonadditive variables of particular importance include classification variables, presence/absence variables, and species richness. The latter, for example, is of interest in connection with the Pennsylvania Breeding Bird Survey (proprietary database, Pennsylvania Academy of Natural Sciences; cf. Brauning, 1992).

## 3. Sample-Based Validation of Spatial Modeling

### 3.1 Overview

Efficacy of the foregoing ideas for a multi-stage spatial modeling strategy requires observational economy in validating each stage of actual modeling to determine the degree and localization of departure in model projections from field conditions. Without such validation, modeling errors can be propagated and magnified from one stage to the next. Since field measurements are expensive, we need to incorporate the most cost effective sampling methods that are possible. Such methods should exploit any available information about the area that can aid in maximizing sampling efficiency through the balance of cost with necessary precision, while maintaining unbiasedness.

Sampling for validation of spatial models should distribute effort according to apparent localization of criticality in order to detect false positives. It should also, however, provide estimates of extent for false negatives and give indications of probable location for the more substantial instances of critical-area non-detection on the part of the model. This phase of the investigation may be directed at adapting and refining available sampling methodologies of observational economy to serve the needs of validation for spatial modeling in the context of critical area analysis on watersheds at varying scales. The following methods have particular appeal in this validation context.

## 3.2 Composite Sampling

When risk is characterized by exposure to chemical or pathogenic contaminants, ground truthing may require the measurement of such contaminants in soil, water, air and biological tissue. Analytical chemistry can become the most expensive component of a sampling and analysis plan. Composite sampling offers a way to obtain the needed information while remaining within budgetary constraints. Specifically, compositing consists of physically mixing $k$ sample units, such as soil cores, followed by analysis of the single composite instead of all $k$ individual samples.

Lovison, Gore and Patil (1994) have identified the three primary applications of composite sampling as follows:

1. Estimate a population mean. Here, the increased efficiency that results from compositing allows one either to reduce the number of analytical tests while achieving the same level of precision, or to increase precision with the same number of analytical tests.

2. Classify each of the constituent sample units with respect to being above or below a numerical criterion, or with respect to having the presence or absence of a trait. Here, Johnson and Patil (1994) have found that compositing can be very cost effective, even after including the additional effort and costs associated with compositing and retesting, provided the prevalence (or exceedance probability) is less than 30 percent for most situations or less than 10 percent for all situations studied.

3. Identify those constituent samples that comprise a specified upper percentile of the distribution. Here, a sweep-out retesting protocol has been developed by Gore and Patil (1994) and has been shown to be

quite efficient, particularly for situations of high sample skewness or of high sample heterogeneity.

## 3.3 Ranked Set Sampling

The design of sampling plans for ground truthing should exploit available auxiliary information to enhance sampling efficiency and reduce survey cost. Several such methods are available, including ratio estimation, regression estimation, and stratification (Thompson, 1992). But, these classical methods depend upon assumptions that limit their applicability or their effectiveness.

A method known as Ranked Set Sampling (RSS), originally proposed by McIntyre (1952), and recently reviewed by Patil, Sinha and Taillie (1994), provides a type of double sampling estimator that is extremely flexible and robust in its use of covariate information. While RSS can utilize quantitative covariates, a strength of the method is that covariate information can also be categorical or mixed categorical-quantitative. In addition, judgment and expert opinion can be components of the auxiliary information without introducing bias into the final estimates.

Ranked set sampling works by randomly allocating $m^2$ prospective sample units into $m$ sets, each of size $m$. The units are then rank ordered within each set with respect to the perceived relative magnitude of the variable of interest. The lowest ranked sample unit may be chosen from the first set, the second lowest rank chosen from the second set, and so on until the $m^{th}$ ranked unit is chosen from the $m^{th}$ set. This whole process is repeated $r$ times in order to obtain the desired sample size $n = rm$. From measurements obtained in this manner, one can obtain unbiased estimates of the mean and of any other characteristic that is representable as a mathematical expectation.

A generalized form of ranked set sampling allocates different numbers of measurement to different rank orders and can achieve high efficiency for skew distributions while focusing sampling intensity upon targeted portions of the population.

Myers, Johnson and Patil (1995) have suggested that information from sources such as aerial photography, satellite imagery, and other spatially referenced databases could be successfully exploited for improved sample design through ranked set sampling.

## 3.4 Adaptive Cluster Sampling

Many natural populations are spatially distributed in a clumped fashion, and are not sampled efficiently by conventional methods. Adaptive cluster sampling has been developed to improve sampling efficiency for such situations (Thompson, 1992).

An initial sample is obtained in a manner that may be either probability based or systematic. After obtaining this initial set of measurements, $\{y_i : i = 1, 2, \cdots, n\}$, a chosen criterion is applied to each measurement to determine if neighboring units should be sampled. A common criterion is to include neighboring units if $y_i > c$ for some constant $c$. The procedure is repeated sequentially until no new units satisfy the criterion. An additional stopping rule is often imposed to keep total sampling intensity within limits.

While the simple arithmetic mean of the initial sample of $n$ units is unbiased, the arithmetic mean of the final sample obtained through adaptive cluster sampling is biased upwards. Thompson (1992) describes two estimators that are unbiased for the population mean or total. One is a modification of the Hansen-Hurwitz estimator when sampling with replacement, and the other is a modification of the Horvitz-Thompson estimator.

One may expect to have fairly extensive information about an area, once it is identified as a critical area, and actually exploit such information, much of which may come from GIS analysis. On one hand, GIS-based information may indicate where the clusters are, thus reducing the need for "adaptive" cluster sampling. On the other hand, even when clusters are identified from an initial sample, the GIS-based information can be used to help select neighbors for subsequent sampling.

### 3.5 Prospective Approach

Each stage of modeling involves two spatial scales and, for ease of exposition, we refer to the larger scale units as "blue units" and the smaller scale units as "green units." For this stage, the critical area assessment proposed earlier has identified a set of blue units as likely candidates for containing critical areas. Model projections have been obtained for each of the green cells within these candidate blue units. In addition, the spatial predictions from Component B are computable for each green cell, although computer resources will place limits on the number of such predictions. Finally, on the basis of the model projections, the green cells have been partitioned into two regimes: the P-regime (for "positive") consists of those green cells currently declared as hotspots, while the N-regime (for "negative") contains the remaining green cells, declared to be not hotspots.

The false positive rate for the hotspot declarations is the conditional probability

(1)          $\Pr(\text{green unit is not a hotspot} \,|\, \text{P-regime})$,

and, similarly, the false negative rate is

(2)                    Pr( green unit is a hotspot | N-regime).

These rates are to be estimated from field measurements, for which an appropriate sampling protocol involves two steps. The first step is the selection of green cells from each of the two regimes for field measurement, and the second step is the actual field measurements within the selected green cells. In most instances, the field measurements yield an *estimate* of the true value, i.e., ground "truthing" is really ground "estimating." An inaccurate ground estimate can produce a classification error in determining the true hot-spot status for the green cell and induce a corresponding inaccuracy in determining the conditional probability given by expression (1) or (2). Efficient sample design for step 2 is intended to yield ground estimates that are close to ground truth and reduce the foregoing source of inaccuracy. Any of the sampling designs described earlier in this section are potentially available for achieving observational economy for step 2. However, there is a point of diminishing returns at which the above classification error becomes negligibly small.

Precise determination of the two error rates is largely controlled by the design of step 1. Efficiency for this step can reduce the number of green cells needing field measurement and hold down overall costs. Each of the three sampling methods (compositing, ranked set sampling, and adaptive cluster sampling) described earlier might be evaluated for their effectiveness in step 1 sampling. Ranked set sampling, particularly with unequal allocations, appears particularly promising since model projections and spatial predictions are both available for ranking purposes. Adaptive sampling could be useful in delineating regions with high error rates. The use of compositing is more problematic since it would involve aggregating green cells. There is an extensive literature on estimating individual sample proportions from composite measurements. However, these methods deal with true categorical responses (e.g., presence/absence rather than dichotomized continuous responses) and assume that the individual sample responses are independent rather than spatially correlated.

The preceding approach does not make full use of available spatial information. One might also examine the possibility of a Bayesian approach, treating the model projections and spatial predictions with suitable error distributions, as the prior. After conditioning on the field measurements, the posterior provides an area-wide estimate of ground truth that is fairly accurate in neighborhoods of the field measurements and less so in more distant regions. The posterior allows assessment of the *magnitude* of error in model projections as well as delineating probable locations of false positives and of false negatives.

# 4. Hierarchies in Watershed Assessment

## 4.1 Overview

The spatial-organizational views of watersheds for purposes of management, analysis, and modeling have been substantially simplistic and subjective. The lineage of such views arises from manual/analog technologies of mapping and airphoto interpretation which prevailed in prior decades. Digital successors to the manual/analog technologies are now well established, but unfortunately have been incarnated essentially as computer-assisted mapping systems to facilitate generation of conventional information structures that have gained little in sophistication relative to their predecessors. The organizational depth of informational representations of watersheds in contemporary (GIS) is still determined primarily by the visual spatial insights of human analysts and rendered as digital layer equivalents of conventional maps. Such layers are organizationally flat despite their derivation from complex surface data and intrinsically hierarchical environmental factor influences.

Simplistic views geared to human visual comprehension cannot add substantially to the efficacy of analysis just by virtue of being computer-facilitated. Generalized delineations of watersheds and/or hydrologic units according to expert intuitive judgment are essentially informational dead ends. Aggregate characterizations of internal variability for heuristically defined units can be compiled, but such compilations only indicate possibilities for improvement - not how optimal organization at a given level of detail can be progressively achieved in working across a spectrum of scales.

A future capability component the prospective approach might look toward intelligent development of information structures from digital data pertaining to watershed processes in a progressive sequence of generalization that optimizes information value of representations at successive levels of detail. This may be motivated by the nature of watersheds as gravity-induced tree structures which may or may not be self-similar, but it also recognizes that factor influences for management/remediation may span watershed boundaries at multiple hierarchical levels. The hierarchical information structures thus generated are conducive to formulation and application of tree metrics as generalized computer analogs of classical photointerpretive recognition and differentiation of drainage patterns.

This approach has a foundation of exploratory theoretical investigation in the Environmental Resources Research Institute at Penn State University under the USGS state water center allotment program, wherein it was pursued under the acronym ESCALATR (Expressing Space Complexes As Logically Abstracted Thematic Regions). Arising from that research is an object-oriented concept of meta-networks as nodal linkages jointly spanning complexes in both coordinate and relational spaces (Myers, 1993). The full concept encompassing parallel computing capability of nodal objects with methods is still beyond the horizon of even working prototype software. However, realization of the structuring capability in meta-networks has recently been determined to be within the scope of suitably enhanced geographic information systems. Future research may build on this initial effort in terms of both software and statistical theory. The goal will be to achieve capability for computer-generated hierarchical representations of watersheds, while also exploring statistical theory that would support comparative analysis of the resulting structures as stochastic networks in the graph theoretic domain.

### References

[1] Aragon, M. E. D., Patil, G. P., and Taillie, C. (1994), Use of best linear unbiased prediction for hotspot identification in two-way compositing, Technical Report Number 94-0416, Center for Statistical Ecology and Environmental Statistics, Pennsylvania State University.

[2] Brauning, D. W. (1992), Atlas of Breeding Birds in Pennsylvania. University of Pittsburgh Press, Pittsburgh. 484 pp.

[3] Costanza, R. and Maxwell, T. (1994), Resolution and predictability: an approach to the scaling problem. *Landscape Ecol.*, **9**, 47–57.

[4] Cressie, N. (1991), *Statistics for Spatial Data*, Wiley, New York, 900 pp.

[5] Gore, S. D. and Patil, G. P. (1994), Identifying extremely large values using composite sample data. *Environmental and Ecological Statistics*, **1**, 227–245.

[6] Gore, S. D., Patil, G. P., and Taillie, C. (1995a), Identification of the largest individual sample values using composite sample data and certain modifications of the sweep-out method. Technical Report Number 94-0701, Center for Statistical Ecology and Environmental Statistics, Pennsylvania State University.

[7] Gore, S. D., Patil, G. P., and Taillie, C. (1995b), Identifying the largest individual sample value from a two-way composite sampling design. Technical Report Number 95-0102, Center for Statistical Ecology and Environmental Statistics, Pennsylvania State University.

[8] Graham, R. L., Hunsaker, C. T., O'Neill, R. V. and Jackson, B. L. (1991), Ecological risk assessment at the regional scale. *Ecological Applications*, **1**, 196–206.

[9] Grieg-Smith, P. (1952), The use of random and continuous quadrats in the study of the structure of plant communities. *Annals of Botany*, **16**, 293–316.

[10] Gy, P. M. (1982), Sampling of Particulate Materials–Theory and Practice. Second edition. Elsevier, Amsterdam.

[11] Isaaks, E. H. and Srivastava, R. M. (1989), *An Introduction to Applied Geostatistics*, Oxford University Press, New York, 561 pp.

[12] Johnson, G. D. and Patil, G. P. (1994), Cost Analysis of Composite Sampling for Classification. Technical Report Number 95-0202, Center for Statistical Ecology and Environmental Statistics, Pennsylvania State University.

[13] Kiester, A. R., *et al.*(1993), Research Plan of the Biodiversity Research Consortium.

[14] Lovison, G., Gore, S. D. and Patil, G. P. (1994), Design and analysis of composite sampling procedures: a review. In *Handbook of Statistics, Volume 12: Environmental Statistics*, G. P. Patil and C. R. Rao, eds. North-Holland, Amsterdam, pp. 103-166.

[15] Maurer, B. A. (1994), *Geographical Population Analysis: Tools for the Analysis of Biodiversity*, Blackwell Scientific Publications, 130 pp.

[16] McIntyre, G. A. (1952), A method for unbiased selective sampling, using ranked sets. *Australian Journal of Agricultural Research*, **3**, 385–390.

[17] Myers, W. L. (1993), A meta-network approach to higher-order spatial constructs and scale effects, Chapter 18. In: *Multivariate Environmental Statistics*, G. P. Patil and C. R. Rao, eds. North Holland.

[18] Myers, W. L. (1994), Environmental remote sensing and geographic information systems-based modeling. In: *Handbook of Statistics, Volume 12: Environmental Statistics*, G. P. Patil and C. R. Rao, eds. North Holland, Amsterdam. pp. 615-642.

[19] Myers, W. L., Johnson, G. D., and Patil, G. P. (1995), Rapid mobilization of spatial/temporal information in the context of natural catastrophes. *COENOSES*, **10**, 89–94.

[20] Myers, W. L., Patil, G. P. and Joly, K. (1997), Echelon approach to areas of concern in synoptic regional monitoring. *Environmental and Ecological Statistics*, **4**, 131–152.

[21] Novotny, V. and Chester, G. (1989), Delivery of sediment and pollutants from nonpoint sources: a water quality perspective. *Soil Water Cons.*, **44**, 568–576.

[22] Patil, G. P., Sinha, A. K., and Taillie, C. (1994), Ranked set sampling. In: *Handbook of Statistics, Volume 12: Environmental Statistics*, G.

P. Patil and C. R. Rao, eds. North-Holland, Amsterdam. pp. 167–200.

[23] Petersen, G., Hamlett, J., Baumer, G., Miller, D., Day, R. and Russo, J. (1991), Evaluation of agricultural nonpoint pollution potential in Pennsylvania using a geographic information system. Report ER-9105, Environmental Resources Research Institute, The Pennsylvania State University, University Park, PA.

[24] Pitard, F. (1993), Pierre Gy's Sampling Theory and Sampling Practice, Second edition. CRC Press, Ann Arbor. 488 pp.

[25] Rastetter, E. B., King, A. W., Cosby, B. J., Hornberger, G. M., O'Neill, R. V. and Hobbie, J. E. (1992), Aggregating fine-scale ecological knowledge to model coarser-scale attributes of ecosystems. *Ecological Applications*, 2, 55–70.

[26] Schneider, D. C. (1994), *Quantitative Ecology: Spatial and Temporal Scaling*. Academic Press, San Diego, 395 pp.

[27] Smith, H. F. (1938), An empirical law describing heterogeneity in the yields of agricultural crops. *Journal of Agricultural Science*, 28, 1–23.

[28] Sugihara, G. and May, R. (1990), Applications of fractals in ecology. *Trends in Research in Ecology and Evolution*, 5, 79–87.

[29] Thompson, S. K. (1992), *Sampling*. John Wiley and Sons Inc., New York, 343 pp.

[30] Turner, M. G., Dale, V. H. and Gardner, R. H. (1989), Predicting across scales:theory development and testing. *Landscape Ecology*, 3, 245–252.

[31] Turner, S. J. (1994), Scale, observation and measurement: critical choices for biodiversity research. (Paper presented at: Measuring and Monitoring Biodiversity in Tropical and Temperate Forests, August 28, 1994. submitted for jury review for publication of conference proceedings.)

[32] Vecchia, A. V. (1992), A new method of prediction for spatial regression models with correlated errors. *Journal of the Royal Statistical Society, Series B,*, 54, 813–830.

[33] Wiens, J. A. (1989), Spatial scaling in ecology. *Functional Ecology*, 3, 385–397.

# R.A. Fisher: The Founder of Modern Statistics*

## C. Radhakrishna Rao[1]

Department of Statistics
Pennsylvania State University
University Park, Pennsylvania

**Abstract.** Before the beginning of this century, statistics meant observed data and descriptive summary figures, such as means, variances, indices, etc., computed from data. With the introduction of the $\chi^2$ test for goodness of fit (specification) by Karl Pearson (1900) and the $t$ test by Gosset (Student, 1908) for drawing inference on the mean of a normal population, statistics started acquiring new meaning as a method of processing data to determine the amount of uncertainty in various generalizations we may make from observed data (sample) to the source of the data (population).

The major steps that led to the establishment and recognition of statistics as a separate scientific discipline and an inevitable tool in improving natural knowledge were made by R.A. Fisher during the decade 1915-1925. Most of the concepts and methods introduced by Fisher are fundamental and continue to provide the framework for the discussion of statistical theory. Fisher's work is monumental, both in richness and variety of ideas, and provided the inspiration for phenomenal developments in statistical methodology for applications in all areas of human endeavor during the last 75 years.

---

* This paper was previously published (without Addendum) in: *Statistical Science* **7**(1), February 1992; pp. 34–48. Reprinted with the permission of the Institute of Mathematical Statistics

[1] C. Radhakrishna Rao is Eberly Professor of Statistics and Director of the Center for Multivariate Analysis at Pennsylvania State University, 417 Thomas Building, University Park, PA 16802

Some of Fisher's pioneering works have raised bitter controversies that still continue. These controversies have indeed helped in highlighting the intrinsic difficulties in inductive reasoning and seeking refinements in statistical methodology.

*Key words and phrases:* Ancillary statistics, Bayes theorem, confounding, consistency, efficiency, $F$-test, factorial experiments, fiducial probability, Fisher information, Fisher optimal scores, likelihood, local control, maximum likelihood, nonparametric tests, randomization, regression, replication, roots of determinantal equation, sufficiency.

## 1. Introduction

In his stimulating address to the members of the Institute of Mathematical Statistics, the American Statistical Association and the Biometric Society, entitled *Rereading of R.A. Fisher,* the late L.J. Savage (1981, pages 678-720) expressed his admiration at the deep and diverse nature of Fisher's contributions and the richness of his ideas, but he was critical of some of the claims made by Fisher as lacking in mathematical rigor and/or logical content. He concluded:

Fisher is at once very near and very far from modern statistical thought generally.

Savage referred to the bitter controversies on some of Fisher's contributions and said:

Of course, Fisher was by no means without friends and admirers too. Indeed, we are all his admirers. Yet he has a few articulate partisans in his controversies onthe foundations of statistical inference, the closest, perhaps being Barnard (e.g., 1963) and Rao (e.g., 1961).

I must emphasize what Savage did not record was that whatever Fisher did was strongly motivated by practical applications. Fisher's research papers look quite different from those we find in current statistical journals. Fisher started off with a description of some live data presented to him for analysis, formulated the questions to be answered and developed the appropriate statistical methodology for the analysis of data. His style of writing was aphoristic and cryptic; often, intermediate mathematical steps are skipped, which may be annoying to the reader. This way, he introduced new ideas and expanded the scope of statistical methodology instead of dissipating his energy and talents in pursuing narrow theoretical concepts. Commenting on Fisher's contributions, Neyman (1951) says:

A very able mathematician, Fisher enjoys a real mastery in evaluating multiple integrals. In addition, he has remarkable talent in the most difficult field of approaching problems of empirical research.

I have been associated with Fisher from 1940 until his death in 1962. I was one of the very few students (perhaps the only student) who did Ph.D. work in statistics under his guidance. I tried to understand his contributions by reading and rereading his writings, and writings on his writings, and above all by numerous personal discussions with him. My own research work is influenced largely by Fisher's ideas. I shall try to comment on different aspects of Fisher's work in the following sections.

We are indebted to Savage for his critical rereading of Fisher. It is important for us to know that some of Fisher's results specially in estimation are not valid in the wide generality claimed by him, some of his conjectures need slight modifications for their validity, and there are minor slips here and there in the mathematical treatment of some of the problems. These are matters of details that need to be examined and the findings put on record. But Savage's criticism does not detract Fisher's pioneering contributions from their usefulness and the motivation they provided for research in statistical theory and applications. They only highlight, what we are aware of today, that there are difficulties in building up a coherent structure for statistical inference, and the search for a monolithic structure for extracting information from data is bound to fail. Fisher himself expressed this view in his last book on *Statistical Methods and Scientific Inference*, a view also held by Savage (1981, page 734):

> The foundations of statistics are shifting, not only in the sense that they have been, and will continue to be changing, but also in the idiomatic sense that no known system is quite solid.

Section 2 gives a broad survey of Fisher's contributions toward the development of statistics over a period of 50 years, starting from 1912, which may be called the Fisherian era of statistics. Some comments on Fisher's book on *Statistical Methods for Research Workers* are given in Section 3.

In the remainder of the paper, references to Fisher's papers will be indicated by the year of publication followed by the volume and paper number, and sometimes the page number in the collected papers edited by J.H. Bennett and published by The University of Adelaide. The titles of books, *Statistical Methods for Research Workers, Statistical Methods and Scientific Inference* and *Calcutta University Lectures* will be abbreviated as SMRW, SMSI and CUL, respectively.

## 2. Fisherian Era of Statistics: 1912 - 1962

The earliest contribution to statistics by Fisher was (1912, 1-1), where the method of maximum likelihood was first used to estimate the unknown parameters. This was followed by a series of papers in the next 12 years, which laid the foundations of statistics as a separate, full-fledged scientific discipline with a great potential for applications in all sciences. He pointed out that a distinction should be drawn between a sample and the population that gave rise to the sample and defined statistics as inductive reasoning of generalizing from a sample to the population. Inductive reasoning has baffled philosophers for a long time, and although its codification started with the writings of Fisher, controversies are bound to raise and continue. It is now generally recognized that there is no monolithic structure for statistical inference, and Fisher's contributions have been of great help in our search for refinements and in introducing new ideas on the subject. In order to understand and evaluate Fisher's work, let us look at the status of statistics before 1912.

### 2.1 Statistics before 1912

*Descriptive statistics.* There was what is now described as descriptive statistics dealing with presentation of data through histograms, bar charts etc., and computation of measures of location, dispersion and association (such as correlation, partial correlation and regression in the case of continuous variables and coefficient of contingency in the case of discrete variables). No distinction was drawn between a sample and the population, and what was calculated from the sample was attributed to the population.

*Curve fitting.* Adolph Quetelet popularized the normal curve of error introduced by Gauss and Laplace by fitting normal curves to all sorts of biological data. Karl Pearson introduced a system of frequency curves to accommodate curves differing in shape from the normal in symmetry and kurtosis. He developed a method of choosing an appropriate curve for given data and also subjecting the choice to an objective test. This test, called the $\chi^2$ test, is the forerunner of all methods of statistical inference and has been rightly hailed as one of the top 20 discoveries of this century, considering all fields of science and technology.

*Testing of hypotheses.* There have been a few attempts at testing of hypotheses, but a major breakthrough was the introduction of the $t$ test, with an exact distribution in finite samples, by Gosset in 1908.

*Estimation.* The method of least squares due to Gauss and the method of least absolute deviations due to Laplace were generally used to estimate parameters in a linear model. Later, Karl Pearson introduced the method of moments to estimate parameters of frequency curves. There was also the concept of standard error (asymptotic standard deviation) to express the precision of an estimate.

*Bayes' theorem.* This was the first major attempt to quantify uncertainty based on observed data. Bayes recommended the use of a uniform prior distribution for the unknown parameters in the absence of any other knowledge and expressing the conclusions in the form of a posterior distribution. Bayes' theorem had engaged the attention of mathematicians like Boole, Venn, Chrystal, Laplace and Poisson, but was not actively pursued.

## 2.2 Fisher's Contributions: 1912 - 1962

*Exact sampling distributions.* In the words of Savage (1981, page 686), "In the art of calculating explicit sampling distributions, Fisher led statistics out of its infancy, and he may never have been excelled in this skill." Fisher obtained the exact null and nonnull distributions of the correlation (1915, 1-4), partial correlation (1924, 1-35) and multiple correlation (1928, 2-61) coefficients, and the $F$ statistic ($z$ statistic) that arises in tests of hypotheses in regression (1922, 1-20; 1924, 1-36).

*Correct use of $\chi^2$.* Although Karl Pearson (1900) introduced the $\chi^2$ statistic and found its asymptotic distribution when the parameters are known, he thought that the same distribution will hold when estimates are substituted for the unknown parameters in computing $\chi^2$. Fisher (1922, 1-19) introduced the concept of degrees of freedom, which depends on the number of unknown parameters estimated and specifies the appropriate asymptotic distributionto be used. He showed that, for the validity of the $\chi^2$ distribution, the parameters may have to be estimated by a more efficient method than that of moments, like maximum likelihood. He also pointed out the modification needed when an observed frequency in any cell is small or even zero.

*Estimation.* Fisher made a major contribution to theoretical statistics in (1922, 1-18), where he considered estimation as a method of reduction of data. He discussed the associated problems of specification (mathematical form of the population involving unknown parameters), estimation (of unknown parameters) and distribution of statistics (estimates) computed from the sample. He recommended maximum likelihood as a general method of estimation, which he used earlier in (1912, 1-1).

*Regression.* Fisher (1922, 1-20) developed the statistical methodology for testing goodness of fit of a regression function and for testing the significance of the individual coefficients. He also considered the problem

of selection of variables (1938, 4-157) based on tests of significance on individual coefficients. Unfortunately, this procedure does not seem to be
satisfactory (see Rao, 1984). In SMRW (7th edition, page 305), Fisher
recommended the regression method in disaggregation of data:

It often happens that the statistician is provided with data
on aggregates which it is required to allocate to different items.
Thus, we may have data on the total consumption of different
households, without knowing how this consumption is allocated
between a man and his wife, or among children of different ages.
If the composition of each household is known, the relative importance of each class of consumer may be obtained by minimizing the
deviation between the consumption recorded, and that expected,
*on assigned scores*, from the composition of the family.

Fisher's suggestion was tried on different sets of data, but the results
were very discouraging (like a negative score for the consumption of rice
by the housewife in a Calcutta study). The flaw appears to be in the
assumption that the *scores* for children, wife and husband remain the same
for all compositions of households.

*Design of experiments.* Fisher introduced a whole new area of research, the design of experiments with a wealth of new ideas on scientific
experimentation, analysis and interpretation of data. Design of experiments also gave a new impetus for research in combinatorial mathematics.
Some fundamental contributions to combinatorics were made in searching
for combinatorial arrangements for design of experiments. [See, e.g., the
contributions of Bose and Shrikhande (1959, disproof of Euler's conjecture)
and Rao (1949, orthogonal arrays).]

*Discriminant function.* In my paper (Rao, 1964) on R.A. Fisher -
"The Architect of Multivariate Analysis", I have described in some detail
Fisher's contributions to multivariate analysis, of which special mention
may be made of discriminant functions. The linear discriminant function
for the classification of an individual into one of two alternative populations proposed by Fisher (Martin, 1936) has become an important tool in
numerical taxonomy and medical diagnosis. The discriminant function for
genetic selection suggested by Fisher (Fairfield Smith, 1936) is an important tool in the selection of plants and animals for genetic improvement.
This is, perhaps, the first attempt at using an empirical Bayes method,
placing a prior distribution on the genetic parameters to be estimated simultaneously for several individuals. This is also an early example where
Stein-type shrinkage estimates emerge when several parameters to be estimated are considered as a random sample from a population of parameters

(see Rao, 1953). However, the Stein phenomenon of reduction in compound mean square loss for fixed values of the parameters was not noticed.

*Roots of determinantal equation.* Generalizing ANOVA techniques to multiple variables, Fisher (1939, 4-163) introduced nonlinear statistics obtained as roots of a determinantal equation of the type $|W - \lambda T| = 0$, where $W$ and $T$ are positive definite random matrices. Originally intended to study the dimensionality of the configuration of a set of points in a high-dimensional space, the distribution of the roots has received new applications in solving problems in physics (Mehta, 1967), signal detection, etc.

*Dispersion on a sphere.* Fisher (1953, 5-249) initiated a new area of research when observations are in the form of direction cosines specifying the orientation of an object, such as the direction of remnant magnetism in the lava flows. Fisher considered the direction cosines in three dimensions as points on a sphere and suggested the density function of the form, $\exp(k \cos \theta)$, where $k$ is positive and $\theta$ is the angular displacement from the true position, $\theta = 0$, at which the dens ity is a maximum. $k$ is a measure of precision; a large value $k$ implies a high concentration of points around a central point and a small value of $k$ implies wide dispersion of points on the sphere. Fisher showed how to estimate the mean direction and the parameter $k$ by the method of maximum likelihood. He also worked out the sampling distributions of the estimates. There is now a considerable body of literature on the statistical analysis of directional data (see Mardia, 1972).

Fisher (1953, 5-249) used his model to estimate the true direction ($\theta$) of remnant rock magnetism in lava flow assuming that the observations collected over a geographical area are independent. He did not consider the possibility of spatial correlations which may have some effect on estimation as shown in Rao (1975).

*Quantification of categorical variables.* In the seventh edition of SMRW (1938, page 299), Fisher gave an example of 12 samples of human blood tested with 12 different sera, giving reactions represented by five symbols, -, ?, $w$, (+) and +, indicating different strengths of reactions. The data were arranged in a $12 \times 12$ two-way table, with the observed reaction given in each cell. Fisher showed how the qualitative reactions can be scored in such a way that the row and column effects are close to additivity, and estimated scores used in a standard ANOVA of two-way classification to test for the row and column effects. Gower (1990) points out that Fisher's optimal scoring methods come close to methods of multiple correspondence analysis and related methods developed several years later by the French school (see Greenacre, 1984).

In a different context of the analysis of contingency tables, Fisher (see Maung, 1941) suggested a singular value decomposition of the matrix of cell frequencies, which is the basis of modern correspondence analysis (Greenacre, 1984). The singular vectors corresponding to the largest singular value provide scores for the row and column categories in such a way that the product moment correlation is maximized. Fisher (1940, 4-175) discussed what is now known as the reciprocal averaging method by which the scores can be computed. Fisher's approach can be extended to the general case of the construction of test criteria for comparison of treatments, etc., when the response variable is qualitative [see Mathen (1954), where I have suggested such an approach].

*Multiplicative model.* Fisher (1923, 1-32) introduced a multiplicative model in a two-way table to explain the responses of 12 varieties to 6 manures, as the usual additive model was found to be unsuitable. The method of fitting the multiplicative model given in the paper is the first attempt to approximate a given matrix by a matrix of unit rank, anticipating the matrix-approximation theorem of Eckart and Young (1936). The paper also provides an extended use of the ANOVA technique (based on nonlinear statistics) to test manural and varietal effects and deviations from the product formula. Perhaps, the first example using regular ANOVA is in 1918 (1-9).

*What else did Fisher do?* In variety and depth of scientific contributions, Fisher has no equal. He touched almost every aspect of current statistical research. Mention may be made of transformation of statistics to stabilize asymptotic variance and/or to induce a higher rate of approach to asymptotic normality, as in the case of inverse hyperbolic tangent transformation of the correlation coefficient (1921, 1-14); analysis of residuals in fitting a regression function (1921, 1-15; 1922, 1-19; 1924, 1-37); distribution of extreme values (1928, 2-63); branching processes in genetics (1930, 2-87); maximum-likelihood estimation in probit analysis (1935, 3-126); specification based on methods of ascertainment of genetic data (1934, 3-113) that led to the investigation of weighted distributions (Rao, 1965); nonparametric inference, such as the sign test, run test, use of order statistics, Fisher-Yates normal score test (see *Statistical Tables* by Fisher and Yates) and treatment of outliers (1922, 1-18); and so on. The list is almost endless. Amazed at Fisher's scholarly and massive achievements, Savage (1981, page 686) says that it would have been more economical to list a few statistical topics in which Fisher displayed no interest than those in which he did.

## 3. Statistical Methods for Research Workers

This is the first full-length book on statistical methods; the first edition was published in 1925 and since then it has gone into numerous editions, with little revision but with some additional material or comments in each edition. In her biography of Fisher, Joan Box says, "The objective of the book was shaped on the anvil of Fisher's scientific thought under the hammer of empirical problems. He conceived of statistics as a tool for research workers and shaped it here to their ends." It is a guide to data analysis, lively and thought-provoking in its presentation, cautioning against misinterpretation of data, questioning the validity of given data to test specified hypotheses, providing reasons for the choice of particular statistical methodology and stressing that the aim of statistical analysis is not just answering the questions raised by the client but to extract all the information from it to answer possibly a wider set of questions and to guide future research. The book is based on his own experience of working with biologists and with the intimate knowledge of the data arising in actual research work. The book also demonstrates the need for an interface between statistics and other sciences for the development of meaningful methodology for analysis of data. It is not a textbook on the mathematical aspects of statistical theory. It is not statistical theory illustrated with made-up examples. It is statistics as it should be used and developed. Each section starts with a live data set, takes the reader step-by-step through various stages of statistical analysis and gradually introduces modern statistical concepts. Fisher himself says the following in the introductory chapter of SMRW:

> ..many will wish to use the book for laboratory reference and not as a connected course of study. ...The great part of the book is occupied by numerical examples. ...The examples have rather been chosen to exemplify a particular process. ...By a study of the processes exemplified, the student should be able to ascertain to what questions, in his material, such processes are able to give a definite answer; and, equally important, what further observations would be necessary to settle other outstanding questions.

It is interesting to note that the SMRW is the only advanced book in statistics that emphasizes the importance of graphical technique in the analysis of the data. There is a full chapter entitled, *Diagrams*, where Fisher says:

> The preliminary examination of most data is facilitated by the use of diagrams. Diagrams prove nothing, but bring out outstanding features readily to the eye; they are therefore no substi-

tute for such critical tests as may be applied to the data, but are valuable in suggesting such tests and in explaining the conclusions based on them.

The following complaints about the book are, perhaps, not without some justification, but appear to be based on an insufficient appreciation of what it is meant to convey.

A prerequisite for reading *Statistical Methods for Research Workers* is that you must have a Master's degree in statistics.

Attributed to M.G. Kendall

In modern mathematical education there is great repugnance to transmitting a mathematical fact without its demonstration. ...Fisher freely pours out mathematical facts in his didactic works without even a bow in the direction of demonstration. I have encountered relatively unmathematical scholars of intelligence and perseverance who are able to learn much from these books, but for most people, time out for some mathematical demonstration seems indispensable to mastery.

Savage (1981, page 685)

Fisher is not against the use of mathematics in teaching statistical methodology. Fisher (1938, 4-159) says:

I want to insist on the important moral that the responsibility for the teaching of statistical methods in our universities must be entrusted, certainly to highly trained mathematicians, but only to such mathematicians as have had sufficiently prolonged experience of practical research and of responsibility for drawing conclusions from actual data, upon which practical action is to be taken. Mathematical acuteness is not enough.

SMRW was the only text on statistical methodology and inference until the Second World War and has inspired many generations of statisticians, and there are plenty of new ideas in the book from which one can still learn.

## 4. Foundations of Theoretical Statistics

Fisher (1922, 1-18), on mathematical foundations of theoretical statistics, is a classic in many ways. For the first time, a clear distinction is drawn between a population and a random sample from it, and the fundamental problems of statistics are stated (see page 280):

(i) Specification (population model specified as a family of probability distributions $P_\theta$ indexed by a parameter $\theta$).

(ii) Estimation (choosing a value of $\theta$ or a member of $P_\theta$, based on the sample, as the appropriate population distribution).

(iii) Distribution (for expressing the precision of the estimate of $\theta$ or the uncertainty in the choice of $\theta$).

The formulation of these problems mark the beginning of the development of statistics as a full-fledged discipline.

Fisher did not give any general discussion of the problem of specification. He was appreciative of Karl Pearson's work on his system of frequency curves and the $\chi^2$ goodness of fit:

> We may instance the development by Pearson of a very extensive system of skew curves, ...Nor is the introduction of the Pearsonian system of frequency curves the only contribution which their author has made to the problem of specification: of greater importance is the introduction of an objective criterion of fit.
>
> Fisher (1922, 1-18, page 281)

A great part of Fisher's paper is devoted to the introduction of new ideas and concepts to provide a logical framework for discussing problems of estimation. For the first time, a clear distinction is made between a *parameter* and an *estimate*. Criteria of *consistency, efficiency* and *sufficiency* were introduced to study properties of estimates and to compare alternative estimates. A measure of *information* on unknown parameters in observed data from an experiment was introduced. The concept of *ancillary statistics* for conditional inference was put forward. *Likelihood*, as a function of the parameters given the data, was defined and the method of *maximum likelihood* was proposed as a general method of estimation.

The terminology and the methods introduced by Fisher form the core of modern estimation theory. No doubt, certain refinements and modifications were made, and will continue to be made, to provide rigor and to meet new situations. We shall examine some of the concepts, results and conjectures made by Fisher and controversies surrounding them, in the light of modern developments.

### 4.1 Consistency

In Fisher (1922, 1-18, page 276), consistency of an estimate is defined as follows:

> A statistic satisfies the criterion of consistency, if, when it is calculated from the whole population, it is equal to the required parameter.

To interpret this definition, we may look at the example on page 283, where Fisher says that

$$\sigma^2 = \left[n^{-1}\Sigma(x_i - \bar{x})^2\right]^{1/2}$$

when calculated from the whole population will lead to the correct value of the standard deviation $\sigma$. Perhaps, the implication is that a statistic $T_n$, as an estimate of a parameter of $\theta$ is defined for all sample sizes and that as $n \to \infty$, $T_n \to \theta_0$ (true value of $\theta$) in probability. We call this CP (consistent in probability). All the editions of SMRW up to the eleventh (1950, page 11) and papers on estimation written subsequent to the 1922 paper carry the above definition of consistency. In a series of lectures delivered at Calcutta University in 1938, Fisher (CUL, page 42) says:

.. If $T$ tends to a limiting value, it is easily recognized by inserting for the frequencies in our sample their mathematical expectations.

In the same publication, Fisher also mentions that an estimate is a function defined on the sample space (presumably of observed relative frequencies) and it is consistent if its value at the expected frequencies is the required parameter. Thus, by consistency, Fisher had in mind both the properties: the estimate tending to the true value in probability and the estimate, considered as a function of observed relative frequencies, taking the true value when the expected values are substituted for the relative frequencies. [The meaning of Fisher's statements is clear when sampling is from a multinomial distribution where we have observed and expected relative frequencies and the estimating function is defined for all vectors $(x_1, \ldots, x_k), x_i \geq 0, \Sigma x_i = 1$. Fisher regarded a continuous distribution as an infinite multinomial, and thus included the continuous case in his definition. Kallianpur and Rao (1955) clarified the situation.]

In 1954, I was preparing a paper on asymptotic efficiency for presentation at the International Statistical Conference, to be held in Rio de Janeiro. I wanted to prove Fisher's bound for the asymptotic variance of a consistent estimate rigorously using conditions under which Hodges-Le Cam phenomenon of superefficiency will not hold. Fisher was visiting Calcutta at that time, and I asked him what exactly did he mean by consistency. He referred to the multinomial case, where the expected cell probabilities ar e $\pi_1(\theta), \ldots, \pi_k(\theta)$ which are functions of, say, a single parameter $\theta$. He said that an estimate $T(p_1, \ldots, p_k)$ of $\theta$ based on observed relative frequencies $p_1, \ldots, p_k$ only, is consistent if

$$(1) \qquad\qquad T(\pi_1(\theta), \ldots, \pi_k(\theta)) \equiv \theta$$

and no limiting property is involved.

Choosing (1) as the definition of consistency, *which I termed as Fisher consistency* (FC), and assuming differentiability of $T$ as a function of $k$ variables and of $\pi_i(\theta)$ as functions of $\theta$, I gave a simple proof of Fisher's inequality in Rao (1955) on the following lines. Differentiating both sides of (1) and the relationship $\Sigma \pi_i(\theta) = 1$, we have

(2)

$$\sum_{i=1}^{k} \frac{\partial T}{\partial \pi_i} \frac{d\pi_i}{d\theta} = 1, \quad \sum \frac{d\pi_i}{d\theta} = 0,$$

$$\Rightarrow \sum_{i=1}^{k} \sqrt{\pi_i} \left( \frac{\partial T}{\partial \pi_i} - \sum_{i=1}^{k} \pi_i \frac{\partial T}{\partial \pi_i} \right) \frac{1}{\sqrt{\pi_i}} \frac{d\pi_i}{d\theta} = 1.$$

Applying the Cauchy-Schwarz inequality to (2) yields

(3)

$$a^2 = \sum \pi_i \left( \frac{\partial T}{\partial \pi_i} \right)^2 - \left( \sum \pi_i \frac{\partial T}{\partial \pi_i} \right)^2 \geq \frac{1}{I}$$

where

$$I = \sum \frac{1}{\pi_i} \left( \frac{d\pi_i}{d\theta} \right)^2$$

is the Fisher information. Notice that under the assumed differentiability condition it is well known that the asymptotic distribution of $\sqrt{n}(T(p_1, \ldots, p_k) - \theta)$ is $N(0, a^2)$ where $a^2$ is the desired asymptotic variance. Equation (3) provides a simple and rigorous demonstration of Fisher's inequality. The proof is similar to that of Cramér-Rao lower bound. In my paper, I have considered the more general inequalities arising in the multiparameter case.

In a subsequent paper, in collaboration with Kallianpur (Kallianpur and Rao, 1955), the concept of FC was extended to the continuous case by considering an estimating function defined on the space of all distribution functions (d.f.'s). If $F_n$ is the empirical d.f. based on a fixed sample size $n$ and $F(\theta)$ is the corresponding true d.f., then an estimate $T_n = T(F_n)$ is FC if $T(F(\theta)) \equiv \theta$. We were able to establish Fisher's inequality by considering Frechet differentiability of $T$.

In the 12th edition of SMRW published in 1954, Fisher, for the first time, added the following paragraph (page 12), while retaining the earlier definition:

> The foregoing paragraphs specify the notion of consistency in terms suitable to the theory of large samples, i.e., by means of the properties required as the sample is increased without limit. Logically it is important that consistency can also be defined strictly for small (i.e. finite) samples by the stipulation that if for each frequency observed its expectation were substituted, then consistent

statistics would be equal identically to the parameters of which they are estimates.

In SMSI (pages 144-146), Fisher says that his definition of consistency as a limiting property is unsatisfactory, and the alternative definition applicable to finite samples is the appropriate one. However, such a definition of consistency may be too restrictive for application in many problems.

## 4.2 Maximum Likelihood

Early references to the method of maximum likelihood (m.l.) for estimation can be found in the works of Gauss, Laplace and Edgeworth, but it was Fisher who saw its great potential for universal use and started studying the properties of m.l. estimates (m.l.e.'s). Encouraged by the nice properties of the m.l.e.'s, judged by criteria such as consistency, efficiency and sufficiency, in the numerous examples he examined, Fisher suggested the m.l. method for universal use. He proved some propositions claiming optimum properties for the m.l.'s, which we now know are not universally true. [See the counter examples by Bahadur (1958), Basu (1988) and Savage (1981). Most of the results can, however, be established rigorously under certain conditions. Fisher was aware of the shortcomings in the mathematical proofs. He suggested the m.l. method only as a "formal solution of problems of estimation" with the following note:

> For my own part I should gladly have withheld publication until a rigorously complete proof could have been formulated; but the number and variety of the new results which the method discloses press for publication.

> Fisher (1922, 1-18, page 290)

Fisher has been accused of priesthood in advocating the m.l. method, because it fails to give acceptable estimates in certain situations. Of course, there is no known method in statistics which is universally optimal, and different methods may have to be used for different purposes and in different situations. The m.l. method is no exception, but it remains as the main stay, which everyone tries when confronted with a new situation. When the m.l. method fails to give acceptable estimates, other methods are sought for.

What can be said in support of m.l. estimates? Under some smoothness conditions, the following are true:

(i) They are Fisher efficient in the sense of having minimum asymptotic variance.

(ii) They are second-order efficient in a decision theoretic sense of having a minimum value for the terms up to the order $(1/n^2)$ in the expansion of expected loss under a bowl-shaped loss function (Rao, 1961b,

1963; Ghosh and Subramaniam, 1974; Efron, 1975; and Akahira and Takeuchi, 1981).

(iii) Suppose that the unknown density function $g$ does not lie in a specified parametric family $f(\theta)$ and an m.l. estimate $\hat{\theta}$ of $\theta$ is obtained under the wrong assumption that it does. In such a case, $\hat{\theta}$ estimates $\theta_g$ defined by

$$I(g; f(\theta)) = \min_{\theta} I(g; f(\theta))$$

where

$$I(g; f(\theta)) = \int g(x) \log\{g(x)/f(x|\theta)\}\, dx$$

is the Kullback-Liebler information number. Or, in other words, the m.l. estimate provides a close approximation to the true density (see Foutz and Srivastava, 1977; White, 1982; and Nishi, 1988).

Can m.l. be used for model selection? Suppose that the class of possible regression in a given problem is the set

$$\{\beta_0, \beta_0 + \beta_1 x, \beta_0 + \beta_1 x + \beta_2 x^2\}$$

of zero, first- and second-degree polynomials. Then, the m.l. method will always choose the second-degree polynomial. Fisher did not consider model selections as a problem of estimation, but, perhaps, as a problem in testing of hypothesis. In recent work, the m.l. principle is extended to cover model selection by using m.l. with an appropriate penalty function depending on the number of parameters (see Akaike, 1973; Rissanen, 1978; Schwarz, 1978; Zhao, Krishnaiah and Bai, 1986; Nishi, 1988; and Rao and Wu, 1989.)

Fisher's work on estimation is of a pioneering nature. The basic concepts and the terminology introduced by him are now routinely used in discussing problems of estimation. Some propositions in estimation proved by Fisher may lack in rigor, but this does not detract their value in their logical content. As stated earlier, Fisher himself was aware of the imperfections in his mathematical treatment. Commenting on the early work of Fisher, Mahalanobis (1938) says:

> Mechanical drill in the technique of rigorous statement was abhorrent to him, partly for its pedantry, and partly as an inhibition to the active use of the mind. He felt it was more important to think actively, even at the expense of occasional errors from which an alert intelligence would soon recover, than to proceed with perfect safety at a snail's pace along well-known paths with the aid of the most perfectly designed mechanical crutches. ...Fisher himself

thinks he was merely a very willful and impatie nt young man. This is no doubt true, but he was impatient not because he was young but because he was a creative genius.

## 4.3 Estimation as Reduction of Data

Fisher viewed the problem of estimation as that of reduction of data. If

$$\underset{\sim}{x} = (x_1, \dots, x_n)$$

constitutes the data, the problem may be posed as that of finding a $k$-vector statistic, with $k \ll n$,

$$\underset{\sim}{T} = (T_1(\underset{\sim}{x}), \dots, T_k(\underset{\sim}{x}))$$

such that "$\underset{\sim}{T}$ contains as much as possible, ideally the whole, of the relevant information" (1922, 1-18, page 278). Reduction of data ("which is usually by its bulk is incapable of entering the mind") in the form of summary figures may offer some convenience in understanding the data and drawing inferences, in addition to the economy in recording only the summary figures for future use, instead of preserving the entire mass of data, much of which may be irrelevant. These considerations are important, especially in scientific research. If, therefore, we define the purpose of estimation as condensation of data, what are the appropriate criteria for choosing the statistic $\underset{\sim}{T}$ to replace the whole sample $\underset{\sim}{x}$? These criteria are developed in Rao (1961a), of which a brief outline is given below. Under these wider criteria, the m.l. estimates seem to provide a good summary of given data provided some regularity conditions hold.

The log-likelihood ratios

$$l(\theta_1, \theta_2 | \underset{\sim}{x}) = \log L(\theta_1 | \underset{\sim}{x}) - \log L(\theta_2 | \underset{\sim}{x})$$
$$= l(\theta_1 | \underset{\sim}{x}) - l(\theta_2 | \underset{\sim}{x})$$

play an important role in statistical inference, whether it is Bayesian or frequentist. If there exists $\underset{\sim}{T}$ such that

$$l(\theta_1, \theta_2 | \underset{\sim}{x}) = l(\theta_1, \theta_2 | \underset{\sim}{T}),$$

then no information is lost, in which case $\underset{\sim}{T}$ is said to be sufficient. If no such $\underset{\sim}{T}$ exists, then we need some criteria for choosing $\underset{\sim}{T}$.

*(i) Wider consistency:* In general, as $n \to \infty$, $l(\theta_1, \theta_2|\underset{\sim}{x}) \to \infty$ or $-\infty$ according as $\theta_1$ or $\theta_2$ is the true value so that complete discrimination is possible in large samples between any two alternative parameter values $\theta_1$ and $\theta_2$. If the same holds when $\underset{\sim}{x}$ is replaced by $\underset{\sim}{T}$, i.e.,

$$l(\theta_1, \theta_2|\underset{\sim}{T}) \to \infty \quad \text{or} \quad -\infty,$$

then $\underset{\sim}{T}$ is said to be consistent in the wider sense.

*(ii) First-order efficiency:* Let us consider discrimination between two close alternatives $\theta$ and $\theta + \delta\theta$. Then expanding $l(\theta + \delta\theta, \theta|\underset{\sim}{x})$, we have up to the first-order term

$$l(\theta + \delta\theta, \theta|\underset{\sim}{x}) = \frac{L'(\theta|\underset{\sim}{x})}{L(\theta|\underset{\sim}{x})}\delta\theta = S_{\underset{\sim}{x}}(\theta)\delta\theta.$$

Similar expansion with $\underset{\sim}{T}$ gives

$$l(\theta + \delta\theta, \theta|\underset{\sim}{T}) = \frac{L'(\theta|\underset{\sim}{T})}{L(\theta|\underset{\sim}{T})}\delta\theta = S_{\underset{\sim}{T}}(\theta)\delta\theta,$$

where $S_{\underset{\sim}{x}}(\theta)$ and $S_{\underset{\sim}{T}}(\theta)$ are the score functions based on $\underset{\sim}{x}$ and $\underset{\sim}{T}$ respectively. If $S_{\underset{\sim}{x}}(\theta)$ and $S_{\underset{\sim}{T}}(\theta)$ are equivalent in some sense, then $\underset{\sim}{T}$ is a good summary of $\underset{\sim}{x}$. Let $U = S_{\underset{\sim}{x}}(\theta) - S_{\underset{\sim}{T}}(\theta)$. We define $\underset{\sim}{T}$ to be first-order efficient if $n^{-1/2}U \to 0$ in probability or in terms of the variance of $n^{-1/2}U$

$$(4) \qquad V(n^{-1/2}U) = \frac{1}{n}(I_{\underset{\sim}{x}}(\theta) - I_{\underset{\sim}{T}}(\theta)) \to 0$$

as $n \to \infty$, where $I_{\underset{\sim}{x}}(\theta)$ is Fisher information based on the whole sample and $I_{\underset{\sim}{T}}(\theta)$, that based on $\underset{\sim}{T}$. Equation (4) means that the information per observation tends to be the same for $\underset{\sim}{x}$ and $\underset{\sim}{T}$ as $n \to \infty$.

*(iii) Second-order efficiency:* Let us consider

$$(5) \qquad V(U) = I_{\underset{\sim}{x}}(\theta) - I_{T}(\theta).$$

Generally, both the terms on the right-hand side diverge, but the difference may tend to a constant as $n \to \infty$. The limiting constant is defined as second-order efficiency, and $\underset{\sim}{T}$ for which this constant is a minimum has the maximum second-order efficiency.

We see that Fisher information comes in a natural way as a criterion for distinguishing between estimates.

Fisher (1925, 2-42) found a general expression for the limit of (5) as $n \to \infty$. However, the proof is not rigorous. Rao (1961b) and Efron (1975) have given a different interpretation of Fisher's expression.

## 5. Conditional Inference

Fisher maintained that statistical methodology developed for data analysis in one area may not be directly applicable in another, since the objectives in different areas may be different. For a business concern, profit or loss in the long run, aggregated over all of its diversified activities, is relevant. In such a case, it is appropriate that decisions taken over a period of time and activities are such that the expected compound loss is minimized. In this process, it may happen that losses in some activities are heavy, but are compensated by larger gain in others. But the situation is different in scientific research. The concept of aggregate loss over a number of different scientific projects or loss in the long run in taking scientific decisions is not meaningful. The data arising from each ex periment to estimate an unknown parameter or to test a hypothesis have to be considered separately, and the amount of uncertainty in the best possible decision taken in each has to be specified. Further, in each case, we need a summary of the data to communicate to others or to place on record for future use.

For statistical analysis of sample data, we need to know how the data are generated, which is specified by a model $(\mathcal{X}, \mathcal{B}, \Theta)$ with $\mathcal{X}$ as the space of all samples, $\mathcal{B}$ as Borel sets and $\Theta$ as identifying the family of probability distributions indexed by a parameter $\theta \in \Theta$. In the terminology of Dawid (1991), this is called the production model. In the Bayesian setup, a hypothetical probability model, called the prior, is imposed on the space $\Theta$ in the form $(\Theta, \mathcal{B}^*, p)$, where $p$ is a known probability measure, in which case, statistical analysis can be done in the standard Bayesian way. No other complications arise, except for the pertinent question, what $p$? [Bayesians may not agree, but I believe, that Bayesian decision theory is essentially based on minimizing loss in the long run (expected) with respect to a chosen prior distribution.]

In the frequentist approach, some new methodological problems arise. Suppose that a certain parameter $\theta$ is estimated by $T(X)$ when $X \in \mathcal{X}$ is observed. In the frequentist approach, the distribution of $T(X)$ is used for drawing inference on $\theta$. Fisher suggested that the distribution of $T(X)$ should be obtained not with respect to the production model, but with reference to a restricted model $(\mathcal{X}_A, \mathcal{B}_A, \Theta)$ where

$$\mathcal{X}_A = \{X \in \mathcal{X} : A(X) = A(x)\}$$

and $x$ is the observed value of $X$, and $A(X)$ is an ancillary statistic (i.e., whose distribution is independent of $\theta$). Fisher considered a number of examples using particular ancillary statistics but did not lay down rules for choosing them.

There has been considerable debate on conditional inference of the type previously described. There are examples like mixtures of experiments (Cox, 1958) and simple random sampling with replacement (Basu, 1988), where there is broad agreement on the need and choice of an ancillary for conditioning. But the general recommendation of Fisher, which is logically of the same status as in the previously described problems, has run into some difficulties, mainly because there can be many choices of an ancillary statistic, each leading to a different kind of inference (Basu, 1988). Some further research appears to be necessary on the choice of an ancillary.

I shall give some examples to show how conditioning on certain features of the observed data can be of help in refining statistical inference.

**Example 1.** The first is a finite sample version of Fisher's example (1925, 2-42, page 26). Suppose that we have two independent samples $X$ and $Y$, giving information on the same parameter $\theta$, from which the estimates $T_1(X)$ and $T_2(Y)$ are obtained are such that

$$(6) \qquad\qquad E[T_1(X)] = E[T_2(Y)] = \theta,$$
$$(7) \qquad\qquad V[T_1(x)] = v_1, \quad V[T_2(Y)] = v_2,$$

where $v_1$ and $v_2$ are independent of $\theta$. Further, suppose that there exist statistics $A_1(X)$ and $A_2(Y)$ such that

$$E[T_1|A_1(X) = A_1(x)] = \theta,$$
$$(8) \qquad\qquad E[T_2|A_2(Y) = A_2(y)] = \theta,$$
$$V[T_1|A_1(X) = A_1(x)] = \nu_1(x),$$
$$(9) \qquad\qquad V[T_2|A_2(Y) = A_2(y)] = \nu_2(y),$$

where $x$ and $y$ are observed values of $X$ and $Y$, respectively, and $\nu_1(x)$ and $\nu_2(y)$ are independent of $\theta$. Then, we might consider the conditional distributions of $T_1$ and $T_2$ given $A_1$ and $A_2$ at the observed values and report the variances of $T_1$ and $T_2$ as $\nu_1(x)$ and $\nu_2(y)$, respectively, as an alternative to (7). What is the right thing to do?

Now, consider the problem of combining the estimates $T_1$ and $T_2$ using the reciprocals of $\nu_1, \nu_2$ and $\nu_1(x), \nu_2(y)$ as alternative sets of weights:

$$(10) \qquad t_1 = \left(\frac{T_1}{\nu_1} + \frac{T_2}{\nu_2}\right) \Big/ \left(\frac{1}{\nu_1} + \frac{1}{\nu_2}\right),$$

$$(11) \qquad t_2 = \left(\frac{T_1}{\nu_1(x)} + \frac{T_2}{\nu_2(y)}\right) \Big/ \left(\frac{1}{\nu_1(x)} + \frac{1}{\nu_2(y)}\right).$$

It is easy to see that the unconditional variances of $t_1$ and $t_2$ satisfy the relation

$$(12) \qquad V(t_1) \geq V(t_2)$$

so that $t_1$ is inadmissible. Does this not mean that $\nu_1(x)$ and $\nu_2(y)$ are more appropriate measures of precision than $\nu_1$ and $\nu_2$ of $T_1$ and $T_2$?

Note that in the previous example, $A_1$ and $A_2$ need not be strictly ancillaries. We need only the conditions in equations (6) to (9) to be satisfied.

**Example 2.** Suppose that a random sample of size 3 has been taken from a row of plants to estimate their average height, and we observe:

| height | $h_1$ | $h_2$ | $h_3$ |
|---|---|---|---|
| position of the plant | 10 | 30 | 31. |

We note that in the observed sample, two values are from two contiguous plants, and there is likely to be a high correlation between the heights of successive plants. In such a case, we could refer the observed sample to the set of samples, where two units out of the three chosen are contiguous and estimate the average height as

$$\frac{2^{-1}(h_2 + h_3) + h_1}{2}$$

which is better than the traditional estimate

$$\frac{h_1 + h_2 + h_3}{3}$$

if, in fact, there is high correlation between contiguous plants.

**Example 3.** Even after a half century of debate, there still seems to be no consensus on when to treat the margins of a $2 \times 2$ contingency table as fixed when conducting a significance test. In a recent article, Greenland (1991) provides a logical justification of conditional tests without appealing to ancillarity, conditionality or marginal information.

**Example 4.** Suppose that we want to estimate the population of the state of West Bengal (India) by random sampling from the complete list of $N$ towns and cities, on which we do not have any prior information on the relative sizes. Theory says that if $x_1, \ldots, x_k$ are the population sizes of $k$ observed units, then an estimate of the population of West Bengal is $N\bar{x}$. Rao (1971) argues that if, among the observed $x_1, \ldots, x_k$, there is one value say, $x_3$ (perhaps the population of Calcutta), which is much larger than the rest, then a better estimate would be $x_3 + (N-1)\bar{x}'$ where $\bar{x}'$ is the average excluding $x_3$. Here, the sample is referred to the set where $x_3$ is observed.

These examples show that the configuration of the observed sample provides some information on appropriate analysis of the data. We owe it to Fisher for introducing this useful concept and the few examples I have given show the need for further discussion and research on conditional inference or choice of a frame of reference. It appears that different samples of the same size from the same population have different information on the unknown parameters, depending on the configuration of the observations in the sample.

## 6. Design of Experiments

In their biographical account of Fisher, Yates and Mather (1963) say the following about Fisher's contributions to design of experiments:

> ..the new ideas of experimental design and analysis soon came to be accepted by research workers. ...The recent spectacular advances in agricultural production owe much to their consistent use. They certainly rank as one of Fisher's greatest contributions to practical statistics.

The subject of experimental designs was developed by Fisher during the years 1921-1923, while he was working at the Rothamsted Experimental Station. He saw the need to collect data in such a way that differences between effects of treatments (or yields of varieties) can be estimated unbiasedly and in the most efficient way under the given constraints on resources. He laid down three fundamental principles, *randomization, replication* and *local control* to be followed in designing an experiment to ensure validity of statistical analysis, to provide an estimate of error for estimated treatment comparisons and to minimize the variance of estimates. Fisher (1931) expressed the roles played by these principles in the form of a diagram (see Figure 1). Describing the importance of experimental design in the collection of data, Fisher (1938, 4-159, page 163) said:

A competent overhauling of the process of collection, or of experimental design may often increase the yield (precision of results) ten or twelve fold, for the same cost in time and labor. To consult a statistician after an experiment is finished is often merely to ask him to conduct a post-mortem examination. He can perhaps say what the experiment died of.

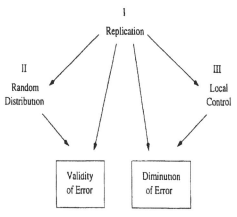

Figure 1.  Fisher's Diagram: "Principles of Field Experiments."

Fisher introduced the concept of factorial designs where each treatment is formed by combining a number of factors at different levels. The aim of an experiment in such a case is to study the effects of individual factors and the interactions which would be of help in determining the optimum mix of factors. Such designs are now routinely used in agriculture and industrial experimentation.

Design of experiments is the most outstanding contribution of Fisher to statistics. G.E.P.Box says, "It is, perhaps, the only tool in statistics which had the greatest impact on analytic and investigative studies in all scientific disciplines and given a status to the statistician as a valued member of a scientific mean." It should also be mentioned that Fisher's ideas of incomplete block designs, confounding and the single replicate factorial designs have inspired a considerable volume of research in combinatorial mathematics. As examples, reference may be made to contributions by the Indian School of Statisticians led by R.C. Bose on finite geometries and mutually orthogonal latin squares, partially balanced and intra- and inter-group balanced incomplete block designs, disproof of Euler's conjecture on the nonexistence of orthogonal latin squares of the order $4n + 2$ (Bose and Shrikande, 1959), coding theory, Kirkman's school girl problem, orthogonal arrays (Rao, 1949), etc.

The importance of replication and local control is well understood, but the principle of randomization and its role in providing valid tests of significance and estimates of error have been questioned from time to time and the controversy still continues. At one end, we have outright rejection of the randomization principle by Basu (1988), and on the other the contributions by Neyman (1935), Rao (1959, 1960), Yates (1964), Youden (1972), Preece (1990), Bailey and Rowley (1987), Bailey (1991) and others who see the need for randomization and examine its role with reference to "validity of the estimation of error". Fisher did not elaborate an the question of validity, but from his writing and comments on papers by others, it is clear, that by "no treatment difference" (null hypothesis), he meant that "each treatment has the same effect on each experimental unit" and by "validity", that "under the null hypothesis, the expected mean square for treatments is equal to that for error" in the appropriate analysis of variance table. A randomization scheme that achieves Fisher's requirement is called *weakly valid*. For instance, it is known (Rao, 1959) that in the case of incomplete block designs such as BIBD, linked block designs and special cases of PBIBD, weak validity is achieved by the randomization rules:

($R_1$) The subsets of treatments are assigned to blocks at random.

($R_2$) Within each block, the varieties of a subset are assigned to the plots at random.

It is also true that, under the nonnull hypothesis, the expected mean square for treatments (say $M_1$) exceeds that for the error (say $M_2$) if an additive model holds (i.e., when an observed yield can be expressed as the sum of treatment effect and plot effect). But, under a nonadditive model, $M_1$ can be less than $M_2$ under the nonnull hypothesis and also under the wider null hypothesis that on the total, over all the experimental units, there are no treatment differences (Neyman, 1935; Rao, 1959).

Weak validity does not ensure the more desirable property that the expected mean square for testing a null hypothesis on subsets of treatments or subsets of contrasts is equal to that for the error. A randomization that ensures this is said to be *strongly valid*, a concept introduced by Grundy and Healy (1950). In some cases, the further randomization rule:

($R_3$) Label the treatments randomly

may introduce strong validity. For a general discussion on randomization schemes for strong validity and recent developments, the reader is referred to Bailey and Rowley (1987) and Bailey (1991).

One of the early arguments against randomization is that an experimenter who knew his material could choose arrangements that were more accurate than some of the arrangements that would be arrived at by random chance. Fisher thought that the experimenter's knowledge could be better utilized in stratifying the material into homogeneous clusters that could be used as blocks, but the act of randomization is necessary at some stage to produce a valid estimate of error which is of great importance, and worth a small reduction in accuracy.

However, a few points remain to be resolved in the practice and theory of randomization. What should one do if a design arrived at by random choice exhibits systematic features? Should one reject this and make another random choice? Any design of experiment must specify the set of designs from which one may be chosen at random. The only condition the set has to satisfy is that the act of randomization provides unbiased estimates of treatment comparisons and valid estimates of error. If there is more than one set with these properties, what further criteria should be used in choosing an appropriate set? To what extent randomization can be sacrificed when some random assignments of treatments to experimental units are difficult to implement in practice as in the case of Youden's (1972) example? Some discussion clarifying these issues will be useful.

## 7. Forms of Quantitative Inference

In Chapter 3 of his book, SMSI, Fisher describes at some length different forms of making inference from observational data. He thinks that a monolithic structure of statistical inference, requiring statements of probability about alternative hypotheses given the observed data, may not always be possible or necessary in taking decisions in experimental sciences. He discusses different types of inference depending on the nature of problems and available data. Rereading this book, one gets the impression that Fisher was reviewing his own work, modifying some statements he had made earlier, answering the criticisms leveled against his contrthe contents of this chapter and comment on some of Fisher's statements.

### 7.1 Tests of Significance

In his early writings (SMRW and DOE), Fisher laid great emphasis on tests of significance. Given a null hypothesis, $H_0$, a test statistic is chosen and its 95% or 99% percentile point is computed. If the observed value of the test statistic exceeds this value, the $H_0$ is rejected. The decision is based on the logical disjunction: Either an intrinsically improbable event has occurred, or the offered hypothesis is not correct. "The level of significance in such a case fulfills the conditions of a measure of the rational grounds for the disbelief it engenders" (SMSI, second edition, page 43). Such a prescription was, perhaps, necessary at a time when statistical concepts were not fully understood and the exact level of significance attained by a test statistic could not be calculated due to lack of computational power. Fisher clarified his views in SMSI both in respect to the level of significance to be used and the situations in which a test of significance is relevant and useful:

> ..for in fact no scientific worker has a fixed level of significance at which from year to year, and in all circumstances, he rejects hypotheses; he rather gives his mind to each particular case in the light of his evidence and his ideas. ...Further, the calculation is based solely on a hypothesis, which, in the light of the evidence, if often not believed to be true at all, so that the actual probability of erroneous decision, supposing such a phrase to have any meaning, may be much less than the frequency specifying the level of significance. ...It (the level of significance) is more primitive or elemental than, and does not justify, any exact probability statement about the proposition.
>
> (SMSI, 2nd edition, pages 42-43)

Fisher thinks that tests of significance have a role to play in scientific research, although they result in a weak form of inference. Thus, when one wants to know whether a normal distribution fits a given data, a general test like the $\chi^2$ goodness of fit is appropriate. If the hypothesis is rejected, an alternative model is sought. The test, by itself, does not indicate what the alternative is. However, once the specification such as the normal family is accepted, then the problem is that of discriminating between alternative values of the parameters of a normal distribution, which falls within the realm of estimation.

It appears from reading SMSI that Fisher gives a limited role to tests of significance in statistical inference, only useful in situations where alternative hypotheses are not specified. He does not recommend any fixed level of significance, but suggests that the observed level of significance has to be used with other evidence that the experimenter may have in making a decision.

However, Fisher's emphasis on testing of null hypotheses in his earlier writings has probably misled the statistical practitioners in the interpretation of significance tests in research work and motivated much theoretical research and publication of text books on a statistical methodology of "limited utility and applicability". [See Wolfowitz (1967) for further remarks.]

## 7.2 Mathematical Likelihood

In the beginning, Fisher introduced likelihood as a quantity "to designate the state of our information with respect to the parameters of the hypothetical population" (Fisher 1922, 1-18, page 334), and more specifically as "measuring our order of preference among different possible populations" (SMRW, 12th edition, page 30). He used these concepts to introduce the m.l. estimation, but found the need to derive the distribution of m.l. estimates, sometimes together with some ancillary statistics, for making inferential statements. He did not state the *pure likelihood principle* as later discussed by Barnard and Birnbaum. However, in SMSI, he referred to likelihood as a "measure of rational belief" in some well-defined sense and proceeded to make inferential statements based on the likelihood function only, keeping the observations fixed. He measured the plausibility of a given value of the parameter by the ratio of its likelihood to the m.l. It is not clear how such a measure can be of help in guiding research investigations, apart from the fact that the likelihood function cannot always be defined, and there are certain other difficulties in dealing with m.l. estimates when there are nuisance parameters (see Cox, 1978).

## 7.3 Fiducial Distribution and Bayes Theorem

There was an attempt by Fisher to make probability statements about the unknown parameters of the Bayesian type without using a prior distribution. In the words of L.J. Savage (1981):

> Fisher's fiducial argument is a gallant but unsuccessful attempt to make the Bayesian omelette without breaking the Bayesian egg.

Fisher was aware of the usefulness of prior probabilities in statistical inference, only when they are inherent and can be ascertained by prior knowledge or introduced through a random mechanism for the choice of a population to be sampled, or estimable from data (empirical Bayes). In fact he uses these concepts in genetic work, as in the construction of discriminant function for genetic selection (Fairfield Smith, 1936; Rao, 1953). But he thought that, in some situations, prior probabilities do not exist (e.g., when the atomic weight of a chemical has to be determined). He describes the axiomatic and personalistic approaches to choose a prior distribution as "bogus" (see 1959, 5-273, for further discussion). However, Fisher's approach for making probability statements about unknown parameters by using the information provided by pivotal statistics is strewn with logical difficulties.

## 7.4 The Final Thought

In a series of lectures given at the Indian Statistical Institute in 1954-1955, Fisher wrote on the blackboard different forms of quantitative inference in the form of an incomplete list (the explanation within parenthesis is mine)

1. Tests of significance (logical disjunction)
2. Mathematical likelihood (measure of rational belief)
3. Fiducial probability (inversion of a pivotal quantity)
4. Bayes theorem (when there is an inherent prior)
5. ...
6. ...

When I asked him what he meant by 5 and 6, he said, these represent other ways which have yet to be discovered, and it is up to the younger generation like you to think about it.

## 8. Concluding Remarks

Fisher is the author of about 300 research publications (reproduced in 5 volumes of his collected papers) and six books, of which four are on statistics and two on genetics. The originality of his papers, their thought-provoking contents, and many suggestions for further development should, in spite of the lack of mathematical rigor of some of his contributions, provide a stimulus and challenge to research workers for many years to come.

The recognition of statistics as a separate scientific discipline came only after the theoretical foundations of the subject were laid and its applications to scientific research was demonstrated by Fisher.

The basis of statistics is inductive logic, which remained uncodified until the beginning of the present century because of inherent difficulties in generalizing from the particular. Attempts at quantifying uncertainty in hypothesis testing through levels of significance provided the initial breakthrough, but there are bound to be controversies in the development of the subject. Fisher may have been wrong in some of the statistical methods he advocated. I say, "may", because there are inherent difficulties in judging the merits of any rule of procedure in inductive logic. Some of Fisher's ideas are still being debated. But, undoubtedly, Fisher was the founder of modern statistical theory and an innovator of various topics that are in the main stream of current research.

I would like to point out that, in the ultimate analysis, statistics as practiced by Fisher and some of his predecessors (see 1947, 4-214; 1938, 4-159; 1953, 5-251) is a way of thinking. Statistical methodology is a process by which we analyze data to provide insight into the phenomenon under investigation rather than a prescription for final decision. There is no fixed rule for answering all questions. Search for new methods will continue. I recall what Fisher said in his preface to his 1950 volume of papers, *Contributions to Mathematical Statistics*:

> In each of these fields there is still much to be done. I am still too often confronted by problems, even in my own research, to which I cannot confidently offer a solution, even to be tempted to imply that finality has been reached (or to take very seriously this claim when made by others)!

## References

[1] Akahira, M. and Takeuchi, K. (1981) *Asymptotic Efficiency of Statistical Estimators: Concepts of Higher Order Asymptotic Efficiency*, Springer, New York.

[2] Akaike, H. (1973) Information theory and extension of the maximum likelihood principle, in: *Proceedings, 2nd International Symposium on Information Theory* (B.N. Petrov et.al., eds.), Akademiai Kiado, Budapest, 267-281.

[3] Bahadur, R.R. (1958), Examples of inconsistency of maximum likelihood estimates. *Sankhyā*, **20**, 2070–210.

[4] Bailey, R.A. (1991), Strata for randomized experiments (with discussion). *J. Roy. Statist. Soc. Ser.* B, **53**, 27–78.

[5] Bailey, R.A. and Rowley, C.A. (1987), Valid randomization. *Proc. Roy. Soc. London Ser.* A, **410**, 105–124.

[6] Barnard, G.A. (1963), Fisher's contributions to mathematical statistics. *J. Roy. Statist. Soc. Ser.* A, **126**, 162–166.

[7] Basu, D. (1988) Statistical Information and Likelihood: A Collection of Critical Essays. Lectures Notes in Statist. **43**, Springer, New York.

[8] Bose, R.C. and Shrikande, S.S. (1959), On the falsity of Euler conjecture about the non-existence of two orthogonal Latin squares of order $4t + 2$. *Proc. Nat. Acad. Sci. USA*, **43**, 734–737.

[9] Cox, D.R. (1958), Some problems connected with statistical inference. *Ann. Math. Statist.*, **29**, 357–372.

[10] Cox, D.R. (1978), Foundations of statistical inference: The case for eclecticism. *Austral. J. Statist.*, **20**, 43–59.

[11] Dawid, A.P. (1991), Fisherian inference in likelihood and prequential frames of reference (with discussion). *J. Roy. Statist. Soc. Ser.* B, **53**, 79–110.

[12] Eckart, C. and Young, G. (1936), The approximation of one matrix by another of lower rank. *Psychometrika*, **1**, 211–218.

[13] Efron, B. (1975), Defining the curvature of a statistical problem,

with applications to second order efficiency (with discussion). *Ann. Statist.*, **3**, 1189–1242.

[14] Fairfield Smith, H. (1936), A discriminant function for plant selection. *Annals of Eugenics*, **7**, 240–250.

[15] Fisher, R.A. (1931) Principles of plot experimentation in relation to the statistical interpretation of the results, in: *Rothamsted Conferences*, **13**, 11–13.

[16] Foutz, R.V. and Srivastava, R.C. (1977), The performance of the likelihood ratio test when the model is incorrect. *Ann. Statist.*, **5**, 1183–1194.

[17] Gosh, J.K. and Subramaniam, K. (1974), Second order efficiency of maximum likelihood estimators. *Sankhyā* Ser. A, **36**, 325–358.

[18] Gower, J.C. (1990), Fisher's optimal scores and multiple correspondence analysis. *Biometrics*, **46**, 947–961.

[19] Greenland, S. (1991), On the logical justification of conditional tests for the two-by-two contingency tables. *Amer. Statist.*, **43**, 248–251.

[20] Greenacre, M.J. (1984) *Theory and Applications of Multiple Correspondence Analysis,* Academic, London.

[21] Grundy, P.M. and Healy, M.J.R. (1950), Restricted randomization and quasi-latin squares. *J. Roy. Statist. Soc. Ser.* B, **12**, 286–291.

[22] Kallianpur, G. and Rao, C.R. (1955), On Fisher's lower bound to asymptotic variance of a consistent estimate. *Sankhyā*, **16**, 331–342.

[23] Mahalanobis, P.C. (1938), Professor Ronald Aylmer Fisher. *Sankhyā*, **4**, 265–272.

[24] Mardia, K.V. (1972) *Statistics of Directional Data,* Academic, London.

[25] Martin, E.S. (1936), A study of an Egyptian series of mandibles with special reference to mathematical methods of sexing. *Biometrika*, **28**, 149.

[26] Mathen, K.K. (1954), Note on design of experiments and testing the efficiency of drugs having local healing power. *Sankhyā*, **14**, 175–180.

[27] Maung, K. (1941), Measurement of association in a contingency table with special reference to pigmentation of hair and eye colour in Scottish school children. *Annals of Eugenics*, **11**, 189–223.

[28] Mehta, M.L. (1967) *Random Matrices and Statistical Theory of Energy Levels,* Academic, New York.

[29] Neyman, J. (1935), Statistical problems in agricultural experimentation. *J. Roy. Statist. Soc. Suppl.*, **2**, 107–154.

[30] Neyman, J. (1951), Review of Fisher's Collected Papers. *Scientific Monthly*, **LXii**, 406–408.

[31] Nishi, R. (1988), Maximum likelihood principle and model selection

when the true model is unspecified. *J. Multivariate Anal.*, **27**, 392–403.

[32] Pearson, K. (1900), On the criterion that a given system of deviations from the probable in a correlated system of variables is such that it can be reasonably supposed to have arisen from random sampling. *Philosophical Magazine*, Series **5**, 157–175.

[33] Preece, D.A. (1990), R.A. Fisher and experimental design: A review. *Biometrics*, **46**, 925–935.

[34] Rao, C.R. (1949), On a class of arrangements. *Proc. Edinburgh Math. Soc. (2)*, **8**, 119–125.

[35] Rao, C.R. (1953), Discriminant function for genetic differentiation and selection. *Sankhyā*, **12**, 229–246.

[36] Rao, C.R. (1955), Theory of the method of estimation by minimum chi-square. *Bull. Inst. Internat. Statist.*, **35**, 25–32.

[37] Rao, C.R. (1959), Expected values of mean squares in the analysis of incomplete block experiments and some comments based on them. *Sankhyā*, **21**, 327–336.

[38] Rao, C.R. (1960), Experimental designs with restricted randomization. *Bull. Inst. Internat. Statist.*, **37**, 394–404.

[39] Rao, C.R. (1961a), Apparent anomolies and irregularities in MLE. *Sankhyā Ser. A*, **24**, 73–101.

[40] Rao, C.R. (1961b) Asymptotic efficiency and limiting information, in: *Proc. Fourth Berkeley Symp. Math. Statist. Probab.*, **1**, Univ. Califormia Press, Berkeley, 531–546.

[41] Rao, C.R. (1963), Criteria of estimation in large samples. *Sankhyā Ser. A*, **25**, 189–206.

[42] Rao, C.R. (1964), R.A. Fisher - The architect of multivariate analysis. *Biometrics*, **20**, 286–300.

[43] Rao, C.R. (1965), On discrete distributions arising out of methods of ascertainment. *Sankhyā Ser. A*, **27**, 311–324.

[44] Rao, C.R. (1971) Some aspects of statistical inference in problems of sampling from finite populations, in: *Foundations of Statistical Inference* (V.P. Godambe and D.A. Sprott, eds.), Holt, Rinehart and Winston, Toronto, 177–202.

[45] Rao, C.R. (1975), Some problems of sample surveys. *Adv. in Appl. Probab.*, **7**, 50–61.

[46] Rao, C.R. (1984), Discussion on "Present position and potential developments: Some personal views, multivariate analysis" by R. Sibson. *J. Roy. Statist. Soc. Ser. A*, **147**, 205–207.

[47] Rao, C.R. and Wu, Y. (1989), A strongly consistent procedure for model selection in regression problem. *Biometrika*, **76**, 369–374.

[48] Rissanen, J. (1978), Modeling by shortest data description. *Automatica*, **14**, 465–471.

[49] Savage, L.J. (1981) *The Writings of Leonard Jimmie Savage. A Memorial Selection,* Amer. Statist. Asoc. and IMS, Hayward, Calif.

[50] Schwarz, G. (1978), Estimating the dimension of a model. *Ann. Statist.*, **6**, 461–464.

[51] Student (1908), The probable error of a mean. *Biometrika*, **6**, 1–25.

[52] White, H. (1982), Maximum likelihood estimation of misspecified models. *Econometrika*, **50**, 1–25.

[53] Wolfowitz, J. (1967), Remarks on the theory of testing of hypotheses. *New York Statistician*, **18**, 1–3.

[54] Yates, F. (1964), Sir Ronald Fisher and the design of experiments. *Biometrics*, **20**, 307–321.

[55] Yates, F. and Mather, K. (1963), Ronald Aylmer Fisher, 1890-1962. *Bibliographic Memoirs of Fellows of the Royal Society of London*, **9**, 91-129.

[56] Youden, W.J. (1972), Randomization and experimentation. *Technometrics*, **14**, 13–22.

[57] Zhao, L.C., Krishnaiah, P.R., and Bai, Z.D. (1986), On the detection of the number of signals in the presence of white noise. *J. Multivariate Anal.*, **20**, 1–25.

## Addendum

In the preceding pages reproduced from *Statistical Science*, 7, 34-48, 1992, I have described Fisher's contributions to the theory of statistical inference and design of experiments in some detail. His contributions to statistical methods in some applied areas were referred to in a general way. In this addendum, I want to describe in some detail Fisher's pioneering work on multivariate statistical methods. Reference may also be made to an earlier paper of mine (Rao, 1964) on "R.A. Fisher: The architect of multivariate analysis".

## Correlation and Regression

The starting point of Fisher's work in MA (multivariate analysis) is the exact sampling distribution of the product moment correlation coefficient $r$ in 1915 (Fisher, 1915). The distribution involving the population value $\rho$ is somewhat complicated and not suitable for practical work without extensive tabulation of the distribution function. Fisher (1921) struck upon the extremely useful transformation $z = tanh^{-1}r$, whose distribution is close to normal for samples of size over 8, with mean $\xi = tanh^{-1}\rho$ and variance $(n - 3)$, so that inference on observed $r$ could be made using the normal theory. See for example Rao (1973, pp.432-436), where a test for the equality of several correlations is developed and a method for obtaining a pooled estimate of $\rho$ from $k$ independent estimates is described.

In a series of papers, Fisher obtained the distributions of the intraclass correlation coefficient (1921), the null distribution of the multiple correlation coefficient (1922b) and the partial correlation coefficient (1924a), all of which provided the tools for a proper interpretation of observed correlation coefficients in studying association between measurements. He pointed out how spurious correlation can arise from a trend in the observations, or by mixing data from different sources. He advocated correlating the residuals after removing the trend, as in the computation of correlation between characters on twins eliminating the effect of age by cubic regression (1924c). A summary of these discussions and the proper use and interpretation of partial correlation coefficients are given in *Statistical Methods for Research Workers* (hereafter referred to as SMRW), first published in 1925.

Fisher (1918) tried to interpret the correlation between characters of relatives in the light of Mendelian inheritance and derived correlations between relatives of different orders under the hypothesis of independent effects of a large number of Mendelian factors and random mating. The effect of deviations from the assumed conditions on the correlation coefficients was also investigated. This paper, with the problems posed and techniques developed in it, laid the foundations of Biometrical Genetics.

In the 1915 paper, Fisher used a geometrical representation of bivariate data in deriving the joint sampling distribution of the estimated variances and covariances. At the suggestion of Fisher, Wishart (1928) used the same method in deriving the joint distribution of the estimated variances and covariances in samples from a general $p$-variate normal distribution, which is known in statistical literature as the Wishart distribution. No attempt was made for a long time to find the joint distribution of the $p(p - 1)/2$ correlations $(r_{ij})$ by integrating the Wishart distribution over the estimated variances. Fisher (1962) obtained the probability density of $(r_{ij})$ in the explicit form

$$\frac{\pi^{-p(p-1)/4} 2^{-p(n-3)/2}}{\left(\frac{n-3}{2}\right)! \ldots \left(\frac{n-p-2}{2}\right)!} |\rho_{ij}^*|^{(n-1)/2} |r_{ij}|^{(n-p-2)/2} F_{n-2}(\gamma_{ij})$$

where

$$\rho_{ij}^* = \rho^{ij} / (\rho^{ii} \rho^{jj})^{1/2},$$

$(\rho^{ij})$ is the matrix reciprocal to $(\rho_{ij})$, the matrix of population correlation coefficients,

$$\gamma_{ij} = -r_{ij}\rho_{ij}^*$$

and

$$F_{n-2}(\gamma_{ij}) = \int_0^\infty \ldots \int_0^\infty \exp\{-\frac{1}{2}(u_1^2 + \ldots + u_t^2$$

$$-2\gamma_{12}u_1u_2 = \ldots\}(u_1 \ldots u_t)^{n-2}\Pi\, du.$$

Fisher wanted to use this distribution in deriving the fiducial distribution of the population correlations ($\rho_{ij}$) given the observed correlations ($r_{ij}$). This problem was engaging the attention of Fisher and his associates at the University of Adelaide shortly before his death.

Fisher (1922a) stressed the importance of regression as a more general concept than correlation by saying that "the regression coefficients are of interest and scientific importance in many classes of data where the correlation coefficient, if used at all, is an artificial concept of no real utility" (SMRW (1967, 129-174)). He obtained the distribution of the regression coefficients for fixed values of the independent variables, and demonstrated how valid inferences could be drawn on regression coefficien ts independently of the sampling distribution of the independent variables. The only distributional assumption to be made in such a case is about the conditional distribution of the **response** variable given the **regressor** variables. Fisher's method introduced a new concept of using conditional tests, which was later used by others as a technique for getting rid of nuisance parameters. It also opened the possibility of deliberately choosing the values of independent variables in an experiment in order to obtain more efficient estimates of the regression coefficients.

The distribution of the multiple correlation coefficient in the null case, which is used in testing the joint significance of all the regression coefficients, was given in (1924b); the distribution in the non-null case, however, appeared a few years later (1928). The 1928 paper made an important advance in the derivation of non-null distributions, but it did not receive much attention as the paper was hard to follow in the absence of several details in the proofs. It is a general characteristic of Fisher's writings as Mahalanobis (1938) says: "he does not attempt to write down the analysis until the problem is solved in his mind, and sometimes, he confesses, after the key to the solution has been forgotten".

Fisher wrote a number of papers during the twenties and thirties illustrating the use of regression in different situations. Fisher (1938a) was the first to discuss the problem of redundancy of independent variables in predicting a response variable, a phenomenon now referred to as the "curse of dimensionality". He suggested some methods for selection of variables, which has been an active area of research during the last 50 or 60 years. We now have cross validation, model selection procedures such as $AIC, BIC$ and $GIC$, and techniques such as partial least squares and projection pursuit regression. We do not yet have a satisfactory solution to the problem.

The possibility of using the regression equation for inverse estimation, i.e., for obtaining confidence (or fiducial) limits to the independent variables for a given value of the dependent variable, was also considered by Fisher. The determination of the age at which boys begin to be taller than girls given in SMRW (1967, p.145) isan example of such estimation. The validity of such a procedure needs further investigation as there is some controversy on the relative efficiencies of direct and inverse regress ion estimates.

Fisher suggested in SMRW (1967, p.298) the use of regression in allocating the total cost of an operation to individual items, such as the estimation of the average cost of maintaining a cow, a buffalo and a calf based on records of total expenditure $(t)$ on cattle incurred by farmers and the composition of the farms (in terms of the numbers of cows (c), buffaloes (b) and calves (v)). For instance if the regression of $t$ on $c, b$ and $v$ is fitted in the form

$$t = a_1 c + a_2 b + a_3 v$$

then $a_1, a_2$ and $a_3$ are the costs of maintaining a cow, a buffalo and a calf. An application made of this method in India gave a negative value for $a_3$ which was attributed to sampling errors! But the explanation is not so simple. Such negative estimates can arise if data on small and big farms are mixed together, and if farms having large numbers of cattle have a smaller proportion of calves and feed the cattle better. Some caution is necessary in using the regression method for such purposes.

Some research work inspired by Fisher's ideas on regression are worth mentioning. In the 1928 paper, Fisher interpreted multiple correlation as the maximum possible correlation between the dependent variable and a linear function of the independent variables. This concept was generalized by Hotelling (1936) to study the relationship between two sets of variables $x_1, \ldots, x_p$ and $y_1, \ldots, y_q$ by maximizing the correlation coefficient between two linear functions $a_1 x_1 + \ldots + a_p x_p$ and $b_1 y_1 + \ldots + b_q y_q$ with respect to $a_i$ and $b_j$. The maximum, minimum and intermediate stable values of the correlation were named by Hotelling as canonical correlations.

Rao (1945) considered the situations where the two sets of variables $x_1, \ldots, x_p$ and $y_1, \ldots, y_p$ are measurements of homologous characters on, say, two brothers in a family and derived what are called familial correlations by maximizing the $intraclass$ correlation between homologous linear functions $a_1 x_1 + \ldots + a_p x_p$ and $a_1 y_1 + \ldots + a_p y_p$. In another papers Rao and

Rao (1987) derived what are called homologous canonical correlations by maximizing the interclass correlation coefficient between $a_1x_1 + \ldots + a_px_p$ and $a_1y_1 + \ldots + a_py_p$. Further research is needed in the interpretation of canonical, familial and homologous canonical correlations.

Fisher (1924a) raised the question of comparing the predictive efficiencies of two sets of independent variables in predicting a dependent variable. Hotelling (1940) provided a partial solution to the problem, and a satisfactory solution has yet to be found.

A new line of research started by Rao (1962b) is an extension of Fisher's use of regression in genetic selection, where a function of concomitant variables is sought such that its correlation with the variable under selection is a maximum. In some problems it may be necessary to maximize the correlation with a desired variable subject to the condition that its correlation with another desired variable is not negative. For instance, if we want to select a parental plant such that the yield of its progeny is high, it is also desirable to see that the resistance to pest does not diminish. The estimation of such a function, called "restricted regression" is based on quadratic programming (see Rao (1964b) for further details). The sampling problems associated with restricted regression have yet to be investigated.

## Discriminant Function

Fisher (1936a) opened a new area of research called discriminant function in answering a question posed by an anthropologist (Martin (1936)) whether there is a method of deciding on the sex of the owner of a jaw bone (recovered from a grave) on the basis of a specified set of measurements made on the jaw bone. Fisher suggested the computation of a linear function of the measurements called the LDF (linear discriminant function) which provided the maximum separation between the distributions of the measurements on male and female jaw bones. Such a function was obtained by determining the compounding coefficients of the measurements in such a way that the ratio of the difference in means to the common standard deviation of the linear function is maximized. This is the origin of the LDF, which was later shown by Welch (1939) to be equivalent to the ratio of the probability densities of the measurements in the two populations. Extension of discriminant analysis when there is more than two alternative populations is given in Rao (1948, 1952, 1973).

The importance of LDF was established by Smith (1947) by showing that it is sufficient with respect to the two populations under consideration, so that no loss of information results in reducing the multiple measurements to a single linear function. Rao (1962a) generalized Smith's result by establishing that the LDF derived from two populations is sufficient not only with respect to the two populations but for a wider set of populations whose mean values lie on the line joining the mean values of the original two populations. This result is useful in testing the appropriateness of the LDF in the classification of an individual when we admit the possibility of the individual belonging to a third unknown group. For an application of such an approach, the reader is referred to Rao (1973, pp.577-579).

At the suggestion of Fisher, a discriminant function for genetic selection was constructed by Fairfield Smith (1936) by regressing the hypothetical genetic value of an individual on observable characteristics. The interpretation of the predicted values in terms of empirical Bayes approach is given by Rao (1953).

Fisher (1938b, 1940) also developed a test criterion for testing the adequacy of a given discriminant function. Rao (1946) extended the test to examine whether a given subset of measurements or a given set of linear functions is sufficient for discrimination.

## Multivariate Analysis of Variance or Analysis of Dispersion

Hotelling (1931) introduced his $T^2$ statistic as a generalization of Students' $t$ for testing hypotheses on the mean values of several correlated variables. The nonnull distribution of $T^2$, which is the same as Mahalanobis $D^2$ introduced for a different purpose, was obtained by Bose and Roy (1938). The nonnull distribution of $D^2$ has the same form as that of the multiple correlation derived by Fisher ten years earlier.

The generalization by Fisher (1939) of Hotelling's $T^2$ to test the differences in mean values of several multivariate populations led to test criteria based on the roots of the determinantal equation

$$|W - \theta(W + B)| = 0$$

where $B$ and $W$ represent the sum of squares and products between and within populations. The distribution of the roots $\theta_1, \theta_2, \ldots$ in the null case is similar to that of canonical correlations of Hotelling (1936) and of test criteria introduced by Roy in 1939 (see Roy (1957)) to test the equality of the dispersion matrices of two multivariate populations.

A general discussion of multivariate tests based on Analysis of Dispersion or MANOVA as it is now generally known is given in papers and books by Anderson (1958), Bartlett (1947, 1951), Rao (1952, 1973), Roy (1957), and others.

Fisher's work on roots of a determinantal equation has extended the theory of canonical correlations of Hotelling (1936) for measuring association between two sets of random variables to cases where one set consists of random and the other dummy variables. The intermediate problem of studying the association between hypothetical (or latent) and observable variables as in problems of genetic selection and factor analysis has been considered by Rao (1953, 1955).

## Canonical Representation of a Discrete Distribution in Two Dimensions

Consider the frequencies $f_{ij}$, $i = 1, \ldots, m$ and $j = 1, \ldots, n$ in a contingency table with row marginal totals $f_{1.}, \ldots, f_{m.}$, and column marginal totals $f_{.1}, \ldots, f_{.n}$, and define the matrix

$$B = (b_{ij}) = (f_{ij} f_{i.}^{-1/2} f_{.j}^{-1/2})$$

with the singular value decomposition

(1) $$B = \sum \rho_i p_i q_i'$$

where $1 = \rho_1 \geq \rho_2 \geq \ldots \geq 0$ are the singular values. At the suggestion of Fisher, Maung (1942) used the singular values $\rho_2, \rho_3, \ldots$ to study the association between two qualitative characters (hair color and eye color) representing the rows and columns of a contingency table. From this study, we have what may be called Fisher identity

(2) $$f_{ij} = f_{i.} f_{.j} (1 + \rho_2 x_i^{(2)} y_i^{(2)} + \rho_3 x_i^{(3)} y_i^{(3)} + \ldots)$$

where $x_i^{(k)}$ and $y_i^{(k)}$ are derived from vectors $p_i$ and $q_i$ in (1). Truncating the series (2) to the first few terms provides a good approximation to the frequencies in a contingency table. The $x_i^{(k)}$ and $y_i^{(k)}$ values together with $\rho_k$ provide the coordinates for biplots in what is currently called correspondence analysis (see Anderson (1996), Benzécri (1973) and Rao (1995)).

Not much was known in the field of multivariate analysis by way of theory or applications before Fisher did his pioneering researches. Many important contributions by others have been inspired by his work. He forged a number of important tools and demonstrated their use in applied research, especially in biometric research, all of which are "bound to stay on the books and used continuously". Truly Fisher was the architect of Multivariate Analysis.

## References

### I. Papers by R.A. Fisher on Multivariate Analysis

[1] Fisher, R.A. (1915), Frequency distribution of the values of the correlation coefficient in samples from an indefinitely large population. *Biometrika*, **10**, 507–521.

[2] Fisher, R.A. (1918), The correlation between relatives on the supposition of Mendelian inheritance. *Trans. Roy. Soc.*, Edinburgh, **52**, 399–433.

[3] Fisher, R.A. (1921), On the "probable error" of a coefficient of correlation deduced from a small sample. *Metron*, **1**, 3–32.

[4] Fisher, R.A. (1922a), The goodness of fit of regression formulae, and the distribution of regression coefficients. *J. R. Statist. Soc.*, **85**, 597–612.

[5] Fisher, R.A. and Mackenzie, W.A. (1922b), The correlation of weekly rainfall. *Quart. J.R. Met. Soc.*, **48**, 234–245.

[6] Fisher, R.A. (1924a), The distribution of the partial correlation coefficient. *Metron*, **3**, 329–332.

[7] Fisher, R.A. (1924b), The influence of rainfall on the yield of wheat at Rothamsted. *Phil Trans.*, B, **213**, 89–142.

[8] Fisher, R.A. (1924c), The resemblance between twins, a statistical examination of Lauaterbach's measurements. *Genetics*, **10**, 569–579.

[9] Fisher, R.A. (1928), The general sampling distribution of the multiple correlation coefficient. *Proc. Roy. Soc.*, A, **121**, 654–673.

[10] Fisher, R.A. (1936), The use of multiple measurements in taxonomic problems. *Ann. Eugen.*, **7**, 179–188.

[11] Fisher, R.A. (1937), The relation between variability and abundance shown by the measurements of the eggs of British-nesting birds. *Proc. Roy. Soc.* B, **122**, 1–26.

[12] Fisher, R.A. (1938a), On the statistical treatment of the relation between sea-level characteristics and high-altitude acclimatization. *Proc. Roy. Soc.*, B, **126**, 25–29.

[13] Fisher, R.A. (1938b), The statistical utilization of multiple measurements. *Ann. Eugen.*, **8**, 376–386.

[14] Fisher, R.A. (1939), The sampling distribution of some statistics obtained from non-linear equations. *Ann. Eugen.*, **9**, 238–249.

[15] Fisher, R.A. (1940), The precision of discriminant function. *Ann. Eugen.*, **10**, 422–429.

[16] Fisher, R.A. (1958) *Statistical Methods for Research Workers*, Thirteenth Edition, Oliver and Boyd, Edinburgh.

[17] Fisher, R.A. (1962), The simultaneous distribution of correlation coefficients. *Sankhyā*, A, **24**, 1–8.

## II. Other Publications Referred to in the Text

[18] Anderson, T.W. (1958) *An Introduction to Multivariate Analysis,* Wiley, New York.

[19] Anderson, T.W. (1996), R.A. Fisher and Multivariate Analysis. *Statistical Science,* **11**, 20–35.

[20] Benzécri, J.P. (1973) *L'Analyse de Données.* Tome 1: La Taxinomie. Tome 2: L'Analise des correspondances. Dunod, Paris.

[21] Bartlett, M.S. (1947), Multivariate analysis. *J. R. Statist. Soc.,* B, **9**, 176–197.

[22] Bartlett, M.S. (1951), The goodness of fit of a single hypothetical discriminant in the case of several groups. *Ann. Eugen.,* **16**, 199–214.

[23] Bose, R.C. and Roy, S.N. (1938), The exact distribution of the Studentized $D^2$-statistic. *Sankhyā,* **4**, 19–38.

[24] Day, B.B. (1937), A suggested method for allocating logging costs to log sizes. *J. For.,* , 69–71.

[25] Fairfield Smith, H. (1936), A discriminant function for plant selection. *Ann. Eugen.,* **7**, 240–250.

[26] Hotelling, H. (1931), The generalization of Student's ratio. *Ann. Math. Stat.,* **2**, 360–378.

[27] Hotelling, H. (1936), Relations between two sets of variates. *Biometrika,* **28**, 321–377.

[28] Hotelling, H. (1940), The selection of variates for use in prediction with some comments on the general problem of nuisance parameters. *Ann. Math. Statist.,* **11**, 271–283.

[29] Mahalanobis, P.C. (1938), Professor Ronald Aylmer Fisher. *Sankhyā,* **4**, 265–272.

[30] Maung, K. (1942), Measurement of association in a contingency table with special reference to the pigmentation of hair and eye colors of Scottish school children. *Ann. Eugen.* London, **11**, 189–223.

[31] Martin, E.S. (1936), A study of an Egyptian series of mandibles with special reference to mathematical methods of sexing. *Biometrika,* **28**, 149–172.

[32] Rao, C.R. (1945), Familial correlations or the multivariate generalization of the intraclass correlation. *Current Science,* **4**, 66.

[33] Rao, C.R. (1946), Tests with discriminant functions in multivariate analysis. *Sankhyā,* **7**, 407–414.

[34] Rao, C.R. (1948), Utilization of multiple measurements in problems of biological classification. *J. Roy. Statist. Soc.,* **10**, 159–203.

[35] Rao, C.R. (1952) *Advanced Statistical Method in Biometric Research,* Wiley, New York.

[36] Rao, C.R. (1953), Discriminant function for genetic differentiation and selection. *Sankhyā*, **12**, 229–246.

[37] Rao, C.R. (1955), Estimation and tests of significance in factor analysis. *Psychometrika*, **20**, 93–111.

[38] Rao, C.R. (1960), Multivariate analysis. An indispensable statistical aid in applied research. *Sankhyā*, **22**, 317–338.

[39] Rao, C.R. (1961), Some observations on multivariate statistical methods in anthropological research. *Bull. Int. Stat. Inst.*, **33**, 99–109.

[40] Rao, C.R. (1962a), Use of discriminant and allied functions in multivariate analysis. *Sankhyā*, A, **24**, 149–154.

[41] Rao, C.R. (1962b), Selection with restrictions. *J. R. Statist. Soc.* B, **24**, 401–405.

[42] Rao, C.R. (1964a), Sir Ronald Fisher - The architect of multivariate analysis. *Biometrics*, **20**, 286–300.

[43] Rao, C.R. (1964b) Problems of selection involving programming techniques, in: *Proceedings of IBM Scientific Computing Symposium on Statistics*, 29-51.

[44] Rao, C.R. (1973) *Linear Statistical Inference and its Applications*, John Wiley.

[45] Rao, C.R. and Rao, C.V. (1987) Stationary values of the product of two Raleigh coefficients: homologous canonical variates, Sankhyā, B. *49*, **113–125,**

[46] Rao, C.R. (1995) *A review of canonical coordinates and an alternative to correspondence analysis using Hellinger distance*, **Qüestio**, 19.15–63

[47] Roy, S.N. (1957) *Some Aspects of Multivariate Analysis*, Wiley, New York.

[48] Smith, C.A. (1947), Some examples of discrimination. *Ann. Eugen.*, **18**, 272–282.

[49] Welch, B.L. (1939), Note on discriminant functions. *Biometrika*, **31**, 218–220.

[50] Wishart, J. (1928), The generalized product moment distribution in samples from a normal multivariate population. *Biometrika*, **20 A**, 32–52.

# Higher Order Asymptotics: Costs and Benefits

N. Reid and D. A. S. Fraser

Department of Statistics
University of Toronto
Toronto, Ontario, Canada

**Abstract.** The theory of higher order asymptotics provides a method for very accurate approximation of $p$-values and confidence limits in parametric inference. These approximations are not widely used in practice, for various reasons, including lack of knowledge, lack of software, and lack of appreciation. This paper describes some results in higher order asymptotics, with a view to the practice and the theory of statistics.

## 1. Introduction

Let $Y_1, Y_2, \ldots, Y_n$ be independent, identically distributed random variables from a density $f(y; \theta)$, where $\theta \in R^k$. We assume that $\theta = (\psi, \lambda)$, where $\psi$ is a scalar parameter of interest and $\lambda$ is a $(k-1)$-dimensional nuisance parameter. Our goal is to construct a *significance function* $p(\psi)$, where $p(\psi_0)$ is the $p$-value for a test of the composite null hypothesis $H_0 : \psi = \psi_0$. The significance function provides confidence limits at any level of confidence by inversion, as well as providing an ordinary $p$-value for particular values of $\psi$.

This problem is a considerable abstraction from what is typically needed in practical work, and some discussion of the possibility for extending the work more generally is given in Section 4. The advantage of considering such an abstraction is that very accurate approximations to the significance function are available, and the structure of these approximations provides insight into the theory of inference, as well as suggesting ways to extend the results to more realistic settings.

351

The approximations are obtained from the theory of higher order asymptotics, which in essence means using the first few terms of an asymptotic expansion to suggest an approximation. In Section 2 we illustrate these with some numerical examples. In Section 3 we describe several results now available for accurate approximation of $p$-values in various settings. In Section 4 we consider some of the costs, or limitations of the results of Section 2, and suggest some benefits of higher order asymptotics for theory and practice. In the remainder of this Section we introduce necessary notation, review first order asymptotic theory, and introduce a higher order approximation.

We denote the log likelihood function based on the sample $y = (y_1, \ldots, y_n)$ by $\ell(\theta; y) = \log f(y; \theta)$; where needed we write $\ell(\psi, \lambda; y)$, and sometimes abbreviate this by $\ell(\theta)$ or $\ell(\psi, \lambda)$. The maximum likelihood estimate $\hat{\theta}$ is assumed to be a unique solution of the score equation $\ell'(\theta; y) = 0$, and usually we will need the restricted maximum likelihood estimate $\hat{\lambda}_\psi$, of the nuisance parameter $\lambda$, which we assume satisfies $\partial \ell(\psi, \hat{\lambda}_\psi)/\partial \lambda = 0$. The profile log likelihood function is $\ell_p(\psi) = \ell(\psi, \hat{\lambda}_\psi)$. The Fisher information function for $\theta$ is denoted by $j(\theta)$:

$$j(\theta) = -\frac{\partial^2 \ell(\theta)}{\partial \theta \partial \theta^T} .$$

This matrix is often partitioned to conform to the partition of $\theta$

$$j(\theta) = \begin{pmatrix} j_{\psi\psi} & j_{\psi\lambda} \\ j_{\lambda\psi} & j_{\lambda\lambda} \end{pmatrix} \qquad j^{-1}(\theta) = \begin{pmatrix} j^{\psi\psi} & j^{\psi\lambda} \\ j^{\lambda\psi} & j^{\lambda\lambda} \end{pmatrix}$$

and the following identity is useful

(1) $$|j_p(\hat{\psi})| = |j(\hat{\psi}, \hat{\lambda})|/|j_{\lambda\lambda}(\hat{\psi}, \hat{\lambda})|$$

where $j_p(\hat{\psi}) = -\ell_p''(\hat{\psi})$ is the Fisher information from the profile log likelihood function.

Under regularity conditions on the model, outlined for example in Lehmann (1983, Ch. 6), we can apply a central limit theorem to the score function $\ell'(\theta) = \ell'(\theta; y)$, which is a sum of independent, identically distributed random variables, and use this to obtain the result

(2) $$\ell_p'(\psi)\{j_p(\hat{\psi})\}^{-1/2}d \longrightarrow N(0, 1)$$

as $n \to \infty$. Similarly by Taylor series expansion of $\ell(\theta)$ about $\hat{\theta}$ we can obtained the related results

(3) $$(\hat{\psi} - \psi)\{j_p(\hat{\psi})\}^{1/2}d \longrightarrow N(0, 1)$$

(4) $$r(\psi) = \pm[2\{\ell_p(\hat{\psi}) - \ell_p(\psi)\}]^{1/2}d \longrightarrow N(0,1) \ .$$

These results are derived in Cox and Hinkley (1974, Ch. 9), without detailed attention to regularity conditions. The results lead to three possible approximate significance functions:

(5) $$p(\psi) \doteq \Phi\{(\hat{\psi} - \psi)j_p^{1/2}(\hat{\psi})\}$$

(6) $$\doteq \Phi\{r(\psi)\}$$

(7) $$\doteq \Phi\{\ell_p'(\psi)j_p^{-1/2}(\hat{\psi})\}$$

where $\Phi(\cdot)$ is the cumulative distribution function for the standard normal distribution. Approximations (5), (6) and (7) are called first order approximations because the relative error in the approximation is $O(n^{-1/2})$, which means in (5), say, that the ratio of the approximate to the exact cumulative distribution function for $\hat{\psi}$ is $1 + O(n^{-1/2})$ for each $\hat{\psi}$ in the so-called *normal deviation region*: $\{\psi : |\hat{\psi} - \psi| \le \delta n^{-1/2}\}$.

The approximations that we consider in this paper are of the form

(8) $$p(\psi) \doteq \Phi(r) + \phi(r)\left(\frac{1}{r} - \frac{1}{u}\right)$$

(9) $$\doteq \Phi\left(r + \frac{1}{r}\log\frac{u}{r}\right) \equiv \Phi(r^*)$$

where $r = r(\psi)$ is defined in (eq4) and $u = u(\psi)$ is to be described below.

Approximation (8) is due to Lugannani and Rice (1980) in the context of independent, identically distributed sampling from a one-parameter exponential family. Approximation (9), often called the $r^*$ approximation, is due to Barndorff-Nielsen (1986). Both approximations are third order approximations, in the sense that under independent, identically distributed sampling they have relative error $O(n^{-3/2})$ in normal deviation regions, under suitable regularity conditions. They are asymptotically equivalent to each other to the same order.

There are a great many examples in the literature indicating the surprising accuracy of (8), some of which are given in Section 3. We also note that given $r$ and $u$, the calculations in (8) or (9) are a fairly trivial extension of the first order approximation (6). The main difficulty in applying (8) or (9) is the determination of a suitable expression for $u = u(\psi)$. A fairly complicated version was given in Barndorff-Nielsen (1986), which was somewhat simplified in Barndorff-Nielsen (1991): these can be used when the model can be embedded in a model with a parameter of fixed dimension. Construction of $u$ is the main focus of the Section 3; for the moment we note only that we will have

$$u = r + O_p(n^{-1/2})$$

for each value of $\psi$, and $u(\hat{\psi}) = r(\hat{\psi}) = 0$. The latter means that that approximation (8) or (9) is computationally unstable near $\psi = \hat{\psi}$ and needs be replaced there by a limiting value if a $p$-value is needed at that central point.

## 2. Numerical examples

### 2.1 Normal parabola

Assume the model is a normal distribution with mean $\psi$ and variance $\psi^2$. This is a $(2, 1)$ curved exponential family, that is, it has a two-dimensional sufficient statistic, and a one-dimensional parameter of interest. The maximum likelihood estimate $\hat{\psi}$ solves a quadratic equation, and there is an exact ancillary statistic (Hinkley, 1977), so that the conditional distribution of $\hat{\psi}$ given $a$ contains all the information about $\psi$. Table 1, a smaller version of Table 4.2 in Fraser, Reid and Wu (1999) compares the exact distribution function for $\hat{\psi}$, conditional on $a$ with approximation (8), based on a sample of size 5 from a $N(1, 1)$ distribution.

| $\psi$ | 0.6 | 0.8 | 1.0 | 1.2 | 1.4 | 1.6 | 1.8 | 2.0 |
|---|---|---|---|---|---|---|---|---|
| exact $p(\psi)$ | 0.9976 | 0.8855 | 0.5802 | 0.3110 | 0.1566 | 0.0794 | 0.0416 | 0.0228 |
| approximate | 0.9976 | 0.8863 | 0.5784 | 0.3094 | 0.1555 | 0.0707 | 0.0412 | 0.0225 |
| first order | 0.9960 | 0.8465 | 0.5014 | 0.2414 | 0.1102 | 0.0513 | 0.0249 | 0.0128 |

Table 1. Adapted from Fraser, Reid and Wu (1999)

### 2.2 A $(2, 1)$ exponential family

This example is discussed in Barndorff-Nielsen and Chamberlin (1991) and in Fraser, Reid and Wu (1999). We assume a simple observation $y = (y_1, y_2)$ has a bivariate distribution $f(y_1, y_2; \psi, \lambda) = \lambda \psi e^{-\psi} e^{-\lambda y_1 - \psi y_2}$, where $\lambda = \lambda(\psi) = \psi^{-1} e^{-\psi}$. This is also a $(2, 1)$ curved exponential family, but no exact ancillary statistic exists. Table 2 gives the approximation to the conditional distribution of $\hat{\psi}$, gives an approximate ancillary statistic. This is compared to the exact distribution, obtained by simulation.

| exact | 0.0013 | 0.0250 | 0.251 | 0.500 | 0.749 | 0.9750 | 0.9987 |
|---|---|---|---|---|---|---|---|
| approximate | 0.0012 | 0.0229 | 0.236 | 0.485 | 0.742 | 0.9759 | 0.9987 |
| first order | 0.0009 | 0.0208 | 0.311 | 0.593 | 0.824 | 0.9874 | 0.9993 |

Table 2. Adapted from Fraser, Reid and Wu (1999)

## 2.3 Extreme value regression

This example is adapted from DiCiccio, Field and Fraser (1990). The model is $y_i = x_i^T \beta + \sigma e_i$, where $x_i$ is a $6 \times 1$ vector of known constants, $e_i$ follows the extreme value distribution $f(e) = \exp\{e - \exp(e)\}$, and $\theta = (\beta, \sigma)$ is of dimension 7. Table 3 compares the exact and approximate $p$-value for testing $\psi = \beta_1$, based on a sample of size 10.

| | | | | | | |
|---|---|---|---|---|---|---|
| exact | 0.995 | 0.99 | 0.975 | 0.95 | 0.90 | 0.75 |
| approximate | 0.994 | 0.991 | 0.981 | 0.947 | 0.887 | 0.714 |
| first order | 0.923 | 0.906 | 0.872 | 0.823 | 0.763 | 0.665 |
| | | | | | | |
| exact | 0.25 | 0.10 | 0.05 | 0.025 | 0.010 | 0.005 |
| approximate | 0.226 | 0.094 | 0.053 | 0.031 | 0.012 | 0.006 |
| first order | 0.320 | 0.192 | 0.141 | 0.109 | 0.067 | 0.049 |

Table 3. Adapted from DiCiccio, Field and Fraser (1990)

## 2.4 Discussion

The examples above are necessarily limited, and chosen to illustrate the accuracy of approximation (8) or (9).[1] There are many similar examples in the published literature, in a variety of regression and other models: a partial survey is given in Reid (1996). In the tables above approximation (8) is recorded, and there are small differences between (8) and (9) in these examples. Often (9) is slightly more accurate, although examples exist for the contrary situation as well. Note in all these cases the substantial improvement over the first order approximation, which in these tables is always $\Phi(r)$. The other two asymptotically equivalent first order approximations given by (6) and (7) are typically much less accurate.

## 3. The structure of the third order approximation

### 3.1 Introduction

There are two aspects to the calculation of $u$ for use in approximation (8) or (9). The first is detailed specification of the statistic $u$ needed for calculation. The second question, left out of the description of the examples in the previous section, is to clarify exactly what the expression $p(\psi)$ is approximating.

---

[1] In fact inference for $\log \sigma$ in Example 2.3 is considerably less accurate than for $\beta_1$ (DiCiccio, Field and Fraser, 1990).

We will illustrate in this section the answers to these questions in a number of special cases, and then refer briefly to more general solutions available in the literature.

### 3.2 Bayes marginal posterior

The easiest derivation for $u$ is obtained in a Bayesian framework, in which case the resulting "p-value" is actually a posterior marginal probability:

$$p(\psi) = pr(\Psi \geq \psi|y)$$
$$= \int_\psi^\infty \pi_m(t|y)dt$$

where

$$\pi_m(\psi|y) = \int \pi(\psi, \lambda|y)d\lambda$$
$$= \int f(y; \psi, \lambda)\pi(\psi, \lambda)d\lambda .$$

By using the Laplace approximation (Tierney and Kadane, 1986) to $\pi_m(\psi|y)$ and an integration by parts, the posterior probability is approximated by (8,9) with

$$(10) \qquad u = \ell_p'(\psi)\{j_p(\hat\psi)\}^{-1/2} \frac{|j_{\lambda\lambda}(\psi, \hat\lambda_\psi)|^{1/2}}{|j_{\lambda\lambda}(\hat\psi, \hat\lambda)|^{1/2}} \frac{\pi(\hat\psi, \hat\lambda)}{\pi(\psi, \hat\lambda_\psi)} .$$

The derivation of this result is sketched in the Appendix. Generalizations of this result to the setting where $\theta$ is not available as a partition $(\psi, \lambda)$ and $\psi$ is specified as a function of $\theta$ are given in DiCiccio and Martin (1993) and Fraser, Reid and Wu (1999).

### 3.3 Canonical exponential family

Suppose the model is $f(y; \theta) = \exp\{\psi s + \lambda^T t - c(\psi, \lambda) - d(y)\}$ where $(s, t) = \{s(y), t(y)\} = \{\Sigma s(y_i)), \Sigma t(y_i)\}$ is the minimal sufficient statistic in independent, identically distributed sampling from a canonical exponential family. Then the conditional density of $s$, given $t$, is again an exponential family, and this conditional density depends only on the parameter of interest $\psi$. This is the basis for the construction of similar or unbiased tests, described in many textbooks on inference, such as Cox and Hinkley (1974, Ch. 5), or Lehmann (1986, Ch. 4). A saddlepoint approximation to this conditional density can be integrated using very similar arguments to those in the Bayesian case, and the result is

$$p(\psi) = pr\{\hat{\Psi} \leq \hat{\psi}|t; \psi\}$$

is approximated by (8) or (9) with

$$(11) \qquad u = (\hat{\psi} - \psi)\{j_p(\hat{\psi})\}^{1/2} \frac{|j_{\lambda\lambda}(\hat{\psi}, \hat{\lambda})|^{1/2}}{|j_{\lambda\lambda}(\psi, \hat{\lambda}_\psi)|^{1/2}} .$$

The details are provided, for example, in the appendix of Pierce and Peters (1992). In fact there are two possible saddlepoint approximations to the conditional density: one obtained from the ratio of the saddlepoint approximation to $f(s,t)$ and the saddlepoint approximation to $f(t)$, and another obtained by approximating the conditional density directly (Davison, 1986; Fraser, Reid and Wong, 1990). The first of these has been used to obtain (11), but there are indications from numerical work in Pierce and Peters (1992) and in Butler, Huzurbazar and Booth (1992) that the second leads to better approximate tail areas. The Bayesian version of the second approach is sketched in the appendix.

### 3.4 Non-normal regression

Suppose $y_i = x_i^T\beta + \sigma e_i$, where $e_i$ has a known distribution $f(e)$. In this model an exact ancillary statistic is given by $a = (a_1, \ldots, a_n)$, $a_i = (y_i - x_i^T\hat{\beta})/\hat{\sigma}$, where $(\hat{\beta}, \hat{\sigma})$ is an equivariant estimate of $(\beta, \sigma)$, for example the maximum likelihood estimate. The conditional density of $(\hat{\beta}, \hat{\sigma})$, given $a$, can be obtained exactly by renormalizing the log likelihood function (Fraser, 1979, Ch. 4 ; Barndorff-Nielsen, 1980):

$$f(\hat{\beta}, \hat{\sigma}|a; \beta, \sigma) = c|j(\hat{\beta}, \hat{\sigma})|^{1/2} \exp\{\ell(\beta, \sigma) - \ell(\hat{\beta}, \hat{\sigma})\}$$

where $c$ is a renormalizing constant, and the dependence of $\ell(\beta, \sigma)$ on $(\hat{\beta}, \hat{\sigma}, a)$ has been suppressed. As outlined in Fraser (1979, Ch.4), inference about $\psi = \beta_i$, say, is obtained from the marginal distribution of the $t$-statistic $t_i = (\hat{\beta}_i - \beta_i)/\hat{\sigma}$, and that for $\sigma$ from the marginal distribution of $\log\hat{\sigma} - \log\sigma$. A Laplace approximation to this marginal density is readily obtained, and has the same form as the Tierney and Kadane (1986) approximation to the posterior marginal, with a flat prior for $(\beta, \log\sigma)$. The result is that, for $\psi = \beta_1$, say,

$$p(\psi) = pr(T_1 \leq t_1; \psi|a)$$

is approximated by (8) or (9) with

(12) $$u = \ell'_p(\psi)\{j_p(\hat{\psi})\}^{-1/2} \frac{|\hat{j}_{\lambda\lambda}(\psi, \hat{\lambda}_\psi)|^{1/2}}{|\hat{j}_{\lambda\lambda}(\hat{\psi}, \hat{\lambda})|^{1/2}} .$$

### 3.5 Scalar parameter models

Suppose that the model has just a one-dimensional parameter $\theta$. Unless $\hat{\theta} = \hat{\theta}(y)$ is sufficient for $\theta$, as for example with an exponential family, we need some means of reducing the dimension from that of $y$, or the minimal sufficient statistic, to that of $\theta$. The most general approach is to seek a statistic $a = a(y)$ that has a distribution exactly or approximately free of $\theta$, that is, an exactly or approximately *ancillary statistic*. In a location model, the vector of residuals $(y_1 - \hat{\theta}, \ldots, y_n - \hat{\theta})$ is exactly ancillary. In a curved exponential model, an approximately ancillary statistic can be constructed by considering the full exponential family embedding model; see, e.g. Barndorff-Nielsen and Cox (1994, Ch.7) or Barndorff-Nielsen and Wood (1998). Assume this is possible, and that the resulting $p^*$-approximation of Barndorff-Nielsen (1980, 1993) is accurate to $O(n^{-3/2})$

(13) $$f(\hat{\theta}; \theta|a) = p^*(\hat{\theta}; \theta|a)\{1 + O(n^{-3/2})\}$$
$$= c\{j(\hat{\theta})\}^{1/2} \exp\{\ell(\theta) - \ell(\hat{\theta})\}\{1 + O(n^{-3/2})\} .$$

Then the $p$-value
$$p(\theta) = pr\{\hat{\Theta} \leq \hat{\theta}|a; \theta\}$$

is again approximated by (8,9) with

(14) $$u = \{\ell_{;\hat{\theta}}(\hat{\theta}) - \ell_{;\hat{\theta}}(\theta)\}\{j(\hat{\theta})\}^{-1/2}$$

where $\ell_{;\hat{\theta}}(\hat{\theta}) = \partial\ell(\theta; \hat{\theta}, a)/\partial\hat{\theta}$ is a sample space derivative of the log likelihood function for fixed $a$. The derivation of (8) or (9) from (13) is very similar to the Bayesian derivation outlined in the Appendix, and it is indicated there how sample space derivatives of the log likelihood arise in the change of variable in the integration from $\hat{\theta}$ to $r$. The approximate ancillary suggested by Barndorff-Nielsen (see, for example, Barndorff-Nielsen and Wood, 1998) is a succession of likelihood ratio statistics bridging the gap from the given model to an embedding model. An alternative approach is developed in Fraser and Reid (1995) by using a coordinate-by-coordinate pivotal quantity.

## 3.6 Discussion

The structure of the solution in the cases outlined above is identical: find an appropriate density to approximate, and approximate the cumulative distribution function of this density by an integration by parts. The density approximation is obtained from two somewhat distinct steps: first a reduction in dimension from that of the data $y$ to that of the full parameter $\theta$. In the Bayesian case this is provided by the posterior. In non-Bayesian settings it is typically obtained by conditioning on an exactly or approximately ancillary statistic. In the location regression model for example an exact ancillary is available. (An exception is the exponential family in which the initial reduction is made by sufficiency.)

The second step is the elimination of nuisance parameters, obtained in the Bayesian and location-regression model by marginalizing to the density of a one-dimensional function of $y$ and $\psi$: simply $\psi$ in the Bayesian version and a $t$-statistic in the regression model. (Again the exponential family model is the exception; the nuisance parameter is eliminated by conditioning.)

Barndorff-Nielsen (1986, 1991) gave a rather general expression for $u$, which assumes that the first reduction in dimension is achieved by conditioning on an approximately ancillary statistic, and the second reduction achieved by finding the marginal distribution of $r_\psi^*$, within this conditioning, where $r_\psi^* = r + r^{-1}\log(u/r)$ is analogous to the $t$-statistic pivotal in the location regression model. The resulting expression for $u$ is given for example as (6.108) in Barndorff-Nielsen and Cox (1994).

One further point that is most easily seen in the scalar parameter case (14) is that for calculating a $p$-value the only dependence on the approximate ancillary is through $\partial\ell(\theta; \hat{\theta}, a)/\partial\hat{\theta}$, and it is not necessary to specify the actual form of $a = a(y)$, as long as it is possible to compute this partial derivative.

This observation was used in Fraser & Reid (1995) to develop an expression for $u$ that does not require specification of the approximate ancillary statistic explicitly. This achieves the first step, of reducing the dimension from $n$ to $k$, and the second step is achieved by finding the marginal distribution of a $t$-type statistic, which is equivalent to $r_\psi^*$. This approach is described and compared to Barndorff-Nielsen's approach in Fraser, Reid and Wu (1999).

A related development is the availability of several general expressions for $u$ that ensure relative error $O(n^{-1})$ in the $p$-value approximation; two recent useful references are Barndorff-Nielsen and Wood (1998) and Skovgaard (1996).

## 4. Costs and Benefits

The biggest hurdle in computing approximation (8) or (9) is that current literature on the theory of these approximations is rather heavy going. In particular it is fairly difficult to implement general expressions for $u$ without using quite detailed knowledge about the model. Once $u$ is computed the calculation of expression (8) or (9) is quite simple. The only further computational stumbling block is the need, when computing the significance function, to obtain both the profile log likelihood function and its curvature at $\hat{\theta}_\psi$, but there are some reasonably general programs for finding restricted maximum likelihood estimates.

In special classes of models, it is possible to have general software available for computing $u$. For example Splus code for implementing (8) in binomial and Poisson regression is given in Brazzale (1998).

Of course having computable approximations is not the same as having useful approximations, and in many applications assumptions of independence, choice of model, treatment of potential outliers, sampling and so on have a much larger effect on the inference than the accuracy of the approximation. Hovever, when accurate approximations are routinely available for practical use, marked discrepancies of those from first order approximations would suggest that careful examination of modelling assumptions may be needed. And of course these are some applications where relatively few data points are available, and it is sensible in these settings to make as precise an inference as possible.

From a theoretical point of view, the similarity in structure of the various special cases leads not only to the general results of Barndorff-Nielsen (1986) and Fraser and Reid (1995), but also to insight about the essential ingredients for accurate inference. For example, in a frequentist approach to approximating $p$-values, sample space derivatives of the log likelihood function are unavoidable. This is in direct contrast to the Bayesian approach, which is conditional on all the data, and suggests that it is not possible in general to choose the prior to ensure that frequentist and Bayesian inferences agree to $O(n^{-3/2})$. This point was raised by Lindley (1992) in the discussion of Pierce and Peters (1992) and expanded in Pierce and Peters (1994). However the only sample space information about the log likelihood function that is needed for third order approximation is the log likelihood function at the data point, and its first (sample space) derivative at the data point. Other sampling information about the likelihood function is not needed, so the Bayesian and frequentist solutions are not 'very different'.

A limitation of the current work is that it is not at all clear how to accommodate models for discrete data. Detailed consideration of the role of continuity corrections is given in Pierce and Peters (1992) for discrete generalized linear models. The sample space derivative approach of Fraser and Reid (1995) requires continuous data $y$.

The assumption of independent, identically distributed data that was made in Section 2 can be relaxed to an assumption of independence, with the individual components $y_i$ depending on a common parameter through a regression type relation as in Section 3.4. It seems likely that some types of short-range dependence can also be accommodated, but we are not aware of any systematic development of this.

Another area largely unexplored is the relation of higher order asymptotic techniques to those based on sampling. DiCiccio and Efron (1996) discuss connections between bootstrap sampling and higher order approximations, but it seems likely that there is a role for higher order asymptotics in other types of sampling, such as Markov chain Monte Carlo sampling, and in designs for evaluating high dimensional integrals. The use of Laplace approximations as an adjunct to MCMC in Bayesian inference is discussed in Kass, Tierney and Kadane (1988).

The surprising accuracy of the approximations in many examples, such as those in Section 2, and in many other published examples, suggests that it is useful to continue to make these methods as widely accessible as possible.

### Acknowledgement

This work was partially supported by the Natural Sciences and Engineering Research Council.

### Appendix

*Derivation of (10):* The posterior marginal density for $\psi$ is

$$\pi_m(\psi|y) = \int \pi(\psi, \lambda|y) d\lambda$$

$$= \int \exp\{\ell(\psi, \lambda)\}\pi(\psi, \lambda) d\lambda \Big/ \int \exp\{\ell(\psi, \lambda)\}\pi(\psi, \lambda) d\psi d\lambda.$$

Laplace approximation to the integrals in both numerator and denominator lead to

$$\pi_m(\psi|y)$$

$$\doteq c \exp\{\ell(\hat{\psi}, \hat{\lambda}) - \ell(\psi, \hat{\lambda}_\psi)\}|j(\hat{\psi}, \hat{\lambda})|^{1/2}|j_{\lambda\lambda}(\psi, \hat{\lambda}_\psi)|^{-1/2}$$

(A1)

$$= c \exp\{\ell_p(\hat{\psi}) - \ell_p(\psi)\}\{j_p(\hat{\psi})\}^{1/2}|j_{\lambda\lambda}(\hat{\psi}, \hat{\lambda})|^{1/2}|j_{\lambda\lambda}(\psi, \hat{\lambda}_\psi)|^{-1/2}.$$

This result is derived in Tierney and Kadane (1986), where they show that the relative error in the approximation is $O(n^{-3/2})$. The second equality uses (1).

We now let

(A2)
$$-\frac{1}{2}r_p^2 = \ell_p(\psi) - \ell_p(\hat{\psi})$$

from which we have

$$-r_p dr_p = \ell_p'(\psi)d\psi.$$

Thus

(A3)
$$\int_{\psi^0}^{\infty} \pi_m(\psi|y)d\psi = c\int_{-\infty}^{r^0} \exp(-\frac{1}{2}r_p^2)r_p u^{-1} dr_p$$

where $u$ is as defined in (10). Both $u$ and $r_p$ are functions of $\psi$, and by expanding (10) and (A2) in Taylor series about $\psi = \hat{\psi}$ we can verify that

(A4)
$$r_p = u + \frac{A}{\sqrt{n}}u^2 + \frac{B}{n}u^3 + O(n^{-3/2}).$$

We can then complete the integration of (A3) by:

$$= \int_{-\infty}^{r_p^0} c'\phi(r_p)\{1 + r_p(u^{-1} - r_p^{-1})\}dr_p$$

(A5)
$$= c'\left\{\Phi(r_p^0) + \phi(r_p^0)\left(\frac{1}{r_p^0} - \frac{1}{u^0}\right) + \int \phi(r_p)d\left(\frac{1}{r_p} - \frac{1}{u}\right)\right\}.$$

From (A4) if follows that $d(r_p^{-1} - u^{-1}) = \{(B - A^2)/n\}dr + O(n^{-1})$, and from Tierney and Kadane (1986) that $c' = 1 + O(n^{-1})$. Taking the limit as $r_p^0 \longrightarrow \infty$ shows that $c' = 1 + (B - A^2)n^{-1} + O(n^{-3/2})$ and (A5) simplifies to

$$\left\{\Phi(r_p^0) + \phi(r_p^0)\left(\frac{1}{r_p^0} - \frac{1}{u^0}\right)\right\}\left\{1 + O(n^{-3/2})\right\}$$

where $r_p^0 = r_p(\psi^0) = \pm[2\{\ell_p(\hat{\psi}) - \ell_p(\psi^0)\}]^{1/2}$ and $u = u(\psi^0)$.

Various versions of this derivation appear in several papers in the literature, including DiCiccio, Field and Fraser (1990), Barndorff-Nielsen (1991), Pierce and Peters (1992), and Reid (1996). The Bayesian version is given in Sweeting (1995) as well as DiCiccio and Martin (1993). The integration by parts step leading to (A4) was originally proposed by Temme (1982); see also Barndorff-Nielsen and Cox (1990, Ch. 3).

A related approximation is obtained by rewriting the integrand in (A3) as

$$\exp\left\{\ell_p(\psi) - \ell_p(\hat\psi) - \frac{1}{2}\log|j_{\lambda\lambda}(\psi,\hat\lambda_\psi)| \right.$$
$$\left. + \frac{1}{2}\log|j_{\lambda\lambda}(\hat\psi,\hat\lambda)|\right\} = \exp\{\ell_a(\psi) - \ell_a(\hat\psi)\},$$

say. We now use the results $\hat\psi_a = \hat\psi + O_p(n^{-1})$, $j_p(\hat\psi) = j_a(\hat\psi) + O_p(n^{-1})$ to obtain

(A6)
$$\int_{\psi^0}^\infty \pi_m(\psi|y)d\psi = \left\{\Phi(r_a^0) + \phi(r_a^0)\left(\frac{1}{r_a^0} - \frac{1}{u_a^0}\right)\right\}\{1 + O(n^{-3/2})\}$$

where

$$-\frac{1}{2}r_a^2(\psi) = \ell_a(\psi) - \ell_a(\hat\psi_a)$$

$$u_a(\psi) = \ell_a'(\psi)\{j_a(\hat\psi_a)\}^{-1/2}\frac{\pi(\hat\psi,\hat\lambda)}{\pi(\psi,\hat\lambda_\psi)}.$$

This is the Bayesian version of the asymptotically equivalent version mentioned at the end of Section 3.3. It is more difficult to compute, as $j_a(\psi) = -\ell_a''(\psi)$ involves a fourth derivative of the full loglikelihood. Fraser, Reid and Wong (1990) treat the exponential family case using numerical derivatives. Numerical evidence indicates though that (A6) is more accurate in problems with many nuisance parameters. Pierce and Peters (1992) first clarified the difference between (A5) and (A6), and gave several numerical examples for the exponential family version.

To obtain (14) from (13) involves a similar integration by parts, but now we compute

$$\int p^*(\hat\theta;\theta|a)d\hat\theta$$

using the change of variable

$$-\frac{1}{2}r^2 = \ell(\theta;\hat\theta,a) - \ell(\hat\theta;\hat\theta,a)$$

leading to

$$-rdr = \{\ell_{;\hat\theta}(\theta;\hat\theta,a) - \ell_{;\hat\theta}(\hat\theta;\hat\theta,a)\}d\hat\theta.$$

## References

[1] Barndorff-Nielsen, O.E. (1980), Conditionality resolutions. *Biometrika*, **67**, 293–310.

[2] Barndorff-Nielsen, O.E. (1983), On a formula for the distribution of the maximum likelihood estimator. *Biometrika*, **70**, 343–365.

[3] Barndorff-Nielsen, O.E. (1986), Inference on full or partial parameters based on the standardized signed log likelihood ratio. *Biometrika*, **73**, 307–322.

[4] Barndorff-Nielsen, O.E. (1991), Modified signed log likelihood ratio. *Biometrika*, **78**, 557–564.

[5] Barndorff-Nielsen, O.E. and Chamberlin, S.R. (1991), An ancillary invariant modification of the signed log likelihood ratio. *Scand. J. Statist.*, **18**, 341–352.

[6] Barndorff-Nielsen, O.E. and Cox, D.R. (1990) *Asymptotic Techniques for Use in Statistics*, Chapman and Hall, London.

[7] Barndorff-Nielsen, O.E. and Cox, D.R. (1994) *Inference and Asymptotics*, Chapman and Hall, London.

[8] Barndorff-Nielsen, O.E. and Wood, A.T. (1998), On large deviations and choice of ancillary for $p^*$ and $r^*$. *Bernoulli*, **4**, 35–63.

[9] Brazzale, A. (1998) Approximate conditional inference in logistic and loglinear models, *J. Comp. Graph. Statist.*, to appear.

[10] Butler, R.W., Huzurbazar, S., and Booth, J.G. (1992), Saddlepoint approximations for the Bartlett-Nanda-Pillai trace statistic in multivariate analysis. *Biometrika*, **79**, 705–716.

[11] Cox, D.R. and Hinkley, D.V. (1974) *Theoretical Statistics*, Chapman and Hall, London.

[12] Davison, A.C. (1986). *Biometrika*

[13] DiCiccio, T.J. and Efron, B. (1995)

[14] DiCiccio, T.J., Field, C.A., and Fraser, D.A.S. (1990), Approximations of marginal tail probabilities and inference for scalar parameters.. *Biometrika*, **77**, 77–95.

[15] DiCiccio, T.J. and Martin, M.A. (1993), Simple modifications for signed roots of likelihood ratio statistics. *J. R. Statist. Soc. B*, **55**, 305–316.

[16] Fraser, D.A.S. (1979) *Inference and Linear Models*, McGraw Hill, New York.

[17] Fraser, D.A.S. and Reid, N. (1995), Ancillaries and third order significance. *Util. Math.*, **47**, 33–53.

[18] Fraser, D.A.S., Reid, N. and Wong, A. (1990) *J. R. Statist. Soc. B*

[19] Fraser, D.A.S., Reid, N. and Wu, J. (1999) A simple general formula for tail probabilities for frequentist and Bayesian inference. *Biometrika*, to appear.

[20] Kass, R.E., Tierney, L.J. and Kadane, J.B. (1988) Asymptotics in Bayesian computation, in: *Bayesian Statistics* **III**, Bernardo, J.M., DeGroot, M.H., Lindley, D.V. and Smith, A.F.M. eds. Clarendon Press, Oxford.

[21] Lehmann, E.L. (1983) *Theory of Point Estimation*, J. Wiley & Sons, New York.

[22] Lehmann, E.L.(1986) *Testing Statistical Hypotheses 2nd ed.*, J. Wiley & Sons, New York.

[23] Lindley, D.V. (1992), in Practical use of higher order asymptotics for multiparameter exponential families (with discussion). *J. R. Statist. Soc. B*, **54**, 728.

[24] Lugannani, R. and Rice, S.O. (1980), Saddlepoint approximation for the distribution of the sum of independent random variables. *Adv. Appl. Probab.*, **12**, 475–490.

[25] Pierce, D.A. and Peters, D. (1992), Practical use of higher order asymptotics for multiparameter exponential families (with discussion). *J. R. Statist. Soc. B*, **54**, 701–738.

[26] Pierce, D. A. & Peters, D. (1994), Higher order asymptotics and the likelihood principle: one parameter models. *Biometrika*, **81**, 1–10.

[27] Reid, N. (1995), The roles of conditioning in inference (with discussion). *Statist. Sci.*, **10**, 138–157.

[28] Reid, N. (1996), Likelihood and higher-order approximations to tail areas: A review and annotated bibliography. *Canad. J. Statist.*, **24**, 141–166.

[29] Skovgaard, I.M. (1996), An explicit large-deviation approximation to one parameter tests. *Bernoulli*, **2**, 145–165.

[30] Sweeting, T.J. (1995) Approximate Bayesian computation based on signed roots of log-density ratios, in: J. M. Bernardo, J. O. Berger, A. P. Dawid & A. F. M. Smith, eds, *Bayesian Statistics V* Oxford University Press., 427–444.

[31] Temme, N.M. (1982), The uniform asymptotic expansion of a class of integrals related to cumulative distribution functions. *SIAM J. Math. Anal.*, **3**, 221–249.

[32] Tierney, L.J. and Kadane, J.B. (1986), Accurate approximation for posterior moments and marginal densities. *J. Am. Statist. Assoc.*, **81**, 82–87.

# The Analysis of Subject Specific Agreement

D. A. Sprott

Department of Statistics and Actuarial Science
University of Waterloo
Waterlo, Ontario, Canada
and
Centro de Investigación en Matemáticas
Guanajuato, Mexico


V. T. Farewell

Department of Statistical Science
University College
London, England

**Abstract.** Two raters independently assign $n$ subjects to $q$ categories. The subject specific agreement between the two raters is defined to be the extent to which the probability of the assignment of a given subject to a category by one rater depends upon, or is determined by, the category to which the same subject is assigned by the other rater. A measure $\nu$ of subject specific agreement is proposed, based on conditional probabilities and related to the log odds ratio, for which there is a conditional likelihood function. This last point is of paramount importance for making quantitative statements of scientific inference. The foregoing is compared with the use of the traditional measure of agreement $\kappa$. The proposed procedures are an example of how the increasing power of computers, characteristic of the late 20th and presumably 21st century, should affect statistical methods and scientific inference.

## 1. Introduction

Consider two judges or raters who assign $n$ items or subjects independently to $q$ categories $C_1, \ldots, C_q$. The resulting observations form a $q \times q$ contingency table of frequencies $\{x_{ij}\}$, $i, j = 1, \ldots, q$, where $x_{ij}$ is the number of subjects assigned to categories $C_i$, $C_j$ simultaneously by Raters 1 and 2 respectively. Suppose the $\{x_{ij}\}$ have the multinomial distribution $\{p_{ij}\}$, where $\sum_{i,j} p_{ij} = 1$, and that the marginal row and column probabilities are $\{p_{i.}\} = \{\sum_j p_{ij}\}$, $\{p_{.j}\} = \{\sum_i p_{ij}\}$, respectively. The problem is to assess the agreement between the raters in the assignment of subjects to categories.

However, the term "agreement" is rather broad. In the present context there are two principal kinds of agreement, which can be termed "subject specific" and "marginal".

Subject specific agreement is the extent to which the probability of the assignment of a *specific* subject to a category by one rater is determined by the category to which the *same* subject is assigned by the other rater. This form of agreement determines predictability – the extent to which the assignment of a subject to a category by one rater can be predicted from the assignment of the *same* subject by the other rater. Subject specific agreement leads to a measure $\nu$ based on the *conditional* probabilities

(1) $$p_{j|i} = P(C_j \text{ by Rater } 2|C_i \text{ by Rater } 1) = P(j|i) = p_{ij}/p_{i.}.$$

When the raters have different status, the asymmetry implied by conditional probabilities is natural. Examples are disease screening and clinical testing. Often, however, the raters are of equal status and hence symmetrically related. In this case a measure of subject specific agreement should be invariant under an interchange of the raters.

Marginal agreement is the extent to which the marginal row probabilities $p_{i.}$ differ from the corresponding marginal column probabilities $p_{.i}$, $i = 1 \ldots q$. This form of agreement determines the extent to which the raters classify subjects similarly on the average.

The most widely used measure of agreement is kappa introduced and defined by Cohen (1960). Kappa is defined as

$$\kappa = \frac{\sum p_{ii} - \sum p_{i.}p_{.i}}{1 - \sum p_{i.}p_{.i}},$$

This definition was initially designed to ensure that chance agreement ($p_{ij} = p_{i.}p_{.j}$) produces $\kappa = 0$, and to ensure that perfect agreement, $\sum p_{ii} = 1$, produces $\kappa = 1$ which is its maximum value. It is well known that kappa depends on both subject specific and marginal agreement, (unless $\kappa = 0$), as will be exemplified shortly.

In developing measures of agreement such as $\kappa$, the emphasis has been essentially empirical, that is, on the ability of a population parameter and its estimate, possibly supplemented by an estimated asymptotic standard error, to capture the essence of what is meant by agreement. Notably lacking has been a recognition of the principles of statistical inference applied to the structure of a statistical model, in the present case the multinomial likelihood, and to the various substructures to which it can lead depending on the circumstances. This can only impair the ability to handle the practical consequences of this type of data. In particular, cases of scientific interest often yield tables with small or zero off-diagonal frequencies indicative of high agreement, but where the agreement may not be as high as expected. See, for example Sackett et al. (1991, pp. 25-26).

The purpose of this paper is to discuss a measure $\nu$ of subject specific agreement separate from marginal agreement. The principles of statistical inference related to this problem are emphasized, utilizing the structure of the multinomial model. This leads to a likelihood function of $\nu$ based on the observed data, a point of paramount importance for the making of efficient quantitative statements of plausibility, (Fisher 1973, pp. 71-76) *and for the combination of evidence from repeatable experiments.* The shape of the likelihood function indicates the form estimation statements should take, and the adequacy of large sample procedures based on asymptotic considerations.

The above points are exemplified initially by the case of $q = 2$ categories. Then a generalization to $q$ categories is discussed.

## 2. Two Categories

### 2.1 Subject Specific Agreement

The observed frequencies $\{x_{ij}\}$ are assumed to have the $2 \times 2$ multinomial distribution with index $\sum x_{ij} = n$ and probabilities $p_{ij}$, $\sum p_{ij} = 1$, $i, j = 1, 2$,

Subject specific agreement focuses on the conditional odds corresponding to the conditional probabilities (1)

$$\alpha_{ij} = \log\{\text{odds}[C_i : C_j \text{ for Rater 2}|C_i \text{ by Rater 1}]\}$$

(2)
$$= \log(p_{i|i}/p_{j|i}), \quad \alpha_{ii} = 0, \quad -\infty \leq \alpha_{ij} \leq \infty,$$

so that

$$\exp(\alpha_{12}) = \text{odds}\{(C_1 : C_2) \text{ for Rater 2} \mid C_1 \text{ by Rater 1}\}$$
$$\exp(-\alpha_{21}) = \text{odds}\{(C_1 : C_2) \text{ for Rater 2} \mid C_2 \text{ by Rater 1}\}.$$

A high degree of subject specific agreement requires that $\exp(\alpha_{12})$ be large and $\exp(-\alpha_{21})$ be small. The reverse is also possible, yielding strong negative agreement, that is, strong disagreement, which equally allows Rater 2 to be predicted from Rater 1. The simplest and most convenient parameter measuring this form of agreement is

$$\nu = \alpha_{12} + \alpha_{21} = \log[p_{1|1}p_{2|2}/(1 - p_{1|1})(1 - p_{2|2})]$$
$$= \log(p_{1|1}p_{2|2}/p_{2|1}p_{1|2})$$
$$(3) \qquad\qquad = \log(p_{11}p_{22}/p_{21}p_{12}).$$

the log odds or cross ratio. Since $\nu$ is invariant under interchange of the raters, its operational interpretation is that the odds of a specific subject being classified in category $C_1$ versus $C_2$ by one rater are $\exp(\nu)$ times greater if the same subject was also classified in $C_1$ rather than in $C_2$ by the other rater.

The important feature of $\nu$ is the existence of a likelihood function. Inferences about $\nu$ will be based on the conditional likelihood $L_c(\nu)$ obtained from the conditional probability of $x_{11}$ given the row and column totals $r = x_{11} + x_{12}$ and $c = x_{11} + x_{21}$,

$$(4) \qquad L_c(\nu; x_{11}, r, c) \propto P(x_{11}; \nu | r, c) =$$
$$\binom{r}{x_{11}}\binom{n - r}{c - x_{11}} e^{\nu x_{11}} \Big/ \sum_i \binom{r}{i}\binom{n - r}{c - i} e^{i\nu}$$

obtained by Fisher (1935, p. 50). It is convenient to standardize the likelihood with respect to its maximum to give the relative conditional likelihood $R_c(\nu) = L_c(\nu)/L_c(\hat{\nu})$, so that $0 \leq R_c(\nu) \leq 1$, and $\hat{\nu}$ is the conditional maximum likelihood estimate. The likelihood "supplies a natural order of preference among the possibilities under consideration" [in this case possible values of $\nu$], Fisher (1973, p. 73). In addition likelihoods based on (4) can be given a probabilistic interpretation in terms of $P$-values or confidence intervals based on tail probabilities.

The likelihood (4) is exact for cell frequencies of any size. This last point is important since, as mentioned earlier, strong agreement, which is the case of highest scientific interest, results in small or zero frequencies off the diagonal. This leads to highly non-normal likelihoods which impair the inferential use of standard maximum likelihood estimation and other procedures depending on asymptotic normal approximations. Of equal importance, information from independent samples pertaining to the same $\nu$, independently of the other parameters in (2), can easily be combined via $\sum \log L_c(\nu; x_{11i}, r_i, c_i)$.

Inferences about $\nu$ can be transformed into inferences about any 1-1 function of $\nu$ by straightforward algebraic substitution. Two parameters of particular interest are the odds ratio, $\exp \nu$, and $\tau = (e^\nu - 1)/(e^\nu + 1)$. The reason for $\tau$ is that its range is $(-1, 1)$. It is one of a family of such measures described in detail by Cook and Farewell (1995). Thus the log odds ratio satisfies all the desiderata discussed in Section 1 for the practical use of a measure of subject specific agreement.

These points are illustrated in the following examples.

**Example 1.** The data in Table 1 are taken from Sackett et al. (1991, p. 26), on the test-retest reliability of a clinician classifying the same set of 100 fundus photographs on two occasions three months apart.

| First examination | Second examination | | Total |
|---|---|---|---|
| | Little or no retinopathy | Moderate or severe retinopathy | |
| Little or no retinopathy | 69 | 11 | 80 |
| Moderate or severe retinopathy | 1 | 19 | 20 |
| Total | 70 | 30 | 100 |

Table 1. Test-Retest Reliability, Example 1.

The conditional likelihood (4) is

$$L_c(\nu; x_{11} = 69, r = 80, c = 70) \propto P(x_{11} = 1; \nu | r = 80, c = 70)$$

$$= \binom{80}{69}\binom{20}{1} e^{69\nu} \bigg/ \sum_{i=50}^{70} \binom{80}{i}\binom{20}{70-i} e^{i\nu}.$$

The resulting conditional relative likelihood, $R_c(\nu)$, is shown in Figure 1.

This is an example where the agreement or reliability is expected to be high, but is not as high as expected, considering that this was a medical expert being tested against himself, and who knew that a reliability assessment was in progress, as discussed by Sackett et al. (1991 pp. 25-26). Under this circumstance, as mentioned in Section 1, a set of lower bounds on plausible values of $\nu$ in the light of the data is preferable to a single number such as the maximum conditional likelihood estimate $\hat{\nu} = 4.70$. For example, $\nu = 3.13, 2.88, 2.67$ are lower 0.20, 0.10, 0.05 conditional relative likelihood bounds, having $P$-values $P(x_{12} \geq 69; \nu = 3.13, 2.88, 2.67) = .093, .045, .022$. While $\hat{\nu}$ gives the odds of a specific subject being classified as $C_1$ or $C_2$ on the second examination as $\exp(4.70) = 110$ times greater if he was so classified on the first examination, the lower bounds suggest that with reasonable plausibility this ratio could be as low as $\exp(3.13) = 22.9$, $\exp(2.88) = 17.8$, or even only $\exp(2.67) = 14.4$ times greater. The corresponding values of $\tau$ are .92, .89, .87. These last three facts seem of more interest than the maximum likelihood or some other single estimate, at least to a patient who is about to be diagnosed. However it is the likelihood function $R_c(\nu)$ shown in Figure 1. that portrays all the sample information about $\nu$. The above facts are merely a condensation of some of its salient features. The asymmetry of $R_c(\nu)$ is apparent.

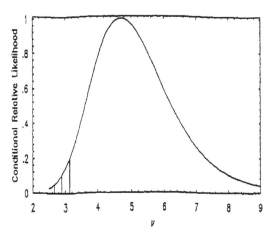

Figure 1.   Relative conditional likelihood $R_c(\nu)$, Example 1

The usual treatment of this type of problem is to quote the maximum likelihood estimate of $\kappa$, as in Sackett et al. (1991 p. 30), which is $\hat{\kappa} = .684$ in Table 1. Sometimes its estimated asymptotic standard error is also given, which is $s = .0896$ in Table 1. But the small frequency of 1 in the third cell suggests that the same skewness that affects $\nu$ also affects $\kappa$, so that $(\hat{\kappa}, s)$ does not provide an adequate summary of the sample information.

**Example 2.** The data in Table 2 are the frequencies with which a given subject classified a given relative as a problem drinker (PD) or as a non-problem drinker (NPD) on the presentation of a family tree diagram on two separate occasions six months apart, Sprott and Vogel-Sprott (1987). There are 24 subjects each yielding a $2 \times 2$ table

In row $i$ of Table 2, column 1 gives the number $N_i$ of subjects yielding the $2 \times 2$ table designated by columns 2 to 5. Column 6 gives the resulting conditional likelihood $R_{c_i}(\nu)$ (4) based on this table. The combined likelihood based on all 24 subjects, assuming that $\nu$ is the same for all subjects, is $R(\nu) = \prod R_{c_i}^{N_i}$. This is shown in Figure 2. Also shown are the nine distinct conditional likelihoods $R_{c_i}(\nu)$ listed in Table 2. These are closely grouped, supporting the assumption that $\nu$ is the same for all subjects.

|           | Trial 2 | |
|-----------|---------|-----------------|
| Trial 1   | PD      | NPD             |
| PD        | $x_i$   | $r_i - x_i$     |
| NPD       | $c_i - x_i$ | $n_i - c_i + x_i$ |

| $N_i$ | $x_i$ | $r_i - x_i$ | $c_i - x_i$ | $n_i - c_i + x_i$ | $R_{c_i}(\nu)$ |
|-------|-------|-------------|-------------|-------------------|----------------|
| 1 | 1 | 0 | 0 | 1 | $e^\nu/(1 + e^\nu)$ |
| 5 | 1 | 0 | 0 | 2 | $e^\nu/(2 + e^\nu)$ |
| 4 | 1 | 0 | 0 | 3 | $e^\nu/(3 + e^\nu)$ |
| 1 | 3 | 0 | 0 | 1 | $e^\nu/(3 + e^\nu)$ |
| 4 | 1 | 0 | 0 | 4 | $e^\nu/(4 + e^\nu)$ |
| 2 | 1 | 0 | 0 | 5 | $e^\nu/(5 + e^\nu)$ |
| 1 | 1 | 0 | 0 | 6 | $e^\nu/(6 + e^\nu)$ |
| 2 | 1 | 0 | 0 | 7 | $e^\nu/(7 + e^\nu)$ |
| 1 | 2 | 0 | 0 | 3 | $e^{2\nu}/(3 + 6e^\nu + e^{2\nu})$ |
| 1 | 1 | 0 | 1 | 3 | $2e^\nu/(3 + 2e^\nu)$ |
| 1 | 1 | 1 | 0 | 3 | $2e^\nu/(3 + 2e^\nu)$ |
| 1 | 1 | 0 | 1 | 2 | $e^\nu/(1 + e^\nu)$ |

Table 2. Test-Retest Reliability, Example

In this case the relative likelihoods $R_c(\nu)$ are themselves tail probabilities, since for each subject, conditional on the marginal totals, the observed table is the most extreme table possible. The tails therefore consist of the observed table only.

From Figure 2. it can be seen that each individual subject contributes little information about reliability. It is the accumulated evidence from all 24 subjects that is crucial. For example, the hypothesis of no reliability based on the accumulated evidence is highly implausible, $R(\nu = 0) = 1.59 \times 10^{-15}$. This means that either $\nu > 0$ or the combined experiments have a probability less than $1.59 \times 10^{-15}$. But no single likelihood in Figure 2. suggests that $\nu = 0$ is implausible.

Similarly, $R(\nu = 2.74, 3.21, 3.49) = .01, .05, .10$ give a set of lower plausible bounds for reliability. For example, at the level of 5% plausibility the odds of a given subject classifying any given relative as a problem versus a non-problem drinker on the second occasion are at least $\exp(3.21) \approx 25$ times greater if the same relative was also classified as a problem rather than as a non-problem drinker on the first occasion. A probabilistic interpretation is: either the above is true or the observed experiments have a probability less than 5%. The corresponding values of $\tau$ are .88, .92, .94.

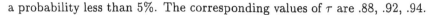

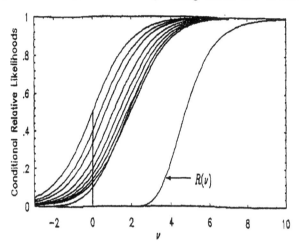

Figure 2.    Relative conditional likelihoods $R_c(\nu)$ and combined relative likelihood $R(\nu)$, Example 2

The combination of evidence here cannot be achieved by pooling the data for all subjects into one table. To do this would require assuming additionally that the remaining "nuisance" parameters, one for each subject in (3), are the same for all subjects. This makes it impossible to apply $\kappa$ to this problem. For each subject the estimate $\hat{\kappa}$ is 1 with an estimated asymptotic standard error 0.

The effect of the marginal frequencies on inferences about $\nu$ will be discussed next.

## 2.2 Marginal Frequencies:  Marginal Agreement and Marginal Balance

**Example 3.** The three samples of Table 3 were chosen to have the same log odds ratio, the maximum conditional likelihood estimates being from (4) $\hat{\nu} \approx 1.93$. The samples are of size 500 in order to mitigate the extreme discreteness in Table 1, so that the margins can be varied while keeping $\hat{\nu}$ essentially constant. Thus the samples in Table 3 differ only in respect of their marginal rates of classification. Tables 3(a) and 3(b) have the same total number $r + c = 800$ of marginal row 1 plus column 1 frequencies; they differ only in their disposition, 474:326 versus 400:400. Table 3(c)

| | Second exam | | | | | | | | |
|---|---|---|---|---|---|---|---|---|---|
| | (a) Sample 1 | | | (b) Sample 2 | | | (c) Sample 3 | | |
| First exam | − | + | Total | − | + | Total | − | + | Total |
| − | 320 | 154 | 474 | 350 | 50 | 400 | 181 | 69 | 250 |
| + | 6 | 20 | 26 | 50 | 50 | 100 | 69 | 181 | 250 |
| Total | 326 | 174 | 500 | 400 | 100 | 500 | 250 | 250 | 500 |

Table 3. Frequencies in Three Samples

has $r = c = n/2 = 250$. Thus Table 3(a) exhibits some marginal disagreement. Tables 3(b) and 3(c) both exhibit perfect marginal agreement; they differ only in respect of their marginal balance, 400:100 and 250:250. Table 3(c) exhibits both perfect marginal agreement and perfect marginal balance.

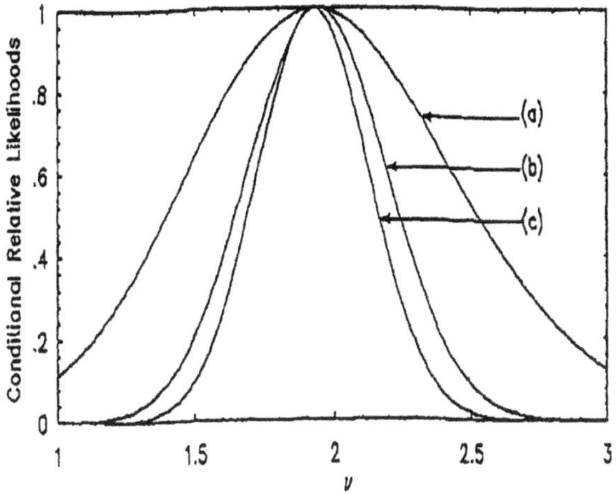


Figure 3. Relative conditional likelihoods of Example 3

The precision or spread of $R_c(\nu)$ about $\hat{\nu}$ is influenced by the margins. This feature of $R_c(\nu)$ is enhanced by both marginal agreement and by marginal balance, as exemplified in the likelihood functions arising from Table 3. These are centered around the common $\hat{\nu}$ and are of decreasing width (that is, nested, with increasing precision) across Tables 3 (a), (b), (c) respectively, as shown in Figure 3.[1]

[1] It can be shown that among all tables having a specified $\hat{\nu}=\log(x_{11}x_{22}/x_{12}x_{21})$ and $r+c$, the local precision in the neighbourhood of $\hat{\nu}$

$$(I^{\nu\nu})^{-1} = \left[\sum(1/x_{ij})\right]^{-1},$$

Thus in the above inferences based on $R_c(\nu)$, the roles of subject specific versus marginal agreement and marginal balance are clearly separated. Subject specific agreement is measured by $\hat{\nu}$, which determines the position of the likelihood $R_c(\nu)$ in the sense that $\hat{\nu}$ is included in all of the likelihood intervals. The observed marginal agreement and marginal balance determine the precision of these intervals.

If marginal agreement is of interest in its own right, a suitable function of the parameters such as $\beta = \log(p_{12}/p_{21})$, could be devised to measure marginal agreement. The marginal distribution of the marginal totals could then be used to assess $\beta$. The relationship between subject specific agreement, marginal agreement and marginal balance is exhibited by the structure of the multinomial likelihood function

$$(5) \quad P(\{x_{ij}\}; \nu, \beta, \rho) = P(x_{11}; \nu | r, t) P(r; \nu, \beta | t) P(t; \nu, \beta, \rho), \quad t = r + c,$$

Cook and Farewell (1994). From (5) it is apparent that subject specific agreement is primary. Inferences about $\nu$ do not depend on the remaining parameters. Inferences about $\beta$ must be conditioned on $\nu$, and about $\rho$ must be conditional on both $\nu$ and $\beta$.

The behaviour of $\kappa$ is quite different. The maximum likelihood estimates are $\hat{\kappa} = .120, .375,$ and $.448$ in Samples 1, 2, and 3, of Tables 3(a), 3(b), and 3(c), respectively. The increase from Table 3(a) to 3(b) can be explained by the dependence of $\kappa$ on marginal agreement. But this can hardly explain the increase from Table 3(b) to 3(c), which differ only in the marginal balance, both exhibiting perfect marginal agreement. This is what can be expected when the likelihood structure (5) is ignored, as is done by kappa. It results in confounding subject specific agreement, marginal agreement, and marginal balance, all without distinction, under the common umbrella of "agreement". Curiously, kappa's proponents defend kappa on this ground and criticize the log odds ratio for not doing so, e. g. Shrout et al. (1987), indicative of the logical level at which agreement has been discussed.

The main difficulty facing $\nu$ has been its limitation to two categories. The following sections extend the foregoing to $q$ categories.

---

where $I^{\nu\nu}$ is the relevant element pertaining to the parameter $\nu$ in the inverse $I^{-1}$ of the information matrix, is a local maximum when there is complete marginal agreement, $r = c$ as exemplified in Tables 3(a) and 3(b). Conditional on this total marginal agreement, the overall maximum is attained when there is complete marginal balance, $r = c = n/2$, as in Table 3(c).

## 3. Generalization to $q$ Categories

### 3.1 The mean $\bar{\nu}$ of the log odds ratios

The following generalization amplifies and extends that of Chamberlin and Sprott (1991). Extend (2) to $i, j = 1, \ldots, q$, and (3) to

(6) $$\nu_{ij} = \alpha_{ij} + \alpha_{ji}, \qquad i < j,$$

the log odds ratio of the $2 \times 2$ table $(x_{ii}, x_{ij}; x_{ji}, x_{jj})$ about the main diagonal. The definition of agreement $\nu$ in (3) can be extended to

(7) $$\bar{\nu} = 2\nu/q(q-1), \quad -\infty < \bar{\nu} < \infty, \qquad \text{where} \quad \nu = \sum \alpha_{ij} = {\sum}' \nu_{ij}.$$

Because of definition (6), all sums involving the $\nu_{ij}$ are over $i < j$. Such sums will be denoted by $\sum'$ as in (7).

The mean $\bar{\nu}$ is the mean of the log odds ratios $\nu_{ij}$ arising from the $q(q-1)/2$ distinct pairs of categories. The quantity $\exp(\bar{\nu})$ is the geometric mean of these odds ratios over all $q(q-1)/2$ pairs of categories. Since $\nu$ is invariant under permutations of the categories and under interchange of the raters, its operational interpretation is that the geometric mean of the odds of a specific subject being classified as $C_i$ versus $C_j$, $i < j$, by one rater is $\exp(\bar{\nu})$ times greater if the same subject was classified in $C_i$ rather than in $C_j$ by the other rater, $i, j = 1$ to $q$, $i < j$.

The analysis is more conveniently expressed in terms of $\nu$. The conditional distribution of $\{x_{ij}\}$ given the row totals, $\{r_i = \sum_j x_{ij}\}$, is the product of $q$ $q$-variate multinomial distributions with indices $\{r_i\}$ and conditional probability parameters given by (1) and (2)

$$P(\{x_{ij}\}; \{\alpha_{ij}\} \mid \{r_i\}) = \prod_{i=1}^{q} r_i! \prod_{i,j=1}^{q} p_{j|i}^{x_{ij}} / x_{ij}!$$

(8) $$= \exp\left(-\sum \alpha_{ij} x_{ij}\right) \prod r_i! p_{i|i}^{r_i} \Big/ \prod x_{ij}!,$$

From (6), the distribution of $\{x_{ij}\}$, $i < j$, and

(9) $$b_{ij} = x_{ji} - x_{ij}, \quad (i < j),$$

is

$$P(\{x_{ij}\}, \{b_{ij}\}; \{\nu_{ij}\}, \{\alpha_{ij}\} \mid \{r_i\}) =$$
$$\exp\left[-{\sum}'(\nu_{ij} x_{ij} + \alpha_{ji} b_{ij})\right] \prod r_i! p_{i|i}^{r_i} \Big/ \prod x_{ij}!$$

The conditional probability of $\{x_{ij}\}$, $(i < j)$, given $\{b_{ij}\}$ and $\{r_i\}$, is thus

$$(10) \quad P(\{x_{ij}\}; \{\nu_{ij}\} \mid \{b_{ij}\}, \{r_i\}) \propto \prod r_i! \exp\left(-\sum\nolimits' \nu_{ij} x_{ij}\right) \Big/ \prod x_{ij}!$$

Letting

$$(11) \qquad\qquad a_{ij} = x_{ij} - x_{12}, \quad (i \neq j, \ a_{12} \equiv 0),$$

so that $b_{ij} = a_{ji} - a_{ij}$, it is easily seen from (10) that the conditional distribution of $x_{12}$ given $\{a_{ij}\}$, $\{r_i\}$, depends only on $\nu$ and so yields the conditional likelihood of $\nu$

$$(12) \qquad L_c(\nu; x) \propto P(x_{12}; \nu \mid \{a_{ij}\}, \{r_i\}) =$$

$$K(x_{12}) \exp(-\nu x_{12}) \Big/ \sum_h K(h) \exp(-\nu h)$$

where

$$(13) \qquad \left.\begin{array}{l} K(h) = 1/\prod d_{ij}(h)!, \\ d_{ij}(h) = a_{ij} + h, \qquad (i \neq j), \\ d_{ii}(h) = r_i - \sum_{j \neq i} d_{ij}(h). \end{array}\right\}$$

The sum in (12) is over all $h$ for which the quantities (13) are non-negative. The $\{d_{ij}(h)\}$ form the conditional reference set of $q \times q$ tables containing the observed table, upon which the conditional probability distribution is based. Since $\{a_{ij}\}$ and $\{r_i\}$ are constant, the column totals of the $q \times q$ table are also constant.

The conditional likelihood $L_c(\bar{\nu})$ can be immediately obtained. When $q = 2$ the results of Section 2.1 are are obtained. With higher dimensional tables a difficulty arises when the resulting conditional reference set contains only a single point and so is uninformative.

## 3.2 Decomposition of $\nu$

A convenient feature of $\nu$ as defined by (7) is the simplicity with which it can be decomposed into its component parts $\nu_{ij}$. These are the individual measures of agreement on the $q(q-1)/2$ pairs of categories $C_i$, $C_j$. They will be of interest when a single measure of agreement is misleading. For example, $\bar{\nu} \approx 0$ can arise because the $\nu_{ij}$ are numerically large, half of them being negative and the remaining half positive. Similarly, any cell $(i, j)$ containing $x_{ij} = 0$ produces an infinite maximum likelihood estimate of the corresponding component $\nu_{ij}$. An isolated zero may be thought to exert an undue effect on the sum, so that it would be desirable to examine other linear functions of subsets of the $\nu_{ij}$.

In principle, any linear function $\delta$ of any subset of the $\nu_{ij}$ can be treated in the same way as $\bar{\nu}$ of Section 3.1. It is merely necessary to eliminate one of the $\nu_{ij}$'s in terms of $\delta$ and the remaining $\nu_{ij}$'s. The unwanted $\nu_{ij}$'s can then be eliminated by conditioning on their coefficients in (10), giving a conditional distribution like (12) depending on $\delta$ only.

For example, the parameter

$$(14) \qquad \delta = \nu_{12} - \frac{1}{2}(\nu_{13} + \nu_{23}),$$

will be relevant in Example 4, and a similar one in Example 5. It is the difference in agreement for categories 1 versus 2 and the mean agreement for categories 1 versus 3 and for categories 2 versus 3. Using (14) to eliminate $\nu_{12}$, the exponent in the conditional distribution (10) can be written

$$-{\sum}' \nu_{ij} x_{ij} = -\delta x_{12} - \nu_{13} t_1 - \nu_{23} t_2 - \sum_{i,j>3}{}' \nu_{ij} x_{ij},$$

where

$$(15) \qquad t_1 = x_{13} + \frac{1}{2} x_{12}, \quad t_2 = x_{23} + \frac{1}{2} x_{12}.$$

Conditioning on $x_{ij}$, $i, j > 3$ eliminates the $\nu_{ij}$, $i, j > 3$. This confines attention to the $3 \times 3$ table containing the first three categories. Conditioning additionally on $t_1$ and $t_2$ eliminates $\nu_{13}$ and $\nu_{23}$. This yields the conditional probability of $x_{12}$ as a function of $\delta$ only

$$(16) \quad P(x_{12}; \delta | \{b_{ij}\}, t_1, t_2) = K(x_{12}) \exp(-\delta x_{12}) \Big/ \sum_h K(h) \exp(-\delta h),$$

| | $C_1$ | $C_2$ | $C_3$ | Total |
|---|---|---|---|---|
| $C_1$ | $--$ | $h$ | $t_1 - \frac{1}{2} h$ | $r_1{}'$ |
| $C_2$ | $b_{12} + h$ | $--$ | $t_2 - \frac{1}{2} h$ | $r_2{}'$ |
| $C_3$ | $b_{13} + t_1 - \frac{1}{2} h$ | $b_{23} + t_2 - \frac{1}{2} h$ | $--$ | $r_3{}'$ |

Table 4.  The Reference Set of Tables for $\delta$

where $K(h)$ are the reciprocals of the product of the factorials of the entries in the bodies of the $3 \times 3$ tables forming the above conditional reference set in Table 4. The $r_i{}'$ are the row totals of the observed $3 \times 3$ subtable.

The diagonal terms are obtained by subtraction. It can be verified that the column totals of the above tables are similarly the column totals of the observed $3 \times 3$ subtable.

## 4. Examples

**Example 4.** The data in Table 5 are taken from Taguchi (1987, pp. 682-684)).

|       | $A_1$ | | | $A_2$ | | | $A_3$ | | | $A_4$ | | |
|-------|-------|-------|-------|-------|-------|-------|-------|-------|-------|-------|-------|-------|
|       | $M_1$ | $M_2$ | $M_3$ | $M_1$ | $M_2$ | $M_3$ | $M_1$ | $M_2$ | $M_3$ | $M_1$ | $M_2$ | $M_3$ |
| $M_1$ | 8 | 0 | 2 | 8 | 1 | 1 | 6 | 0 | 4 | 6 | 2 | 2 |
| $M_2$ | 0 | 10 | 0 | 0 | 9 | 1 | 3 | 7 | 0 | 0 | 8 | 2 |
| $M_3$ | 0 | 0 | 10 | 4 | 1 | 5 | 2 | 3 | 5 | 4 | 3 | 3 |
| Total | 8 | 10 | 12 | 12 | 11 | 7 | 11 | 10 | 9 | 10 | 13 | 7 |

Table 5.  Classification of Musical Chords

They are the frequencies with which a given person classified a given musical chord chosen from $M_1$, $M_2$, $M_3$, as $M_1$, $M_2$, or $M_3$. Each chord was presented ten times. The experiment was performed on four subjects $A_1$, $A_2$, $A_3$, and $A_4$, who varied in musical training from high to none at all.

From (11), for subject $A_1$, $\{a_{ij}\} = (0, 2, 0, 0, 0, 0)$, so that from (13) the possible $3 \times 3$ tables forming the conditional reference set are of the form in Table 6 for $h = 0, 1, 2, 3, 4$, where $h = x_{12} = 0$ is the observed table.

|       | $M_1$ | $M_2$ | $M_3$ | Total |
|-------|-------|-------|-------|-------|
| $M_1$ | $8 - 2h$ | $h$ | $2 + h$ | 10 |
| $M_2$ | $h$ | $10 - 2h$ | $h$ | 10 |
| $M_3$ | $h$ | $h$ | $10 - 2h$ | 10 |
| Total | 8 | 10 | 12 | 30 |

Table 6.  The Reference Set of Tables (13) for $A_1$, Example 4

The resulting probability (12) of the observed table is

$$P(x_{12} = 0; \nu | \{a_{ij}\}, \{r_i\}) = K(0) \left/ \sum_{h=0}^{4} K(h) \exp(-h\nu), \right.$$

where $K(h)$ is given by (13). The maximum likelihood estimate is $\hat{\nu} = \infty$, at which $P(x_{12} = 0; \hat{\nu}) = 1$. This means that the conditional relative likelihoods are again tail probabilities, $R_c(\nu; x) = P(x_{12} = 0; \nu)$ as in Example 2. This varies from a maximum of 1 at $\nu = \infty$ to 0 at $\nu = -\infty$. The infinite maximum likelihood estimate causes no difficulties. It simply means that asymptotic considerations and approximations are inapplicable. The inferences take the form $\bar{\nu}_l \leq \bar{\nu} \leq \infty$. The lower 5% relative conditional likelihood bound in this case is $\bar{\nu}_l = 3.02$, corresponding to the odds ratio $\exp \bar{\nu} = 20.5$, and to $\tau = .91$. This means that unless an event with probability less than 5% has occurred, the geometric mean over all three chords of the odds that a chord is classified as $M_i$ versus $M_j$ is at least $\exp(3.02) = 20.5$ times greater if the chord actually was $M_i$ rather than $M_j$.

Much the same can be said for the remaining three subjects. In all cases the maximum likelihood estimate is infinite, as will always be the case when a zero entry is present. Thus the data cannot discriminate among upper bounds of agreement of the stated chord with the correct chord. But, as in Example 1, of more interest are plausible lower bounds. The lower 5% likelihood bounds for $\bar{\nu}$ are 1.36, 0.601 and $-0.080$ respectively, giving odds ratios $\exp \bar{\nu} = 3.90$, 1.82, .92, and $\tau = .59$, .29, $-.04$. These bounds decrease in accordance with the decreasing musical training of $A_i$. However, the sample information is conveyed by the four resulting relative likelihood functions $R_c(\bar{\nu})$ of (12) shown in Figure 4.

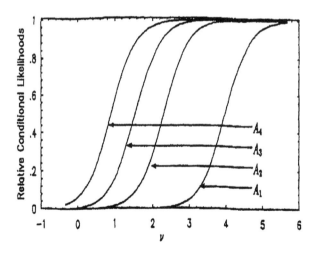

Figure 4.   Relative conditional likelihoods $R_c(\bar{\nu})$ of Example 4

Consider the parameter $\delta$ (14) for subject $A_2$. From (9) and (15) $\{b_{ij}, i < j\} = (-1, 3, 0)$, $t_1 = t_2 = 3/2$. The reference set Table 4 is given in Table 7 with possible values $h = 1,3$. The observed table is $h = x_{12} = 1$.

The probability (16) of the observed table is

$$P(x_{12} = 1; \delta | \{b_{ij}\}, t_1, t_2) =$$
$$\exp(-\delta)/[K(1)\exp(-\delta) + K(3)\exp(-3\delta)] = 1/[1 + 0.57143\exp(-2\delta)].$$

|        | $M_1$              | $M_2$              | $M_3$              | Total |
|--------|--------------------|--------------------|--------------------|-------|
| $M_1$  | $\frac{1}{2}(17-h)$| $h$                | $\frac{1}{2}(3-h)$ | 10    |
| $M_2$  | $-1+h$             | $\frac{1}{2}(19-h)$| $\frac{1}{2}(3-h)$ | 10    |
| $M_3$  | $\frac{1}{2}(9-h)$ | $\frac{1}{2}(3-h)$ | $4+h$              | 10    |
| Total  | 12                 | 11                 | 7                  | 30    |

Table 7.   Reference set Table 4 for $\delta$ of $A_2$, Example 4

Again the maximum likelihood estimate in infinite, so that $R_c(\nu; x)$ = $P(x_{12} = 1; \nu)$. This is shown in Figure 5. along with that of $\bar{\nu}$ for comparison. The value $\delta = 0$ has high plausibility or support by the data, $R_c(\delta = 0) = .636$. This implies that the mean agreement produced by categories 1 versus 3 and 2 versus 3 is comparable to the agreement produced by categories 1 versus 2. Hence the overall agreement is not wholly due to $\nu_{12}$ with its infinite maximum likelihood estimate.

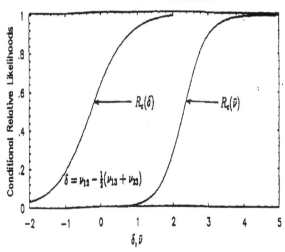

Figure 5.   Relative conditional likelihoods $R_c(\bar{\nu})$, $R_c(\delta)$ for subject $A_2$, Example 4

**Example 5.** The data in Table 8 are taken from Sackett et al. (1991, p. 30). From (11) $\{a_{ij}\} = (0, -3, 2, -2, -3, 3)$, so that from (13) the possible $3 \times 3$ tables forming the conditional reference set are of the form in Table

9 for $h = 3, 4, 5, 6$. The observed table is $h = x_{12} = 3$. The resulting probability (12) of the observed table is

$$P(x_{12} = 3; \nu \mid \{a_{ij}\}, \{r_i\}) = K(3)\exp(-3\nu) \Big/ \sum_{h=3}^{6} K(h)\exp(-h\nu)$$

The maximum likelihood estimate is $\hat{\nu} = \infty$, at which $P(x_{12} = 3; \hat{\nu}) = 1$, so that again the observed conditional relative likelihood is the probability $R_c(\nu; x) = P(x_{12} = 3; \nu)$. This is shown in Figure 6.

|  |  | resident's reading | | | |
| --- | --- | --- | --- | --- | --- |
|  |  | $\leq$ 5cm | 6 – 9cm | $\geq$ 10cm | Total |
| clinical | $\leq$ 5cm | 12 | 3 | 0 | 15 |
| clerk's | 6 – 9cm | 5 | 7 | 1 | 13 |
| reading | $\geq$ 10cm | 0 | 6 | 6 | 12 |
|  | Total | 17 | 16 | 7 | 40 |

Table 8.  Agreement about the level of central venous pressure

|  |  | resident's reading | | | |
| --- | --- | --- | --- | --- | --- |
|  |  | $\leq$ 5cm | 6 – 9cm | $\geq$ 10cm | Total |
| clinical | $\leq$ 5cm | $18 - 2h$ | $h$ | $-3 + h$ | 15 |
| clerk's | 6 – 9cm | $2 + h$ | $13 - 2h$ | $-2 + h$ | 13 |
| reading | $\geq$ 10cm | $-3 + h$ | $3 + h$ | $12 - 2h$ | 12 |
|  | Total | 17 | 16 | 7 | 40 |

Table 9.  The Reference Set of Tables (13) for Example 5

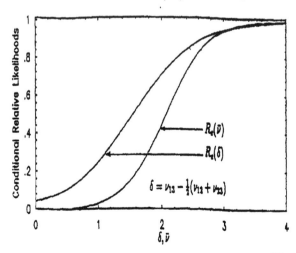

Figure 6.  Relative conditional likelihoods $R_c(\bar{\nu})$, $R_c(\delta)$, Example 5

Since categories are ordered, it might be thought that the agreement between categories 1 versus 3 is greater than the agreement between the intermediate categories 1 versus 2 and 2 versus 3. To examine this consider the parameter $\delta = \nu_{13} - \frac{1}{2}(\nu_{12} + \nu_{23})$, which is (14), with $\nu_{12}$ and $\nu_{13}$ interchanged, so that $t_1 = x_{12} + \frac{1}{2}x_{13}$, $t_2 = x_{23} + \frac{1}{2}x_{13}$. The reference set Table 4 is Table 10 with possible values $h = 0,2$. The observed table is $h = x_{13} = 0$.

| | | resident's reading | | | |
|---|---|---|---|---|---|
| | | $\leq$ 5cm | 6 – 9cm | $\geq$ 10cm | Total |
| clinical | $\leq$ 5cm | $12 - \frac{1}{2}h$ | $3 - \frac{1}{2}h$ | $h$ | 15 |
| clerk's | 6 – 9cm | $5 - \frac{1}{2}h$ | $7 + h$ | $1 - \frac{1}{2}h$ | 13 |
| reading | $\geq$ 10cm | $h$ | $6 - \frac{1}{2}h$ | $6 - \frac{1}{2}h$ | 12 |
| | Total | 17 | 16 | 7 | 40 |

Table 10.    Reference Set Table 4 for $\delta$, Example 5

The probability of the observed table (16) is

$$P(x_{13} = 0; \delta | b_{12} = 2, b_{13} = 0, b_{23} = 5, t_1 = 3, t_2 = 1) =$$
$$K(0)/[K(0) + K(2)\exp(-2\delta)] = 1/[1 + 22.5\exp(-2\delta)] = R_c(\delta).$$

which is also $R_c(\delta)$ as in the preceding examples. This is also shown in Figure 6. The value $\delta = 0$ is implausible, as are negative values of $\delta$. The evidence supports positive values of $\delta$. This means the evidence supports $\nu_{13} > \bar{\nu}_s = (\nu_{12} + \nu_{23})/2$. This in turn suggests that classification in the two extreme categories $\leq$ 5cm versus $\geq$ 10cm is the main source of agreement. The intermediate category 6 – 9cm produces much less agreement. Based on these two raters it appears that a three category scale is too fine for such data to support. A two category scale such as $\leq$ 7.5 versus $>$ 7.5 might be preferable.

## 5. Discussion

The introduction of $\kappa$ in 1960 served the valuable purpose of distinguishing between real agreement and agreement produced by pure chance, as described in Section 1. Further $\hat{\kappa}$ is computationally convenient and its large sample marginal distribution, on which to base inferences, can be approximated.

But one of the effects of the development of computers in the intervening thirty-eight years is that methods of inference that were previously only of theoretical interest are now practical. One of the most obvious examples is the use of conditional distributions, Fisher (1934), (1935, p. 50), and likelihood functions, Fisher (1973, pp. 75-76). The computer is specifically tailored to the use of conditional distributions and likelihood functions rather than the hitherto use of marginal distributions because it replaces high dimensional integrations, requiring large sample approxima- tions or simulations, by conditioning. The latter requires only fixing a large number of specified variables at their observed values in the joint proba- bility (density) function of the observations and renormalizing to integrate to one over a much lower dimension, as exemplified by (4), (12), and (16). The resulting inferences are more relevant since they are conditioned on the salient features of the observed sample, for example the marginal fre- quencies as discussed in Section 2.2. This leads to the methods used in the previous sections.

Therefore the ideas and methods of Section 2 are all implicit in Fisher (1935, p. 50, 1973, pp. 71-76). The only reason for dwelling on them here is that subsequent developments, particularly in the area of measuring agreement, have largely ignored them. For instance, although the use of the odds ratio as a measure of agreement was advocated by Spitzenagel and Helzer (1985) and criticized by Shrout et al. (1987) as cited earlier, the arguments took place at the rather primitive level of comparing the extent to which parameters such as $\kappa$ and $\nu$ and their point estimates succeed in capturing the essence of what is commonly meant by the word "agreement", in particular, their response to marginal agreement. No consideration of scientific inference, likelihood structure such as (5), the combination of data, or the effective utilization of the sample information, was involved. Much the same can be said about more recent developments.

Similarly, although the odds ratio is widely used elsewhere in biostatis- tics, this use is often based on maximum likelihood estimation interpreted as a method of obtaining asymptotically unbiased estimates with minimum variance. This requires that the relevant likelihood functions be approx- imately normal, which is often the case. But when agreement is under consideration this is a questionable point, as the preceding examples all show. To escape this difficulty the methods here are based on the graphing and full exploitation of the conditional likelihood function (4) and general- izations thereof.

Darroch and McCloud (1986) also separate marginal agreement from subject specific agreement as relating to observer differences and to cat- egory distinguishability respectively. Their measure of category distin- guishability, $\delta$, can be written as

$$\delta = 2 \sum_{i<j} [1 - \exp(-\nu_{ij})]/q(q-1)$$

The measure $\delta$ does not submit to inference procedures of the kind discussed here. In particular it does not have a readily accessible likelihood function.

An additional feature of $\nu$ that facilitates its practical use is its highly conditional nature, which implies a robustness against different sampling schemes. The full $q \times q$ multinomial model $\{p_{ij}\}$ of Section 1 is not required. Only the conditional submodel $\{p_{j|i}\}$ (8) is required. There the origin of the marginal row frequencies is irrelevant, allowing its unmodified use when these margins are fixed, as in Example 4, or restricted, as in Jannarone et al. (1987).

Finally, $\nu$ is more closely related to conditional logistic regression models than are such correlational measures as $\kappa$. This may facilitate its use in regression methods to examine the influence of covariates on this form of agreement.

### References

[1] Chamberlin, S. R., and Sprott, D. A. (1991), On a discrete distribution associated with the statistical assessment of nominal scale agreement. *Discrete Mathematics*, **92**, 39–47.

[2] Cook, R. J., and Farewell, V. T. (1994), Conditional inference for subject-specific and marginal agreement. *Canadian Journal of Statistics*, **23**, 333–344.

[3] Cohen, J. (1960), A coefficient of agreement for nominal scales. *Educational and Psychological Measurement*, **20**, 37–46.

[4] Darroch, J. N. and McCloud, P. I. (1986), Category distinguishability and observer agreement. *Australian Journal of Statistics*, **28**, 371–388.

[5] Fisher, R. A. (1934), Two new properties of mathematical likelihood. *Proceedings of the Royal Society of London, A*, **144**, 285–307.

[6] Fisher, R. A. (1935), The logic of inductive inference (with discussion). *Journal of the Royal Statistical Society*, **98**, 39–54.

[7] Fisher, R. A. (1973) *Statistical Methods and Scientific Inference*, Hafner Press, New York.

[8] Jannarone, R. J., Macera, C. A., and Garrison, C. Z. (1987), Evaluating interrater agreement through "case-control" sampling. *Biometrics*, **43**, 433–437.

[9] Sackett, D. L., Haynes, R. B., Guyatt, G. H., and Tugwell, P. (1991) *Clinical Epidemiology: A Basic Science for Clinical Medicine, Second Edition*, Little, Brown and Company, Toronto.

[10] Shrout, P. E., Spitzer, R. L., and Fleiss, J. L. (1987), Quantification of agreement in psychiatric diagnosis revisited. *Arch Gen Psychiatry*, **44**, 172–177.

[11] Spitznagel, E. L. and Helzer, J. E. (1985), A proposed solution to the base rate problem in the kappa statistic. *Arch Gen Psychiatry*, **42**, 725–728.

[12] Sprott, D. A. and Vogel-Sprott, M. D. (1987), Use of the log odds ratio to assess the reliability of dichotomous questionnaire data. *Applied Psychological Measurement*, **11**, 307–316.

[13] Taguchi, G. (1987) *System of Experimental Design, Vol.* **2.**, Kraus International Publications: White Plains, N. Y.

# Some Important Classes of Problems in Statistical Experimental Design, Multivariate Analysis, and Sampling Theory

J.N. Srivastava

Department of Statistics
Colorado State University
Fort Collins, Colorado

**Abstract.** In this paper, some fields of research arising in the sub-disciplines of Statistics described in the title, are presented. The author feels that research on the problems described herein will lead to substantial fundamental advances.

## 1. Introduction

The presentation is brief. No suggestions have been made with respect to corollary type of work. Rather, we describe entirely new areas where research may prove to be significant and fruitful. This paper is a stepping stone for a possibly large number of research papers.

## 2. Statistical Design of Scientific Experiments

In this section, we shall consider three areas of research. One is concerned with the nature of, and the measurement of the information contained in an experiment. For example, we may be concerned with the information available for discriminating between rival models, or given a particular model we may be interested in measuring how much information is there on different sets of parameters within this model.

The second area is that of designs for identification of the model as well as for estimating paramters under the correct model. Much work has already been done in this area. However, even all of that is too little. What we need is a full-fledged and complete theory.

The third area is one where we want to do an experiment with a large number of factors, where high order interactions are present. The task is to find out what is the correct set of high order interactions, and to estimate them.

## 2.1 Information Contained in an Experiment

Let $X$ be a random variable, about which nothing is known. Now, there are infinitely many possible distributions, and $X$ could come from any one of these distributions. Since nothing is known, the information contained in $X$ is infinite. When we say that $X$ comes from a distribution $f(x)$, we are specifying one out of an infinite number of distributions to which $X$ could belong. Clearly, this specification itself has an infinite amount of information in a sense. After we have specified the distribution of $X$, we consider the information contained in $X$ under the knowledge that $X$ comes from $f(x)$. If $f(x)$ is completely known, then $X$ does not give any new information.

If on the other hand, $f(x)$ is not completely known, then $X$ will give some information. This is the information that is measured by various classical information measures. For example, if $f(x) = f(x, \theta)$ where $\theta$ is a scaler or vector parameter which is unknown, and $f$ is known, then $X$ contains information, and that information is on $\theta$. Let $X_i (i = 1, \cdots, 2n)$ be a set of independent random variables each with the distribution $f(x, \theta)$. Then it is clear that any $n$ out of these $2n$ random variables have the same amount of information in them.

Now, for simplicity suppose that $\theta = (\mu, \sigma^2)$ where $\mu$ is a location parameter, and $\sigma^2$ is a scale parameter. We wish to compare the information contained in the set $S_1$ of the $n$ variables $(X_1, \cdots, X_n)$, with the set $S_2$ containing $(X_{n+1}, \cdots, X_{2n})$ At this point, it may be recalled that if $X$ has a normal distribution, then the usual information measure for $X$ does not involve $\mu$ at all. Thus, under that information measure, $S_1$ and $S_2$ will have the same information in them.

Now suppose that we have $\mu_i = \sum_{j=1}^{k} a_{ij}\alpha_j$, for all $i$. Here, the $a_{ij}$ are known real numbers and the $\alpha_j (j = 1, \cdots, k)$ are unknown parameters. Now, suppose that it is known that some of the $\alpha$ are zero, but it is not known which of the $\alpha$'s are actually negligible. Indeed, let $\Delta$ be a family of subsets of $(\alpha_1, \cdots, \alpha_k)$ such that it is known that there is a $\delta \in \Delta$ such that the $\alpha$'s corresponding to $\delta$ are nonnegligible, and the other $\alpha$'s are actually negligible.

The problem then is to compare sets of the kind $S_1$ and $S_2$. In other words, suppose that $S$ is a set of $n$ variables $(X_1, \cdots, X_n)$ which are all independent, with $X_i$ having the distribution $f(\mu_i, \sigma^2)$. Then the question arises, what kind of information does $S$ give concerning the $\alpha$'s and $\sigma^2$. Notice that in this case, $S$ gives information as to which value of $\delta$ inside $\Delta$ is the true value, and also information on the actual numerical value of the parameters contained in $\delta$. This will be illustrated by the following example from the $2^4$ factorial experiment.

In this case, we have $k = 16$, and the $\alpha$'s are the various parameters belonging to the $2^4$ factorial experiment. These shall be denoted by $\mu_0$ (the general mean), $B_i$ (main effect of the $i$th factor), $B_{ij}$ (interaction between factors $i$ and $j$), etc. Now suppose that it is known that at most $\nu(= 4)$ parameters are nonnegligible. Furthermore, suppose that it is known that the set of nonnegligible parameters obey the tree structure. This means that if a particular interaction is significant then at least one lower order effect which involves the factors occuring in the said nonnegligible interaction, is also nonnegligible. (For example, if the interaction $B_{256}$ is nonnegligible, then so is one of the three two-factor interactions $B_{25}, B_{26}$ and $B_{56}$.) It will then be seen that $\Delta$ consists of the following possible sets of parameters:

(2.1)
$$\Delta_1 = (\mu_0, B_i, B_j, B_k); : \Delta_2 = (\mu_0, B_i, B_j, B_{ij});$$

$$\Delta_3 = (\mu_0, B_i, B_j, B_{ik}); : \Delta_4 = (\mu_0, B_i, B_{ij}, B_{ik});$$

and

$$\Delta_5 = (\mu_0, B_i, B_{ij}, B_{ijk}),$$

where $(i \leq i, j, k \leq 4$, and $i, j, k$ are distinct.

Here, $\Delta$'s are a set of mutually inclusive and exhaustive subclasses of $\Delta$, and the $\Delta_i(i = 1, 2, 3, 4, 5)$ contain respectively $4, 6, 24, 12$, and $24$ $\delta$'s respectively.

The question faced by a researcher is how to find out which of the 70 members of $\Delta$ correspond to the true nonnegligible set, and furthermore to estimate the parameters in that set. Henceforth, we shall assume that $\delta_0$ in $\Delta$ corresponds to the set of parameters which are indeed nonnegligible. Thus the problem is to identify $\delta_0$ and estimate the paramters occuring in $\delta_0$.

In the above example, suppose that $\delta_0 = (\mu_0, B_2, B_{12}, B_{24})$. Then, any given set of observations $S$ (which corresponds to a design) gives us information of two kinds. The first kind of information helps us in identifying $\delta_0$, and the second kind of information helps us in estimating the elements of $\delta_0$. Thus, we have a clear cut problem arrising here. The problem is to measure the amount of information given by a general set $S$ concerning the identification of $\delta_0$, and also concerning the parameters contained in $\delta_0$. It appears to the author that in general the total information contained in a set $S$ may roughly be constant, particularly if $\Delta$ has a large number of members. Thus, given two designs $S_1$ and $S_2$ we need to know how much revealing power each design has, as measured by its ability to identify $\delta_0$. Similarly, we would like to compare the two designs with respect to their ability to estimate accurately any given member $\delta$ of $\Delta$.

It may be remarked that the current so called "optimal design theory" is a bit misleading, since it assumes that it knows $\delta_0$ and proceeds from an intermediate stage rather than from the more fundamental stage where we know $\Delta$ and need to identify $\delta_0$ first. Of course, if we have successfully identified $\delta_0$, and need only to estimate the parameters in $\delta_0$ accurately then the current optimal design theory becomes meaningful.

### 2.2 Model Identification in Factorial Experiments

The example discussed above in the last section illustrates the model identification problem in factorial experiments. In the last section, we consider the question of the measurement of information in any given design, both with respect to its ability to discriminate between the competing members of $\Delta$, and also to estimate the parameters accurately in the $\delta$ which happens to correspond to the nonnegligible parameters. Notice that, in general, a wrong $\delta$ may get identified in place of $\delta_0$.

Thus, there remains the question of how to correctly identify $\delta_0$. There is a considerable amount of work available in the literature in this direction. This includes the theory of search designs, and also the more recent work by Srivastava and Chu (1997) on multistage designs for model identification. However, this whole field is wide open. The importance of this field cannot be exaggerated, since it is easy to show that if the correct $\delta_0$ is not identified, and in its place we choose another $\delta$ (say, $\delta^*$), then there will be a bias introduced in our results. How serious the bias will be will depend upon many aspects. In some cases, one may be lucky and the bias may be small. In other cases, it may distort our results very badly. The worst thing about it is that in any given situation we will not know how big the bias-vector is. Thus, it is important to focus on the identification of $\delta_0$. In some cases, where we are interested in finding an optimal level-combination, it may be possible to bypass the problem of first identifying $\delta_0$. Here, the problem can be formulated in a somewhat different way. The new way will ask the question as to how to choose $\delta$ so that the optimum level-combination gets identified, or at least, we identify a level combination whose yield is close to that of the optimum.

There are generally two kinds of problems arising in factorial experiments. One is the estimation of the response surface. Here, using a relatively small number of observations from a design $S$, we try to estimate the yield of each of the $2^m$ level combinations. The other kind is concerned with finding the optimum level-combination i.e., the one whose yield is optimized in some sense. (For example, we may be concerned with finding the level combination whose yield is the maximum.)

In the first category of problems, i.e., where we are concerned with the response surface, the research mentioned in the last section, will be important. Here it seems that we will need to identify $\delta_0$ as accurately as possible. On the other hand, when we want to identify the optimal treatment, or one very close to it, we will need to adopt a different strategy.

## 2.3 Model Identification of Row-Column Designs

In the field of row column designs, traditionally, it has been assumed that the additive model holds. Researchers have concentrated more on the combinatorial aspect. However, as the recent work of Srivatava and Wang (1995) indicates, nonadditivity holds only in exceptional circumstances. Generally, one may have many cells which pollute the results because of being nonadditive.

In this field, the design aspect is less important than the analysis aspect. Given the data from a row column design, it is necessary to analyze it and determine which are the large nonadditive cells. A method of doing that has been developed and used in the above paper of Srivastava and Wang. This method has been used in the context of latin squares. However, in Srivastava (1996), a somewhat more general method has been explained. Refinement of these methods should yield useful results with respect to identification.

In situations where there is too much nonadditivity, i.e., a very large percentage of cells are nonadditive, the problem becomes more severe. In this case, if it is possible to take more than one observation per cell, then the problem becomes simple. However, the case where we can take only one observation per cell remains to be solved.

This last case, however, can be approached by other methods. These methods are developed in various papers of the author (for example, see Srivastava and Beaver (1986), and references therein), dealing with nested multidimensional block designs (NMBD). This approach will reduce the number of cells which contribute large nonadditivity parameters. In many situations it may make nonadditivity quite negligible, or nonadditivity may still remain, but it may be of a minor nature. In other situations, large amounts of nonadditivity may still remain, unless the NMBD is chosen very carefully.

Thus, the theory of nested multidimensional block designs becomes very important. A thorough study of this field vis-a-vis nonadditivity is very desirable.

## 2.4 Factorial Experiments With a Large Number of Factors and a Large Number of Interactions

As the title of this subsection reveals, there are situations where we have a large factorial experiment on the one hand, and on the other hand it is not true that only the small order parameters are nonnegligible. In other words, higher order interactions are common. An example of a field where this would happen is the field of nutrition.

The importance of creating expert systems to monitor human health can not be exaggerated. For this purpose, it will be important to conduct factorial experiments which involve a rather large number of factors. Also, it is quite clear that interactions of high orders can be expected quite commonly. The question is how to approach this field.

Here, a few suggestions will be offered.

One approach is as follows. Suppose there are $m$ factors and $m = m_1 + m_2 + \cdots + m_k$, where the $m_i$ are relatively small. One possibility is to do a small experiment of size $2^{m_i}$ corresponding to the $i$th set of factors. In this experiment, the level of all other factors must be kept constant. By using model-identification theory, for each $i$ we should find out the response surface for the corresponding $2^{m_i}$ factorial experiment.

Next, we should consider the division of $m$ into another set of a group of factors. For example, let $m = m'_1 + m'_2 +, \cdots, + m'_{k'}$. We don't necessarily have $k = k'$. In other words, the second division is a division of a different kind from the first one and the grouping of the factors may be quite different. By using ideas such as those of the intra and inter-group balanced designs, it may be possible to do a set of such experiments, each experiment corresponding to a breakup of $m$ into several parts.

The question is how many such partial experiments are needed in order to reveal the whole response surface to a reasonable degree.

In actual practice, in the field of nutrition there may be one more complication. It may be that instead of one response we may have a very large multitude of responses to cope with. Also, it may not be possible to measure each response variable accurately. For example, the treatments may be various rations to be used on consecutive days and the subject may be able to tell only whether he feels better or worse compared to the previous day.

Needless to say, the use of orthogonal arrays etc. may be totally out of place in the above context.

Now, a theory must first exist, before it can be applied. Thus, before we can apply a particular factorial design theory in the field of development of expert systems for human health, we need to solve the problem to a reasonable degree on the theoretical level. Thus, a clear-cut problem of a theoretical type, described above, arises here.

We illustrate the above idea by considering an experiment with 16 factors each at 2 levels. Let $EG(m, q)$ denote the finite Euclidean Geometry of $m$ dimensions based on the finite field $GF(q)$. We represent the 16 factors by the 16 points in $EG(4, 2)$ each point being of the form $(C_1, C_2, C_3, C_4)$, where $C_i \in GF(2)$.

Now, in $EG(4, 2)$ there are $2^4 = 16$ points. Also there are 140 planes (or two-dimensional spaces). These 140 planes are divisible into 35 sets of 4 planes each. Each plane contains 4 points of $EG(4, 2)$. The 4 planes in any set contain among themselves all the 16 points of the geometry. Thus, we have a block design with $k = 4, v = 16, b = 140, r = 35$. It can be checked that this is a BIBD with $\lambda = (\lambda_2) = 7$. The design is resolvable into 35 replications. Furthermore, it turns out that the design is also a 3-design, so that each of 560 distinct triplets of points in $EG(m, 2)$ occurs in $\lambda_3 (= 1)$ block.

Now, take one of the replications (containing 4 blocks) from the above mentioned 3-design. We will assume that there is absolutely no information available on the different factors, except that it is unlikely to have interactions of order 5 or higher. Then we can proceed as follows. Take anyone replicate of four blocks, and take one block out of this replicate. We shall number the replicates from 1 to 35 and the blocks in the replicate $i$ will be denoted by $i.1$ to $i.4$. Thus, we take block 1.1. Now, consider a factorial experiment with 16 factors in which the levels of all factors are held constant at 1, and only the levels of the factors included in block 1.1 are varied. (We shall assume that from prior considerations, the level combination $(1, 1, \cdots, 1)$ of the 16 factors is considered to be "optimum". In other words, without further information this combination is supposed to be the best. However, the purpose of the experiment is to find out whether there are other combinations which are better, and in general, to study the whole response surface, i.e., to find out the yields of all the $2^{16}$ level combinations.)

We conduct a $2^4$ experiment using factors in block 1.1. If necessary, we make this experiment sequential, or at least multistage. In other words, not all the treatments in this $2^4$ experiment have to be tried at the same time. The author recommends that we begin with the following design $T_1$:

$$T_1 = \begin{bmatrix} 1 & 0 & 0 & 0 & 0 \\ 1 & 1 & 0 & 0 & 0 \\ 1 & 1 & 1 & 0 & 0 \\ 1 & 1 & 1 & 1 & 0 \end{bmatrix}$$

This design has the following feature. It contains the treatment (1111). Also, suppose we consider two treatments to be adjacent if they occur as adjacent columns in $T_1$. Then, it is clear that two adjacent treatments differ only in one factor.

In nutrition experiments, the author feels that changing only the level of one factor at a time may be most conducive to understanding the underlying phenomena. After having tried $T_1$, the data can be analyzed by using the interaction-sieve (IS) and the temporary elimination procedure (TEP) introduced in Srivastava (1987), and further applied in Srivastava and Hveburg (1990), Li (1990), and Chu (1997). If necessary, further treatments can be added to $T_1$ one by one (with level-change in one only factor) until, we feel that all the factorial effects which are significant have been identified, and estimated. Next, using these results, the response surface for the $2^4$ experiment corresponding to block 1.1 can be estimated. It is assumed that the response surface has been verified by trying at least one treatment (not tried yet), and checking that the observed yield for that treatment is close to the expected yield obtained from the estimated response surface.

Now, in general, there can be two purposes behind the conduct of the 16-factor experiment. The first one $(P_1)$ is where we are seeking the "optimum" treatment out of the $2^{16}(= 65536)$ treatments. It is understood that the optimum may occur at an area of the response surface which is quite flat. In other words, there may be a relatively large number of treatments whose true yield is close to that of the optimum. In that case, we will try to find at least one treatment which is reasonably close to the optimum.

The second purpose $(P_2)$ is that of estimating the whole response surface. In other words, here we want to estimate as accurately as possible the yields of all the 65536 treatments.

In general, the purpose will be $P_1$ in situations such as the development of a drug for a particular disease. Here, the 16-factors are certain promising componants, and we are trying to determine their optimum combination. The purpose $P_2$ will occur in the development of the expert system for human health.

Consider first the case $P_2$. Having conducted the $2^4$ experiment with factors in block 1.1, and having estimated the response surface for this case, let $(x_1, x_2, x_3, x_4)$ be the optimum level combination obtained. Here, the

$x$'s denote the levels of the factors in block 1.1. At this stage, we rename the levels of the factors in block 1.1 so that under the estimated response surface, (1111) again becomes the level combination which corresponds to the optimum yield. Having done this, we proceed to block 1.2 and conduct a $2^4$ experiment as in the case of block 1.1. At the end of this experiment, we again estimate the response surface, and find the optimum level combination.

Having completed the experiment with block 1.2, we again rename the levels of the factors in block 1.2 so that for these factors the level combination (1111) is optimum. We repeat this procedure with the factors in block 1.3 and factors in block 1.4, which would complete the first replicate. Note that at the completion of this first replicate, we arrive at a renaming of the the levels of the 16 factors, so that the (new) level-combination $(1, 1, \cdots, 1)$ appears to be optimum. (Obviously this is not necessarily the global optimum.)

Now, based on the information gained so far on the different factors and the responses to various treatments, we should try to judge as to which pairs or triplets of factors could be of more interest at this point of time in our investigation. Having determined that, we select one more replicate out of the 34 replicates not tried so far. We should select a particular replicate in which the blocks seem to be more interesting from the above point of view. We conduct experiments of the $2^4$ type on each successive block of replicate 2, each time renaming the levels so as to correspond to the optimum combination.

Again, taking into account the "experience" obtained so far concerning the yields of various treatment combinations and the factorial effects which turned out to be significant in the successive $2^4$ experiments, one can select a third replicate. Experimentation in the third replicate then is carried on in an analogous manner. We can then proceed, until somehow we are assured that the combination obtained is relatively close to optimum.

Here, an important question arises. When is it that we should stop experimentation. After all, there are 65536 treatments. We do not want to try even a small fraction of these. Also, our "experience" at a particular time in our experimentation may make us feel that the combination that we have obtained at that point is close to the optimum. The question is how can we check that.

Suppose that at any particular moment the "current optimum" is such that there are a total $u$ treatments which are more optimal than this one. Also, suppose that until that time, there is a set of $v$ treatments which have not been tried. Then, if out of these $v$ treatments we select one randomly, the chance is $(u/v)$ that we may come across a treatment which is more optimal. Assuming $v$ to be relatively large, the chance that after trying $k$ treatments randomly, we do not find anyone to be more optimum than the current optimum is approximately $[1 - (u/v)]^k$. Thus, if we randomly select $k$ treatments not tried thus far, and it turns out that none of them is more optimum than the current one, and furthermore if $k$ is relatively large, then it may be safely concluded that we have obtained a combination which is at least close to the optimum.

If, on the other hand, we do come across a treatment which is significantly more optimum than the current one, then we rename the levels of the factors so that this new optimum treatment has levels $(1, 1, \cdots, 1)$. After doing this, we may again try one or more replicates and continue as before. It appears intuitively that this process should rapidly converge.

Research is needed to put the above ideas on a sound theoretical footing, and to develop methodology corresponding to the same. This methodology will of course show what is lacking in the above procedure as well as its fruitfulness.

Now consider the case for $P_1$. In this case, we do the block 1.1 as before, but at the end of the experiment, we do not rename the levels of the factors. We simply go ahead and do the blocks $1, 2, 1.3, 1.4$ in succession, thus completing the first replicate. Based on the experience gained from the first replicate, we select one of the remaining 34 replicates, and proceed as before. We go through this procedure a few times. In this process, we may have conjectures concerning certain 3-factor or 4-factor interactions to be nonnegligible. We try those replicates in which the corresponding 3-factors do occur in a block. Thus, we try a few replicates. At this stage, we conjecture a model. In other words, we conjecture possible sets of nonnegligible factorial effects. Using these, we generate the response surface. Then, we randomly select a treatment and compare its observed and expected yields. We do the same for several treatments. If the difference between the observed and the expected values appear to be large, we go through one or more further replicates, and repeat the process.

This procedure may or may not converge. If there are 4-factor interactions which correspond to a set of 4 factors which do not occur in any block, then such a 4-factor interaction may not be estimable. In that case, of course, further theory is needed.

Theory has to be developed firstly for the case when there is a single (continuous) response $y$. After we have a sufficiently powerful theory for this case, we may consider the situation under paired comparisons where we may be able to know only whether a particular treatment has a $y$-value larger than or smaller than the $y$-value for another treatment.

The above procedures for $P_1$ and $P_2$ have been given under the assumption that there is no knowledge of the various factorial effects. Whatever knowledge we have already gathered by experiment and also by direct human experience in general, will be translated into knowledge on various kinds of factorial effects both in the whole experiment, and also in the smaller subexperiments. This knowledge will have to be handled in many cases using Bayesian techniques. Also, from various biological considerations, results in biomathematics, etc., it will be possible to predict to various degrees of accuracy differences between two treatments in many cases. Thus, for the development of expert systems for human health, what is needed is more statistical science rather than pure mathematical statistics. Experimental design theory, particularly in the Bayesian frame work will have to be blended into the same.

## 3. Multivariate Analysis

There is a vast variety of unsolved problems in the field. However, below we shall outline only three classes. These are motivated from the point of view of application to the health industry.

The first one, dubbed "antimeta-analysis", is analogous to small area estimation problems arising in sampling. The other one deals with the situation where we have a multitude of responses, some of which may even be unknown. Furthermore, the observations are of an extremely incomplete and varied nature. On some responses it is easy to take observations, and a large number of them may be available. On other responses observations may be scanty or incomplete. On still some other responses, observations could be taken only if the responses "come into play".

### 3.1 Antimeta-analysis

We first consider "meta-analysis". Suppose that there are three treatments $A$, $B$, and $C$. We wish to compare them, and find out how effective each one of them is. Suppose that these drugs have been tried at four different centers and the percentage effectiveness is as shown in the table below.

As we notice from this table, the figures for the different drugs vary from center to center. The question arises whether this variation is due to random fluctuations or not. Assuming that it is due to random fluctuations, we wish to combine the results from the different centers. Even if there are other influences besides random fluctuations, we wish to have a combined picture from all the centers comparing the three drugs $A, B$, and $C$. Any technique whose purpose is to come to an integrated decision concerning $A, B$, and $C$, in which the results from the different centers are combined in someway, comes under the general title of "meta-analysis".

An analysis such as the above would be of great interest, for example, to a pharmaceutical company. The interest of such a company lies in finding out which of the three drugs is the most effective in the population as a whole. Whichever drug is the most effective, if this drug is manufactured and marketed, the sale should be expected to be the largest. Thus, it becomes of great interest to somehow integrate the results from the different centers into one overall result. Because of this, during the past decade or so, meta-analysis has grown in importance and interest.

However, consider the situation from the point of view of the consumer. Consider two individuals $i_1$ and $i_2$. It is possible that the drug $A$ is more suitable to $i_1$ relative to drugs $B$ and $C$. However, for $i_2$ the drug $B$ is more effective, than $A$ and then $C$. Thus, each individual needs to know which drug is the most suitable for him/her personally. So, from the point of view of the consumer, what is needed is a stratification of people into several strata such that in one strata the drugs are effective in the order $A > B > C$ in the second strata it might be $B > A > C$, and similarly for other strata with $A > C > B, C > A > B, B > C > A, C > B > A$.

From this point of view, combining the results from the centers may result actually in a loss of information.

Tables such as the one shown above are actually the result of summarization of various kinds of information. Initially, the data would come on each particular individual separately. In the above table, there are 10200 individuals. Such a table is obtained by combining these individuals in various ways and summarizing the information. Meta-analysis aims at further combining the information into one overall comparison of drugs.

As mentioned above, from the individual point of view, it is unimportant to know which drug is more effective on an overall basis. What one needs to know is which drug is most effective for himself as an individual.

Thus, we have a new challenge. It is to start with the initial observations (in the above example, the 10200 cases), and somehow categorize them into six groups as mentioned above. One may say that if each individual is given only one drug, then how can we know how the other two drugs would have worked on him. However, that's a different kind of question which actually deals with the foundations and philosophy of medical experimentation itself. We do not intend to go into that here. But, it may be noted that if the only data that we have on the different subjects is whether they were benefited or not benefited by a particular drug, then, the antimeta-analysis can not be performed. Such analysis, which is indeed useful to people, will be possible if on each person we collect information relevant to the person's response to different drugs. This data could involve various other measurements from time to time on the body of individuals.

The measurements themselves should be selected in such a way that they throw light on the response of a person to the different drugs.

## 3.2 Missing data

A large amount of work is available on missing data, both in uniresponse and multiresponse situations. However, as the work of Srivastava and Zaatar (1973) indicates, even in the very simple situation of estimating a bivariate covariance matrix from a normal distribution, many of the estimators which intuitively appear to be promising, do not do too well. When large bodies of data are missing, the author strongly feels that it is quite improper to use any of the available methods to estimate these and fill in the blanks, and proceed to analyze the data so completed.

What is needed indeed is the study of stochastic processes which cause missing data. This may have to be done to some extent in different disciplines where missing data arise. Of course, the same stochastic process can arise in two different disciplines. Thus, once a study has been made with a particular stochastic process in one discipline, it is possible that it will be usable in some other disciplines.

We illustrate the above idea by considering a situation from the area of keeping weather records. Suppose that in a particular area, there are a number of rain and snow gauges, which have been installed in a number of different places. Normally, each of these gauges will give the observations. However, sometimes the observation from a gauge may be missing. Let us look into the question of what may cause the information to be missing. This may happen, because of a variety of reasons. A gauge itself may have stopped working, because it was too old or somehow it got broken. The storm in a particular locality may be so severe that there may be a blockade, so that people who would read the gauges simply could not do so. The weather may be so severe that it may break a particular gauge in a certain spot. The data for a particular day may have been read from the gauge and recorded, but the record itself may have been lost for some reason. And so on. Thus, it is clear that data in this situation will be missing because of a variety of reasons. These reasons, can be modeled into appropriate stochastic processes involving various parameters. Each of these processes are causes censoring of the data. The censoring process will then be super-imposed on the stochastic model for the general data itself. Then, using the available data various parameters could be estimated, and missing values could also be dealt with.

This is a fascinating and useful field, the importance of which can not be exaggerated.

## 3.3 Too Many Responses with Scanty Data

Consider the problem of building an Expert System for Human Health (ESHH). Clearly, it is a very complex problem, whose solution would profoundly ameliorate our health condition and prolong life. The building up of ESHH would necessarily involve collection of large amounts of data either through (partially) controlled experiments or otherwise. This data would include measurements on a large number of response, such as blood pressure, temperature, pulse rate, various variables included in blood tests, etc. Potentially, however, the number of responses is extremely large.

Indeed, the responses may be grouped into three broad categories. Firstly, there are responses which are considered to be "relevant", and which are not costwise prohibitive.

The second category consists of responses which may be relevant but which are expensive to measure, and also responses which are considered irrelevant. The third category includes responses which are not even explicitly included in the researcher's list of responses (both relevant and irrelevant); these include, for example, various types of pains, discomforts or other abnormal bodily experiences. (The precise nature of body "pain" may vary from time to time, and potentially, there are an infinite number of such pains, discomforts and other experiences.)

For responses in the first category, data may be generally complete, if it is coming from a controlled experiment. If the data is not from such an experiment, but is simply being observed and collected, then there may be missing values, for which a stochastic process approach mentioned in the last section may be necessary.

For the second category, there may be very little data, this being on responses which are very expensive, but which nevertheless are measured sometimes because of their importance.

The third category will produce data on a response if it "arises". For, example, on a given subject, data on a particular type of pain could be obtained, only if firstly the pain did occur, and secondly if the pain was such that some variables (which are measurable) could be correctly associated with the pain. A similar remark holds for various discomforts, or other experiences.

For developing an ESHH a large amount of both macro and micro model building will be required. This will necessitate that the reposes be "co-analyzed", and inter-relationships between them be explored and an accurate model for the same be established. This will have to be done in spite of the fact that the data on some of the most important responses may sometimes be very scanty. The author feels that a predominantly Bayesian approach may be required in much of the work concerning the

ESHH, particularly in the initial and intermediate stages.

The building of these theories will take time, and it is time that the multivariable problems in this field be formulated and tackled. This field may be very different from classical multivariate analysis.

## 4. Sampling

In Srivastava (1985), a general class of estimators in sampling theory was introduced. It is convenient to discuss it relative to the Horvitz-Thompson estimator $\hat{Y}_{HT}$ for the population total $Y$. The new class of estimators is a three-fold generalization of this. Firstly, instead of $Y$, there is an infinite class of functions $Q(\psi)$, whose estimation is considered. Another generalization is with respect to the fact that whereas $\hat{Y}_{HT}$ looks at one unit at a time in the sample, the new class of estimators look at $1, 2$, or more units at a time. The third aspect of generalization is that a new function has been introduced. This function is denoted by $r(\omega)$. The symbol $\omega$ denotes a sample, and thus it can take any value in the class of all subsets of the population $U$ (which is supposed to contain $N$ units.) Thus $\omega$ can take $2^N$ possible values out of which one corresponds to the empty set. The function $r$ maps $\omega$ to real numbers. In general, $r$ can be any function. However, as we shall see some restrictions on the function $r$ lead to nice results. In what follows, the units in the population $U$ will be denoted by the integers $1, 2, \cdots, N$, and the variable of interest $(y)$ will have the value $y_i$ for the unit $i$, where $i = 1, 2, \cdots, N$. In the subject of sampling theory the estimation of $Y$ is a topic of great interest. Out of the new class of estimators introduced in the above paper, there are an infinite number of estimators that become available for estimating $Y$, $\hat{Y}_{HT}$ being one of them. Among the available estimators one estimator denoted by $\hat{Y}_{Sr1}$ has been particularly studied (see for example Srivastava and Ouyang (1992). This class of estimators $\hat{Y}_{Sr1}$ turns out to be very significant, and is pre-eminent (Srivastava (1998)).

The author believes that the above theory can be extended very fruitfully in many directions creating a wealth of results which should be very useful in different circumstances. Below, we shall indicate some of the directions in which extension may be very fruitful. As the research in this field proceeds, the author believes that there will be newer strategies that will become evident which will lead to possible further improvements. The author strongly feels that this theory will be at least as good as any other theory in every field of application.

In what follows, we shall first describe some of the major results already obtained under the above theory. Next, we shall outline many of the outstanding problems. It will be noted that some of these problems are formulated in a way so that one can go to work on them right away. They are clear-cut mathematical problems and do not have philosophical or other considerations attached to them.

Throughout, let $\underline{i}$ denote the set of $k$ units $(i_1, i_2, \cdots i_k)$ with $1 \leq i_1 < i_2 \cdots < i_k \leq N$. Let $\psi$ be any function which maps $\underline{i}$ to real numbers. Thus, $\psi(\underline{i})$ is a real number for all possible $\binom{N}{k}$ values of $\underline{i}$. (Although the elements of $\underline{i}$ are written in an ordered manner, the set of units $\underline{i}$ is considered throughout to be unordered.) Now, define

$$Q(\psi) \equiv \Sigma_1 \psi(\underline{i}),$$

where $\Sigma_1$ denotes the sum over all possible $\binom{N}{k}$ values of $\underline{i}$. In Srivastava (1985), the estimation of $Q(\psi)$ is considered. Although there is some theory relating to biased estimators, most of the paper is concerned with unbiased estimators, which we proceed to describe for later use.

Throughout this section, let $p$ denote the sampling "measure". Thus, if $\omega$ is any sample then $p(\omega)$ is the probability that the sample is drawn under our PST ("physical sampling technique"). Note that there is no restriction on the PST that we use. Thus, all possible PST's are covered by this theory. In this formulation, $\omega$ already stands for a sample whose units are distinct. The case where we have a sample $\omega$ whose units are not distinct is covered by this case. Similarly, the order in which the units are drawn has also been ignored, because that is also indirectly covered by this theory. Now, let $\Omega$ denote the set of all possible $2^N$ samples. We now define a function $r$ from $\Omega$ to real numbers. Thus, $r(\omega)$ is a real number for all $\omega \in \Omega$. The function $r$ satisfies the condition that $r(\omega)p(\omega) = 0$, if $p(\omega) = 0$.

For $1 \leq t \leq N$, let $1 \leq j_1 < j_2 \cdots < j_t \leq N$, and let $\underline{j} = (j_1, j_2, \cdots, j_t)$. Although the elements of $\underline{j}$ are written in an ordered form, we shall always consider $\underline{j} \equiv (\underline{j}(t))$ to be an unordered set. Let

(4.1)                     $\pi_r(\underline{j}(t)) = \Sigma_2 \, p(\omega)r(\omega)$ ,

where $\Sigma_2$ runs over all samples $\omega$ which contain the $t$ units $\underline{j}(t)$. For $1 \leq k \leq N$, recall $\underline{i}$. Let $T_r(\underline{i}, t)$ be the class of all $t$-sets $\underline{j}(t)$, such that the unordered set $\underline{j}$ contains the unordered set $\underline{i}$, and such that $\pi_r(\underline{j}(t)) \neq 0$. Let $\nu_r(\underline{i}, t)$ denote the number of elements in the set $T_r(\underline{i}, t)$. Again, for all $k$ such that $1 \leq k \leq N$, for all $\underline{i}$, and for all $t$ (with $1 \leq t \leq N$), let $\alpha(\underline{i}, t)$ be real numbers such that $\alpha(\underline{i}, t) = 0$, if $\nu_r(\underline{i}, t) = 0$, and such that $\sum_{t=1}^{N} \alpha(\underline{i}, t) = 1$. Also, for all real numbers $x$, we denote the Moore-inverse of $x$ by $x^-$. (Thus, $x^- = 0$ if $x = 0$, and $x^- = 1/x$, if $x \neq 0$.)

We are now able to present an estimator for $Q(\psi)$. For all $\omega \in \Omega$, let

$$(4.2) \qquad \beta_r(\underline{i}, \omega) = r(\omega) \sum_{t=1}^{N} \alpha(\underline{i}, t) \; [\nu_r(\underline{i}, \underline{t})]^{-1} \left\{ \sum_{3} (\pi_r(\underline{j}(t)))^{-} \right\},$$

where $\Sigma_3$ runs over all $t$-sets $\underline{j}$, such that $\omega$ contains $\underline{j}$, and (simultaneously) $\underline{i}$ is contained in $\underline{j}$. Define

$$(4.3) \qquad \hat{Q}^{Sr}(\psi) \equiv \hat{Q}(\psi) = \Sigma_4 \; \psi(\underline{i})\beta_r(\underline{i}, \omega)$$

where $\Sigma_4$ runs over all possible values of $\underline{i}$ which are contained in the sample $\omega$. Notice that $\hat{Q}(\psi)$ depends upon $\omega$. In Srivastava (1985), where this estimator was proposed, some fundamental results concerning this class of estimators were presented. For lack of space, we can not describe the results fully, but the nature of these results is mentioned, for convenience of the reader, in the following.

**Results 4.1** (1) (Theorems 4.1, 4.5, 4.6 and 4.7 of the above paper). For all choices of $r$ and $\alpha$, (subject to the stated conditions), for all choices of the PST's (and, hence $p$), and for all choices of the function $\psi$ (and the value of $k$), it is shown that the estimator $\hat{Q}(\psi)$ at (4.3) is an unbiased estimate of $Q(\psi)$.

(2) (Theorem 4.2) Consider two estimators $\hat{Q}_1, (\psi_1)$ and $\hat{Q}_2(\psi_2)$. Here, $\psi_1$ and $\psi_2$ may be different functions, and $\psi_i(i = 1, 2)$ may involve $k_i$ (as the value of $k$), and $k_1$ and $k_2$ may be different. Also $\hat{Q}_i(i = 1, 2)$, may involve the value $r_i$ of $r$ and the value $\alpha_i$ of $\alpha$, with $r_1 \neq r_2$ and $\alpha_1 \neq \alpha_2$. (In other words, the two elements in any of the pairs $(k_1, k_2), (\psi_1, \psi_2), (r_1, r_2)$, and $(\alpha_1, \alpha_2)$ may be distinct or may be identical). Then, the paper presents an expression, which is quite simplified and computable, for $\mathrm{Cov}(\hat{Q}_1(\psi_1), \hat{Q}_2(\psi_2))$. Notice that by taking $k_1 = k_2$ and $\psi_1 = \psi_2$, this gives the covariance of two (possibly different) estimators of $Q(\psi)$. Also, by further taking $\alpha_1 = \alpha_2$ and $r_1 = r_2$, this gives us the expression for $\mathrm{Var}(\hat{Q}(\psi))$.

(3) (Theorems 4.3 and 4.4) Consider $\hat{Q}_1(\psi_1)$ and $\hat{Q}_2(\psi_2)$ in the same vast generality as above. Then, the paper presents estimates (from a sample) of $\mathrm{Cov}(\hat{Q}_1(\psi_1), \hat{Q}_2(\psi_2))$

(4) (Theorems 6.1, 6.2). When $k = 1$, and $\psi(\underline{i}) = y_i$ for all $i$, we get $Q(\psi) = Y$. By taking $\alpha(i, t) = 0$ for $t \neq 1$ (and, hence, $\alpha(i, 1) = 1$), the estimator $\hat{Q}^{Sr}(\psi)$ reduces to one denoted by $\hat{Y}_{Sr1}$, where

$$(4.4) \qquad \hat{Y}_{Sr1} = r(\omega) \sum_{4} [y_i/\pi_r(i)]$$

where $\sum_4$ sums over all $i$ in $\omega$ and where

$$(4.5) \qquad\qquad \pi_r(i) = \Sigma_5 \; p(\omega) r(\omega),$$

where $\sum_5$ sums over all samples $\omega$ which contain the unit $i$. The paper provides an expression for $\mathrm{Var}(\hat{Y}_{Sr1})$. (Also, $\hat{Y}_{HT}$ is shown to be a special case of $\hat{Y}_{Sr1}$ when $r(\omega) = $ constant, for all $\omega$.)

When sample size is variable, the choice $r(\omega) = |\,\omega\,|^{-1}$ was considered in Srivastava (1988); this leads to the estimator $\hat{Y}_{S2}$ given by

$$(4.6) \qquad\qquad \hat{Y}_{S2} = \Sigma_4 \; y_i/\pi'_i \; , \text{where}$$

$$(4.7) \qquad\qquad \pi'_i = \Sigma_5 \; p(\omega)/\mid \omega \mid \; .$$

The Mickey and Hartley-Ross estimators are specialization of the author's estimator $\hat{Q}^{Sr}(\psi)$ (Srivastava Ouyang (1992)); so also is the Lahiri-Midzuno estimator.

More detailed studies were made on $\hat{Q}_{Sr1}$ in (Srivastava and Ouyang (1992)); we mention some of these below.

**Results 4.2** Assume $y_i > 0$, for all $i$, and let $\underline{y}' = (\underline{y}_1, \cdots, y_N)$.

(1) (Theorems 2.4, 6.1) It was shown that the following "Zero-Variance Equations" (ZVE) constitute a necessary and sufficient condition that $\mathrm{Var}(\hat{Y}_{Sr1}) = 0$.

There exist positive real numbers $r_1, r_2, \cdots, r_N$, such that

$$(4.8a) \qquad y_i \; /Y = \pi_r(i) \; /r_i, \text{ for } i = 1, 2, \cdots, N \; ;$$

and

$$(4.8b) \qquad \Sigma_4(1/r_i) = 1/r(\omega), \text{for all } \omega \; (\text{with } p(\omega) > 0) \; .$$

(It is seen that the theory of $p.p.s.$ ("probability proportional to size") sampling (along with the usage of the estimator $\hat{Y}_{HT}$) arises as a special case of the above, in case the sample size is fixed. When the sample size is variable, $\hat{Y}_{S2}$ satisfies the above but $\hat{Y}_{HT}$ does not. As a result, $\hat{Y}_{HT}$ is inadmissible for variable sampling size whereas $\hat{Y}_{S2}$ is admissible.)

(2) (Theorem 4.1) If we choose $r$ so that (4.8b) is satisfied (for some positive $(r_1, \cdots, r_N)$), then $\hat{Y}_{Sr1}$ is admissible in the class of homogeneous linear unbiased ($hlu$) estimators of $Y$.

(3) (Section 7) An iterative method for solving the ZVE is presented, along with some necessary conditions which the $r_i$ must satisfy if $(r_1, \cdots, r_N)$ is a solution of the ZVE.

(4) The "Principle of Invariance" was propounded, and its usefulness in estimating $Y$ was pointed out. This principle says that $Y$ is invariant under a permutation of $(y_i, \cdots, y_N)$, a fact which can be exploited by $\hat{Y}_{Sr1}$ but not by $p.p.s.$ sampling and $\hat{Y}_{HT}$.

Ouyang and Schreuder (1993) and Schreuder, Gregoire, and Wood (1993), and Ouyang, Schreuder, Max, and Williams (1993) applied the above theory to forestry. Other related results are also available. See, for example, Yusuf, (1986), and Ouyang, Srivastava, and Schreuder (1993).

Further work (Srivastava (1998), yet unpublished, is available. One part of it is concerned with the comparison of two strategies (I and II) for estimating $Y$, assuming that a guess value $\underline{y}^*$ of $\underline{y}$ is available. Strategy I uses $pps$ sampling using $\underline{y}^*$ and $\hat{Y}_{HT}$. Strategy II does SRSWOR (simple random sampling without replacement), and uses $\hat{Y}_{Sr1}$ where the function $r$ is obtained by solving the ZVE in which $\underline{y}^*$ is used in place of $\underline{y}$. When $\underline{y}$ and $\underline{y}^*$ are close to being proportional, the two procedures are close, but in other cases, Strategy II is decidedly better. This fact is then utilized in solving the problem of multiple characters. Suppose $g$ variables are present, and the corresponding $\underline{y}$ vectors are $\underline{y}^{(1)}, \cdots, \underline{y}^{(g)}$. Now, to use $p.p.s.$ sampling, we need to draw $g$ different samples, which is expensive. The alternative is to use some compromise. But when the $\underline{y}^{(i)}(i = 1, \cdots, y)$ differ too much, a good compromise is not possible. Hence, the classical theory is unable to produce an acceptable solution. Hence, using Strategy II, the problem can be satisfactorily solved, since a different $r$ function can be used for each different character.

Note that in the above theory, the function $r$ can be chosen even after the sample is drawn. (All that is required is that the choice of $r$ be not dependent on the sample drawn.)

Now, we are ready to discuss new unsolved problems, whose solution may be quite fruitful.

(i) We start with the estimation of $Y$. Consider solving the ZVE. The nature of dependence of $\underline{r}(= (r_1, \cdots, r_N))$ on the $\underline{y}$ needs to be studied.

(ii) A somewhat biased and nonlinear estimate which may be interesting can be developed as follows. We allow $r$ to depend upon the sample $\omega$ and using $\omega$ we first estimate certain broad features of the population of $y$-values. We then use it to obtain a guess value $\bar{\underline{y}}$ of $\underline{y}$, and find $r$ using the $\bar{\underline{y}}$ in the ZVE. This research will be in the direction of improving upon the usual estimator $\{\frac{N}{n}\Sigma_4 y_i\}$ of $Y$ (under SRSWOR) which is admissible in the hlu class.

(iii) When it is known that $(y_1, \cdots, y_N)$ is a set of $N$ independent variables drawn from a population with some known features, we use this information in $r$, and improve our estimator. This will give rise to "model-assisted" estimators.

(iv) The function $r$ is of a scale type. We can investigate a "location correction" as follows. Let $C > O$ be such that $y_i > C$, for all $i$. Then we can consider a population $U_C$ whose $\underline{y}$-vector is $(y_1 - C, \cdots, y_N - C)$. Solving the ZVE for $U_C$, we can determine the function $r(= r_C,$ say ). If $\hat{Y}_{Sr1}(C)$ is the estimate of the total for $U_C$, we then take the estimate of $Y$ to be $\hat{Y}_{Sr1,c}$ where

$$(4.9) \qquad\qquad \hat{Y}_{Sr1,C} = CN + \hat{Y}_{Sr1}(C) .$$

Notice that $\hat{Y}_{Sr1,C}$ will change with $C$. The question arises, as to what value of $C$ will be best, and how to determine it since $\underline{y}$ is not known.

(v) Study the following technique. Suppose we obtain a guess value for $\underline{y}$ as follows. Using $\omega$, we estimate (using extreme value theory) the value of $\min\limits_i y_i \equiv y_{(1)}$ and $\max\limits_i y_i \equiv y_{(N)}$.

We then consider $\underline{y}$ to be evenly spread out between $y_{(1)}$ and $y_{(N)}$, i.e. We take $y_j^* = y_{(1)} + (j-1)[y_{(N)} - y_{(1)}]/(N-1)$, for $j = 1, 2, \cdots, N$. Using the $\underline{y}^*$, we obtain $r$. If $\omega \equiv (i_1, i_2, \cdots, i_n)$ is the sample, we rename the units $(q_1, \cdots, q_n)$ as below. The unit $i_h$ $(h = 1, \cdots, n)$ is named $q_h$ if $|y_{i_h} - y_j^*|$ is minimum for $j = q_h$. We then use $\hat{Y}_{Sr1}$, and estimate $Y$ using the invariance principle.

(vi) For all $\underline{i}$, assume that $\psi(\underline{i}) > 0$, and $Q(\psi)$ is estimated by $\hat{Q}(\psi)$. Obtain the ZVE for this situation, and find conditions on $r$ so that the ZVE may be satisfiable. Study questions such as admissibility, and other questions of the type outlined above (for the case when $Q(\psi) = Y$).

(vii) Study other estimators of $Y$ corresponding to other choices of the $\alpha$'s. Compare these with whatever is available.

(viii) Let $S^2 = (1/(N-1))\{\sum_{i=1}^N (y_i - \bar{Y})^2\}$, with $\bar{Y} = Y/N$, so that $S^2$ is the population variance. Obviously, using the above theory, a large number of estimators of $S^2$ can be obtained. One important case is as follows. Notice that $S^2 = [1/2N(N-1)]Q(\psi)$, where $k = 2$ and $\psi(\underline{i}) = \psi(i_1, i_2) = (y_{i_1} - y_{i_2})^2$. One simple class of estimators arises by choosing (for all $\underline{i}$)) $\alpha(\underline{i}, 2) = 1$, and $\alpha(\underline{i}, t) = 0$, if $t \neq 2$. Various choices of $r$ are still available. The choice $r(\omega) = 1$ leads to the usual estimate of $S^2$ (under SRSWOR). For a general PST (and hence, a general $p$), the ZVE could be developed, and restrictions on the function $r$ so that the ZVE are satisfiable can be obtained. Then, various functions $r$ can be studied and compared with each other (with respect to their suitability for producing a very accurate estimator $\hat{Q}^{Sr}(\psi)$).

(ix) In the same spirit as in (viii), we can approach the general problem of estimation of the variance of any estimator of $Y$ (or such other quantities) under any PST. Basically, the expression for this variance will be of the form $\{H_1 - H_2\}$, where $H_1 = \sum_1^N d_i y_i^2$, and $H_2 = \sum_{i \neq j=1} d_{ij} y_i y_j$. We can estimate $H_1$ and $H_2$ separately. If the $d_i$ and the $d_{ij}$ and the $y_i$, are positive, we can (for $H_1$) take $k = 1$, and $\psi(\underline{i}) = d_i y_i^2$ and also (for $H_2$) take $k = 2$, and $\psi(\underline{i}) = y_i y_j d_{ij}$, and try various $r$ functions (taking $\alpha(\underline{i}, 1) = 1$, for $H_1$, and $\alpha(\underline{i}, 2) = 1$, for $H_2$, and all other $\alpha$'s to be zero in each case). The problem of producing estimators which are nonnegative can then be studied in this context.

(x) All the above theories should be extended to the more general case where $\psi(\underline{i})$ may take both positive and negative values for different values of $\underline{i}$. For example, if we are estimating $Y$, and the $y_i$ are not necessarily positive, we need the appropriate generalization of the above theories. Also, as suggested in some cases above, estimators which use the sample somewhat, and hence are biased, may also turn out to be very interesting, and sometimes may do better then the unbiased ones.

The author believes that once the above kind of problems are attacked in the right spirit and solved, an amazing array of powerful new estimators will be generated which may change the face of the subject of sampling theory.

### References

[1] Chu, J.Y. (1997) Model Identification in Factorial Experiments. (Ph.D. Dissertation) Colorado State University, Fort Collins, CO.

[2] Li, J.F. (1991) Sequential and Optimal Single Stage Factorial Designs, With Industrial Applications. Ph.D. Dissertation, Colorado State University, Fort Collins, CO.

[3] Ouyang, Z. and Schreuder, H.T. (1993), Approximate Srivastava Estimation in Forestry. *Forest-Science*, **39**, 309–320.

[4] Ouyang, Z., Schreuder, H.T., Max, T. and Williams, M. (1993), Poisson-Poisson and Binomial-Poisson Sampling in Forestry. *Survey Methodology*, **19**, 115–121.

[5] Ouyang, Z., Srivastava, J.N, and Schreuder, H.T. (1993), A General Ratio Estimator, and its Application to Regression Model Sampling in Forestry. *Annals. Inst. Stat. Math.*, **45**, 113–127.

[6] Schreuder, H.T., Gregoire, T.G., and Wood, G.B. (1993) *Sampling Methods for Multiresource Forest Inventory*, Wiley, New York.

[7] Srivastava, J.N. (1987), On the Inadequacy of Customary Orthogonal Arrays in Quality Control and General Scientific Experimentation, and the Need of Probing Designs of Higher Revealing Power. *Comm.*

*Stat.*, **16**, 2901–2940.

[8] Srivastava, J.N. (1988) On a General Theory of Sampling Using Experimental Design Concepts, II. Relation With Arrays, in: *Essays in Probability and Statistics*, North Holland, Amsterdam, 267–283.

[9] Srivastava, J.N. (1993) Multivariate Analysis With Few or Incomplete Observations, in: *Future Directions for Multivariate Analysis*, (ed., C.R. Rao), North-Holland.

[10] Srivastava, J.N. (1993), Nonadditivity in Row-column Designs. *J. Comb. Inf. Sys. Sc.*, **18**, 85–96.

[11] Srivastava, J.N. (1996) A Critique of Some Aspects of Experimental Design, in: *Handbook of Statistics Vol. 13* (S. Ghosh and C.R. Rao, ed), 309–341.

[12] Srivastava, J.N., and Beaver, R.J. (1986), On the Superiority of the Nested Multidimensional Block Designs, Relative to the Classical Incomplete Block Designs. *Jour. Statist. Plann. Inf.*, **13**, 133-150.

[13] Srivastava, J.N., and Chu, J.Y. (1997) Multistage Design Procedures for Identifying Two-factor Interactions, When Higher Effects are Negligible. Submitted for publication. (48 typed pages).

[14] Srivastava, J.N. and Hveberg, R. (1992), Sequential Factorial Probing Designs for Identifying and Estimating Nonnegligible Parameters. *Jour. Statist. Plann. Inf.*, **30**, 141–162.

[15] Srivastava, J.N. and Ouyang, Z. (1992), Studies on the General Estimator in Sampling Theory, Based on the Sample Weight Function. *Jour. Statist. Plann. Inf.*, **31**, 199–218.

[16] Srivastava, J.N. and Ouyang, Z. (1992), Some Properties of a Certain General Estimator in Finite Population Sampling. *Sankhya*, **54**, 261–264.

[17] Srivastava, J.N., and Wang, Y.C. (1996) Row-Column Designs: Nonadditivity Makes Them Hazardous to Use. To appear in Bose Conference Issue of Jour. Statist. Plann. Inf.

[18] Srivastava, J.N. and Zaatar, M.K. (1973), A Monte-Carlo Comparison of Four Estimators of the Dispersion Matrix of a Bivariate Normal Population, Using Incomplete Data. *Jour. Amer. Stat. Assoc.*, **68**, 180-183.

[19] Youssif, G.A. (1986) Some Studies on Srivastava Estimators for the Population Total Under Post-stratification. M.S. Thesis, Colorado State University, Fort Collins, CO.

# Pre-Limit and Post-Limit Theorems for Statistics

Gábor J. Székely

Department of Mathematics and Statistics
Bowling Green State University
Bowling Green, Ohio

and

Alfréd Rényi Institute of Mathematics
Hungarian Academy of Sciences
Budapest, Hungary

**Abstract.** Classical limit theorems of probability theory, especially the central limit theorem are standard tools in statistics. These limit theorems are based on the tail behavior of the component distributions. Finitely many statistical observations, however, can never completely justify any tail behavior. In this paper we introduce the new concept of *pre-limit theorems* relying on the central section, or body, of the distribution rather than the tail. Instead of limiting behavior our proposed 'pre-limit' theorems provide approximations for distributions of compositions when the number $n$ of observations is large but not too large. The first pre-limit result for sums of i.i.d. random variables is joint work with L. Klebanov and S.T. Rachev ([12]).

Even if we are ready to accept inferences on tails, many widely applied statistical models do not really match with theoretical expectations. Some examples are the Pareto-Mandelbrot model of income distributions or the Geometric Brownian Black-Scholes model of stock price fluctuations. The beauty and other inherent values of these models, however, suggest that under the surface there is some essential truth in these mathematically nice self-similar models, even if the data do not really fit in many cases. In this paper a new concept of *post-limit theorems* is introduced based on the idea that the time can be considered to be a random process. Combining this stochastic deformation of time and the theory of algebraic probability ([11], [25]) we can construct a larger class of random-time stable models which includes the classical stable models, and many other more flexible 'randomly selfsimilar' models. Other post-limit type results include theorems on degenerate U-statistics and the algebraic background of infinitely divisible limit distributions.

Several open problems will be proposed, possibly for the 21st century.


## 1. Pre-Limit Approach

There is a considerable debate in the literature about the applicability of $\alpha$-stable distributions which appear in Lévy's limit theorems (see, for instance [18], [4] and the references given there). A serious drawback of Lévy's approach is that, in practice, one can never know whether the underlying distribution is heavy tailed, or just has a long but truncated tail. Limit theorems for stable laws are not robust with respect to truncation of the tail or with respect to any change from 'light' to 'heavy' tail or conversely. We propose a new pre-limiting approach that helps to overcome this main drawback of Lévy-type limit theorems and also of the central limit theorem.

Let us start with an example.

**Example 1.** *Exponential decay.* One of the most popular examples for exponential distributions is the random time for radioactive decay. The exponential distribution is in the domain of attraction of the Gaussian law. In quantum physics it has been shown ([8] ,[32], [21]) that theoretically the radioactive decay is not exactly exponentially distributed, and recent experimental evidence supports that conclusion (see [31]). But then one faces the following paradox.

Let $p(t)$ be the probability density that a physical (quantum mechanical) system is in the initial state at the moment $t \geq 0$. It is known (see, for example, [34] p. 42) that $p(t) = |f(t)|^2$, where

$$f(t) = \int_0^\infty \omega(E) \exp(iEt) dE,$$

and $\omega(E) \geq 0$ is the density of energy of the disintegrating physical system. For a broad class of physical systems,

$$\omega(E) = \frac{A}{(E - E_o)^2 + \Gamma^2}, \ E \geq 0,$$

(see [34] and the references there), where $A$ is a normalizing constant, and $E_o$ and $\Gamma$ are the mode and the measure of dissipation of the system energy (with respect to $E_o$). For typical nonstable physical systems the ratio $\Gamma/E_o$ is very small ( of order $10^{-15}$ or smaller). Therefore we have that

$$f(t) = e^{iE_o t} \frac{A}{\Gamma} \int_{-\frac{E_o}{\Gamma}}^{\infty} \frac{e^{i\Gamma t y}}{y^2 + 1} dy$$

differs by a very small value (of magnitude $10^{-15}$) from

$$f_1(t) = e^{iE_o t} \frac{A}{\Gamma} \int_{-\infty}^{\infty} \frac{e^{i\Gamma t y}}{y^2 + 1} dy = \pi e^{iE_o t} \frac{A}{\Gamma} e^{-t\Gamma}, \ t > 0.$$

That is, $p(t) = |f(t)|^2$ is approximately equal to $\left(\frac{\pi A}{\Gamma}\right)^2 e^{-2t\Gamma}$, which gives (as an approximation) the classical exponential distribution model of the decay time. On the other hand it is equally easy to find the asymptotic representation for $f(t)$ as $t \to \infty$. Namely,

$$\int_{-\frac{E_o}{\Gamma}}^{\infty} \frac{e^{i\Gamma t y}}{y^2 + 1} dy = \int_{-\arctan(\frac{E_o}{\Gamma})}^{\frac{\pi}{2}} e^{i\Gamma t \tan z} dz \sim$$

$$-\frac{\cos^2(\arctan(\frac{E_o}{\Gamma}))}{it\Gamma} e^{-itE_o}.$$

Therefore

$$f(t) \sim i \frac{A}{E_o^2 + \Gamma^2} \frac{1}{t} \ \text{as } t \to \infty,$$

where

$$A = \frac{1}{\int_0^{\infty} \frac{dE}{(E-E_o)^2 + \Gamma^2}},$$

or

(1.1) $$p(t) \sim \frac{A^2}{E_o^2 + \Gamma^2} \frac{1}{t^2} \ , \text{as } t \to \infty.$$

Therefore, $p(t)$ belongs to the domain of attraction of a stable law with index $\alpha = 1$. Thus if $T_j, j \geq 1$, are independent identically distributed (i.i.d.) random variables describing the times of decay of a physical system, then the sum $\frac{1}{\sqrt{n}} \sum_{j=1}^{n} (T_j - c_j)$ with some constants $c_j$ does *not* tend to a Gaussian distribution (as we would have expected under exponential decay with $c_j = E(T_j)$).

Let us describe the situation in a more general setting. Given $X_j, j \geq 1$, i.i.d. random variables the limiting behavior of the normalized partial sums $S_n = n^{-1/\alpha}(X_1 + \ldots + X_n)$ depends on the tail behavior of $X_j$'s. Both the proper normalization, $n^{-1/\alpha}$ in $S_n$, and the corresponding limiting law are extremely sensitive to tail truncation. We claim that in this sense the problem of limiting distributions for the sums of i.i.d. random variables is *ill-posed*. Below we describe a well-posed version and an attempt to solve it in the form of a *pre-limit theorem*.

Let us fix two positive constants $c$ and $\gamma$, and consider the following semi-metric between the random variables $X$ and $Y$:

$$(1.2) \qquad d_{c,\gamma}(X, Y) = \sup_{|t| \geq c} \frac{|f_X(t) - f_Y(t)|}{|t|^\gamma}.$$

(Here and in what follows $F_Y(x)$ and $f_Y(t)$ stand for the cumulative distribution function (cdf) and the characteristic function of $X$, respectively.)

Observe that in the case $c = 0$ (1.2) defines a well-known probability distance (see, for example [22], [33]) in the space of all random variables for which $d_{0,\gamma}(X, Y)$ is finite, see [33].

Next recall that $Y$ is a strictly $\alpha$-stable random variable if for every positive integer $n$,

$$(1.3) \qquad Y_1 d = U_n := \frac{Y_1 + \ldots + Y_n}{n^{1/\alpha}}$$

where $d =$ stands for equality in distribution and $Y_j, j \geq 1$ are i.i.d. $Y_j d = Y$, see [34].

Let $X, X_j, j \geq 1$ be a sequence of i.i.d. random variables such that $d_{0,\gamma}(X, Y)$ is finite for some strictly stable random variable $Y$. Suppose that $Y, Y_j, j \geq 1$ are i.i.d. strictly $\alpha$-stable random variables, and $\gamma > \alpha$. Then

$$d_{0,\gamma}(S_n, Y) = d_{0,\gamma}(S_n, U_n)$$

$$= \sup_t \frac{|f_X^n(t/n^{1/\alpha}) - f_Y^n(t/n^{1/\alpha})|}{|t|^\gamma}$$

$$\leq n \sup_t \frac{|f_X(t/n^{1/\alpha}) - f_Y(t/n^{1/\alpha})|}{|t|^\gamma} = \frac{1}{n^{\gamma/\alpha-1}} d_{0,\gamma}(X, Y),$$

see [34]. From this we can see that $d_{0,\gamma}(S_n, Y)$ tends to zero as $n$ tends to infinity, that is, we have convergence (in $d_{0,\gamma}$ distance) of the normalized sums of $X_j$ to a strictly $\alpha$-stable random variable $Y$ provided that $d_{0,\gamma}(X, Y) < \infty$. However, *any* truncation of the tail of the distribution of $X$ leads to $d_{0,\gamma}(X, Y) = \infty$ . Our goal is to analyze the closeness of the sum $S_n$ to a strictly $\alpha$-stable random variable $Y$ without the assumption on the finiteness of $d_{0,\gamma}(X, Y)$, restricting our assumptions to bounds in terms of $d_{c,\gamma}(X, Y)$ with $c > 0$. Our result is based on the behavior of the characteristic function of the components outside of a small interval around 0. The pre-limiting approach can be illustrated by the following theorem (for more details see [12]).

**Theorem 1.1.** *Take an arbitrary probability distribution $h$ with bounded continuous density function and denote a bound by $c = c(h)$. Introduce the smoothed Kolmogorov distance $k_h(F, G)$ between two arbitrary cumulative distribution functions $F(x)$ and $G(x)$ as*

$$k_h(F, G) = \sup_x |F * h(x) - G * h(x)|,$$

*where $*$ is denotes convolution. Let further $X, X_j, j \geq 1$ be i.i.d. with bounded density function, and $S_n = n^{-1/\alpha} \sum_{j=1}^{n} X_j$. Suppose further that $Y$ is a strictly $\alpha$-stable random variable. Let $\gamma > \alpha$ and $\Delta > \delta$ be arbitrary given positive constants and let $n \leq (\frac{\Delta}{\delta})^\alpha$ be an arbitrary positive integer. Then*

$$k_h(F_{S_n}, F_Y) \leq \inf_{a>0} \left( \sqrt{2\pi} \frac{d_{\delta,\gamma}(X, Y)(2a)^\gamma}{n^{\frac{2}{\alpha}-1}\gamma} + 2\frac{c(h)}{a} + 2\Delta a \right).$$

In this theorem if $\Delta \to 0$ and $\Delta/\delta \to \infty$, then $n$ can be large enough so that the right-hand side becomes sufficiently small, that is, we obtain the classical limit theorem for weak convergence to an $\alpha$-stable law.

**Proposed Problem 1.** Generalize this theorem for U-statistics, and prove local pre-limit and large deviation results.

## 2.   Post-Limit Approach

Suppose we do not have reservations about the tail behavior of the sample and thus we do not need to use our pre-limit approach. Even then several difficulties may arise in statistical applications of classical limit theorems of probability theory. We can fix some of them by further research. I coined the term 'post-limit approach' for these investigations. Let me illustrate three of them.

(A) In the theory of U-statistics if the kernel is degenerate then the limiting distribution is the distribution of a linear combination of squares of i.i.d. normal variables. In this linear combination the coefficients typically depend on the kernel and also on the distribution of the components

(see [26], [15]). In testing problems (and in many other statistical procedures) these distributions are unknown and thus classical limit results on
degenerate kernel U-statistics cannot be applied (although there are some
exceptions where we do know these coefficients, and they do not depend
on unknown distributions). We can fix this problem by estimating the tail
probabilities of the limiting distributions without using any information on
the unknown component distributions.

Even the following crude inequality which does not use any information
on the kernel seems to be useful (for more details see [2]):

(2.1)                $P\{Q \geq x\} \leq P\{Z^2 \geq x\}, \quad \forall x \geq 1 + 2^{-1/2}$

for all quadratic forms $Q$ of normal variables centered at 0, $EQ = 1, Q \geq 0$
and standard normal $Z$.

Inequality (2.1) can be applied to test diagonal symmetry, multivariate normality, homogeneity, just to name a few, such that these tests are
consistent against all alternatives. Using the same inequality we can also
test 'Gaussianity' of stochastic processes in a consistent way against all
alternatives.

**Example 2.** *Testing for diagonal symmetry.* Testing symmetry is
a classical problem with dozens of suggested solutions. The asymptotic
confidence level of Kolmogorov and many other tests is known under the
assumption that the underlying distribution $F$ is continuous. In the multivariate case it is not that easy to eliminate the dependence on $F$, in
other words classical tests are not similar. We can overcome this difficulty
and construct rotation and scale invariant tests by applying the following
lemma (see [29]). (A simple modified test applies if the center of symmetry
is unknown.)

Let X and Y be i.i.d. $\mathbb{R}^d$-valued random vectors with finite expectation.
Then

$$E\left(\|X + Y\| - \|X - Y\|\right) \geq 0$$

with equality if and only if the distributions of $X$ and $-X$ coincide.

Denote by $Z_k$, $k = 1, 2, ...$ a sequence of i.i.d. standard normal random
variables, and put $Q = \sum_k c_k Z_k^2$ where $c_k \geq 0$ and $\sum_k c_k = 1$. The kernel
$f(x, y) = \|x + y\| - \|x - y\|$ satisfies all conditions to see that the sequence
of statistics

$$T_n(X_1, ..., X_n) := \frac{\sum_{1 \leq i < j \leq n}\left(\|X_i + X_j\| - \|X_i - X_j\|\right)}{\sum_{1 \leq i \leq n} \|X_i\|}$$

has a non-degenerate limit $Q - 1$ in distribution as $n$ tends to infinity
if the sample comes from a diagonally symmetric distribution, and goes

to infinity otherwise. This property makes it possible to base a test on $T_n$. Though the coefficients $c_k$ appearing in the limit distribution are not invariant under changes in the distribution of $X_i$, inequality (2.1) can help us with the critical values.

**Remark 2.1**

$T_n$ *is a reasonable empirical measure of multivariate asymmetry. The corresponding theoretical measure is*

$$T = E\left(\|X + Y\| - \|X - Y\|\right) / [2E\left(\|X\|\right)].$$

*$T = 0$ iff the distribution of $X$ is diagonally symmetric, $T = 1$ iff $T$ is a constant different from 0, otherwise $T$ is strictly between 0 and 1.*

**Proposed Problem 2.** Since the special form of $Q$ depends on the kernels it is reasonable to try to find sharper inequalities than (2.1) for given kernels. Find inequalities of this type and apply them to construct less conservative tests than the ones based on (2.1).

(B) Another type of difficulty with statistical applications of classical limit theorems is that the statistical data might not justify our simple (normal, stable, etc.) models. Examples include the 'Pareto-Mandelbrot' model of income distributions or the Geometric Brownian 'Black-Scholes' model of stock price fluctuations. Let us suppose that the time (e.g. market time) is random (depends on the volatility). Combining this stochastic deformation of time and the theory of algebraic probability([25] and [11]) we can obtain more flexible 'randomly selfsimilar' models.

**Example 3.** *Pareto-stable laws.* More than hundred years ago Vilfredo Pareto ([20]) observed that the number of people in the population whose income exceeds a given level $x$ can be satisfactorily approximated by $Cx^{-\alpha}$ for some $C$ and $\alpha > 0$. (See [1], [4], [14] for more details.) Later, Mandelbrot ([16], [17]) argued that stable laws should provide the appropriate model for income distributions. After some statistical studies on income data he made two claims:

(i) the distribution of income for different (but sufficiently long) time periods must be of the same type, in other words the distribution of the income follows a stable law (Lévy's stable law, see [3]),

(ii) the tails of the Gaussian law are too light to describe the distribution of the income in typical situations.

It is known that the variance of any non-Gaussian stable law is infinite, thus an essential condition for a non-Gaussian stable limit distribution for sums of random incomes is that the summands must have heavy tails in the sense that the variance of the summands must be infinite. On the other hand it is obvious that the incomes are always bounded random variables (in view of the finiteness of all available money in the world, and the existence of a smallest monetary unit). Even if we assume that the support of the income distribution is infinite, there exists a considerable number of empirical studies showing that the income distributions have Pareto tails with index $\alpha$ between 3 and 4, thus the variance is finite, see [6]. Thus in practice the underlying distribution *cannot* be heavy tailed.

Before suggesting a solution let us see another example.

**Example 4.** *Liquid financial markets.* Most liquid financial markets show that using Gaussian distribution systematically underestimates the probability of large price fluctuations, an issue of utmost importance in financial risk management. The empirical kurtosis is one indicator, typical values are 60 for US\$ /Swiss Franc exchange rate futures or 74 for US\$/DM. They are not even close to 0, as one would expect from a Gaussian model.

In order to get a simple flexible model define *random stability* as follows.

Let $X, X_1, \ldots, X_n$ be a sequence of i.i.d. $d$-dimensional random vectors. Assume that $\{\nu_p, \ p \in \Delta\}$, $\Delta \subset (0,1)$, is a family of non-negative, integer-valued random variables, independent of $\{X_j \ j \geq 1\}$. Assume further that $\nu_p$ has finite expectation, with $E\nu_p = 1/p$ for all $p$. We shall study the distributions of sums $S_p = \sum_{j=1}^{\nu_p} X_j$, $p \in \Delta$.

*Definition.* A $d$-dimensional random vector $X$ is said to be *randomly stable* if for some $\alpha \in (0, 2]$

$$X_1 d = p^{1/\alpha} \sum_{j=1}^{\nu_p} X_j.$$

This definition is essentially Mandelbrot's claim (i) with random time. It is not difficult to describe families $\{\nu_p \ p \in \Delta\}$ for which randomly stable vectors exist.

Denote by $P_p$ the generating function of the random variable $\nu_p$, and denote by $\wp$ the semigroup with operation of superposition $\circ$ generated by the family $\{P_p, \ p \in \Delta\}$.

We can prove (see [11]) that a necessary and sufficient condition for the existence of a randomly stable random vector $X$ for a given family $\{\nu_p, \ p \in \Delta\}$, is that the semigroup $\wp$ be commutative.

As a consequence we can also deduce that $g(u)$ is the characteristic function of a randomly stable distribution if and only if

$$(2.2) \qquad g(u) = \varphi(\exp(\alpha\tau + \alpha \int_{S^{d-1}} [\ln |(u,x)| - \frac{i\pi}{2} sign(u,x)]d\sigma(x)),$$

where $(u, x)$ is a scalar product, the measure $\sigma$ is defined as

$$\sigma(A) = \left\{ \frac{X}{\|X\|} \in A \right\},$$

$$\tau = \mathbf{C} + \mathbb{E} \ln \|X\| + \frac{1}{\alpha} \left( \int_0^\infty \ln x \, dA(x) - \mathbf{C} \right),$$

and $\mathbf{C} = -\Gamma'(1)$ is the Euler constant.

**Proposed Problem 3.** Find efficient estimators for the unknown parameters and measures in this representation. Find similar random time stable models for more complicated compositions than additions. Random time stability for symmetric polynomials and other U-statistics promise extremely wide applicability.

In all areas where the theory of fractals and chaos have been applicable one can try this new approach. These areas include turbulent motions of fluids, the chaotic rhythm of life, and market prices.

(C) The post limit approach can go beyond stability. Since the publication of papers by B. de Finetti, P.Lévy, A. Kolmogorov, B. Gnedenko and A. Khinchin almost 60 years ago it became clear that infinitely divisible distributions play a fundamental role in statistical modeling. The monograph [25] is a kind of post-limit approach for infinitely divisible limiting distributions in the following sense. In [25] a general theory was built up on the idea that the basic limit theorems remain valid if convolution is replaced by other operations. The convolution semigroup of probability distributions endowed with the topology of weak convergence can be replaced by other commutative topological semigroups with unit element. In this general setting the limit of products of 'small' elements from an arbitrarily small neighborhood of the unit element tend to an infinitely divisible element of the semigroup. Thus a topological property ('infinite smallness') implies an algebraic property ('infinite divisibility'). Thus we could eliminate probability measures from this part of probability theory. Only the spirit of probabilists remained there and a much wider applicability. For more details see [25]. Let me close the paper with an open problem from the field of algebraic probability theory.

**Proposed Problem 4.** Is it possible to decompose the convolution semigroup of probability distributions on the real line into the direct product of two nontrivial subsemigroups = components (nontrivial decomposition means that all components contain non-degenerate distributions)?

If such a decomposition is possible then we can solve problems 'coordinatewise' as in many other branches of mathematics. Let me explain this by an example.

**Example 5.** *Direct decomposition.* Every probability distribution on the real line has a unique decomposition into the convolution of a normal and an anti-normal distribution (anti-normal = has no normal convolution component). This is a direct decomposition but the anti-normal distributions do not form a convolution semigroup as it was observed by R. A. Fisher and D. Dugué [5]. For related aspect of the same problem see [24] and [30].

Interestingly enough in this seemingly remote field of probability and statistics the name of R. A. Fisher appears again. This fact itself suggests that the solution of this problem promises new insight for probabilists and statisticians of the next century.

### References

[1] Arnold, B.C. (1983) *Pareto Distributions* I. C. Publ. House, Fairland, MD.

[2] Bakirov, N., Mori, T.F., and Szekely, G.J. (in preparation) A Unified Method for Testing Multivariate Hypotheses.

[3] Bunge, J. (1996), Composition semigroups and random stability. *Ann. of Probab.*, **24**, 1476–1489.

[4] DuMouchel, W. (1983), Estimating the stable index $\alpha$ in order to measure tail thickness: a critique. *Ann. Statist*, **11**, 1019–1031.

[5] Fisher, R. A., Dugué, D. (1948), Un résultat asséz inattendu d'arithmétique des lois de probabilités. *C.R. Acad. Sci. Paris*, **227**, 1205.

[6] Gnedenko, B.V. and Korolev, V.Yu. (1996) *Random Summation. Limit Theorems and Applications* CRC Press, Boca Raton.

[7] Hahn, M.G., Klass, M.J. (1997), Approximation of partial sums of arbitrary i.i.d. random variables and the precision of the usual exponential upper bound. *Ann. of Probab.*, **25**, 1451-1470.

[8] Khalfin, L.A. (1958), Contribution to the decay theory of a quasi-stationary state. *J. of Experimental and Theoretical Physics*, **6**, 1053–1063.

[9] Klebanov, L.B., Rachev, S.T. (1996), Sums of a random number of random variables and their approximations with $\nu$ - accompanying infinitely divisible laws. *Serdica*, **22**, 471–498.

[10] Klebanov, L.B., Maniya, G.M. and Melamed, J.A. (1984), A problem of Zolotarev and analogs of infinite divisible and stable distributions in a scheme for summing of a random number of random variables. *Theory Probab. Appl.*, **29**, 791–794.

[11] Klebanov, L. B., Kozubowski, T.J., Rachev, S.T., and Székely, G.J. (in preparation) Stability and Random Time.

[12] Klebanov, L.B., Rachev, S.T., Szekely, G. J. *Central Pre-limit Theorem and Its Applications*, to appear.

[13] Korolyuk, V.S. , Borovskih, Ju., V. (1989) *The Theory of U-statistics (in Russian)*. Naukova Dumka, Kijev.

[14] Kozubowski, T. J., Rachev, S. T. (1994), The theory of geometric stable distributions and its use in modeling financial data. *European J. Oper. Res.*, **74**, 310–324.

[15] Lee, A. (1990) *U-statistics. Theory and practice.* Marcel Dekker, Inc., New York.

[16] Mandelbrot, B., (1959), Variables et processus stochastiques de Pareto-Levy, et la repartition des revenus. *C.R. Acad. Sc. Paris*, **23**, 2153–2155.

[17] Mandelbrot, B., (1960), The Pareto-Levy law and the distribution of income. *Internat. Econ. Rev.*, **1**, 79–106.

[18] Mittnik, S., and Rachev, S. T. (1993), Modeling asset returns with alternative stable distributions. *Econometric Rev*, **12**, 261–330.

[19] Móri, T.F., Székely, G.J. (1982), Asymptotic behavior of symmetric polynomial statistics. *Annals of Probability*, **10**, 124–131.

[20] Pareto, V. (1897) *Cours d'Economie Politique.* F. Rouge, Lausanne, Switzerland.

[21] Petrosky, T., Prigogine, I. (1997) *Advances in Chemical Physics* Vol. XCIX, 1-120, Wiley, Chichester, New York.

[22] Rachev S.T. (1991) *Probability Metrics and the Stability of Stochastic Models.* Wiley, Chichester, New York.

[23] Rempala, G., Székely, G.J. (1998), On estimation with elementary symmetric polynomials. *Random Operators and Stochastic Equations*, **6/1**, 77–88.

[24] Ruzsa, I. Z. , Székely, G. J. (1985), No distribution is prime. *Z. Wahrscheinlichkeitstheorie v. Geb.*, **70**, 263–269.

[25] Ruzsa, I. Z. , Székely, G.J. (1988) *Algebraic Probability Theory.* Wiley, Chichester, New York.

[26] Serfling, J. R. (1980) *Approximation Theorems of Mathematical Statistics.* Wiley, Chichester, New York.

[27] Shirjajev, A. N. (1998) *Foundations of Stochastic Financial Mathematics (in Russian).* Fazis, Moscow.

[28] Székely, G. J. (1986) *Paradoxes in Probability Theory and Mathematical Statistics.* Reidel-Kluwer, Dordrecht, Holland.

[29] Székely, G.J. (1996) *Contests in Higher Mathematics.* Springer, New York.

[30] Székely, G. J. , Zempléni, A. (1997), A direct decomposition of the convolution semigroup of probability distributions. *Studia Sci Math Hung.*, **32**, 20–27.

[31] Wilkinson, S.R., Bharucha, C.F., Fisher, M.C. (1997), Experimental evidence for non-exponential decay in quantum tunnelling. *Nature*, **387**, 575–577.

[32] Winter, R.G. (1961), Evolution of a quasi-stationary state. *Phys. Rev.*, **123**, 1503–1507.

[33] Zolotarev V.M. (1986) *Modern Theory of Summation of Independent Random Variables*. Nauka, Moscow.

[34] Zolotarev V.M. (1983) *One-Dimensional Stable Distributions*. Nauka, Moscow (in Russian, English translation: Amer. Math. Soc., Providence, R.I., 1986).

# Regression Modelling with Fixed Effects: Missing Values and Related Problems

Helge Toutenburg, Andreas Fieger and Christian Heumann
Institut für Statistik
Ludwig Maximilians Universität München
Munich, Germany

**Abstract.** The paper considers three problems in the linear regression.

(i)    The predictive performance of (possibly biased) restricted and
mixed regression estimators with respect to a stochastic target function $T(y) = \Lambda y + (I - \Lambda)\mathbb{E}(y)$, with $\Lambda$ a weight matrix is investigated. Under MDEP-matrix superiority, application comes from imputation for missing values resulting in a biased mixed estimator.

(ii)    The weighted mixed regression estimator is a flexible data oriented method to handle the problem of missing values in the $X$-matrix. Alternative strategies to optimize the weights are discussed.

(iii)    The basic question of detecting a non-MCAR process is demonstrated using outlier detection methods. Especially the power of detecting a non-MCAR process is investigated using adaptions of Cooks distance, of DRSS and DXX in a simulation study.

## Problem 1.
## Predictive Performance of Restricted and Mixed Regression Estimators

### 1.1 Introduction

Generally predictions from a linear regression model are made either for the actual values of the study variable or for the average values at a time. However, situations may occur in which one may be required to consider the predictions of both the actual and average values simultaneously. For example, consider the installation of an artificial tooth in patients through a specific device. Here a dentist would like to know the life of a restoration, on the average. On the other hand, a patient would be more interested in knowing the actual life of restoration in his/her case. Thus a dentist is interested in the prediction of average value but he may not completely ignore the interest of patients in the prediction of actual value. The dentist may assign higher weightage to prediction of average values in comparison to the prediction of actual values. Similarly, a patient may give more weightage to prediction of actual values in comparison to that of average values.

This section considers the problem of simultaneous prediction of actual and average values of the study variable in a linear regression model when a set of linear restrictions binding the regression coefficients is available, and analyzes the performance properties of predictors arising from the methods of restricted regression and mixed regression besides least squares.

### 1.2 Specification of Model and Target Function

Let us postulate the following linear regression model:

$$(1.1) \qquad \mathbf{y} = \mathbf{X}\beta + \mathbf{u}$$

where $\mathbf{y}$ is a $n \times 1$ vector of $n$ observations on the study variable, $\mathbf{X}$ is a $n \times K$ full column rank matrix of $n$ observations on $K$ explanatory variables, $\beta$ is a column vector of regression coefficients and $\mathbf{u}$ is an $n \times 1$ vector of disturbances.

It is assumed that the elements of $\mathbf{u}$ are independently and identically distributed with mean zero and variance $\sigma^2$.

If $\hat{\beta}$ denotes an estimator of $\beta$, then the predictor for the values of study variable within the sample is generally formulated as $\hat{\mathbf{T}} = \mathbf{X}\hat{\beta}$ which is used for predicting either the actual values $\mathbf{y}$ or the average values $\mathbb{E}(\mathbf{y}) = \mathbf{X}\beta$ at a time.

When the situation demands prediction of both the actual and average values together, Toutenburg and Shalabh (1996) defined the following stochastic target function

$$(1.2) \qquad T(\mathbf{y}) = \lambda \mathbf{y} + (1 - \lambda)\mathbb{E}(\mathbf{y}) = \mathbf{T}$$

and use $\hat{\mathbf{T}} = \mathbf{X}\hat{\beta}$ for predicting it where $0 \leq \lambda \leq 1$ is a nonstochastic scalar specifying the weightage to be assigned to the prediction of actual and average values of the study variable; see, e. g. Shalabh (1995).

**Remark (i).** In case that $\lambda = 0$, we have $\mathbf{T} = \mathbb{E}(\mathbf{y}) = \mathbf{X}\beta$ and then optimal prediction coincides with optimal estimation of $\beta$, whereas optimality may be defined, e. g., by minimal variance in the class of linear unbiased estimators or by some mean dispersion error criterion if biased estimators are considered. The other extreme case $\lambda = 1$ leads to $\mathbf{T} = \mathbf{y}$. Optimal prediction of $\mathbf{y}$ is then equivalent to optimal estimation of $\mathbf{X}\beta + \mathbf{u}$. If the disturbances are uncorrelated this coincides again with optimal estimation of $\mathbf{X}\beta$, i. e., of $\beta$ itself. If the disturbances are correlated according to $\mathbb{E}(\mathbf{u}\mathbf{u}') = \sigma^2 \mathbf{W}$, then this information leads to solutions $\hat{\mathbf{y}} = \mathbf{X}\hat{\beta} + \hat{\mathbf{u}}$ (cp. Goldberger, 1962).

**Remark (ii).** The two alternative prediction problems—the $\mathbf{X}\beta$-superiority and the y-superiority, respectively—are discussed in full detail in Rao and Toutenburg (1995, Chapter 6). As a central result, we have the fact that the superiority (in the Loewner ordering of definite matrices) of one predictor over another predictor can change if the criterion is changed. This was one of the motivations to define a target as in (1.2) that combines these two risks.

In the following we consider this problem but with the nonstochastic scalar $\lambda$ replaced by a nonstochastic matrix $\Lambda$. The target function is therefore

$$(1.3) \qquad T(\mathbf{y}) = \Lambda \mathbf{y} + (\mathbf{I} - \Lambda)\mathbb{E}(\mathbf{y}) = \mathbf{T}.$$

Our derivation of the results makes no assumption about $\Lambda$, but one may have in mind $\Lambda$ as a diagonal matrix with elements $0 \leq \lambda_i \leq 1$, $i = 1, \ldots, n$.

### 1.3 Exact Linear Restrictions

Let us suppose that we are given a set of $J$ exact linear restrictions binding the regression coefficients:

$$(1.4) \qquad \mathbf{r} = \mathbf{R}\beta$$

where $\mathbf{r}$ is a $J \times 1$ vector and $\mathbf{R}$ is a $J \times K$ full row rank matrix.

If these restrictions are ignored, the least squares estimator of $\beta$ is

$$(1.5) \qquad \mathbf{b} = (\mathbf{X'X})^{-1}\mathbf{X'y}$$

which may not necessarily obey (1.4). Such is, however, not the case with restricted regression estimator given by

$$(1.6) \qquad \mathbf{b_R} = \mathbf{b} + (\mathbf{X'X})^{-1}\mathbf{R'}[\mathbf{R}(\mathbf{X'X})^{-1}\mathbf{R'}]^{-1}(\mathbf{r} - \mathbf{Rb})$$

which invariably satisfies (1.4).

Employing (1.5) and (1.6), we get the following two predictors for the values of the study variable within the sample:

$$(1.7) \qquad \hat{\mathbf{T}} = \mathbf{Xb},$$

$$(1.8) \qquad \hat{\mathbf{T}}_R = \mathbf{Xb_R}.$$

In the following we compare the estimators $\mathbf{b}$ and $\mathbf{b_R}$ with respect to the predictive mean dispersion error (MDEP) of their corresponding predictions $\hat{\mathbf{T}} = \mathbf{Xb}$ and $\hat{\mathbf{T}}_R = \mathbf{Xb_R}$ for the target function $\mathbf{T}$.

From (1.3), and the fact that the ordinary least squares estimator and the restricted estimator are both unbiased, we see that

$$(1.9) \qquad \mathbb{E}_\Lambda(\mathbf{T}) = \mathbb{E}(\mathbf{y}),$$

$$(1.10) \qquad \mathbb{E}_\Lambda(\hat{\mathbf{T}}) = \mathbf{X}\beta = \mathbb{E}(\mathbf{y}),$$

$$(1.11) \qquad \mathbb{E}_\Lambda(\hat{\mathbf{T}}_R) = \mathbf{X}\beta = \mathbb{E}(\mathbf{y}),$$

but

$$(1.12) \qquad \mathbb{E}(\hat{\mathbf{T}}) = \mathbb{E}(\hat{\mathbf{T}}_R) \neq \mathbf{T}.$$

Equation (1.12) reflects the stochastic nature of the target function $\mathbf{T}$, a problem which differs from the common problem of unbiasedness of a statistic for a fixed but unknown (possibly matrix valued) parameter. Therefore both the predictors are only "weakly unbiased" in the sense that

$$(1.13) \qquad \mathbb{E}_\Lambda(\hat{\mathbf{T}} - \mathbf{T}) = 0,$$

$$(1.14) \qquad \mathbb{E}_\Lambda(\hat{\mathbf{T}}_R - \mathbf{T}) = 0.$$

### 1.3.1 MDEP Using Ordinary Least Squares Estimator

To compare alternative predictors, we define the matrix-valued mean-dispersion error for $\tilde{\mathbf{T}} = \mathbf{X}\hat{\beta}$ as follows:

(1.15) $$\text{MDEP}_\Lambda(\tilde{\mathbf{T}}) = \mathbb{E}(\tilde{\mathbf{T}} - \mathbf{T})(\tilde{\mathbf{T}} - \mathbf{T})'.$$

First we note that

$$\mathbf{T} = \Lambda\mathbf{y} + (\mathbf{I} - \Lambda)\mathbb{E}(\mathbf{y})$$

(1.16) $$= \mathbf{X}\beta + \Lambda\mathbf{u},$$

$$\hat{\mathbf{T}} = \mathbf{X}\mathbf{b}$$

(1.17) $$= \mathbf{X}\beta + \mathbf{P}\mathbf{u},$$

with the symmetric and idempotent projection matrix $\mathbf{P} = \mathbf{X}(\mathbf{X}'\mathbf{X})^{-1}\mathbf{X}'$. Hence we get

$$\text{MDEP}_\Lambda(\hat{\mathbf{T}}) = \mathbb{E}(\mathbf{P} - \Lambda)\mathbf{u}\mathbf{u}'(\mathbf{P} - \Lambda)'$$

(1.18) $$= \sigma^2(\mathbf{P} - \Lambda)(\mathbf{P} - \Lambda)',$$

using our previously made assumptions on $\mathbf{u}$.

### 1.3.2 MDEP Using Restricted Estimator

The problem is now solved by calculation of

(1.19) $$\text{MDEP}_\Lambda(\hat{\mathbf{T}}_R) = \mathbb{E}(\hat{\mathbf{T}}_R - \mathbf{T})(\hat{\mathbf{T}}_R - \mathbf{T})'.$$

Using the abbreviation

(1.20) $$\mathbf{F} = \mathbf{X}(\mathbf{X}'\mathbf{X})^{-1}\mathbf{R}'[\mathbf{R}(\mathbf{X}'\mathbf{X})^{-1}\mathbf{R}']^{-1}\mathbf{R}(\mathbf{X}'\mathbf{X})^{-1}\mathbf{X}'$$

and

(1.21) $$\mathbf{r} - \mathbf{R}\mathbf{b} = -\mathbf{R}(\mathbf{X}'\mathbf{X})^{-1}\mathbf{X}'\mathbf{u},$$

we get from (1.6), (1.8), (1.16) and (1.17) the following

$$\hat{\mathbf{T}}_R - \mathbf{T} = \mathbf{X}\mathbf{b}_R - \mathbf{T}$$

(1.22) $$= (\mathbf{P} - \mathbf{F} - \Lambda)\mathbf{u}.$$

As $\mathbf{F} = \mathbf{F}'$, $\mathbf{P} = \mathbf{P}'$ and $\mathbf{P}\mathbf{F} = \mathbf{F}\mathbf{P} = \mathbf{F}$, we have

$$\text{MDEP}_\Lambda(\hat{\mathbf{T}}_R) = \sigma^2(\mathbf{P} - \mathbf{F} - \Lambda)(\mathbf{P} - \mathbf{F} - \Lambda)'$$

(1.23) $$= \sigma^2[(\mathbf{P} - \Lambda)(\mathbf{P} - \Lambda)' - (\mathbf{F} - \Lambda\mathbf{F} - \mathbf{F}\Lambda')].$$

### 1.3.3 MDEP Matrix Comparison

Using the results (1.18) and (1.23), the difference of the MDEP-matrices can be written as

$$\Delta_\Lambda(\hat{\mathbf{T}}; \hat{\mathbf{T}}_R) = \text{MDEP}_\Lambda(\hat{\mathbf{T}}) - \text{MDEP}_\Lambda(\hat{\mathbf{T}}_R)$$
$$= \sigma^2(\mathbf{F} - \Lambda\mathbf{F} - \mathbf{F}\Lambda')$$
(1.24)
$$= \sigma^2\left[(\mathbf{I} - \Lambda)\mathbf{F}(\mathbf{I} - \Lambda)' - \Lambda\mathbf{F}\Lambda'\right].$$

Then $\hat{\mathbf{T}}_R$ becomes MDEP-superior to $\hat{\mathbf{T}}$ if $\Delta_\Lambda(\hat{\mathbf{T}}; \hat{\mathbf{T}}_R) \geq 0$.

For $\Delta_\Lambda(\hat{\mathbf{T}}; \hat{\mathbf{T}}_R)$ to be non-negative definite, it follows from Baksalary, Schipp and Trenkler (1992) that necessary and sufficient conditions are

$$(i) \mathcal{R}(\Lambda\mathbf{F}) \subset ((\mathbf{I} - \Lambda)\mathbf{F}$$
$$(ii) \lambda_1 \leq 1$$

where $\lambda_1$ denotes the largest characteristic root of the matrix $[(\mathbf{I} - \Lambda)\mathbf{F}(\mathbf{I} - \Lambda')]^+\Lambda\mathbf{F}\Lambda'$.

For the simple special case of $\Lambda = \theta\mathbf{I}$, the conditions reduce to $\theta \leq \frac{1}{2}$.

### 1.4 Missing values in the X-Matrix and the Mixed Estimator

An interesting problem in all regression models relates to missing data. In general, we may assume the following structure of data:

(1.25)
$$\begin{pmatrix} \mathbf{y}_{\text{obs}} \\ \mathbf{y}_{\text{mis}} \\ \mathbf{y}_{\text{obs}} \end{pmatrix} = \begin{pmatrix} \mathbf{X}_{\text{obs}} \\ \mathbf{X}_{\text{obs}} \\ \mathbf{X}_{\text{mis}} \end{pmatrix} \beta + \mathbf{u}.$$

Estimation of $\mathbf{y}_{\text{mis}}$ corresponds to the prediction problem discussed in Chapter 6 of Rao and Toutenburg (1995) in full detail. We may therefore confine ourselves to the structure

(1.26)
$$\mathbf{y}_{\text{obs}} = \begin{pmatrix} \mathbf{X}_{\text{obs}} \\ \mathbf{X}_{\text{mis}} \end{pmatrix} \beta + \mathbf{u}$$

and change the notation as follows:

(1.27)
$$\begin{pmatrix} \mathbf{y} \\ \mathbf{y}_* \end{pmatrix} = \begin{pmatrix} \mathbf{X} \\ \mathbf{X}_* \end{pmatrix} \beta + \begin{pmatrix} \mathbf{u} \\ \mathbf{u}_* \end{pmatrix}, \quad \begin{pmatrix} \mathbf{u} \\ \mathbf{u}_* \end{pmatrix} \sim (0, \sigma^2\mathbf{I}).$$

The submodel

(1.28)
$$\mathbf{y} = \mathbf{X}\beta + \mathbf{u}$$

presents the completely observed data and should fulfill the standard assumptions (i. e., $\mathbf{X}$ is nonstochastic of full column rank). The other submodel

$$\mathbf{y}_* = \mathbf{X}_*\beta + \mathbf{u}_*$$

is related to the partially observed $\mathbf{X}$-variables. The dimensions of the two models are $m_C$ and $m_*$, respectively, with $n = m_C + m_*$.

Let $\mathbf{M} = (m_{ij})$ define the missing indicator matrix (c.p. Rubin, 1976) with $m_{ij} = 1$ if $x_{ij}$ is not observed and $m_{ij} = 0$ if $x_{ij}$ is observed. Under the assumption that missingness is independent of $\mathbf{y}$, i.e.,

$$f(\mathbf{M}|\mathbf{y}, \mathbf{X}) = f(\mathbf{M}|\mathbf{X})$$

we have

$$f(\mathbf{y}|\mathbf{M}, \mathbf{X}) = \frac{f(\mathbf{y}, \mathbf{M}|\mathbf{X})}{f(\mathbf{M}|\mathbf{X})} = \frac{f(\mathbf{M}, \mathbf{y}|\mathbf{X})}{f(\mathbf{M}|\mathbf{y}, \mathbf{X})} = f(\mathbf{y}|\mathbf{X})$$

which means that the the CC-estimator (complete case)

$$(1.29) \qquad\qquad \mathbf{b} = (\mathbf{X'X})^{-1}\mathbf{X'y}$$

is consistent for $\beta$.

As an alternative one may impute estimates or fixed values for the missing data so that the partially unknown matrix $\mathbf{X}_*$ is replaced by a known matrix $\mathbf{R}$ resulting in

$$(1.30) \qquad\qquad \mathbf{y}_* = \mathbf{R}\beta + (\mathbf{X}_* - \mathbf{R})\beta + \mathbf{u}_*$$

or, equivalently written in the shape of stochastic linear restrictions,

$$(1.31) \qquad\qquad \mathbf{r} = \mathbf{R}\beta + \delta + \phi, \quad \phi \sim (0, \sigma^2\mathbf{I})$$

with $\delta = (\mathbf{X}_* - \mathbf{R})\beta$ a bias vector. Combining the CC-model (1.28) and the filled-up model (1.31) results in the mixed model (Theil and Goldberger, 1961)

$$(1.32) \qquad\qquad \begin{pmatrix} {}^l\mathbf{y} \\ \mathbf{r} \end{pmatrix} = \begin{pmatrix} \mathbf{X} \\ \mathbf{R} \end{pmatrix} \beta + \begin{pmatrix} {}^l\mathbf{0} \\ \delta \end{pmatrix} + \begin{pmatrix} \mathbf{u} \\ \phi \end{pmatrix}.$$

For $\delta = 0$, the BLUE in (1.32) is given by the mixed estimator

$$(1.33) \qquad \tilde{\mathbf{b}}_\mathbf{R} = \mathbf{b} + (\mathbf{X'X})^{-1}\mathbf{R'}\left[\mathbf{I} + \mathbf{R}(\mathbf{X'X})^{-1}\mathbf{R'}\right]^{-1}(\mathbf{r} - \mathbf{Rb})$$

with dispersion matrix

(1.34)
$$V(\tilde{b}_R) = V(b) - \sigma^2 (X'X)^{-1} R' \left[ I + R(X'X)^{-1} R' \right]^{-1} R(X'X)^{-1}$$
(1.35)
$$= V(b) - D ,$$

say, whence it follows that the variance covariance matrix of b exceeds the variance covariance matrix of $\tilde{b}_R$ by a non-negative definite matrix and thus $\tilde{b}_R$ is more efficient than b.

In case that $\delta \neq 0$, the mixed estimator $\tilde{b}_R$ becomes biased and its bias vector is

(1.36)
$$\text{Bias}(\tilde{b}_R, \beta) = Dd$$

where

(1.37)
$$d = (X'X)R^+ \delta \sigma^{-2}$$
(1.38)
$$R^+ = R'(RR')^{-1} .$$

Therefore $\text{Bias}(\tilde{b}_R, \beta) \in \mathcal{R}(D)$ and we may apply result A1 given in the Appendix to get the following theorem.

**Theorem 1.** Let $M(\hat{\beta}, \beta) = \mathbb{E}(\hat{\beta} - \beta)(\hat{\beta} - \beta)'$ define the MDE matrix of an estimator $\hat{\beta}$ of $\beta$. Then the biased estimator $\tilde{b}_R$ is MDE-superior over the OLSE b in the sense that the variance covariance matrix of b exceeds the mean squared error matrix of $\tilde{b}_R$ by a non-negative definite matrix if and only if

(1.39)
$$\rho = \sigma^{-2} \delta' \left[ I + R(X'X)^{-1} R' \right]^{-1} \delta \leq 1 .$$

If u and $\phi$ are independently normally distributed, then $\rho$ is the noncentrality parameter of the statistic

(1.40)
$$F = \frac{1}{J s^2} (r - Rb)' \left[ I + R(X'X)^{-1} R' \right]^{-1} (r - Rb)$$

which follows a noncentral $F_{J,n-K}(\rho)$-distribution under $\rho \leq 1$.

### 1.4.1 MDEP Using Mixed Estimator

Using the mixed estimator $\tilde{b}_R$, we have $\tilde{T}_R = X\tilde{b}_R$. Hence we have to calculate

(1.41)
$$\text{MDEP}_\Lambda \left( \tilde{T}_R \right) = \mathbb{E}(\tilde{T}_R - T)(\tilde{T}_R - T)' .$$

Using the abbreviation

$$(1.42) \qquad \mathbf{A} = (\mathbf{X'X})^{-1}\mathbf{R'} \left[\mathbf{I} + \mathbf{R}(\mathbf{X'X})^{-1}\mathbf{R'}\right]^{-1}$$

and taking into account that

$$(1.43) \qquad \mathbf{A}\left[\mathbf{I} + \mathbf{R}(\mathbf{X'X})^{-1}\mathbf{R'}\right]\mathbf{A'} = \mathbf{D}$$
$$(1.44) \qquad \mathbf{PX}(\mathbf{X'X})^{-1}\mathbf{R'A'} = \mathbf{D}$$
$$(1.45) \qquad \mathbf{XDX'} = \mathbf{F} \quad \text{in case that } \phi = 0$$

we may write

$$(1.46) \qquad \mathbf{\tilde{T}_R} - \mathbf{T} = (\mathbf{P} - \Lambda)\mathbf{u} + \mathbf{XA}\left[\phi - \mathbf{R}(\mathbf{X'X})^{-1}\mathbf{X'u}\right] + \mathbf{XA}\delta .$$

Therefore, using (1.43), (1.44) and (1.45) we get

$$\mathrm{MDEP}_\Lambda\left(\mathbf{\tilde{T}_R}\right) = \sigma^2 (\mathbf{P} - \Lambda)(\mathbf{P} - \Lambda)'$$
$$- \sigma^2\left(\mathbf{XDX'} - \Lambda\mathbf{XDX'} - \mathbf{XDX'}\Lambda'\right)$$
$$(1.47) \qquad + \mathbf{XA}\delta\delta'\mathbf{A'X'}$$

## 1.4.2 MDEP-Matrix Comparison

The difference of the MDEP-matrices of $\mathbf{\hat{T}}$ and $\mathbf{\tilde{T}_R}$ can be written as

$$(1.48) \qquad \Delta_\Lambda(\mathbf{\hat{T}}; \mathbf{\tilde{T}_R}) =$$
$$\sigma^2\left[(\mathbf{I} - \Lambda)\mathbf{XDX'}(\mathbf{I} - \Lambda)' - \Lambda\mathbf{XDX'}\Lambda'\right] - \mathbf{XA}\delta\delta'\mathbf{A'X'}$$

Then using Baksalary et al. (1992) and the result A1 of the Appendix, we have

$$\Delta_\Lambda(\mathbf{\hat{T}}; \mathbf{\tilde{T}_R}) \geq 0$$

if and only if

$$(1.49)$$
$$\text{(i) } \left[(\mathbf{I} - \Lambda)\mathbf{XDX'}(\mathbf{I} - \Lambda)' - \Lambda\mathbf{XDX'}\Lambda'\right] \geq 0$$
$$(1.50)$$
$$\text{(ii)} \sigma^{-2}\delta'\mathbf{A'X'}\left[(\mathbf{I} - \Lambda)\mathbf{XDX'}(\mathbf{I} - \Lambda)' - \Lambda\mathbf{XDX'}\Lambda'\right]^{-}\mathbf{XA}\delta \leq 1 .$$

**Problem 2.**
**Missing values in the X-matrix and the weighted mixed regression estimator**

In the following we again assume the situation given in equation (1.26), that is missing values in $\mathbf{X}$ only. Filling in replacement values for the missing values leads to the setup of biased mixed estimation as in equations (1.31) and (1.32). Since the additional information is biased, it seems pertinent to use a weight lower than one for this part of the model. This can be achieved by rewriting the target function to be minimized from

$$S(\beta) = (\mathbf{y} - \mathbf{X}\beta)'(\mathbf{y} - \mathbf{X}\beta) + (\mathbf{r} - \mathbf{R}\beta)'(\mathbf{r} - \mathbf{R}\beta)$$

to

$$S(\beta, \lambda) = (\mathbf{y} - \mathbf{X}\beta)'(\mathbf{y} - \mathbf{X}\beta) + \lambda(\mathbf{r} - \mathbf{R}\beta)'(\mathbf{r} - \mathbf{R}\beta) \ ,$$

with $0 \leq \lambda \leq 1$. The solution given by

$$\mathbf{b}(\lambda) = (\mathbf{X}'\mathbf{X} + \lambda\mathbf{R}'\mathbf{R})^{-1}(\mathbf{X}'\mathbf{y} + \lambda\mathbf{R}'\mathbf{r})$$

may be called the weighted mixed regression estimator (WMRE). This estimator may be interpreted as the familiar mixed estimator in the model

$$\begin{pmatrix} \mathbf{y} \\ \sqrt{\lambda}\mathbf{r} \end{pmatrix} = \begin{pmatrix} \mathbf{X} \\ \sqrt{\lambda}\mathbf{R} \end{pmatrix} \beta + \begin{pmatrix} \mathbf{u} \\ \sqrt{\lambda}\phi \end{pmatrix} \ .$$

Using $\mathbf{Z}_\lambda = (\mathbf{X}'\mathbf{X} + \lambda\mathbf{R}'\mathbf{R})$, we have the alternative representation

$$\mathbf{b}(\lambda) = \mathbf{Z}_\lambda^{-1}(\mathbf{X}'\mathbf{X}\beta + \mathbf{X}'\mathbf{u} + \lambda\mathbf{R}'\mathbf{X}_*\beta + \lambda\mathbf{R}'\phi)$$
$$= \beta + \lambda\mathbf{Z}_\lambda^{-1}\mathbf{R}'(\mathbf{X}_* - \mathbf{R})\beta + \mathbf{Z}_\lambda^{-1}(\mathbf{X}'\mathbf{u} + \lambda\mathbf{R}'\phi)$$

from which it follows that the WMRE is biased and its bias vector is given by

$$\text{Bias}\mathbf{b}(\lambda) = \lambda\mathbf{Z}_\lambda^{-1}\mathbf{R}'\delta \ ,$$

with covariance matrix as

$$V(\mathbf{b}(\lambda)) = \sigma^2\mathbf{Z}_\lambda^{-1}(\mathbf{X}'\mathbf{X} + \lambda^2\mathbf{R}'\mathbf{R})\mathbf{Z}_\lambda^{-1} \ .$$

**2.1 Ways of finding an optimal $\lambda$**

One strategy to find an optimal $\lambda$ is to minimize the MDEP. Let $\bar{y} = \bar{\mathbf{x}}'\beta + \sigma\bar{\epsilon}$ be a nonobserved future realisation of the regression model that is to be predicted by $p = \bar{\mathbf{x}}'\mathbf{b}(\lambda)$. Minimizing the MDEP of $p$ given by $\mathbb{E}(p - \bar{y})^2$ with respect to $\lambda$ leads to the relation (Rao and Toutenburg, 1995)

$$\lambda = \frac{1}{1 + \sigma^{-2}\rho_1(\lambda)\rho_2^{-1}(\lambda)}$$
$$\rho_1(\lambda) = \text{tr}[\mathbf{Z}_\lambda^{-1}\mathbf{S}\mathbf{Z}_\lambda^{-1}\mathbf{R}^{-1}\text{[}delta\delta'\mathbf{R}\mathbf{Z}_\lambda^{-1}]$$
$$\rho_2^{-1}(\lambda) = \text{tr}[\mathbf{Z}_\lambda^{-1}\mathbf{S}_\mathbf{R}\mathbf{Z}_\lambda^{-1}\mathbf{S}\mathbf{Z}_\lambda^{-1}] ,$$

with $\mathbf{S} = \mathbf{X}'\mathbf{X}$ and $\mathbf{S}_\mathbf{R} = \mathbf{R}'\mathbf{R}$. In general, the solution has to be found iteratively while $\sigma^2$ and $\delta$ have to be estimated by some procedure, e. g., $\sigma^2$ may be estimated from the complete cases. For the special case that only one observation is missing (i. e., $\mathbf{r}$ and $\delta$ are scalars), an explicit but unknown solution is available as

(2.1)
$$\lambda = \frac{1}{1 + \sigma^{-2}\delta^2} .$$

A second strategy is to minimize the trace of the MDE matrix with respect to $\lambda$, which is given by

$$\text{tr}\,\text{MDE}(\mathbf{b}(\lambda), \boldsymbol{\beta}) = \text{tr}\,[\sigma^2\mathbf{Z}_\lambda^{-1}(\mathbf{S} + \lambda^2\mathbf{S}_\mathbf{R})\mathbf{Z}_\lambda^{-1}] + \lambda^2\mathbf{Z}_\lambda^{-1}\mathbf{R}'\delta\delta'\mathbf{R}\mathbf{Z}_\lambda^{-1}]$$

Note that the solution $\lambda_{\text{tr}}$ has to be found iteratively.

A third way is to compare $\mathbf{b}(\lambda)$ and $\mathbf{b}$ with respect to the MDE MDE-I criterion. This results in the condition that $\mathbf{b}(\lambda)$ is MDE better than $\mathbf{b}$, if

$$\rho_\lambda = \sigma^{-2}\delta'[(2\lambda^{-1} - 1)\mathbf{I} + \mathbf{R}\mathbf{S}^{-1}\mathbf{R}']^{-1}\delta \leq 1 .$$

It can be shown, that the generalized version of (2.1),

$$\lambda_e = \frac{1}{1 + \sigma^{-2}\delta'\delta}$$

always fulfills this condition. Alternatively $\lambda_{\text{max}}$ could be chosen such that $\rho_\lambda = 1$ holds. Again, $\lambda_{\text{max}}$ has to be found iteratively.

## 2.2 A small simulation study

In a small simulation study we compared the estimators $\mathbf{b}$ (complete case estimator, which is the same as $\mathbf{b}(\lambda)$ with $\lambda = 0$), $\mathbf{b}(1) = \mathbf{b}_\mathbf{R}$ and $\mathbf{b}(\lambda)$ (with $0 < \lambda < 1$). The comparison of the respective estimators was conducted using the scalar risk function

$$R(\mathbf{I}) = \mathbb{E}\big(\mathbf{b}(\lambda) - \boldsymbol{\beta}\big)'\big(\mathbf{b}(\lambda) - \boldsymbol{\beta}\big)$$

estimated by its empirical version

$$\hat{R}(\mathbf{b}(\lambda), \beta) = \frac{1}{\#\text{rep}} \sum_{i=1}^{\#\text{rep}} \left(\mathbf{b}(\lambda)_i - \beta\right)' \left(\mathbf{b}(\lambda)_i - \beta\right)$$

where #rep means the number of repeated simulations of the error terms applied to one specific covariate data set. The details of the setup can be obtained from the authors on request.

Using the weights computed from $\delta$ and $\sigma^2$ which are known in the simulation study (generating 100 different covariate data sets), all weighted estimators were found to be better than the complete case estimator $\mathbf{b}$, as expected from the theory. Comparing the weighted estimators using the different $\lambda$-values previously mentioned with the estimator $\mathbf{b}_R$ shows that $\mathbf{b}_{\lambda_{tr}}$ performs best in this comparison ($\mathbf{b}_{\lambda_{tr}}$ was better than $\mathbf{b}_R$ in 91 of 100 runs, while $\mathbf{b}_{\lambda_e}$ was better than $\mathbf{b}_R$ in only 50 of 100 runs). On the other hand, using the weights computed from estimated $\hat{\delta} = \mathbf{r} - \mathbf{R}\mathbf{b}$ and $\hat{\sigma}^2$ (from the complete data), we observed that $\mathbf{b}_{\hat{\lambda}_e}$ was better than $\mathbf{b}$ in 99 of 100 runs, $\mathbf{b}_{\hat{\lambda}_{tr}}$ was better than $\mathbf{b}$ in 97 of 100 runs, while $\mathbf{b}_R$ was better than $\mathbf{b}$ in only 79 of 100 runs, but also that, e.g. $\mathbf{b}_{\hat{\lambda}_e}$ was better than $\mathbf{b}_R$ in only 32 of 100 runs and $\mathbf{b}_{\hat{\lambda}_{tr}}$ was better than $\mathbf{b}_R$ in only 43 of 100 runs. These results yield no transitive ordering of the estimators.

One possible reason for these results could be that the true $\lambda_e$ is typically underestimated by $\hat{\lambda}_e = 1/(1 + \hat{\sigma}^{-2}\hat{\delta}'\hat{\delta})$ (the degree is also depending on $\sigma^2$ and the covariance structure of $\mathbf{X}$), since it can be shown that $\mathbb{E}(\hat{\delta}'\hat{\delta}) = \delta'\delta + \sigma^2(J + \sum_{j=1}^{J} \mu_j)$, where $\mu_j$ are the eigenvalues of $\mathbf{R}\mathbf{S}^{-1}\mathbf{R}'$.

These observations suggest the construction of a bias corrected version of the estimators. An interesting direction is to use bootstrapping techniques to obtain a bias correction (using different resampling techniques). The results of this approach indicate that the estimates can be improved concerning the bias. But there is still noticeable underestimation.

## 2.3 Some concluding remarks

For handling the problem of missing values of some explanatory variables, weighted mixed estimation seems to be a promising approach. However, the determination of the weighing scalar requires careful attention. It will also be interesting to develop suitable procedures for confidence intervals and hypothesis testing. So far the results hold only for $J < p$, i.e., the number of restrictions is smaller than the number of variables. For the missing value context, we also need to investigate the case when $J > p$.

## Problem 3.
## Detection of non-MCAR processes in linear regression models

Missing data values in $\mathbf{X}$ are said to be missing completely at random (MCAR) if

$$f(\mathbf{M}|\mathbf{y}, \mathbf{X}, \phi) = f(\mathbf{M}|\phi) \quad \forall \mathbf{y}, \mathbf{X},$$

using the indicator matrix $\mathbf{M}$, defined in section 1.4.
For a mixed model with missing values in $\mathbf{X}_1$, we have

$$\mathbb{E}(y_i|X_{i1}, \ldots, X_{ip}) = \beta_0 + \sum_{j=1}^{p} \beta_j X_{ij}$$

$$\mathbb{E}(y_i|X_{i2}, \ldots, X_{ip}) = \beta_0 + \beta_1 \tilde{X}_{i1} + \sum_{j=1}^{p} \beta_j X_{ij}$$

with $\tilde{X}_{i1} = \mathbb{E}(X_{i1}|X_{i2}, \ldots, X_{ip})$. This means that imputing conditional means $\tilde{X}_{i1}$ and applying least squares on the completed data produce consistent estimates assuming MCAR (Little, 1992).

### MCAR Diagnosis

There are several approaches to detect missing data, which are non-MCAR. These include

- comparison of the means of $\mathbf{y}$ in the complete subsample (CC-data) and in the partially observed subsample,
- diagnostic plots, as introduced by Simon and Simonoff (1986), or
- the usage of diagnostic measures originally intended for the detection of outliers.

We will discuss the latter ideas in more detail.
Possible diagnostics include

- Cook's distance,
- the change in the residual sum of squares, or
- the change in the determinant of $\mathbf{X}'\mathbf{X}$ where originally the comparison is between the data sets $\mathbf{X}$ and $\mathbf{X}_{(i)}$, the data without case number $i$.

In the context of detecting a non-MCAR mechanism, the CC-data $\mathbf{X}$ and the partially observed data $\mathbf{X}_*$ are compared. Cook's distance now compares the (weighted) difference of the CC-estimator $\mathbf{b} = (\mathbf{X}'\mathbf{X})^{-1}\mathbf{X}'\mathbf{y}$ and the mixed-estimator $\tilde{\mathbf{b}}_R$ from (1.33)

$$\frac{(\tilde{\mathbf{b}}_R - \mathbf{b})'(\mathbf{X}'\mathbf{X} + \mathbf{R}'\mathbf{R})(\tilde{\mathbf{b}}_R - \mathbf{b})}{ps_R^2}.$$

Analogously, the change in the residual sum of squares (DRSS)

$$\frac{(\mathrm{RSS_R} - \mathrm{RSS_C})/m_*}{\mathrm{RSS_C}/(m_C - m_* - K + 1)}$$

and the change in the determinant (DXX)

$$\frac{\det(\mathbf{X'X})}{\det(\mathbf{X'X} + \mathbf{R'R})}$$

are used to gain information on the nature of the missing data mechanism.

**Idea**

The basic idea is to compare CC and 'valid imputation assuming MCAR' (Simonoff, 1988). If MCAR does not hold, then a MCAR-imputation for the missing values in $\mathbf{X}_*$ is not adequate. If we compare the diagnostic measure to it's distribution under $H_0$: "MCAR is valid", we should be able to detect a possible non-MCAR process. This is more general than comparing group means (see above), as this procedure can also detect non-MCAR with $\mathbb{E}(y) = \mathbb{E}(y_*)$.

**Distribution under $H_0$**

The distribution of the diagnostic measures under $H_0$ can be investigated using a Monte-Carlo method. The algorithm is as follows

• compute **b**
• replace $\mathbf{X}_*$ by 'valid imputation assuming MCAR' **R**
• replace $\mathbf{y}_*$ by $\hat{\mathbf{y}}_* = \mathbf{Rb}_C + \epsilon$, $\epsilon \sim N(0, \hat{\sigma}_C^2)$
• produce MCAR samples from the filled-in data repeatedly to generate a Null-distribution

The basic idea here is that, no matter what the true missing data mechanism is, the generated data will always have unobserved values that are MCAR. A basic underlying assumption that has to be fullfilled to keep type-I error under control is that the relationship between the missing values in $\mathbf{X}_*$ and the observed values can adequately be fitted by a linear regression model.

**Simulation Study**

A simulation study was conducted to investigate the properties of the above approach for different imputation methods and different correlation structures of the data matrix $X$. The structure was as follows.

• Generate $\mathbf{X} = (\mathbf{1}, \mathbf{x}_1, \mathbf{x}_2)$ with missing values only in $\mathbf{x}_2$.
• Repeat this step for varying $\rho = \mathrm{corr}(x_1, x_2)$, and
• varying amount of cases with missing values.

- Consider different non-MCAR processes.

The processes generating missing values were a mean split and a variance split process. The mean split process selects a value $x_{i2}$ as missing value with probability $p_1$ if $(x_{i2} - \bar{x}_2)$ exceeds a specified constant $c$. If $(x_{i2} - \bar{x}_2) \leq c$ the value is selected as missing value with probability $p_2$.

The variance split process is alike the mean split, but the absolute difference $|x_{i2} - \bar{x}_2| > c$ is used to decide if a value is selected as suitable for the missing value.

In brief, the simulation studies suggested that Cook's distance performes good for mean split while DRSS and DXX for variance split. Interestingly enough, performance also depended on $\rho$. For low absolute $\rho$, the usage of DRSS seems to perform better, whereas for high $\rho$, DXX gives better results.

In contrast to the simulation study, the missing data mechanism is unknown in real applications so that there is no general ranking of the diagnostic measures, concerning their ability for the detection of non-MCAR processes.

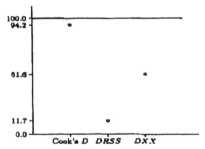

Figure 1. Mean Split, $p_1 = 0.005$, cutoff $= 0.25$, $p_2 = 0.74$, FOR, $\alpha = 0.05$, $\varrho = 0.7$

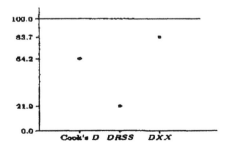

Figure 2. Variance Split, $p_1 = 0.006$, cutoff $= 0.8$, $p_2 = 0.7$, FOR, $\alpha = 0.05$, $\varrho = 0.7$

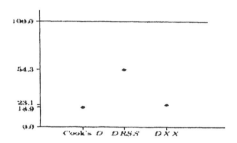

Figure 3. Variance Split, $p_1 = 0.006$, cutoff $= 0.8$, $p_2 = 0.7$, FOR, $\alpha = 0.05$, $\varrho = 0.3$

## Appendix

**Result A1 (Baksalary and Kala, 1983)** Let $\mathbf{A}$ be a non-negative definite matrix and let $\mathbf{a}$ be a column vector. Then $\mathbf{A} - \mathbf{aa}' \geq 0 \Leftrightarrow$

$$\mathbf{a} \in \mathcal{R}(\mathbf{A}) \quad \text{and} \quad \mathbf{a}'\mathbf{A}^-\mathbf{a} \leq 1 \,,$$

where $\mathbf{A}^-$ is any g-inverse of $\mathbf{A}$, that is, $\mathbf{AA}^-\mathbf{A} = \mathbf{A}$.

**Result A2 (Baksalary, Liski and Trenkler, 1989)** Let $\mathbf{A} = \mathbf{C}_1\mathbf{C}_1' - \mathbf{C}_2\mathbf{C}_2'$. Then $\mathbf{A} \geq 0 \Leftrightarrow$

$$\text{(i)} \mathcal{R}(\mathbf{C}_2) \subset \mathcal{R}(\mathbf{C}_1)$$
$$\text{(ii)} \lambda_1(\mathbf{C}_2'(\mathbf{C}_1\mathbf{C}_1')^-\mathbf{C}_2) \leq 1 \,.$$

**Theorem A3 (Baksalary et al., 1992).** *Let $\mathbf{F}$ be a symmetric non-negative definite $n \times n$-matrix. Then*

$$(\mathbf{I} - \Lambda)'\mathbf{F}(\mathbf{I} - \Lambda)' - \Lambda'\mathbf{F}\Lambda \geq 0 \qquad \Leftrightarrow$$

*1.* $\mathcal{R}(\Lambda'\mathbf{F}) \subset \mathcal{R}((\mathbf{I} - \Lambda)'\mathbf{F})$
*2.* $\lambda_1[\{(\mathbf{I} - \Lambda)'\mathbf{F}(\mathbf{I} - \Lambda)\}^+\Lambda'\mathbf{F}\Lambda] \leq 1$.

## References

[1] J. K. Baksalary and R. Kala (1983), Partial orderings between matrices one of which is of rank one. *Bulletin of the Polish Academy of Science, Mathematics*, **31**, 5–7.

[2] J. K. Baksalary, E. P. Liski, and Götz Trenkler (1989), Mean square error matrix improvements and admissibility of linear estimators. *Journal of Statistical Planning and Inference*, **23**, 312–325.

[3] J. K. Baksalary, Bernhard Schipp, and Götz Trenkler (1992), Some further results on hermitian matrix inequalities. *Linear Algebra and Its Applications*, **160**, 119–129. refa4Arthur S. Goldberger (1962) Best linear unbiased prediction in the generalized regression modelJournal of the American Statistical Association57369–375

[5] Roderick J. A. Little (1992), Regression with missing $X$'s: a review. *Journal of the American Statistical Association*, **87**, 1227–1237.

[6] C. Radhakrishna Rao and Helge Toutenburg (1995) *Linear Models: Least Squares and Alternatives,* Springer, New York.

[7] Donald B. Rubin (1976), Inference and missing data. *Biometrika*, **63**, 581–592.

[8] Shalabh (1995), Performance of Stein-rule procedure for simultaneous prediction of actual and average values of study variable in linear regression model. *Bulletin of the International Statistical Institute*, **56**, 1375–1390.

[9] G. A. Simon and Jeffrey S. Simonoff (1986), Diagnostic plots for missing data in least squares regression. *Journal of the American Statistical Association*, **81**, 501–509.

[10] Jeffrey S. Simonoff (1988), Regression diagnostics to detect nonrandom missingness in linear regression. *Technometrics*, **30**, 205–214.

[11] H. Theil and Arthur S. Goldberger (1961), On pure and mixed estimation in econometrics. *International Economic Review*, **2**, 65–78.

[12] Helge Toutenburg and Shalabh (1996), Predictive performance of the methods of restricted and mixed regression estimators. *Biometrical Journal*, **38**, 951–959.

# Some Recent Developments in Nonparametric Inference for Right Censored and Randomly Truncated Data

Grace L. Yang

Department of Mathematics
University of Maryland
College Park, Maryland

**Abstract.** Product-limit estimates $F_n$ of a distribution function $F$ play an essential role in the statistical analysis of censored and randomly truncated data. Prior to the publication of Gill in 1983 on the weak convergence of $F_n$ for right censored data over the entire support of $F$, properties of $F_n$ were studied and known almost exclusively on compact intervals $[0, b]$ for which $F(b) < 1$. Results on compact intervals may be sufficient for computing survival probabilities in biostatistics, but they are insufficient for many other problems, such as regression analysis, and the estimation of moments and mean residual life. Investigation of these problems requires understanding of the upper tail behavior of $F_n$ in the region $[b, \infty)$. In the absence of such knowledge, complicated assumptions on $F$ which are difficult to verify are often used to solve problems.

We discuss some recent results on the asymptotic properties of $F_n$, with special attention to the tail behavior of $F_n$ and the relaxation of stringent variance conditions. A randomization technique that solves many convergence problems for discrete $F$ is presented. We then present some open problems in the censoring and random truncation models.

## 1. Introduction

There is a huge literature on statistical inference about censored data. There are about 4800 titles in the Math. Sci. database that contains keywords "censorship", "censoring" or "censored", see S. Csörgő (1996). The count has not yet included the keyword "random truncation".

My presentation is necessarily narrowly focused and by no means a complete survey. I will highlight several recent results which serve as motivation for some open problems. The references will be made to those that are directly related to the presentation. The reader can easily find related literature in the references therein.

Let me begin with right censored data. Let $X$ be a random variable with a continuous, unknown distribution function $F(t) = P[X \le t]$. Let $C$ be a random variable with a continuous, unknown distribution function $L(t) = P[C \le t]$. A right censored sample of $X$ consists of $n$ pairs of iid random vectors $(Z_j, \delta_j)$, for $j = 1, \dots, n$, where

$$Z_j = \min(X_j, C_j) \quad \text{and} \quad \delta_j = I[X_j \le C_j]$$

and $X_j, C_j$ are independent observations from the distributions $F$ and $L$ respectively. Then the survival function of $Z$ is $S_Z(t) = P[Z > t] = P[X > t]P[C > t]$. Put

$$(1) \qquad\qquad Q(u, 1) = P[Z \le u, \delta = 1].$$

The nonparametric inference about $F$ uses the the ordered $Z_j$, denoted $Z_{k,n}$, and the associated censoring indicators $\delta_{k,n}$ as shown below.

$(Z_{1,n}, \delta_{1,n})$     $(Z_{2,n}, \delta_{2,n})$                                   $(Z_{n,n}, \delta_{n,n})$     $t$

The well known Kaplan-Meier estimate of $F$ or its survival function $S = 1 - F$ is given by

$$(2) \quad \hat{S}(t) = 1 - \hat{F}(t) = \prod_{k=1}^{n} \left[ 1 - \frac{\delta_{k,n}}{n - k + 1} \right]^{I[Z_{k,n} \le t]} \quad for \ \ 0 \le t < \infty.$$

Similar estimates have been used in life table constructions for centuries. What finally caught the attention of many statisticians are the two papers by Breslow & Crowley (1974) and Aalen (1978) based on his 1975 Ph.D. thesis that put the life table methods on solid mathematical ground. Since then, the statistical theory of incomplete data has enjoyed a rapid development with numerous new applications.

There are two basic theorems in the statistical analysis of censored data. To state them, we shall use the symbol $\Longrightarrow$ below to denote the convergence in distribution as $n$ tends to infinity and $\int_0^t = \int_{(0,t]}$, and throughout the paper. Further, although $S(x) = S(x-) = P[X \geq x]$ for a continuous $S$, for estimation purposes it is necessary to distinguish them. Thus $S(x-)$ will be used whenever it occurs.

**Theorem 1.**

$$\frac{\sqrt{n}(\hat{F}(t) - F(t))}{1 - F(t)} \Longrightarrow W(t), \quad for \ t \in [0, T] \quad with \ S_Z(T) > 0,$$

where $W(t)$ is a mean zero Gaussian process with covariance function

$$C(s \wedge t) = EW^2(s) = \int_0^s \frac{dQ(u, 1)}{S_z^2(u-)}$$

$$for \quad 0 \leq s \leq t \leq T \quad and \quad S_Z(T) > 0.$$

The second theorem concerns the cumulative hazard function of $X$,

$$(3) \qquad \Lambda(t) = \int_0^t \frac{1}{1 - F(u-)} dF(u) = \int_0^t \frac{dQ(u, 1)}{S_Z(u-)}, \quad t \geq 0.$$

The first equality defines $\Lambda$ and the second is obtained under the right censoring.

Let $\hat{\Lambda}(t)$ be an estimate of $\Lambda$ obtained by replacing $F$ in eq.(2) by the Kaplan-Meier estimate $\hat{F}$. In fact

$$(4) \qquad d\hat{\Lambda}(t) = \frac{d\hat{Q}(t, 1)}{\hat{S}_Z(t_-)} = \begin{array}{ll} \frac{\delta_{k,n}}{n-k+1}, & if \ t = Z_{k,n} \\ 0, & otherwise, \end{array}$$

where $\hat{Q}(t, 1) = \sum_j I[X_j \leq t, \delta_j = 1]/n$ and $\hat{S}_Z(t-) = \sum_j I[Z_j \geq t]/n$.

**Theorem 2.**

$$\sqrt{n}(\hat{\Lambda}(t) - \Lambda(t)) \Longrightarrow W(t), \quad for \ t \in [0, T] \quad with \ S_Z(T) > 0,$$

where $W(t)$ is as given in Theorem 1.

Both of these theorems are given in Breslow & Crowley (1974) but the proof of Theorem 2. is incomplete and was later corrected and simplified by several authors, see, e.g. Burke, Csörgő & Horváth (1981).

Aalen (1978) formulated the censoring problem in terms of multivariate counting processes and modeled the hazard rate $d\Lambda(t)$. The formulation can be readily seen in eq.(4) in which the middle term is derived from the ratio of two counting processes. Martingale methods are then employed with an appropriate filtration $\{\mathcal{F}_t; t \geq 0\}$, for instance $\mathcal{F}_t$ is the $\sigma$-field of events generated by those $(Z_{j,n}, \delta_{j,n})$ that occur in the time interval $[0, t]$. This particular approach greatly has advanced the statistical theory of censored data and its applications.

Another important approach was first proposed by Lo & Singh (1986). Instead of counting processes, they obtained an iid representation of the Kaplan-Meier estimate in the spirit of Bahadur.

**Theorem 3.** *For $S_Z(T) > 0$,*

$$\sqrt{n}(\hat{F}(t) - F(t)) = \sum_{j}^{n} \eta_j(t)/\sqrt{n} + R_n(t), \quad \text{for } t \in [0, T],$$

*where $\eta_j(t)$ are iid zero mean bounded processes and the remainder term is uniformly bounded over the compact interval $[0, T]$, i.e.,*

$$\sup_{t \in [0,T]} |R_n(t)| = 0((\frac{\log n}{n})^{(3/4)})) \quad a.s.$$

The rates $(\frac{\log n}{n})^{(3/4)}$ have been improved to $\frac{\log n}{n}$ by Major & Rejtö (1988).

Up until now, the overwhelming majority of the papers on the subject are results on compact intervals $[0, T]$, where $T$ must satisfies $S_Z(T) > 0$. However for estimation of functionals such as $\int \phi dF$, the distributional properties of the tail of $\hat{S}$ are needed. The simplest such example is the estimator for the mean $EX$,

$$\widehat{EX} = \int_0^T \hat{S}(u)du + \int_T^\infty \hat{S}(u)du.$$

where the tail is $\hat{S}(u)$ for $u > T$. But the tail of $\hat{S}$ is difficult to study under either censoring or random truncation. The tail problem is often manifested in the censored regression analysis of which estimation of the mean is its simplest special case. Usually, the K-M estimates are used to estimate the distribution of the residual $\epsilon$ in the regression model:

$$X = A'\beta + \epsilon, \quad \text{with } E[X \mid A] = A'\beta,$$

in which the residual $\epsilon = X - A'\beta$ is subject to right censoring.

In the absence of complete understanding of the tail behavior of the K-M estimates, the statistical properties of the estimates of the regression coefficients $\beta$ are often obtained under rather severe assumptions on the tails of $S$ and the survival function of $C$, $S_C$, or by simply ignoring the tails.

With this introduction, I will present some recent results focusing on the following topics:

- Tail development under right censoring
- Random truncation,
- Elimination of the continuity restriction on the underlying distribution functions.
- Some open problems. In order to streamline the presentation, nonnegative random variables will be used throughout.

## 2. The Tails Under Right Censoring

Burke, Csörgő & Horváth (1981, 1988) appear to be the first to study the strong approximation and the weak convergence of the Kaplan-Meier estimate $\hat{F}$ over an expanding interval $[0, T_n]$ where, subject to the following restrictions,

$$T_n < b_z = \sup\{t : S_Z(t) > 0\}, \quad \text{and} \quad S_Z(T_n) \geq (2\epsilon \tfrac{\log n}{n})^{1/2},$$

$T_n$ increases with $n$ at a certain rate as $n \to \infty$. Here $\epsilon$ is a given positive constant.

Note that the result depends on knowing the survival function of $Z$, $S_Z$, at $T_n$ as in the previous theorems. But under the nonparametric assumptions, $S_Z$ is unknown.

Gu & Lai (1990) used a modified K-M estimate and replaced the $T$ by a certain order statistic $Z_{k_n, n}$. Their proof again requires conditions on the unknown $S$ and $S_C = P[C > u]$. The paper also studies similar problems in the random truncation model.

The first result on weak convergence over the entire range of observations, $[0, Z_{n,n}]$, is due to Gill (1983) under

$$Condition \quad C : \qquad \int_0^{b_z} \frac{dF(u)}{S_C(u-)} < \infty,$$

where $S_c(u-) = P[C \geq u]$ and $b_Z = \sup\{u : S_Z(u) > 0\}$.

Let $D[0, b_Z]$ denote the space of right continuous functions with left limits, and endowed with the Skorohod topology.

**Theorem 4.** *Under condition C,*

(5)
$$\sqrt{n}\left(\frac{\hat{F}(t) - F(t)}{1 - F}\right)_{Z_{n,n}} \Longrightarrow W(t), \quad in \ D[0, b_z],$$

*where* $(\xi(t))_{Z_{n,n}}$ *is the process* $\xi$ *stopped at* $Z_{n,n}$ *and* $W(t)$ *is a Gaussian process as defined in Theorem 1.*

Ying (1989) extended the weak convergence to the entire line without stopping the process at $Z_{n,n}$.

Condition C comes from the covariance function $C(t)$ of Theorem 1 with $s = t$. It holds under light censoring. But in general it is too strong as can be seen from the following inequality:

(6)
$$C(t) = \int_0^t \frac{dF(u)}{S_C(u-)} = \int_0^t \frac{d\Lambda(u)}{S(u-)S_C(u-)} \geq \int_0^t \frac{d\Lambda(u)}{S(u-)} \to \infty, \quad as \ t \to \infty$$

For instance, it does not hold for the proportional hazard models $S_C(x) = S^\beta(x)$ if $\beta \geq 1$. Furthermore, results obtained under condition C are not distribution free. It is annoying that in the nonparametric analysis of $F$ one has to verify Condition C that depends on the unknown $F$ and $S_C$. This problem, of course, does not exist in the uncensored case when ordinary empirical distribution functions are used.

On the other hand, a counter example of Chen & Ying (1996) shows that Condition C is necessary for Theorem 4.

The question as to how close to the upper boundary $b_z$ for which the weak convergence of $\sqrt{n}(\hat{F}(t) - F(t))$ holds without Condition C has recently been answered by S. Csörgő (1996). He showed that $Z_{n,n}$ needs to be replaced by another order statistics $Z_{n-\lfloor n^{2/3}a(n)\rfloor,n}$ where the symbol $\lfloor h \rfloor$ denotes the integral part of the value $h$ and $\lfloor n^{2/3}a(n)\rfloor$ is a sequence of positive integers tending to infinity with $a(n)$ satisfying the conditions in section 2 of S. Csörgő (1996). To state the result, let $C_n(t)$ be an estimator of the variance $C(t)$. It can be either the estimate used by Hall & Wellner (1980),

$$C_n(t) = \begin{array}{l} n\sum_{\{j:Z_j \leq t\}} \frac{\delta_j}{(n-j)(n-j+1)}, \qquad \text{for } 0 \leq t \leq Z_{n,n} \\ \infty, \qquad\qquad\qquad\qquad\qquad \text{for } t > Z_{n,n} \end{array}$$

or the one obtained by replacing $S_Z$ and $Q(u,1)$ in $C(t)$ by the empirical distribution functions $\hat{S}_Z$ and $\hat{Q}(u,1)$ as defined in eq.(4). Furthermore, let $K(t) = \frac{C(t)}{1+C(t)}$ be the time-scale transformation introduced by Hall & Wellner (1980). One of the results obtained by S. Csörgő is

**Theorem 5.**

$$\sqrt{n}\left(\frac{\hat{F}(t) - F(t)}{(1 - \hat{F}(t))(1 + C_n(t))}\right)_{Z_{n - n^{2/3}a(n), n}} \Longrightarrow B(K(t)) \quad in \ D[0, b_z]$$

*where B is a Brownian bridge on* $[0, 1]$.

With regard to the functional $\int \phi dF$ under censoring, an entirely different line of attack was introduced by Stute & Wang (1993). Their approach is not to use the formula

$$\int_0^\infty S(u) du$$

but to use

$$\int_0^\infty u dF(u) \quad \text{or the more generally } \int_0^\infty \phi(u) dF(u).$$

The following Strong Law of Large Numbers under right censoring holds.

**Theorem 6.** *Let* $\hat{F}$ *be the Kaplan-Meier estimate of* $F$ *where* $F$ *is a continuous distribution function. Then for any* $F-$*integrable function* $\phi(x)$,

$$S_n = \int_0^\infty \phi(x) d\hat{F}(x) \to \int_{x < b_z} \phi(x) dF(x) \quad a.s. \ as \ n \to \infty.$$

The reader is referred to Stute & Wang (1993) for a more general statement of the theorem without the continuity restriction on $F$.

A departure from the usual counting process formulation with a filtration $\{\mathcal{F}_s, s \geq 0\}$ of increasing $\sigma$-fields is that their proof utilizes a sequence of shrinking $\sigma$-fields $\{\mathcal{F}_n : n = 1, 2, \ldots\}$ generated by the data as follows,

$$(7) \quad \mathcal{F}_n = \sigma\{(Z_{j,n}, \delta_{j,n}), 1 \leq j \leq n, (Z_k, \delta_k), k \geq n + 1\}, n = 1, 2, \cdots.$$

Then $S_n$ written as $S_n = \sum_{j=1}^n W_{j,n} \phi(Z_{j,n})$ where $W_{j,n} = \hat{F}(Z_{j,n}) - \hat{F}(Z_{j,n}-)$, is $\mathcal{F}_n$ measurable and a reverse super martingale in $n$. By the Hewitt-Savage 0-1 law $\mathcal{F}_\infty = \bigcap_{n \geq 1} \mathcal{F}_n$ is trivial, $S_n$ converges to a constant as $n$ tends to infinity.

The proof of the reverse super martingale property of $S_n$ relies on a crucial fact that the conditional rank of $Z_{n+1}$ given the $\sigma$-field $\mathcal{F}_{n+1}$ has a discrete uniform distribution, i.e.,

$$P[Z_{n+1} = Z_{k,n+1} \mid \mathcal{F}_{n+1}] = \frac{1}{n + 1},$$

and that the ranks of $(Z_1, \ldots, Z_n)$ and $n$ vector

(8)                        $(Z_{j,n}, \delta_{j,n}), \quad j = 1, \ldots, n,$

are independent.

This generalizes a standard result that the rank vector of $(Z_1, \ldots, Z_n)$ is independent of the vector of the order statistics $Z_{j,n}, j = 1, \ldots n$. In fact, more is true. The rank vector of $(Z_1, \ldots, Z_n)$ is independent of the vector

$$(Z_{j,n}, \delta_{j,n}, A_{j,n}), \quad j = 1, \ldots, n,$$

where $A_{j,n}$ are any additional independent covariates that are associated with $Z_{j,n}$ (in the usual setup). This approach opened up a new way of studying censoring problems, see e.g., Stute (1993) and some of his recent publications.

To conclude this section, we shall point out a recent result on life expectancy function widely used in life tables, or the so-called mean residual life function in other areas of applications.

Life expectancy function is defined by

(9)

$$M(t) = E[X - t \mid X \geq t] = \frac{\int_0^\infty (x - t)dF(x)}{1 - F(t-)}, \quad \text{provided } 1 - F(t-) > 0.$$

A natural estimator for $M(t)$ is

(10)                        $\hat{M}(t) = \frac{\int_{[x \geq t]} (x - t)d\hat{F}(x)}{1 - \hat{F}(t-)}.$

The weak convergence of the process $\{\sqrt{n}(\hat{M}(t) - M(t)), t \in [0, T]\}$ over compact intervals has been obtained by Yang (1977, 1978) under no censoring and also undercompeting risks. See also Hall & Wellner (1981) for expanding intervals. The problem remains open for convergence over the entire interval $[0, \infty)$ for either censored or randomly truncated data. The uncensored case has only recently been solved by M. Csörgő and Zitikis (1996). Note that the uncensored case corresponds to $P[C = \infty] = 1$. Then $\hat{F}$ reduces to the usual empirical distribution function of $X$ and the $Z_{n,n}$ is replaced by the largest order statistic $X_{n,n}$. Csörgő and Zitikis show that for strong consistency and weak approximation over $[0, \infty)$, it is necessary to introduce a certain weight function $q(t)$. Under the condition of finite second moment of $X$, the following result holds.

**Theorem 7.**

$$P\left[M(t) \in \left[\hat{M}(t) - \frac{z_\alpha}{\sqrt{n}q(\hat{F}(t))}, \hat{M}(t) + \frac{z_\alpha}{\sqrt{n}q(\hat{F}(t))}\right], t \in [0, X_{n,n}]\right]$$
$$= 1 - \alpha + o(1),$$

*as* $n \longrightarrow \infty$, *where* $z_\alpha$ *is a constant that depends on* $F$ *through* $q$ *and the confidence coefficient* $\alpha$.

## 3. Random Truncation

The randomly truncated data refers to bivariate observations of $(X, Y)$ restricted to the lower half plane as shown in the following Fig. 1.

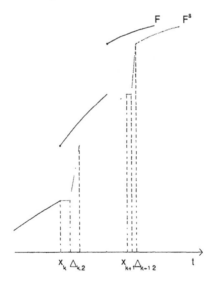

Figure 1. Randomly Truncated Data

The data of this kind is abundant in astronomy in which $X$ represents the absolute luminosity of a galaxy and $Y$ a function of the red shift, and any pair $(X, Y)$ with $X < Y$ is not observable. In applications, it is of importance to estimate the distribution function of $X$ as well as the proportion of missing data in the region $X > Y$, an added feature of the random truncation problem. This proportion cannot, of course, be computed by "deterministic" means. However, stochastic methods offer a solution.

In the stochastic formulation, $X$ and $Y$ are assumed to be independent random variables (the assumption is accordance with the cosmological principle) with unknown continuous distribution functions $F$ and $G$ respectively. The random truncation model for the data is defined by the joint conditional distribution function $H$ of $X$ and $Y$ given $X \geq Y$ where

$$(11) \qquad H(x,y) = P[X \leq x, Y \leq y | X \geq Y].$$

This model can also be derived from cross-sectional sampling of lifetimes or event times ( see e.g. Patil & Rao, 1978, He & Yang 1992), from the length-biased consideration ( see, e.g., Gill, Vardi and Wellner 1985), and from measuring the strength and stress of a component in the reliability theory.

Given a sample of $n$ iid pairs $(X_j, Y_j), j = 1, \ldots, n$ from $H$, we wish to estimate $F$, $G$ and the proportion $\alpha = P[X \geq Y] = \int GdF$ (or the missing proportion $1 - \alpha$). Here, the Lynden-Bell estimates $\tilde{F}_n$ and $\tilde{G}_n$ play the same fundamental role as the K-M estimate in right censoring. It is informative to introduce these estimates by way of model identification of the hazard rates of $F$ and $G$.

For notational convenience, the randomly truncated observation of $(X, Y)$ will be denoted by $(U, V)$. Thus $U_j$ is the observed $X_j$ and $V_j$ is the observed $Y_j$, subject to $X_j \geq Y_j$. The marginal distributions of $U$ and $V$ are thus given by

$$F^*(x) = P[U \leq x] = P[X \leq x \mid X \geq Y] = \int_0^x G(s)dF(s)/\alpha,$$

$$G^*(x) = P[V \leq x] = P[Y \leq x \mid X \geq Y] = \int_0^x (1 - F(s-))dG(s)/\alpha,$$

with

$$(12) \qquad \alpha = P[X \geq Y] = \int GdF.$$

Note that we have considered $X$ and $Y$ as non negative random variables for convenience.

The coverage probability $R(x) = P[V \leq x \leq U]$ that the random interval $[V, U]$ covers any chosen point $x$ plays a pivotal role in the statistical inference. Under the independence of $X$ and $Y$, the following fundamental relations hold:

$$(13) \qquad R(x) = P[V \leq x \leq U] = G^*(x) - F^*(x-) = \frac{(1 - F(x-))G(x)}{\alpha}.$$

The interval $(a_Y, b_X)$ for which $R(x) > 0$ determines the estimable range of $F$, where

$$a_Y = \inf\{y : G(y) > 0\} \quad \text{and} \quad b_X = \sup\{x : F(x) < 1\}.$$

For $x \in (a_Y, b_X)$, the following ratios are well defined:

$$\frac{dF^*(x)}{R(x)} = \frac{G(x)dF(x)/\alpha}{G(x)[1 - F(x-)]/\alpha} = \frac{dF(x)}{1 - F(x-)} = d\Lambda(x),$$

$$\frac{dG^*(x)}{R(x)} = \frac{[1 - F(x-)]dG(x)/\alpha}{G(x)[1 - F(x-)]/\alpha} = \frac{dG(x)}{G(x)}.$$

These equalities show that the hazard rate $d\Lambda(x)$ of $X$ can be identified by $\frac{dF^*(x)}{R(x)}$, the distribution $G$ can be identified through the equality $\frac{dG^*(x)}{R(x)} = \frac{dG(x)}{G(x)}$, and, moreover, the nuisance parameter $\alpha$ gets canceled out.

$F^*(x)$ and $R(x)$ can be estimated by the corresponding empirical distribution functions $F_n^*(x) = \frac{\sum_i I[U_i \le x]}{n}$ and $R_n(x) = \frac{\sum_j I[V_j \le x \le U_j]}{n}$ of the data. Thus a natural estimate of the hazard rate $d\Lambda(x)$ is

$$d\tilde{\Lambda}(x) = \frac{dF_n^*(x)}{R_n(x)} = \frac{F_n^*(x) - F_n^*(x-)}{\sum_{i=1}^n I(V_i \le x \le U_i)/n},$$

provided the denominator is positive.

Inverting $d\tilde{\Lambda}$ to its distribution function by the Doléans-Dade exponential formula yields the Lynden-Bell (1971) estimate $\tilde{F}_n$, and similarly for the construction of the estimate $\tilde{G}_n$. We have

$$\tilde{F}_n(x) = 1 - \prod_{s \le x} \left[1 - \frac{dF_n^*(x)}{R_n(x)}\right], \quad \tilde{G}_n(x) = \prod_{s > x} \left[1 - \frac{dG_n^*(x)}{R_n(x)}\right]$$

These estimates were studied by Woodroofe (1985) who obtained the weak convergence of the L-B estimates on compact intervals. Keiding & Gill (1990) generalized the result to the maximum allowable interval. Mandrekar & Thelen (1990) obtained similar results by introducing a reverse martingale in $t$ to study $\tilde{G}_n(t)$.

**Theorem 8.** Set $a_Y = a_X = 0$ *for simplicity. Under*

$$\text{Condition B:} \qquad \int_{a_X}^\infty \frac{dF}{G} < \infty.$$

*for $t \in [0, b]$   with $0 < b < b_X$,*

$$\sqrt{n}[\Lambda_n(t) - \Lambda(t)] \Longrightarrow W(t), \quad in \ D[0, \infty)$$

$$\frac{\sqrt{n}[\bar{F}_n(t) - F(t)]}{1 - F(t)} \Longrightarrow W(t), \quad in \ D[0, \infty]$$

*where $\{W(t); t \in [0, \infty)\}$ is a zero mean Gaussian martingale with covariance function*

(14)
$$\begin{aligned}
\Gamma(s \wedge t) &= (1 - F(s))^2 \int_{a_X}^{s} \frac{dF^*(u)}{R^2(u)} \\
&= (1 - F(s))^2 \alpha \int_{a_X}^{s} \frac{dF(u)}{(1 - F(u-))^2 G(u)} \quad for \ s \le t.
\end{aligned}$$

If Condition B is violated other limiting distributions are possible. Woodroofe (1985) showed that if $G = F^c$, $1 < c < \infty$, then $n^\theta[\bar{F}_n(t) - F(t)]$ converges weakly to a stable distribution for all $t$ such that $F(t) < 1$, where $\theta = \frac{1}{1+c}$.

From the asymptotic variances given in (14), one sees that the Lynden-Bell estimate is much less stable than the Kaplan-Meier estimate .

The following strong law under random truncation has been proved by He & Yang (1998a).

**Theorem 9.** *For arbitrary $F$ and $G$ subject to $\alpha > 0$,*

$$\int_0^\infty \phi(x) d\bar{F}_n(x) \to \int \phi(x) dF_a(x) \quad a.s. \ as \ n \to \infty,$$

*for any $F$-integrable function $\phi(x)$, where*

$$F_a(x) = \frac{F(x) - F(a_Y)}{1 - F(a_Y)}.$$

In the proof, it is shown that $M_n = \int \phi(x) d\bar{\Lambda}_n(x); n = 1, 2, \ldots$, but not the integrals $\int \phi(x) d\bar{F}_n(x); n = 1, 2, \ldots$, forms a reverse super martingale with respect to the sequence of shrinking $\sigma$-fields, $\mathcal{F}_n = \sigma\{U_{jn}, V_{jn}, 1 \le j \le n, (U_k, V_k), k \ge n\}, n = 1, 2, \ldots,$.

While the construction of the estimates $\bar{F}_n$ and $\tilde{G}_n$ can be carried out without knowing $\alpha$, our proof of the strong law (in the general case) does require estimation of $\alpha$. In fact, the proof relies on an interesting representation of a standard estimate $\alpha_n$ of $\alpha$ given by

$$\alpha_n = \int \tilde{G}_n d\tilde{F}_n$$

A better representation of $\alpha_n$ is given by

$$\hat{\alpha}_n = \frac{\tilde{G}_n(x)[1 - \tilde{F}_n(x-)]}{R_n(x)}$$

and that $\hat{\alpha}_n$ is independent of $x$ provided that $R_n(x) > 0$.

The estimate $\alpha_n$ is interesting in its own right. If $G_n$ and $F_n$ were the empirical distribution functions of two independent, non truncated samples of the $X$ and the $Y$, then $\alpha_n$ is nothing but the Wilcoxon statistic in the Mann-Whitney form.

## 4. Elimination of the Continuity Restriction on $F, G$, and $S_C$ by Randomization

Order statistics, ranks and counts appear almost everywhere in the proofs of weak convergence and the strong laws under either censoring or random truncation. It is therefore most convenient to impose continuity assumption $F, G$, and $S_C$. By employing a randomization trick, it can be shown that these results can be extended to arbitrary distribution functions $F$ and $G$ without much difficulty. This technique is known to some, but it does not seem to be widely used. The technique is discussed in Govindarajulu & Le Cam & Ragavachari (1965) in the study of rank tests based on complete data. A specific construction is given in Major & Rejtő (1988) for right-censored data. Here we shall illustrate the technique for randomly truncated data used in proving Theorem 9.

The procedure consists of two steps. First shift the distribution of $F$ to the right at each jump point of $F$ (the amount of shift varies with the jump point) and then randomizes the shifted $F$ locally at each jump point, so that the resulting distribution, say $F^s$, is a continuous distribution function. See Fig. 2. Do a similar shift and randomization for $G$. Specifically, let $C_F$ and $C_G$ be the continuity sets of $F$ and $G$ respectively. Let $A = \{x_k : k = 1, 2, \ldots\} = C_F^c \cup C_G^c$, the combined set of discontinuity points of $F$ and $G$, i.e., $x_k$ is either a jump point of $F$ or $G$ or both. Let $\varepsilon_k, \eta_k : k = 1, 2, \ldots$ be independent uniform random variables to be specified below. Define a left continuous and non decreasing function $h$ by

$$(15) \qquad h(x) = x + \sum_{k : x_k < x} \frac{1}{k^2}, \qquad x \in (-\infty, \infty)$$

Transform $X$ to $X^s$ as follows:

$$
\begin{aligned}
X^s = x^s &= h(x) && \text{if } X = x \in C_F, \\
&= h(x_k) + \tfrac{1+\epsilon_k}{2k^2}, && \text{if } X = x_k \in A,
\end{aligned}
$$

where $\epsilon_k$ is uniformly distributed over the interval $\Delta_{k,2} = (h(x_k) + \tfrac{1}{2k^2}, h(x_k+)]$.

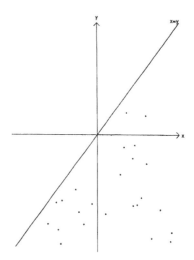

Figure 2. Randomization of $F$

Likewise, transform $Y$ to $Y^s$ as :

$$
\begin{aligned}
Y^s = y^s &= h(y) && \text{if } Y = y \in C_G, \\
&= h(x_k) + \tfrac{\eta_k}{2k^2}, && \text{if } Y = x_k \in A,
\end{aligned}
$$

where $\eta_k$ is uniformly distributed over the interval $\Delta_{k,1} = [h(x_k), h(x_k) + \tfrac{1}{2k^2})$. The random variables $X^s$ and $Y^s$ have continuous distribution functions $F^s$ and $G^s$. Note that $\Delta_{k,1}$ and $\Delta_{k,2}$ are disjoint. They are so constructed that the events $[X \geq Y] = [X^s = Y^s]$. See Fig. 2. It can be shown that, for every $x \in (-\infty, \infty)$,

$$
F(x) = F^s(h(x+)), \quad \tilde{F}_n(x) = F_n^s(h(x+))
$$

(16)      $$
G(x) = G^s(h(x+)), \quad \tilde{G}_n(x) = G_n^s(h(x+)),
$$

where $F_n^s$ and $G_n^s$ are the Lynden-Bell estimates computed from the transformed data $X_j^s, Y_j^s, j = 1, \ldots, n$. Therefore results stated in the previous section for continuous $F^s, G^s$ can be carried over to arbitrary $F, G, S_C$ with only minor modifications to include jumps of $F$ and $G$.

## 5. Some Open Problems

This presentation points out two particular areas in censoring and random truncation for future research. One pertains to the problem of tails of the estimated distribution and the other is to find ways to eliminate the dependence on conditions like C and B with the goal of making the nonparametric inference fully "distribution free". For easy reference, we restate these conditions below:

$$Condition\ C: \int_0^{bz} \frac{dF(u)}{S_C(u-)} < \infty, \qquad Condition\ B: \int_{a_X}^{\infty} \frac{dF(u)}{G(u)} < \infty,$$

as used in Theorems 4 and 8. In this general direction, several specific problems can be posed.

1. Obtain similar results like Theorem 5 for randomly truncated data.
2. Investigate the weak convergence of the mean-residual life process over the entire line under either right censoring or random truncation.
3. Rao & Zhao (1993, 1996) studied the special type of censored regression model:

$$Y_i = x_i'\beta + \epsilon_i, \quad i = 1, \dots, n,$$

only $Y_i^+ = Y_i I[Y_i \geq 0]$ and $x_i$ are observable. Under some mild conditions, they established the asymptotic normality of the LAD (least absolute deviation) estimates of the regression coefficients under some mild conditions. Could this approach be applied to the general right censored regression model ?

4. Unify the treatment of the problems discussed in this paper by considering the data which are obtained under both censoring and random truncation.

## References

[1] Aalen, O. (1978), Nonparametric inference for a family of counting processes. *Ann. Statist.*, **6**, 701-726.

[2] Burke, M. D. Csörgő, S. & Horváth, L. (1981), Strong approximations of some biometric estimates under random censorship. *Z. Wahrsch. Verv. Gebiete*, **56**, 87-112.

[3] Burke, M. D. Csörgő, S. & Horváth, L. ( 1988), Correction to and improvement of 'Strong approximations of some biometric estimates under random censorship'. *Probab. Theory & Related Fields*, **79**, 51-57.

[4] Chen, K. & Ying, Z. (1996), A counterexample to conjecture concerning the Hall-Wellner band. *Ann. Statist.*, **24**, 641–646.

[5] Csörgő, M. & Zitikis, R. (1996), Mean residual life processes. *Ann. Statist.*, **24**, 1717–1739.

[6] Csörgő, S. (1996), Universal Gaussian approximations under random censorship. *Ann. Statist.*, **24**, 2744–2778.

[7] Gill, R. D. (1983), Large sample behavior of the product-limit estimator on the whole line. *Ann. Statist.*, **11**, 49–58.

[8] Gill, R. D., Vardi, Y. and Wellner, J. (1985), Large sample theory of empirical distributions in biased sampling models., **16**, 1069–1112.

[9] Gill, R. D. (1994) *Lectures on Survival Analysis Ecole d'Eté de Probabilités de Saint Flour XXII-1992,* Lecture Notes in Mathematics **1581** Springer, Berlin.

[10] Govindarajulu, Z. & Le Cam, L. & Raghavachari, M. (1965) Generalizations of theorems of Chernoff and Savage on the asymptotic normality of test statistics *Proc. of the Fifth Berkeley Symp. on Math. Stat. and Prob.* eds: Le Cam, L. & Neyman, J., Univ. of Calif. Press, Berkeley and Los Angeles, 1966, **1** 609–638.

[11] Gu, M. G. & Lai, T. L. (1990), Functional laws of the iterated logarithm for the product limit estimator of a distribution function under random censorship or truncation. *Ann. Probab.*, **18**, 160–189.

[12] Hall, W. J. & Wellner, J. (1980), Confidence bands for a survival curve from censored data. *Biometrika*, **67**, 133–143.

[13] Hall, W. J. & Wellner, J. (1981) Mean residual life. In *Statistics and Related Topics,* eds. Csörgő, M. et al., North-Holland, Amsterdam, 169–184.

[14] He, S. and Yang, G. L. (1994) Estimating a lifetime distribution under different sampling plans. In *Statistical Decision Theory and Related Topics, V,* eds: Gupta, S. S. & Berger, J. O., Springer-Verlag, New York, 73–85.

[15] He, S. and Yang, G. L. (1998a), The strong law under random truncation. To appear in. *Ann. Statist.*, **26**, n.3.

[16] He, S. and Yang, G. L. (1998b), Estimation of the truncation probability in the random truncation model. To appear in. *Ann. Statist.*, **26**, n.3.

[17] Keiding, N. and Gill, R. D. (1990), Random truncation models and Markov processes. *Ann. Statist.*, **18**, 582–602.

[18] Lo, S. H. and Singh, K. (1986), The product-limit estimator and the bootstrap: Some asymptotic representations. *Probab. Theory Related Fields*, **71**, 455–465.

[19] Major, P. and Rejtö, L. (1988), Strong embedding of the estimator of the distribution function under random censorship. *Ann. Statist.*, **16**, 1113-1132.

[20] Mandrekar, V. and Thelen, B. (1990) Joint weak convergence on the whole line in the truncation model *Proc. of the R. C. Bose Symposium on Probability, Statistics and Design of Experiments* Wiley Eastern, New Delhi 495-515

[21] Rao, C. R. & Patil, G. P. (1978), Weighted distributions and size-biased sampling with applications to wildlife populations and human families. *Biometrics*, **34**, 179-189.

[22] Rao, C. R. & Zhao, L. C. (1993), Asymptotic normality of LAD estimator in censored regression models. *Math. Methods of Statist.*, **2**, 228-239.

[23] Rao, C. R. & Zhao, L. C. (1996), Recent contributions to censored regression models. *Tatra Mountains Math. Publ.*, **7**, 133-137.

[24] Stute, W. and Wang, J. L. (1993), The strong law under random censorship. *Ann. Statist.*, **21**, 1591-1607.

[25] Stute, W. (1993), Consistent estimation under random censorship when covariables are present. *J. of Multivariate Analysis*, **45**, 89-103.

[26] Stute, W. (1996), The jackknife estimate of variance of a Kaplan-Meier integral. *Ann. Statist.*, **24**, 2679-2704.

[27] Woodroofe, M. (1985), Estimating a distribution function with truncated data. *Ann. Statist.*, **13**, 163-177.

[28] Yang, G. (1978), Estimation of a biometric function. *Ann. Statist.*, **6**, 112-116.

[29] Yang, G. (1977), Life expectancy under random censorship. *Stoch. Processes Appl.*, **6**, 33-39.

[30] Ying, Z. L. (1989), A note on the asymptotic properties of the product-limit estimator on the whole line. *Statist. Probab. Lett.*, **7**, 311-314.

# Estimation of Parameters in Nonlinear Regression Models

## Juan Zhang[1]
Roth Associates
Rockville, Maryland

## Shyamal D. Peddada[2]
Division of Statistics
University of Virginia
Charlottesville, Virginia

## Alan D. Rogol[3]
Department of Pediatrics and Pharmacology
University of Virginia Health Sciences Center
Charlottesville, Virginia

**Abstract.** This article is motivated by two important applications; growth of a boy during puberty and the motion of polar ice. We address the problem of drawing inferences on an unknown parameter $\theta$ in a nonlinear regression model where the regression function is a "smooth" nonlinear function of $\theta$. Two types of parameter spaces are considered, namely, the $p$ dimensional Euclidean space $\mathcal{R}^p$, and $\mathcal{R}^p \otimes SO(p)$, where $SO(p)$ is the special orthogonal group of $p \times p$ orthogonal matrices. In the case of $\mathcal{R}^p$

[1] Research was partially supported by a grant from Genentech foundation for growth and development.

[2] Research was partially supported by ONR grant N 00014-92-J-1009.

[3] Research was partially supported by grants from Genentech foundation for growth and development and the NIH grant RR 00847 to the General Clinical Research Center at the University of Virginia.

we review some of the existing methodology and introduce a few new ones along with several open research problems. Research in the case of $\mathcal{R}^p \otimes SO(p)$ is very recent. We shall review some of the recent developments in this area and suggest several important open research problems for the $21^{st}$ century.

## Introduction

Parametric nonlinear regression models are becoming increasingly popular among practitioners for building models between a response variable $Y$ and an explanatory variable $X$. For instance, Preece and Baines [17] modeled height of a boy at puberty using

$$(1) \qquad H = \theta_{1h} - \frac{2(\theta_{1h} - \theta_{2h})}{\exp(\theta_{3h}[x - \theta_{4h}]) + \exp(\theta_{5h}[x - \theta_{4h}])},$$

where $H$ is height at age $x$ and $\theta_h = (\theta_{1h}, \theta_{2h}, \dots, \theta_{5h})'$ are parameters of the model. Similarly, Zhang et al [30] modeled velocity of growth $(V)$ of a boy at puberty as a function of the levels of growth hormone $(G)$ and serum testosterone $(T)$ using the following nonlinear model.

$$(2) \qquad V = \sqrt{G} T^{\theta_{1v}} \exp(-\theta_{2v}(\frac{T - \theta_{3v}}{\sqrt{G}})^2), \quad \theta_{1v}, \theta_{2v}, \theta_{3v} > 0.$$

In the above examples the parameter space $\Theta$ of interest is $\mathcal{R}^p$. Estimation and test of hypotheses concerning $\theta \in \mathcal{R}^p$ has been an active area of research during the past two decades. For a review on this topic one may refer to [1, 6, 22, 23]. Though volumes have been written on this subject, several issues remain to be addressed. For instance sometimes, depending upon the shape of the function, it may be a numerical challenge to obtain estimate parameters in a nonlinear regression model. Though this is an important issue, we shall not address it in the present article. We shall focus on some of the statistical issues concerning the analysis of nonlinear regression models.

Often the $p$ values and confidence regions reported in a nonlinear regression analysis are inaccurate. In the case of fixed effects nonlinear models, several articles and books (cf [1, 3, 22, 23, 27] have addressed this issue. However, very little is discussed in the case of nonlinear random and mixed effects models. Simulation results presented in Section 2 suggest that some of the existing procedures for nonlinear regression models (both fixed as well as random effects) may perform disastrously in terms of coverage probabilities and hence there is a need for developing new and improved procedures for performing inference on $\theta$. One of the reasons for poor coverage probability is because the standard errors are usually underestimated. Consequently, a new estimator of the covariance matrix of ordinary least squares estimator (OLSE) is introduced in Section 2.1. Using the Minimum Norm Quadratic Estimation (MINQE) principle, in Section 2.2. and in Section 2.3. we introduce new variance estimators for nonlinear random effects models under homoscedastic errors and under heteroscedastic errors, respectively. The proposed methodology is illustrated using an example from Pediatrics.

There are other forms of parameter spaces $\Theta$ that are encountered in applications. One such space is $\Theta = \mathcal{R}^p \times SO(p)$, where $SO(p)$ is the group of $p \times p$ orthogonal matrices with determinant $+1$. This parameter space describes motion of a rigid body in a $p$ dimensional space, where the unknown translation vector belongs to $\mathcal{R}^p$ and the rotation is an orthogonal matrix belonging to $SO(p)$. Research in this field is relatively new. For homologous data, Chang [4], Chang and Ko [5], and Ko and Chang [11] developed a nearly complete theory for estimating various motion parameters along with errors in estimation. Peddada and Chang [14] and Peddada and McDevitt [15] developed methodology for nonhomologous data, including bootstrap confidence regions for the motion parameters ([14]). In Section 3 we shall describe several important open research problems in this field.

### Nonlinear regression models in $\mathcal{R}^p$

### Fixed effects model

Let

$$(3) \qquad Y_i = f(\mathbf{X}_i, \theta) + \epsilon_i, \quad i = 1, 2, \dots, n,$$

be a nonlinear regression model, where $Y_i$ is the response on the $i^{th}$ individual with $q \times 1$ explanatory variable $\mathbf{X}_i$, and $f(\cdot, \theta)$ is a twice differentiable nonlinear function of the $p \times 1$ parameter vector $\theta$. We shall assume that $\epsilon_i$ are unobservable $i.i.d$ random errors with mean 0 and variance $\sigma^2$. We may rewrite (3) in a vector notation as:

$$\mathbf{Y} = \mathbf{f}(\mathbf{X}, \theta) + \epsilon,$$

where $\mathbf{Y} = (Y_1, Y_2, \dots, Y_n)'$, $\mathbf{f}(\mathbf{X}, \theta) = (f(\mathbf{X}_1, \theta), f(\mathbf{X}_2, \theta), \dots, f(\mathbf{X}_n, \theta))'$, and $\epsilon = (\epsilon_1, \epsilon_2, \dots, \epsilon_n)'$.

Assuming $\epsilon \sim N(0, \sigma^2 I)$, a common approach for performing inference regarding $\theta$ is to obtain the OLSE $\hat{\theta}_{olse}$ and use the approximation

$$(4) \qquad \frac{(\hat{\theta}_{olse} - \theta)' \hat{\mathbf{V}}'_{olse} \hat{\mathbf{V}}_{olse} (\hat{\theta}_{olse} - \theta)}{p\hat{\sigma}^2} \sim \text{approximately } F(p, n - p),$$

where $(i,j)^{th}$ element of $\hat{\mathbf{V}}_{olse}$ is $v_{ij} = \partial f(\mathbf{x}_i, \boldsymbol{\theta})/\partial \theta_j$ evaluated at $\hat{\boldsymbol{\theta}}_{olse}$. Throughout we shall use the notation $\hat{\mathbf{V}}_{olse} = (\mathbf{V}_1, \mathbf{V}_2, \dots, \mathbf{V}_n)'$, $\mathbf{V}_i = (v_{i1}, v_{i2}, \dots, v_{ip})'$. $F(p, n-p)$ denotes central $F$ distribution with $(p, n-p)$ degrees of freedom. Unfortunately the above approximation may not be reasonable if the sample size is small and if curvature effects are profound. Box and Coutie [3] suggested a modification to (4) which accounts for the curvature effects. From our simulations we have noticed that this correction need not improve the coverage probability substantially.

Some authors [7, 8] explored jackknife methodology to deal with curvature effects, nonnormality of random errors, and influential observations.

For any array $\mathbf{W}$, let $\mathbf{W}_{(i)}$ denote $\mathbf{W}$ after removing the $i^{th}$ observation from the calculation. Thus $\hat{\boldsymbol{\theta}}_{(i)}$ denotes $\hat{\boldsymbol{\theta}}_{olse}$ when the $i^{th}$ observation is omitted from the sample. Then the $i^{th}$ pseudo-value corresponding to standard jackknife is:

$$(5) \qquad \mathbf{P}_{1i} = n\hat{\boldsymbol{\theta}}_{olse} - (n-1)\hat{\boldsymbol{\theta}}_{(i)},$$

and the jackknife estimate of $\theta$ and its variance estimate are given by

$$(6) \qquad \hat{\boldsymbol{\theta}}_{J_1} = \frac{1}{n}\sum_{i=1}^{n}\mathbf{P}_{1i}, \quad \mathbf{V}_{J_1} = \frac{1}{n(n-p)}\sum_{i=1}^{n}(\mathbf{P}_{1i} - \hat{\boldsymbol{\theta}}_{J_1})(\mathbf{P}_{1i} - \hat{\boldsymbol{\theta}}_{J_1})'.$$

Fox et al [8] considered the following pseudo-values

$$(7) \qquad \begin{aligned} \mathbf{P}_{2i} &= \hat{\boldsymbol{\theta}}_{olse} + n(\hat{\mathbf{V}}'_{olse}\hat{\mathbf{V}}_{olse})^{-1}\frac{\hat{\mathbf{V}}_i\hat{\epsilon}_i}{(1-\hat{h}_{ii})}, \\ \mathbf{P}_{3i} &= \hat{\boldsymbol{\theta}}_{olse} + n(\hat{\mathbf{V}}'_{olse}\hat{\mathbf{V}}_{olse})^{-1}\hat{\mathbf{V}}_i\hat{\epsilon}_i, \qquad i = 1, 2, \dots, n, \end{aligned}$$

where $\hat{h}_{ii} = \hat{\mathbf{V}}'_i(\hat{\mathbf{V}}'_{olse}\hat{\mathbf{V}}_{olse})^{-1}\hat{\mathbf{V}}_i$. Simonoff and Tsai [27] introduced three new jackknife procedures based on the following pseudo-values. Of these, $\mathbf{P}_{5i}$ and $\mathbf{P}_{6i}$ take into account the curvature effects of the nonlinear regression function:

$$\mathbf{P}_{4i} = \hat{\boldsymbol{\theta}}_{olse} + n(\hat{\mathbf{V}}'_{olse}\hat{\mathbf{V}}_{olse})^{-1}\hat{\mathbf{V}}_i\hat{\epsilon}_i(1 - h_{ii}), \quad \mathbf{P}_{5i} = \hat{\boldsymbol{\theta}}_{olse} + \frac{n\mathbf{T}_i^{-1}\hat{\mathbf{V}}_i\hat{\epsilon}_i}{1 - \hat{h}_{ii}^*},$$

$$(8) \qquad \mathbf{P}_{6i} = \hat{\boldsymbol{\theta}}_{olse} + n\mathbf{T}_i^{-1}\hat{\mathbf{V}}_i\hat{\epsilon}_i.$$

Here $\mathbf{T}_i = \hat{\mathbf{V}}'_{olse}\hat{\mathbf{V}}_{olse} - [\hat{\epsilon}'_{(i)}][\hat{\mathbf{V}}_{(i)}..]$, $\hat{\mathbf{V}}_{(i)}..$ is $(n-1) \times p \times p$ array with $i^{th}$ face deleted from $\hat{\mathbf{V}}..$. $\hat{\mathbf{V}}..$ is an $n \times p \times p$ second derivative array where

$v_{ist} = \partial^2 f(X_i, \theta)/\partial\theta_s\theta_t$, and $\hat{h}_{ii}^* = \hat{\mathbf{V}}_i' \mathbf{T}_i^{-1} \hat{\mathbf{V}}_i$. For a $n_1 \times n_2$ matrix $\mathbf{B}$ and a $n_2 \times p \times p$ array $\mathbf{C} = \{(\mathbf{c}_{rs})\}$, we define $[\mathbf{B}][\mathbf{C}] = \{(\mathbf{Bc}_{rs})\}$.

The jackknife point estimator $\hat{\theta}_{Jj}$ and its variance estimator $\mathbf{V}_{Jj}$ are obtained from the formulae given in (6) using the pseudo-values $\mathbf{P}_{ji}$, $j = 1, 2, \ldots, 6$, and $i = 1, 2, \ldots, n$. Each of the above jackknife procedures are asymptotically equivalent to the OLSE ([31]).

Suppose $Var(Y_i) = \sigma_i^2$, $i = 1, 2, \ldots, n$, then following [13, 16, 24, 25, 26], one may consider the following class of variance estimators for estimating the covariance matrix of $\hat{\theta}_{olse}$:

$$(9) \qquad \hat{\mathbf{V}}_\omega = (\hat{\mathbf{V}}_{olse}' \hat{\mathbf{V}}_{olse})^{-1} \sum_{i=1}^{n} \frac{\hat{\epsilon}_i^2}{\omega_{ii}} \hat{\mathbf{V}}_i \hat{\mathbf{V}}_i' (\hat{\mathbf{V}}_{olse}' \hat{\mathbf{V}}_{olse})^{-1},$$

where $\omega_{ii}$ is some suitable function of $\hat{h}_{ii}$ such that $\omega_{ii} \to^{probability} 1$ as $n \to \infty$. Taking $\omega_{ii} = 1 - \hat{h}_{ii}$, Shao [24] proposed

$$(10) \qquad \hat{\mathbf{V}}_0 = (\hat{\mathbf{V}}_{olse}' \hat{\mathbf{V}}_{olse})^{-1} \sum_{i=1}^{n} \frac{\hat{\epsilon}_i^2}{1 - \hat{h}_{ii}} \hat{\mathbf{V}}_i \hat{\mathbf{V}}_i' (\hat{\mathbf{V}}_{olse}' \hat{\mathbf{V}}_{olse})^{-1}.$$

Under certain conditions on the nonlinear regression model, he proved that

$$(11) \qquad n(\hat{\mathbf{V}}_0 - (\mathbf{V}'\mathbf{V})^{-1} \sum_{i=1}^{n} \sigma_i^2 \mathbf{V}_i \mathbf{V}_i' (\mathbf{V}'\mathbf{V})^{-1}) \to 0 \quad a.s.,$$

where $\mathbf{V}$ is the matrix of partial derivatives of $f(\cdot, \theta)$ evaluated at the true value of the parameter $\theta$. From our simulations we found the above procedures may underestimate the true covariance matrix when the sample sizes are small and the curvature effects are pronounced. Hence we introduce the following new member in the class of estimators (9)

$$(12) \qquad \hat{\mathbf{V}}_1 = \frac{n}{n - p} (\hat{\mathbf{V}}_{olse}' \hat{\mathbf{V}}_{olse})^{-1} \sum_{i=1}^{n} \frac{\hat{\epsilon}_i^2}{(1 - \hat{h}_{ii})^2} \hat{\mathbf{V}}_i \hat{\mathbf{V}}_i' (\hat{\mathbf{V}}_{olse}' \hat{\mathbf{V}}_{olse})^{-1}.$$

Under the conditions stated in [24] and using the proofs given therein it can be deduced that

$$(13) \qquad n(\hat{\mathbf{V}}_1 - (\mathbf{V}'\mathbf{V})^{-1} \sum_{i=1}^{n} \sigma_i^2 \mathbf{V}_i \mathbf{V}_i' (\mathbf{V}'\mathbf{V})^{-1}) \to 0 \quad a.s.$$

**Remark 2.1**

*If the errors are homoscedastic, that is $Var(Y_i) = \sigma^2$, for all $i = 1, 2, \ldots, n$, then we propose the following natural modification to (12):*

$$(14) \qquad \hat{\mathbf{V}}_2 = \frac{n\hat{\sigma}^2}{n-p}(\hat{\mathbf{V}}'_{olse}\hat{\mathbf{V}}_{olse})^{-1} \sum_{i=1}^{n} \frac{1}{1-\hat{h}_{ii}} \hat{\mathbf{V}}_i \hat{\mathbf{V}}'_i (\hat{\mathbf{V}}'_{olse}\hat{\mathbf{V}}_{olse})^{-1},$$

*where $\hat{\sigma}^2 = \frac{1}{n-p}\sum_{i=1}^{n} \hat{\epsilon}_i^2$.*

**Remark 2.2**

*Suppose $\mathbf{Y}_i = (Y_{i1}, Y_{i2}, \ldots, Y_{in_i})'$ with*

$$Y_{ij} = f(\mathbf{X}_{ij}, \boldsymbol{\theta}) + \epsilon_{ij}, \quad j = 1, 2, \ldots, n_i, i = 1, 2, \ldots, k,$$

*where $\epsilon_{ij}$ are i.i.d with $E(\epsilon_{ij}) = 0$ and $Var(\epsilon_{ij}) = \sigma_i^2$. Then as in [13, 16] we propose the following class of variance estimators:*

$$(15) \qquad \hat{\mathbf{V}}_3 = (\hat{\mathbf{V}}'_{olse}\hat{\mathbf{V}}_{olse})^{-1} \sum_{i=1}^{k} \frac{\hat{\epsilon}'_i \hat{\epsilon}_i}{n_i - \delta_{ii}} \hat{\mathbf{V}}'_i \hat{\mathbf{V}}_i (\hat{\mathbf{V}}'_{olse}\hat{\mathbf{V}}_{olse})^{-1},$$

*where $\delta_{ii}$ is some function of $Tr(\hat{\mathbf{V}}'_i(\hat{\mathbf{V}}'_{olse}\hat{\mathbf{V}}_{olse})^{-1}\hat{\mathbf{V}}_i)$ such that $\delta_{ii} \to 0$ as $\sum_i^k n_i \to \infty$, $\hat{\mathbf{V}}_i$ is a $n_i \times p$ matrix of partial derivatives of $f(\mathbf{X}_{ij}, \boldsymbol{\theta})$ with respect to $\boldsymbol{\theta}$, and $\hat{\epsilon}_i = (\hat{\epsilon}_{i1}, \hat{\epsilon}_{i2}, \ldots, \hat{\epsilon}_{in_i})'$ with $\hat{\epsilon}_{ij} = \mathbf{Y}_{ij} - f(\mathbf{X}_{ij}, \hat{\boldsymbol{\theta}}_{olse})$.*

A small simulation study was performed using the two nonlinear models (1) and (2), to investigate the performance of procedures described above. For model (1) we considered $\boldsymbol{\theta} = (158.165, 170.069, 0.113, 1.430, 13.988)'$, and the values of $x$ to be 10.575, 11.244, 11.534, 12.205, 12.627, 13.247, 13.548, 14.222, 14.600, 15.241, 15.559, 15.923. In the case of model (2), we performed simulations by taking $\boldsymbol{\theta} = (1.933, 0.058, 3.932)'$. The values of serum testosterone (in log scale) and growth hormone levels were taken to be $(T, G) = (4.968, .630), (5.969, 1.00), (5.733, .682), (6.194, .446), (6.170, .461), (6.365, .972), (6.407, .309), (6.501, .604), (6.288, .253), (6.612, .190), (6.50, .324), (6.284, .233)$. Simulation results are based on 1000 simulation runs. The random errors were generated from a normal distribution with mean zero and standard deviation $\sigma^2 = .32$. Since the random errors are assumed to be homoscedastic, we used $\hat{\mathbf{V}}_2$ as an estimator of the covariance matrix of $\hat{\boldsymbol{\theta}}_{olse}$ in the simulation study. The parameters used in the simulation study are based on the estimates obtained from a real data on one of the children in our sample.

In each case we constructed confidence regions for $\theta$ using the approximation

$$(16) \qquad (\hat{\theta} - \theta)'[\widehat{Var}(\hat{\theta})]^{-1}(\hat{\theta} - \theta) \sim^{approximately} pF_{p,n-p},$$

where $\hat{\theta}$ is the point estimator and $\widehat{Var}(\hat{\theta})$ is the appropriate variance estimator. We compare the above estimation procedures in terms of coverage probability and the volume of the ellipsoid in (16).

Results of the simulation study are reported in Table 1. It seems that the approximation given in (4), which is commonly used, tends to be very liberal (Row number 1). For instance in case of the velocity model (2) it attains a coverage probability of .844 while the nominal level was .95. In this particular model none of the jackknife procedures do any better. However, the pivotal statistic based on the new proposed variance estimator $\hat{V}_2$ seems to perform very well for both models (Row number 8). It is more accurate than rest of the methods. Rather than using the $F$ approximation in (4), one may bootstrap the pivotal statistic in (4) (Row number 9). Results reported in this simulation study are based on 1000 bootstrap samples. In the case of Preece-Bains model the bootstrap methodology approximately achieves the desired coverage probability at the expense of the volume of the ellipsoid. Volume of the bootstrap ellipsoid is substantially larger than that of the proposed methodology based on the new variance estimator $\hat{V}_2$ (Row number 8). In the case of the velocity model it achieves almost the same coverage probability as the new procedure. Thus it seems that the new procedure performs better than some of the available alternative procedures.

| Volume of the ellipsoid is given within parenthesis | | |
|---|---|---|
| Method | Model | |
| Row number | $\hat{\theta} : \widehat{Var}(\hat{\theta})$ | Preece-Baines model | Velocity model |
| 1 | $\hat{\theta}_{olse} : \hat{V}_{olse}$ | .872(.0069) | .844(.0002) |
| 2 | $\hat{\theta}_{J_2} : \hat{V}_{J_2}$ | .895(.0864) | .849(.0004) |
| 3 | $\hat{\theta}_{J_3} : \hat{V}_{J_3}$ | .564(.0019) | .709(.0001) |
| 4 | $\hat{\theta}_{J_4} : \hat{V}_{J_4}$ | .208(.0001) | .509(.00001) |
| 5 | $\hat{\theta}_{J_5} : \hat{V}_{J_5}$ | .866(3.3741) | .803(.0065) |
| 6 | $\hat{\theta}_{J_6} : \hat{V}_{J_6}$ | .574(.0193) | .712(.0065) |
| 7 | $\hat{\theta}_{J_4} : \hat{V}_{J_5}$ | .904(3.3741) | .845(.0023) |
| 8 | $\hat{\theta}_{olse} : \hat{V}_2$ | .974(.2091) | .926(.0005) |
| 9 | Bootstrap $\hat{\theta}_{olse} : \hat{V}_{olse}$ | .943(3.5965) | .930(.0011) |

Table 1. Nominal coverage probability $= .95$

## Random effects models under homoscedastic errors

In some situations an investigator may collect several repeated measurements on the same individual. A common method of modeling such a data is to use random effects models. Analysis of random and mixed effects nonlinear models has been an active area of research during the past two decades. For a review on this subject one may refer to [6].

For the $i^{th}$ individual, let $Y_{ij}$ denote the $j^{th}$ response corresponding to the $q \times 1$ covariate vector $\mathbf{X}_{ij}$, $j = 1, 2, \ldots, n_i$, $i = 1, 2, \ldots, k$. Let

$$\mathbf{Y}_i = \mathbf{f}(\mathbf{X}_i : \boldsymbol{\theta}_i) + \boldsymbol{\epsilon}_i,$$

(17) $\qquad E(\boldsymbol{\epsilon}_i | \boldsymbol{\theta}_i) = \mathbf{0}$ and $Var(\boldsymbol{\epsilon}_i | \boldsymbol{\theta}_i) = \sigma^2 \mathbf{I}_{n_i}$, $\quad i = 1, 2, \ldots, k,$

where $\mathbf{Y}_i = (Y_{i1}, Y_{i2}, \ldots, Y_{in_i})'$, $\mathbf{f}(\mathbf{X}_i : \boldsymbol{\theta}_i) = (f(\mathbf{X}_{i1} : \boldsymbol{\theta}_i), f(\mathbf{X}_{i2} : \boldsymbol{\theta}_i), \ldots$
$\ldots, f(\mathbf{X}_{in_i} : \boldsymbol{\theta}_i))'$, $f$ is some continuous function which is twice differentiable with respect to $\boldsymbol{\theta}$, and $\boldsymbol{\epsilon}_i = (\epsilon_{i1}, \epsilon_{i2}, \ldots, \epsilon_{in_i})'$. We shall assume that $\boldsymbol{\epsilon}_i$ are mutually independently distributed. For simplicity of notation we shall drop $n_i$ from the subscript of the $n_i \times n_i$ identity matrix $\mathbf{I}_{n_i}$.

We shall assume that $\boldsymbol{\theta}_i$ are $i.i.d.$ random variables with

(18) $\qquad\qquad\qquad E(\boldsymbol{\theta}_i) = \boldsymbol{\theta}$ and $Var(\boldsymbol{\theta}_i) = \boldsymbol{\Psi}.$

Alternatively, we write

(19) $\quad \boldsymbol{\theta}_i = \boldsymbol{\theta} + \boldsymbol{\delta}_i$, where $\boldsymbol{\delta}_i$ are i.i.d with $E(\boldsymbol{\delta}_i) = 0$, $V(\boldsymbol{\delta}_i) = \boldsymbol{\Psi}.$

Thus the $p$ dimensional vector $\boldsymbol{\theta}_i$ may be regarded as the parameter specific to $i^{th}$ individual while $\boldsymbol{\theta}$ is the parameter corresponding to the population from which the $i^{th}$ individual was selected.

Construction of the true maximum likelihood estimates for $\boldsymbol{\theta}$ is computationally very intensive and is nearly impossible. Hence a variety of alternative procedures have been proposed in the literature. These procedures can be classified into two general types of methods.

One approach is to linearize $f(\mathbf{X}_i, \boldsymbol{\theta} + \boldsymbol{\delta}_i)$ by performing Taylor's series expansion in terms of $\boldsymbol{\delta}_i$ and then to estimate the parameters under the new model where random effects enter the model in a linear form (cf [12, 29]). Performance of these methods relies heavily on the validity of the first order Taylor's series expansion of the nonlinear function $f$. This can be a serious problem if the curvature effects are pronounced. This issue has not been addressed adequately in the literature. There is a need for a careful and detailed analysis on how the intrinsic and parametric curvature affect the confidence region estimation and test of hypothesis regarding $\theta$.

A second approach is to obtain the OLSE of each $\theta_i$, which are then used to construct estimators for $\theta$ and $\Psi$. These procedures are usually referred to as the "two stage procedures." There are several variations to the two stage procedures. Among these, the "Global Two Stage (GTS)" procedure introduced by Steimer et al. [28] is the most popular one. Though the two stage procedures do not rely on the Taylor series expansion of $f$, they do, however, depend upon the asymptotic behavior of the OLSE of $\theta_i$.

In this article we propose a new two stage procedure derived from the Minimum Norm Quadratic Estimation (MINQE) theory that was introduced by Rao in a series of articles [18, 19, 20]. Justification to the following methodology provided in the Appendix.

Let $\hat{\theta}_i$ denote the OLSE of $\theta_i$ under the model (17). Then, conditional on $\theta_i$, $\hat{\theta}_i$ are independently distributed with conditional variance covariance matrix denoted by $Var(\hat{\theta}_i|\theta_i)$. As in Section 2.1. let $\hat{\mathbf{V}}_{i,olse}$ denote the matrix of partial derivatives of $\mathbf{f}(\mathbf{X}_i : \theta_i)$ with respect to $\theta_i$, evaluated at the ordinary least squares estimator $\hat{\theta}_i$. Also, let $\hat{\sigma}^2 = \sum_{i=1}^{k} \|\mathbf{Y}_i - \mathbf{f}(\mathbf{X}_i : \hat{\theta}_i)\|_2^2/(n - p)$, where $n = \sum_{i=1}^{k} n_i$. Then a commonly used estimator of $Var(\hat{\theta}_i|\theta_i)$ is $\mathbf{C}_i = \hat{\sigma}^2(\hat{\mathbf{V}}_{i,olse}' \hat{\mathbf{V}}_{i,olse})^{-1}$. From Section 2.1., a reasonable alternative estimator for $Var(\hat{\theta}_i|\theta_i)$ is

(20)
$$\hat{\mathbf{V}}_{2_i} = \frac{n_i\hat{\sigma}^2}{n_i - p}(\hat{\mathbf{V}}_{i,olse}' \hat{\mathbf{V}}_{i,olse})^{-1} \sum_{j=1}^{n_i} \frac{1}{1 - \hat{h}_{i,jj}} \hat{\mathbf{V}}_{i,j} \hat{\mathbf{V}}_{i,j}' (\hat{\mathbf{V}}_{i,olse}' \hat{\mathbf{V}}_{i,olse})^{-1}.$$

Expressions in (20) are defined identical to those in (14) with the exception of a subscript $i$ which is used to denote the $i^{th}$ individual.

Then we obtain MINQE for $\Psi$ and $\theta$ by the following iterative process: Starting with the initial estimates

(21)    $$\hat{\Psi}_{(0)} = \frac{1}{k}\sum_{i=1}^{k}(\hat{\theta}_i - \hat{\theta}_{(0)})(\hat{\theta}_i - \hat{\theta}_{(0)})', \quad \hat{\theta}_{(0)} = \frac{1}{k}\sum_{i=1}^{k}\hat{\theta}_i,$$

we iterate between the following two steps until convergence.

*Step 1:* Produce refined estimates of $\theta_i$ as follows

$$\hat{\theta}_{i,(c+1)} = (\hat{\Sigma}_i^{-1} + \hat{\Psi}_{(c)}^{-1})^{-1}(\hat{\Sigma}_i^{-1}\hat{\theta}_i + \hat{\Psi}_{(c)}^{-1}\hat{\theta}_{(c)}),$$

where $\hat{\boldsymbol{\theta}}_{(c)}$ and $\hat{\boldsymbol{\Psi}}_{(c)}$ are the estimates of the population parameters at the $c^{th}$ iteration.

<u>Step 2:</u> Obtain updated estimates of the population parameters as follows

$$\hat{\boldsymbol{\theta}}_{(c+1)} = \frac{1}{k}\sum_{i=1}^{k}\hat{\boldsymbol{\theta}}_{i,(c+1)}, \quad \hat{\boldsymbol{\Psi}}_{(c+1)} = \frac{1}{k}\sum_{i=1}^{k}(\hat{\boldsymbol{\theta}}_i - \hat{\boldsymbol{\theta}}_{(c+1)})(\hat{\boldsymbol{\theta}}_i - \hat{\boldsymbol{\theta}}_{(c+1)})'.$$

In the above expressions we consider two possible choices for $\hat{\boldsymbol{\Sigma}}_i$, namely, $\mathbf{C}_i$ and $\mathbf{V}_{2_i}$. The resulting estimators of $\theta$ and $\Psi$ are denoted by $\hat{\boldsymbol{\theta}}_{AMINQE}$, $\hat{\boldsymbol{\Psi}}_{AMINQE}$ and $\hat{\boldsymbol{\theta}}_{BMINQE}$, $\hat{\boldsymbol{\Psi}}_{BMINQE}$ respectively. Accordingly, we have

$$(22) \qquad \widehat{\mathrm{Var}}_{AMINQE}(\hat{\boldsymbol{\theta}}) = \left\{\sum_{i=1}^{k}(\hat{\boldsymbol{\Psi}}_{AMINQE} + \mathbf{C}_i)^{-1}\right\}^{-1},$$

and

$$(23) \qquad \widehat{\mathrm{Var}}_{BMINQE}(\hat{\boldsymbol{\theta}}) = \left\{\sum_{i=1}^{k}(\hat{\boldsymbol{\Psi}}_{BMINQE} + \mathbf{V}_{2_i})^{-1}\right\}^{-1}.$$

**Simulation**

We compared the performance of the following three confidence regions for $\theta$ in terms of the coverage probability:

$$(24) \qquad \begin{aligned} \mathcal{C}_{GTS} = \Big\{&\theta \in \mathcal{R}^p : \\ &(\hat{\boldsymbol{\theta}}_{GTS} - \theta)'(\widehat{\mathrm{Var}}_{GTS}(\hat{\boldsymbol{\theta}}))^{-1}(\hat{\boldsymbol{\theta}}_{GTS} - \theta) \le c_\alpha\Big\}, \end{aligned}$$

$$(25) \qquad \begin{aligned} \mathcal{C}_{AMINQE} = \Big\{&\theta \in \mathcal{R}^p : \\ &(\hat{\boldsymbol{\theta}}_{AMINQE} - \theta)'(\widehat{\mathrm{Var}}_{AMINQE}(\hat{\boldsymbol{\theta}}))^{-1}(\hat{\boldsymbol{\theta}}_{AMINQE} - \theta) \le c_\alpha\Big\}, \end{aligned}$$

$$(26) \qquad \begin{aligned} \mathcal{C}_{BMINQE} = \Big\{&\theta \in \mathcal{R}^p : \\ &(\hat{\boldsymbol{\theta}}_{BMINQE} - \theta)'(\widehat{\mathrm{Var}}_{BMINQE}(\hat{\boldsymbol{\theta}}))^{-1}(\hat{\boldsymbol{\theta}}_{BMINQE} - \theta) \le c_\alpha\Big\}, \end{aligned}$$

where $c_\alpha = \frac{p(n-1)}{n-kp} F_{p,n-kp}(\alpha)$, $F_{p,n-kp}(\alpha)$ is the upper $\alpha^{th}$ percentile of central $F$ distribution with $p, n - kp$ degrees of freedom. In (24), $\hat{\theta}_{\text{GTS}}$ and $\widehat{\text{Var}}_{\text{GTS}}(\hat{\theta})$ are the respective estimates of $\theta$ and $Var(\hat{\theta}_{\text{GTS}})$ using the GTS methodology.

All simulations were performed using the following two models. The first model is same as (1) mentioned in Section 1 while the second model was introduced by Zhang et al [30] for describing the daily levels of serum testosterone in boys at puberty:

$$
(27) \qquad y_{ij} = \theta_{2i} - \frac{2(\theta_{2i} - \theta_{1i})}{\exp[\theta_{3i}(x_{ij} - \theta_{5i})] + \exp[\theta_{4i}(x_{ij} - \theta_{5i})]} + \epsilon_{ij},
$$
$$
j = 1, 2, ..., m, \quad i = 1, 2, ..., k,
$$

$$
(28) \qquad y_{ij} = \frac{\theta_{1t}}{1 + \exp[-\theta_{2t}(ij - \theta_{3t})]} + \epsilon_{ij}, j = 1, 2, ..., m, \ i = 1, 2, ..., k,
$$

where $k$ is the number of individuals and $m$ is the number of observations per individual. We generated $i.i.d.$ normal random vectors $\theta_i$ with $E(\theta_i) = \theta$ and covariance matrix $\Psi$. The random error vectors $\epsilon_{ij}$ were chosen to be $i.i.d.$ multivariate normal with mean vector $0$ and covariance matrix $\sigma^2 I$ where $\sigma^2 = 0.32$.

For (27), we chose $\theta = (165.8758, 178.4199, 0.1130, 1.4361, 13.6987)'$, and

$$
(29) \qquad \Psi = \begin{pmatrix} 52.8688 & 45.9304 & 0.1575 & 0.5354 & 2.3831 \\ 45.9304 & 42.7685 & 0.1029 & 0.1313 & 1.4709 \\ 0.1575 & 0.1029 & 0.0010 & 0.0042 & 0.0121 \\ 0.5354 & 0.1313 & 0.0042 & 0.0649 & 0.1035 \\ 2.3831 & 1.4709 & 0.0121 & 0.1035 & 0.5691 \end{pmatrix}
$$

and for (28) we chose $\theta = (6.4802, 1.0502, 11.2544)'$, and

$$
(30) \qquad \Psi = \begin{pmatrix} 0.0109 & -0.0015 & 0.0150 \\ -0.0015 & 0.0291 & -0.0597 \\ 0.0150 & -0.0597 & 0.5112 \end{pmatrix}
$$

The choices of $\theta$, $\Psi$ and $\sigma^2$ were obtained by applying the GTS method to a real data set available to us on 23 boys in puberty. We assume $\theta_i$ and $\epsilon_{ij}$ are mutually independent.

We considered two patterns of $k$ and $m$, namely, $k = 10$, $m = 12$, and $k = 20$, $m = 30$. From the simulations reported in Table 2, it is apparent GTS procedure can be substantially improved using AMINQE and BMINQE, with BMINQE being the best. The point of this modest simulation study is to demonstrate that the well accepted GTS methodology can be disastrous in some situations and there is some hope and need for developing better alternative procedures.

| Volume of the ellipsoid is given within parenthesis | | | |
|---|---|---|---|
| Pattern of | Method | Preece-Baines | Testosterone |
| $k$ and $n$ | | model | model |
| $k = 10, m = 12$ | GTS | .764(.1736) | .842(.5082) |
| | AMINQE | .862(.3524) | .939(.7359) |
| | BMINQE | .918(.5741) | .949(.9340) |
| $k = 20, m = 30$ | GTS | .905(.0307) | .922(.1537) |
| | AMINQE | .952(.0528) | .960(.2091) |
| | BMINQE | .958(.0670) | .961(.2391) |

Table 2. Coverage probabilities. Nominal rate $= .95$

## Random effect models under heteroscedastic errors

We now consider
$$\mathbf{Y}_i = \mathbf{f}(\mathbf{X}_i : \boldsymbol{\theta}_i) + \epsilon_i,$$

(31)    $E(\epsilon_i|\theta_i) = \mathbf{0}$, and $Var(\epsilon_i|\theta_i) = \text{Diag}[\sigma_1^2 \mathbf{I}_{n_i} : \sigma_2^2 \mathbf{I}_{n_i} : \cdots : \sigma_r^2 \mathbf{I}_{n_i}]$,

where $\sigma_1^2, \sigma_2^2, \ldots, \sigma_r^2$, are $r$ unknown parameters. Such a model may arise if there are, say, $r$ sets of response variables collected on the $i^{th}$ individual with different experimental errors associated with each response variable.

Following is an example from Pediatrics that motivated us to consider the above model.

**Example 2.1** Pediatricians are interested in understanding the relationship between the age at maximal change in serum testosterone levels and the age at maximal velocity of growth (also known as peak height velocity, PHV). It is hypothesized that, for "normally growing children", the PHV is preceded by maximal change in testosterone.

Zhang et al [30] modeled velocity of growth as a function of age using

(32) $$V = \theta_{1v} \exp[-\theta_{2v}(x - \theta_{3v})^2], \quad \theta_{1v}, \theta_{2v}, \theta_{3v} > 0,$$

and modeled testosterone as a function of age using

(33) $$T = \frac{\theta_{1t}}{1 + \exp[-\theta_{2t}(x - \theta_{3t})]}, \quad \theta_{1t}, \theta_{2t}, \theta_{3t} > 0.$$

Notice from (32) that the age at PHV is $\theta_{3v}$, and from (33) note that the age at maximal change in testosterone is $\theta_{3t}$. Hence, for the overall population of boys growing normally, the hypothesis of interest is $H_0 : \theta_{3v} = \theta_{3t}$ against the alternative $H_a : \theta_{3v} > \theta_{3t}$.

The data consists of repeated measurements on 23 "normally growing boys," taken every four to six months. A variety of measurements are collected including age of the boy, growth velocity and serum testosterone. Age of the boys ranged between 9 years to 17 years.

Two important conditions that need to considered in our analysis are:

$C1$. Growth velocity and serum testosterone are not independent but they are correlated variables.

$C2$. Variability in the measurement error associated with growth velocity is smaller than the variability in the measurement error associated with serum testosterone.

There are several ways of modeling the above data. We propose to use random effects model for modeling $C1$ and use heteroscedastic error structure to model $C2$.

For the $i^{th}$ individual let $n_i \times 1$ vector of growth velocity measurements be denoted by $Y_{i1}$ and serum testosterone measurements be given by $Y_{i2}$. Let the nonlinear models corresponding to these response variables be given by

(34) $$Y_{i1} = f_1(X_i, \theta_{i1}) + \epsilon_{i1},$$

and

(35) $$Y_{i2} = f_2(X_i, \theta_{i2}) + \epsilon_{i2},$$

where $X_i$ is the $n_i \times 1$ vector of ages of the $i^{th}$ individual and $f_1$ and $f_2$ are respectively the nonlinear functions defined in (32) and (33). Further, as per $C2$, we shall assume that $Var(\epsilon_{il}|\theta_i) = \sigma_l^2 I$, $l = 1, 2$, where $\theta_i = (\theta_{i1}', \theta_{i2}')'$. Stacking (34) and (35) into one long model we have

(36) $$Y_i = f(X_i, \theta_i) + \epsilon_i,$$

where $\mathbf{Y}_i = (\mathbf{Y}'_{i1}, \mathbf{Y}'_{i2})'$ and $f(\mathbf{X}_i, \theta_i) = (f_1(X_i, \theta_{i1})', f_2(X_i, \theta_{i2})')'$, and $\epsilon_i = (\epsilon'_{i1}, \epsilon'_{i2})'$. We shall assume that $\theta_i$ are *i.i.d.* random vectors with a mean vector $\theta$ and a covariance matrix $\Psi$. Thus correlations between the vectors $\mathbf{Y}_{i1}$ and $\mathbf{Y}_{i2}$ are modeled through using covariance matrix $\Psi$. This is exactly in the form of (31).

In this article we construct a MINQE like procedure by replacing $\Sigma_i$ by $\hat{\mathbf{V}}_{3_i}$ in Step 1 and Step 2 of Section 2.2., where $\hat{\mathbf{V}}_3$ is as defined in (15). Let the resulting estimators of $\theta$ and $\Psi$ be denoted by $\hat{\theta}_{HMINQE}$ and $\hat{\Psi}_{HMINQE}$, respectively. Then an estimator of $Var(\hat{\theta}_{HMINQE})$ is given by

$$(37) \qquad \widehat{Var}_{HMINQE}(\hat{\theta}_{HMINQE}) = \{\sum_{i=1}^{k}(\hat{\Psi}_{HMINQE} + \mathbf{V}_{3_i})^{-1}\}^{-1}.$$

Hence, for any arbitrary linear combination $\mathbf{C}\theta$, where $\mathbf{C}$ is a $s \times p$ matrix of rank $s$, one may use the following Wald statistic for performing inference on $\mathbf{C}\theta$:

$$(38) \qquad (\mathbf{C}\hat{\theta} - \mathbf{C}\theta)'[Var(\mathbf{C}\hat{\theta})]^{-1}(\mathbf{C}\hat{\theta} - \mathbf{C}\theta) \sim^{\text{approximately}} sF_{s,rn-p},$$

where $F_{a,b}$ denotes central $F$ distribution with $(a, b)$ degrees of freedom.

**Example 2.1 (continued)** In this example there are two response variables and each response variable has three unknown parameters. Thus we have $r = 2$ and $p = 6$. We may rewrite our hypothesis of interest as

$$H_0 : (0, 0, 1, 0, 0, -1)\theta = 0 \text{ Versus } H_a : (0, 0, 1, 0, 0, -1)\theta > 0.$$

Based on the data on $k = 23$ boys on two variables (growth velocity and testosterone) we find that

$$\frac{c'\hat{\theta}}{s.d.(c'\hat{\theta})} = 17.0975.$$

Total sample size was 451 and hence the $p - value$ is less than 0.001. Thus we conclude that there is evidence to believe that peak height velocity is preceded by the maximal change in testosterone.

**Some open research problems**

As seen in the previous two sections, existing methodology for analyzing nonlinear fixed and random effects models is not very satisfactory. This is particularly true with respect to coverage probabilities of confidence regions and the $p$ values for tests of hypotheses. There are two major reasons for this problem. Firstly, the covariance matrix of $\hat{\theta}$ is often underestimated. There is a need for developing new robust variance estimators which have small bias. A second problem is to derive better approximations to the distribution of pivotal statistics. The usual $F$ or $t$ approximations are not very satisfactory. This problem is particularly serious when the errors are heteroscedastic.

As observed in Example 2.1, in some situations an experimenter may collect repeated measurements on several correlated response variables. In such situations, as an alternative to the model described in Section 2.3., one may build a multivariate nonlinear random effects models with a suitable covariance structure.

For the $i^{th}$ individual, let $\mathbf{Y}_{ij}$ be $n_i \times 1$ vector of responses measured on $j^{th}$ variable, $j = 1, 2, \ldots, r$. Let $\mathbf{Y}_i$ denote the $n_i \times r$ matrix obtained by stacking the column vectors $\mathbf{Y}_{ij}$ side by side. For the $i^{th}$ individual, let $f_j(\mathbf{x}_i, \theta_{ij})$ denote the nonlinear function associated with the $j^{th}$ response variable with associated parameter $\theta_{ij}$, a $p \times 1$ vector. Let $\epsilon_{ij}$ be the random error term associated with $\mathbf{Y}_{ij}$. Stack the columns $f_j(\mathbf{x}_i, \theta_{ij})$ and $\epsilon_{ij}$, to form $n_i \times r$ matrices $F_i(\mathbf{x}_i, \theta_i)$ and $\mathbf{E}_i$, respectively. Thus we obtain the following multivariate nonlinear random effects model:

$$(39) \qquad \mathbf{Y}_{n_i \times r} = F_i(\mathbf{x}_i, \theta_i) + \mathbf{E}_i, \ i = 1, 2, \ldots, k,$$

Here $\theta_i = (\theta_{i1}, \ldots, \theta_{ir})'$. We may model the correlations between variables by imposing a covariance structure on the columns of $\mathbf{E}_i$. We may take $\theta_i$ to be random variables with mean vector $\theta$ and covariance structure that is block diagonal. In other words, we may assume that $Var(\theta_{ij}) = \Psi_j$, and $Var(\theta_{ij_1}, \theta_{ij_2}) = 0$. We shall of course assume that observations on different individuals are independent.

Thus we may model the data mentioned in Example 2.1 using the above multivariate model as well. Model (39) provides a richer covariance structure between the response variables than the heteroscedastic random effects model described in Section 2.3.. A disadvantage with (39) is that it has many more parameters than the heteroscedastic model described in Section 2.3.. As far as we know multivariate models such as (39) have not been discussed in the literature and there is a need for developing statistical procedures for analyzing such models.

## Nonlinear regression models in $\mathcal{R}^p \times SO(p)$: Some open research problems

Research in this field is fairly recent [4, 5, 11, 14, 15] with applications to polar ice tracking using satellite images and estimation of the motion of tectonic plates. This is a practically important and a growing field of research. In this section we shall describe various important open research problems in this field.

Consider a rigid body moving in a $p$ dimensional space with a translation vector $\alpha \in \mathcal{R}^p$ and a rotation matrix $\beta \in SO(p)$, a group of $p \times p$ orthogonal matrices with determinant value $+1$. Let $\mathcal{P} = \{P_1, P_2, \ldots, P_m\}$ and $\mathcal{Q} = \{Q_1, Q_2, \ldots, Q_n\}$ denote a collection of $m + n$ points on a rigid body. Suppose at time $t_0$ we observe the collection $\mathcal{P}$, with coordinates of $P_i$ recorded as $\mathbf{x}_i \in \mathcal{R}^p$, and at time $t_1$ we observe the collection $\mathcal{Q}$, with coordinates of $Q_j$ recorded as $\mathbf{y}_j \in \mathcal{R}^p$. If $\mathcal{P} \equiv \mathcal{Q}$ and $m = n$, then we say that we have a *homologous* data, else the observed data is *nonhomologous*.

In the case of homologous data there is a one-one correspondence between the $\mathbf{x}$ and $\mathbf{y}$ values. Consequently, each $\mathbf{y}_i$ can be modeled as

$$(40) \qquad \mathbf{y}_i = \alpha + \beta \mathbf{x}_i + \epsilon_i, \quad i = 1, 2, \ldots, n,$$

where $\epsilon_i$ is a random disturbance vector. In this case the parameters $\alpha$ and $\beta$ can be estimated using the OLSE.

When dealing with greyscale images, such as medical and satellite images, it is unlikely to obtain homologous pairs of data. As the demand for image analysis of moving objects increases, it will become important to develop suitable tracking algorithms to track objects between a sequence of images and to estimate the motion parameters along with any changes in areas and volumes of the objects.

Recently [14, 15] introduced a methodology to estimate motion parameters when the data are nonhomologous. Due to nonhomology, a nonlinear model of the form (40) is not meaningful. Consequently, Peddada and Chang [14] estimate the motion parameters $\alpha$ and $\beta$ by iteratively solving the following pair of minimization problems.

$$(41) \qquad \min_{\alpha, \beta} S(\alpha, \beta) = \min_{\alpha, \beta} \sum_{i=1}^{n} \|\mathbf{y}_i - \alpha - \beta \hat{\xi}_i\|$$

where, for a given iterative value of $\alpha$ and $\beta$, corresponding to each $\mathbf{y}_i$, $i = 1, 2, \ldots, n$, $\hat{\xi}_i$ is chosen to be a point on the $Polygon\{\alpha + \beta\mathbf{x}_1, \alpha + \beta\mathbf{x}_2, \ldots, \alpha + \beta\mathbf{x}_m\}$, such that

$$(42) \qquad \min_{1 \leq i \leq m} \rho(\mathbf{y}_i, Polygon\{\alpha + \beta\mathbf{x}_1, \alpha + \beta\mathbf{x}_2, \ldots, \alpha + \beta\mathbf{x}_m\})$$

where, $\rho$ denotes a suitable distance measure. Methodology described in [15] is a slight variation to the above procedure.

The numerical and statistical properties of the estimates derived from (41) are completely unknown. It is important to determine the conditions under which the algorithm converges to a solution. It is intuitive to believe that shape of the body will play an important role. For instance there will be multiple solutions if the body is a symmetric polygon. The geometry of the body may also play an important role for proving consistency and asymptotic normality of the point estimators.

Suppose $k$ rigid bodies are moving together in a particular region. For the $i^{th}$ body, let us suppose $\mathbf{x}_{ij_1}$ denotes the $j_1^{th}$ coordinate at time $t_0$ and suppose $\mathbf{y}_{ij_2}$ denotes the $j_2^{th}$ coordinate at time $t_1$. Often scientists are interested in obtaining the "mean" direction and speed of motion of these rigid bodies. For instance, a polar scientist may be interested in knowing how a pack of ice is moving rather than how a specific ice floe is moving. This problem may be formulated in terms of nonlinear random effects model on $\mathcal{R}^p \otimes SO(p)$. As far as we know, no research has been done in this area. We shall present a variety of important open problems which need to be studied.

For simplicity of exposition we shall describe the problems in terms of motion on a plane and assume that the data are homologous. If the data are nonhomologous and if the body are moving higher dimensional space then the problems are accordingly generalized.

Since the motion is on a two dimensional plane, the rotation matrix $\beta$ is completely characterized by the angle of rotation $\theta \in (0, 2\pi)$. Thus $\beta_{11} = cos\theta$, $\beta_{12} = -sin\theta$, $\beta_{21} = sin\theta$, and $\beta_{22} = cos\theta$. Motion of the $i^{th}$ body can be described using

$$(43) \qquad \mathbf{y}_{ij} = \alpha_i + \beta_i \mathbf{x}_{ij} + \epsilon_{ij}, \quad j = 1, 2, \ldots, n_i, i = 1, 2, \ldots, k,$$

where $\epsilon_{ij}$ is the random disturbance associated with the $j^{th}$ coordinate. Let $\delta_i = (\alpha_{1i}, \alpha_{2i}, \theta_i)'$. Since all $k$ bodies are traveling in the same region, it may be reasonable to believe that $\delta_1, \delta_2, \ldots, \delta_k$ are random vectors with a common distribution function $F$ and a common motion parameter vector $\delta = (\alpha', \theta)'$. Thus the ultimate problem of interest is to estimate $\delta = (\alpha', \theta)'$ along with its "covariance matrix." Of course when $p > 2$ the parameters of interest are the common translation vector $\alpha$ and the common rotation matrix $\beta$.

Any estimation procedure that one may propose, must take into account various types of correlations that exist in the data.

Marginally, it may be reasonable to assume that $\alpha_i = (\alpha_{1i}, \alpha_{2i})'$ is bivariate normally distributed with mean vector $\alpha = (\alpha_1, \alpha_2)'$ and covariance matrix $\Sigma$. Also marginally, it may be reasonable to assume that $\theta_i$

is distributed according to von Mises distribution with a concentration parameter $\kappa$ and a location parameter $\theta$. It is unreasonable to assume that $\alpha_i$ and $\theta_i$ are independently distributed. Further, when a collection of bodies, such as ice floes, are moving together it is important to model the spatial dependence between the various bodies. Further, as noted in Peddada and Chang [14] and Peddada and McDevitt [15], the residuals within the $i^{th}$ body, for example, $\|\epsilon_{ij_1}\|_2^2$, $\|\epsilon_{ij_2}\|_2^2$ at locations $j_1$ and $j_2$, can not be taken to be independently distributed.

As mentioned earlier, when dealing with satellite data it is rare to obtain homologous observations. Hence it will be important to develop a new statistical estimation procedures for estimating $\alpha$ and $\beta$ when the underlying data are nonhomologous and $p \geq 2$.

## Appendix

### The derivation of MINQE

Expressing $\hat{\theta}_i$ in the form of a linear model we have

$$(44) \qquad \hat{\theta}_i \approx \theta + \mathbf{b}_i + \delta_i, \ i = 1, 2, \ldots, k,$$

where $\mathbf{b}_i$ are $i.i.d$ with mean $E(\mathbf{b}_i) = 0$, $Var(\mathbf{b}_i) = \Psi$, and $\delta_i$ are independently distributed with $E(\delta_i) = 0$ and $Var(\delta_i|\mathbf{b}_i) = V(\hat{\theta}_i|\theta_i)$. Stacking the $k$ linear models we have

$$(45) \qquad \hat{\eta} \approx [\mathbf{I} : \mathbf{I} : \ldots : \mathbf{I}]'\theta + \mathbf{b} + \sum_{i=1}^{k} \mathbf{L}_i\delta_i,$$

where $\hat{\eta} = (\hat{\theta}_1', \hat{\theta}_2', \ldots, \hat{\theta}_k')'$, $\mathbf{b} = (\mathbf{b}_1' : \mathbf{b}_2' : \ldots : \mathbf{b}_k')'$, $\mathbf{L}_i$ is a $pk \times p$ matrix of the form $[\mathbf{0}' : \mathbf{0}' : \cdots : \mathbf{I}' : \cdots : \mathbf{0}']'$, where each $\mathbf{0}$ is a $p \times p$ null matrix and $\mathbf{I}$ is a $p \times p$ identity matrix located at the $i^{th}$ location. Though the left hand side of (45) is approximately equal to the right hand side for notational simplicity we shall pretend that they are exactly equal and proceed further. The model (45) is a special case of the following general mixed effects model:

$$(46) \qquad \mathbf{W} = \mathbf{Z}\beta + \mathbf{L}\,\mathbf{b} + \sum_{i=1}^{k} \mathbf{L}_i\delta_i.$$

In the above model $\mathbf{W} = (\mathbf{W}_1', \mathbf{W}_2', \ldots, \mathbf{W}_k')'$ with $\mathbf{W}_i$ a $p \times 1$ observable random vector, $\mathbf{Z}$ is a known $kp \times p$ design matrix, $\mathbf{L}$ and $\mathbf{L}_i's$ are known design matrices of orders $kp \times kp$ and $kp \times p$, respectively. Further, $\beta = (\beta_1, \beta_2, \ldots, \beta_p)'$ is a $p \times 1$ unknown parameter vector and $b = (b_1', b_2', \ldots, b_k')'$ with $b_i$ a $p \times 1$ unobservable random vector. Finally, $\delta = (\delta_1', \delta_2', \ldots, \delta_k')'$ is a $kp \times 1$ with $\delta_i's$ a $p \times 1$ unobservable random vector. We assume that $E(b) = 0$, $Var(b_i) = \Psi$, $i = 1, 2, \ldots, k$, so $Var(b) = I_k \otimes \Psi$, where $I_k$ is a $k \times k$ identity matrix; $E(\delta_i) = 0$, $Var(\delta_i) = \Sigma_i$ of order $p \times p$, $i = 1, 2, \ldots, k$; $Var(b_i, b_j) = 0$, $Var(\delta_i, \delta_j) = 0$, if $i \neq j$, and also $Var(b_s, \delta_t) = 0$ for all $s$ and $t$.

In terms of $W$, this leads to

$$E(W) = Z\beta,$$

$$Var(W) = L(I \otimes \Psi)L' + \sum_{i=1}^{k} L_i \Sigma_i L_i'.$$

We now derive the MINQE of $\Psi$ in the model (46) and later apply the results to (45).

Let $M$ be any $p \times p$ symmetric matrix, a suitable choice of $M$ would lead to different elements of $\Psi$ through $Tr(M\Psi)$. Let $W'AW$ be an estimator of $Tr(M\Psi)$. We shall obtain the MINQE of $\Psi$ by choosing $A$ such that the quadratic form $W'AW$ is (a) invariant of all translations of the data, (b) "closest" to a "natural" estimator of $Tr(M\Psi)$ if $b$ and $\delta$ were observable. The condition (a) implies that $A$ must satisfy $AZ = 0$. Thus under translation invariance we have

(47)

$$W'A\,W = b'L'ALb + b'L'A\sum_{i=1}^{k} L_i\delta_i + \sum_{i=1}^{k}\delta_i'L_i'ALb + \sum_{i=1}^{k}\sum_{j=1}^{k}\delta_i'L_i'AL_j\delta_j$$

(48) $\quad = (b', \delta_1', \delta_2', \ldots, \delta_k')' \begin{pmatrix} L'AL & L'AL_1 & \ldots & L'AL_k \\ L_1'AL & L_1'AL_1 & \ldots & L_1'AL_k \\ \vdots & \vdots & & \vdots \\ L_k'AL & L_k'AL_1 & \ldots & L_k'AL_k \end{pmatrix} \begin{pmatrix} b \\ \delta_1 \\ \delta_2 \\ \vdots \\ \delta_k \end{pmatrix}.$

Let

(49) $\quad B = \begin{pmatrix} L'AL & L'AL_1 & \ldots & L'AL_k \\ L_1'AL & L_1'AL_1 & \ldots & L_1'AL_k \\ \vdots & \vdots & & \vdots \\ L_k'AL & L_k'AL_1 & \ldots & L_k'AL_k \end{pmatrix},$

and $\bar{b} = \frac{1}{k} \sum_{i=1}^{k} \mathbf{b}_i \mathbf{b}_i'$. If $\mathbf{b}_i$ are observable $r.v$'s, then $\mathbf{I}_k \otimes \bar{b}$ would be a natural estimator of $\mathbf{I}_k \otimes \Psi$. So $(\mathbf{I}_k \otimes \mathbf{M})(\mathbf{I}_k \otimes \bar{b})$ is a natural estimator of $(\mathbf{I}_k \otimes \mathbf{M})(\mathbf{I}_k \otimes \Psi) = \mathrm{diag}(\mathbf{M}\Psi, \dots, \mathbf{M}\Psi)$. Hence $Tr(\mathbf{M}\bar{b})$ is a natural estimator of $Tr(\mathbf{M}\Psi)$. Note that $Tr(\mathbf{M}\bar{b}) = Tr(\mathbf{M}(\frac{1}{k} \sum_{i=1}^{k} \mathbf{b}_i \mathbf{b}_i')) = \frac{1}{k} \sum_{i=1}^{k} \mathbf{b}_i' \mathbf{M} \mathbf{b}_i$, which can be rewritten in terms of $\mathbf{b}, \delta_1, \dots, \delta_k$ as

$$(50) \quad (\mathbf{b}', \delta_1', \delta_2', \dots, \delta_k') \begin{pmatrix} \mathrm{diag}(\dfrac{\mathbf{M}}{k}, \dots, \dfrac{\mathbf{M}}{k})_{kp \times kp} & \mathbf{0}_{kp \times kp} \\ \mathbf{0}_{kp \times kp} & \mathbf{0}_{kp \times kp} \end{pmatrix} \begin{pmatrix} \mathbf{b}_{kp \times kp} \\ \delta_{1 p \times 1} \\ \delta_{2 p \times 1} \\ \vdots \\ \delta_{kp \times 1} \end{pmatrix}.$$

Denote $\mathbf{B}_1 = \begin{pmatrix} \mathrm{diag}(\frac{\mathbf{M}}{k}, \dots, \frac{\mathbf{M}}{k}) & \mathbf{0} \\ \mathbf{0} & \mathbf{0} \end{pmatrix} = \begin{pmatrix} \mathbf{M}^* & \mathbf{0} \\ \mathbf{0} & \mathbf{0} \end{pmatrix}$, where $\mathbf{M}^* = \mathrm{diag}(\frac{\mathbf{M}}{k}, \dots, \frac{\mathbf{M}}{k})$. Then it follows that the MINQE of $Tr(\mathbf{M}\Psi)$ is obtained by minimizing the norm $\|\mathbf{B} - \mathbf{B}_1\|$. Let $\mathbf{U} = (\mathbf{L}, \mathbf{L}_1, \dots, \mathbf{L}_k)$, it can be verified that

$$\mathbf{U}'\mathbf{A}\mathbf{U} = \mathbf{B}.$$

Let $\Gamma^*$ be a $2kp \times 2kp$ symmetric weight matrix with the first diagonal block being $\Gamma$ a $kp \times kp$ weight matrix and the rest being null matrices. For different weight matrices $\Gamma$, we different MINQE procedures. Thus we we are obtaining a family of estimators. For instance, from the following calculations it can be proved that if $\Gamma = \mathbf{I}$, then MINQE procedure will be identical to the well known Standard Two Stage (STS) procedure. The weighted MINQE of $Tr(\mathbf{M}\Psi)$ is obtained by minimizing

$$Tr((\mathbf{B} - \mathbf{B}_1)\Gamma^*(\mathbf{B} - \mathbf{B}_1)\Gamma^*)$$

$$= Tr\left( \left( \mathbf{U}'\mathbf{A}\mathbf{U} - \begin{pmatrix} \mathbf{M}^* & \mathbf{0} \\ \mathbf{0} & \mathbf{0} \end{pmatrix} \right) \Gamma^*(\mathbf{U}'\mathbf{A}\mathbf{U} - \begin{pmatrix} \mathbf{M}^* & \mathbf{0} \\ \mathbf{0} & \mathbf{0} \end{pmatrix})\Gamma^* \right)$$

$$= Tr(\mathbf{A}\mathbf{V}\mathbf{A}\mathbf{V}) - 2Tr(\mathbf{U}'\mathbf{A}\mathbf{U}\Gamma^* \begin{pmatrix} \mathbf{M}^* & \mathbf{0} \\ \mathbf{0} & \mathbf{0} \end{pmatrix} \Gamma^*),$$

where $\mathbf{V} = \mathbf{U}\Gamma^*\mathbf{U}'$. Note that

$$Tr(\mathbf{U}'\mathbf{A}\mathbf{U}\Gamma^* \begin{pmatrix} \mathbf{M}^* & \mathbf{0} \\ \mathbf{0} & \mathbf{0} \end{pmatrix} \Gamma^*) = Tr(\mathbf{B}\Gamma^* \begin{pmatrix} \mathbf{M}^* & \mathbf{0} \\ \mathbf{0} & \mathbf{0} \end{pmatrix} \Gamma^*)$$

$$= Tr(\mathbf{L}'\mathbf{A}\mathbf{L}\Gamma\mathbf{M}^*\Gamma)$$

$$= Tr(\mathbf{A}\mathbf{L}\Gamma\mathbf{M}^*(\mathbf{L}\Gamma)').$$

Thus the problem reduces to

$$\min_{AZ=0} (Tr(\mathbf{AVAV}) - 2Tr(\mathbf{AL\Gamma M^*(L\Gamma)'})),$$

which is solved in the following theorem.

**Theorem 4.1** *The unique solution for the problem*

$$\min_{AZ=0} (Tr(\mathbf{AVAV}) - 2Tr(\mathbf{A(L\Gamma)M^*(L\Gamma)'}))$$

*is* $\mathbf{A^* = R(L\Gamma)M^*(L\Gamma)'R}$, *where* $\mathbf{R}_{kp \times kp} = V^{-1}(\mathbf{I} - \mathbf{Z(Z'V^{-1}Z)^- Z'V^{-1})}.$

*Proof:*The proof of the theorem follows exactly along the same lines as in Rao [20]. Hence we omit the proof of this theorem.     •

Let $\omega_{kp \times 1} = \mathbf{RW} = V^{-1}(I - Z(Z'V^{-1}Z)^- Z'V^{-1})W = V^{-1}(W - \hat{W})$. Then

$$\widehat{Tr(\mathbf{M\Psi})} = \mathbf{W}'A^*W = \mathbf{W}'RL\Gamma M^*(L\Gamma)'RW$$

$$= Tr(\mathbf{M^*(L\Gamma)'RWW'RL\Gamma}) = Tr(\mathbf{M^*(L\Gamma)'\omega\omega'L\Gamma}).$$

Let

$$(51) \qquad (\mathbf{L\Gamma})'\omega\omega'(\mathbf{L\Gamma}) = \begin{pmatrix} \mathbf{F}_{11} & \mathbf{F}_{12} & \cdots & \mathbf{F}_{1k} \\ \mathbf{F}_{21} & \mathbf{F}_{22} & \cdots & \mathbf{F}_{2k} \\ \vdots & \vdots & & \vdots \\ \mathbf{F}_{k1} & \mathbf{F}_{k2} & \cdots & \mathbf{F}_{kk} \end{pmatrix}.$$

Then

$$\widehat{Tr(\mathbf{M\Psi})} = Tr(\mathbf{M^*(L\Gamma)'\omega\omega'L\Gamma})$$

$$= Tr\left( \operatorname{diag}\left( \frac{\mathbf{M}}{k}, \ldots, \frac{\mathbf{M}}{k} \right) \operatorname{diag}(\mathbf{F}_{11}, \ldots, \mathbf{F}_{kk}) \right)$$

$$= Tr(\frac{\mathbf{M}}{k} \sum_{i=1}^{k} \mathbf{F}_{ii}).$$

Since $\mathbf{M}$ is an arbitrary symmetric matrix, so the MINQE for $\Psi$ would be $\frac{1}{k} \sum_{i=1}^{k} \mathbf{F}_{ii}$.

By taking the weight matrix $\Gamma = \operatorname{diag}((\Psi + \Sigma_1), \ldots, (\Psi + \Sigma_k))$, which is the covariance matrix of $(\hat{\theta}'_1, \hat{\theta}'_2, \ldots, \hat{\theta}'_k)'$, and applying the above procedure to our special case (45), we have

$$(52) \quad \hat{\Psi} = \frac{1}{k} \sum_{i=1}^{k} \mathbf{F}_{ii}$$

$$= \frac{1}{k} \sum_{i=1}^{k} \left[ \hat{\theta}_i - \left\{ \sum_{i=1}^{k} (\hat{\Psi} + \hat{\Sigma}_i)^{-1} \right\}^{-1} \sum_{i=1}^{k} (\hat{\Psi} + \hat{\Sigma}_i)^{-1} \hat{\theta}_i \right]$$

$$\left[ \hat{\theta}_i - \left\{ \sum_{i=1}^{k} (\hat{\Psi} + \hat{\Sigma}_i)^{-1} \right\}^{-1} \sum_{i=1}^{k} (\hat{\Psi} + \hat{\Sigma}_i)^{-1} \hat{\theta}_i \right]'.$$

By solving the above implicit equation we obtain the MINQE for $\hat{\Psi}$. This can be accomplished by iteratively solving the following pair of equations

$$(53) \qquad \hat{\Psi} = \frac{1}{k} \sum_{i=1}^{k} (\hat{\theta}_i - \hat{\theta})(\hat{\theta}_i - \hat{\theta})',$$

$$(54) \qquad \hat{\theta} = \left( \sum_{i=1}^{k} (\hat{\Psi} + \hat{\Sigma}_i)^{-1} \right)^{-1} \sum_{i=1}^{k} (\hat{\Psi} + \hat{\Sigma}_i)^{-1} \hat{\theta}_i.$$

In the above expression $\hat{\Sigma}_i$ is some suitable estimator of $\Sigma_i$ and $\hat{\theta}$ is the weighted least squares estimator of $\theta$.

It is easy to verify that the iterative process in (53) and (54) is equivalent to the following two step iterative process starting with (21) as the initial guess for $\hat{\Psi}$ and $\hat{\theta}$.

*Step 1:* Produce refined estimates of $\theta_i$ as follows

$$\hat{\theta}_{i,(c+1)} = (\hat{\Sigma}_i^{-1} + \hat{\Psi}_{(c)}^{-1})^{-1} (\hat{\Sigma}_i^{-1} \hat{\theta}_i + \hat{\Psi}_{(c)}^{-1} \hat{\theta}_{(c)}),$$

where $\hat{\theta}_{(c)}$ and $\hat{\Psi}_{(c)}$ are the population parameters at the $c^{th}$ iteration.

*Step 2:* Obtain updated estimates of the population parameters as follows

$$\hat{\theta}_{(c+1)} = \frac{1}{k} \sum_{i=1}^{k} \hat{\theta}_{i,(c+1)}, \quad \hat{\Psi}_{(c+1)} = \frac{1}{k} \sum_{i=1}^{k} (\hat{\theta}_i - \hat{\theta}_{(c+1)})(\hat{\theta}_i - \hat{\theta}_{(c+1)})'.$$

The algorithm is iterated until convergence.

In (54) we consider two possible choices for $\hat{\Sigma}_i$, namely, $\mathbf{C}_i$ and $\mathbf{V}_{J_i}$.

## References

[1] Bates, D., and Watts, D. G. (1988) *Nonlinear Regression Analysis and its Applications,* Wiley, New York, NY.

[2] Beal, S.L., and Sheiner, L.B. (1988), Heteroscedastic nonlinear regression. *Technometrics,* **30**, 327–338.

[3] Box, G. E. P. and Coutie, G. A. (1956) Application of digital computers in the exploration of functional relationships, *Proc. I. E. E.* **103** Part B, Suppl. 1, 100–107.

[4] Chang (1986), Spherical regression. *Annals of Statistics,* **14**, 907–924.

[5] Chang and Ko (1995), *M* - estimates of rigid body motion on sphere and in Euclidean space. *Annals of Statistics,* **23**, 1823–1847.

[6] Davidian, M., and Giltinan, D. (1995) *Nonlinear Models for Repeated Measurement Data,* London: Chapman and Hall.

[7] Duncan, G. T. (1978), An Empirical study of Jackknife constructed confidence regions in nonlinear regression. *Technometrics,* **20**, 123–129.

[8] Fox, T., Hinkley, D., and Larntz, K. (1980), Jackknifing in Nonlinear Regression. *Technometrics,* **22**, 29–33.

[9] Gallant, A. R. (1971) Statistical Inference for nonlinear regression models, *Ph. D. Thesis,* Iowa State University.

[10] Hartley, H. O. (1964), Exact confidence regions for the parameters in nonlinear regression laws. *Biometrika,* **51**, 347–353.

[11] Ko, D. and Chang (1993), Robust *M* estimators on spheres. *Journal of Multivariate Analysis,* **45**, 104–136.

[12] Lindstrom, M.J. and Bates, D.M. (1990), Nonlinear mixed effects models for repeated measures data. *Biometrics,* **46**, 673–687.

[13] Peddada, S. D. (1993) Jackknife variance estimation and bias reduction, *Handbook of Statistics,* **9**, 723–744; C. R. Rao, ed., Elsevier Science Publishers.

[14] Peddada, S. D., and Chang, T. C. (1996), Bootstrap confidence region estimation of the motion of rigid bodies. *Journal of American Statistical Association,* **91**, 231–241.

[15] Peddada, S. D., and McDevitt, R. J. (1996), Least Average Residual Algorithm (LARA) for tracking the motion of arctic sea ice. *IEEE Transaction on Geoscience and Remote Sensing,* **34**, 915–926.

[16] Peddada, S. D., and Patwardhan, G. (1992), Jackknife variance estimators in linear models. *Biometrika,* **79**, 654–657.

[17] Preece, M.A., and Baines, M.J. (1978), A new family of mathematical models describing the human growth curve. *Annals of Human Biology,* **5**, 1–24.

[18] Rao, C.R. (1970), Estimation of heteroscedastic variances in linear models. *Journal of American Statistical Association*, **65**, 161–172.

[19] Rao, C.R. (1971), Estimation of Variance covariance components - MINQUE theory. *Journal of Multivariate Analysis*, **1**, 257–275.

[20] Rao, C.R. (1972), Estimation of variance and covariance components in linear models. *Journal of American Statistical Association*, **67**, 112–115.

[21] Rao, C.R., and Kleffe, J. (1987) *Estimation of variance components in linear models with applications*, North-Holland, New York, NY

[22] Ratkowsky, D. A. (1990) *Handbook of Nonlinear Regression Models*, STATISTICS: textbooks and monographs, Marcel Decker Inc., New York, NY

[23] Seber, G. A. F., and Wild, C. J. (1989) *Nonlinear Regression*, Wiley, New York, NY

[24] Shao, J (1990), Asymptotic theory in heteroscedastic nonlinear models. *Statistics and Probability Letters*, **10**, 77-85.

[25] Shao, J (1992), Consistency of least squares estimator and its jackknife variance estimator in nonlinear models. *The Canadian Journal of Statistics*, **20**, 415–428.

[26] Shao, J and Tu, D. (1995) *The jackknife and bootstrap*, Springer, New York NY

[27] Simonoff, J. S., and Tsai. C. (1986), Jackknife-based Estimators and Confidence Regions in Nonlinear Regression. *Technometrics*, **28**, 103–112.

[28] Steimer, J.L., Mallet, A., Golmard, J.L. and Boisvieux, J.F. (1984), Alternative approaches to estimation of population pharmacokinetic parameters: Comparison with the nonlinear mixed effect model. *Drug Metabolism Reviews*, **15**, 265–292.

[29] Vonesh, E. F. and Carter, R. L. (1992), Mixed effects nonlinear regression for unbalanced repeated measures. *Biometrics*, **48**, 1–17.

[30] Zhang, J., Peddada, S. D., Malina, R. M., Preece, M. A. and Rogol, A. (1997) A longitudinal assessment of hormonal and physical alterations during normal puberty in boys - VI: Modeling of linear growth, serum testosterone (T) and mean growth hormone (GHmean) concentrations and growth velocity as a function of T and GHmean, *American Journal of Human Biology*, submitted.

[31] Zhang, J. (1997) Analysis of Nonlinear Fixed and Random Effects Models with Applications to Statural Growth and Hormonal Changes in Boys at Puberty, *Ph.D. dissertation*, University of Virginia, Charlottesville, Virginia.

Milton Keynes UK
Ingram Content Group UK Ltd.
UKHW020322111024
449327UK00041B/2697